QUANTITATIVE CONSERVATION BIOLOGY

THEORY AND PRACTICE OF POPULATION VIABILITY ANALYSIS

QUANTITATIVE CONSERVATION BIOLOGY

THEORY AND PRACTICE OF POPULATION VIABILITY ANALYSIS

William F. Morris
Duke University

Daniel F. Doak
University of California, Santa Cruz

THE COVER
Emperor goose (*Chen canagica*). Photograph courtesy of Joel Schmutz, U.S. Geological Survey.

QUANTITATIVE CONSERVATION BIOLOGY

THEORY AND PRACTICE OF POPULATION VIABILITY ANALYSIS

Sinauer Associates Inc.
23 Plumtree Road
Sunderland, MA 01375 USA

FAX: 413-549-1118
EMAIL: orders@sinauer.com; publish@sinauer.com
WEB SITE: http://**www.sinauer.com**

Downloadable computer programs associated with the research herein described can be found at http://www.sinauer.com/PVA/

Library of Congress Cataloging-in-Publication Data

Morris, William F., 1961-
Quantitative conservation biology : theory and practice of population viability analysis / William F. Morris and Daniel F. Doak.
p. cm.
Includes bibliographical references and index.
ISBN 0-87893-546-0 (pbk.)
1. Population viability analysis—Mathematical models. 2. Conservation biology. I. Doak, Daniel F., 1961- II. Title.

QH352.5 . M67 2002
577.8′8—dc21 2002003860

Printed in U.S.A.
5 4 3 2 1

WFM

To Kerstin Ohlander, and to family and friends who have helped make it possible to finish this book in difficult times.

DFD

To inspiring teachers I have had: in mathematics, C. Mel and G. Iversen; in writing, N. Bruce, W. Knight, and G. Moore; in biology, R. Terwilliger, S. Kinsman, M. Jacobs, R.T. Paine, and P. Kareiva; and in life, Barbara Doak.

Contents

SAS and MATLAB Programs

Preface

Conserving biological diversity will often require us to ask hard questions and make difficult decisions regarding populations of threatened and endangered species. How much traditional harvesting of a population can be allowed without endangering it? Is it worth the effort to try to preserve a particular population, or is it so likely to go extinct that limited resources would be better invested in trying to save other populations? To improve the outlook for a particular population, would it be more effective to enhance the survival rate of abundant juveniles or of adults that make up a smaller proportion of the population, and by how much do we need to do so? How many populations of a species do we need to preserve to ensure reasonable protection against both "run-of-the-mill" local extinctions and widespread but infrequent catastrophic events?

These are all *quantitative* questions (how much? how many? how likely?) and they demand quantitative answers. Providing these answers is the goal of population viability analysis (PVA), a broad suite of population modeling and data-fitting methods for assessing extinction risk and guiding the management of rare or threatened species. The promise that PVA holds as a tool for guiding conservation decision-making has been recognized by governmental science advisory boards (National Research Council 1995), by professional organizations such as the Ecological Society of America (Carroll et al. 1996), and by nongovernmental conservation organizations such as The Nature Conservancy (Morris et al. 1999).

Unfortunately, a largely unbridged chasm separates the promised benefits of PVA and its actual use by conservation planners. For example, PVA would ideally contribute to the two major conservation planning processes mandated by the U.S. Endangered Species Act: *Habitat conservation plans*, which are proposals to assess and ameliorate the consequences of "taking" (i.e., harvesting, killing, or destroying critical habitat of) endangered species; and *recovery plans*, which propose concrete steps for improving the status of endangered species so that they can eventually be de-listed. Yet recent analyses have found that PVA is rarely utilized in either of these contexts (Harding et al. 2001, Morris et al. 2002). Instead of seeing PVA as a valuable tool to aid their decision making, most field-oriented conservation biologists retain the misperception that PVA models can only be constructed and understood by an elite priesthood of mathematical population ecologists. As a result, PVA remains largely the purview of academic conservation biologists, who are often removed from the concerns of managing threatened species in the "real" world. This situation deprives conservation practitioners of a set of important decision-making tools, and also shields those who advocate PVA methods from an understanding of the complexities and limitations imposed by real-world deadlines and data limitations.

We believe that, at least in large part, the gap between conservation practitioners and population modelers is due to the lack of good guides to PVA for managers and other empirical ecologists. In particular, there have been no easy-to-understand introductions that describe population modeling for biologists and conservationists with little prior exposure to mathematical modeling, while simultaneously providing deeper coverage of the theory that underlies these method and their application. Our goal in writing this book is to rectify this situation, making population modeling methods clear and usable for conservation biologists in particular and for ecologists in general. We have tailored the book to provide intelligible, intuitive, comprehensive, and practical explanations of population modeling to empirical ecologists and conservation biologists—that is to say, our goal has been to explain concepts and methods for the majority of field biologists, not to the minority of theoreticians who already understand them. We assume readers have no more mathematical background than is typically required for an undergraduate degree in biology. Although at times we use more advanced mathematical concepts (e.g., partial derivatives and probability densities), we try to give intuitive explanations and graphical illustrations to back up the mathematical symbolism.

Because the complexities of population dynamics often preclude simple mathematical expressions for extinction risk, many methods for assessing population viability rely on computer simulation. To make these simulation-based methods (and also several computationally intensive data analysis methods) clear and to make it easier for readers to use them, we have included in the book many fully functional computer programs written in the mathematical language MATLAB. These programs can also be downloaded at **www.sinauer.com/PVA/**. Our hope is that the book and programs together will make the methods we illustrate not just understandable, but also eminently usable by a wide range of ecologists.

The focus of our book is linking modeling methods to real data. Therefore, as an important pedagogical feature of the book, we illustrate all key analyses by means of extensive worked examples that proceed from raw data on actual endangered species to estimating parameters in PVA models, and finally to generating predictions about population viability and best management practices. The book proceeds from simple methods that rely on easily obtained data to more complex, "data-hungry" methods—but always with an eye toward what is practical given realistic constraints on the amount and kinds of data typically available for rare and endangered species. Thus we have omitted many sophisticated population models that are simply too complex to ever be practical in a world of severe data limitations.

Indeed, a shortage of data is a perennial problem in conservation biology. One of the most difficult challenges that conservation scientists face is how to use limited and uncertain data to improve decision-making. While managers ideally want good information and near-certain results ($p = 0.05$) to make firm decisions, in conservation, as in other applied scientific fields (engineering and, especially, medicine), often demands are such that decisions must be made with the information at hand, however bad it may be. The PVA methods we present are ways to use data to improve decisions by reducing and quantifying (but not eliminating) uncertainty, thereby improving conservation practice.

While data uncertainties are frequently used as a reason to rely solely on expert opinion—or on simple political expediency—when deciding difficult issues, we believe that use of more formal analyses can frequently benefit conservation practice. Quantifying and specifying both the likelihoods of different outcomes and the uncertainty about these likelihoods will inevitably sharpen our understanding of current population health, as well as highlighting what we can and cannot conclude given the data on hand. In the absence of such scientific analysis of conservation situations, personalities, politics, and dollars will drive what actions are and are not taken, often with little or no regard to their real conservation value.

This book, and indeed PVA as a whole, focuses on populations, but conservation biologists also seek to maintain intact communities, landscapes, ecosystems, and entire ecoregions. While methods to analyze the viability of populations can be complex, the basic goal is not: our aim is simply to prevent extinction. In contrast, the health or functionality of communities and ecosystems are not as easily defined, let alone analyzed, in a straightforward way. However, with the protection of biodiversity as the underlying goal of all conservation efforts, the viability of the individual populations that make up a community or ecosystem is an ultimate measure of conservation success. Therefore we hope that, despite its exclusive focus on populations, population viability analysis will be of value in the struggle to maintain ecological diversity and integrity at higher levels of biological organization.

ACKNOWLEDGMENTS

We have many people to thank for helping us to improve the clarity and accuracy of this book. For reading and providing editorial suggestions on portions of the manuscript, we thank Susan Alberts, Elizabeth Crone, Greg Dwyer, Susan Harrison, L. Scott Mills, Ingrid Parker, Michael Wisdom, and the students in graduate seminars at Duke University, the University of California at Santa Cruz, the University of California at Santa Barbara, and the University of Montana. Cecil Frost and Susan Harrison kindly provided us with raw data for reanalysis. Special thanks go to Bill Fagan and Bruce Kendall, who extensively reviewed all chapters of the book. This book had its origin in a workshop on PVA methods that we helped to teach for staff of The Nature Conservancy (TNC), which was conducted at the National Center for Ecological Analysis and Synthesis (NCEAS) in Santa Barbara, CA, in 1998. We thank Craig Groves for helping to organize that workshop, out of which a TNC handbook (Morris et al. 1999) and, eventually, this book grew, as well as our co-authors on the handbook (Martha Groom, Peter Kareiva, John Fieberg, Leah Gerber, Peter Murphy, and Diane Thomson), and the staff of NCEAS for logistical assistance.

Finally, we thank both the theoretical population biologists whose work we have summarized in these pages and the many field conservation biologists whose ongoing efforts to monitor rare and endangered species provide the data that have made past PVAs possible, and will form the basis of improved ones in the future.

William F. Morris
Durham, North Carolina

Daniel F. Doak
Santa Cruz, California

1

What Is Population Viability Analysis, and How Can It Be Used in Conservation Decision Making?

Broadly defined, population viability analysis (PVA) is the use of quantitative methods to predict the likely future status of a population or collection of populations of conservation concern. By "future status" we typically mean the likelihood that the population (or the total number of individuals across all populations) will be above some minimum size at a given future time. That size may be zero if we simply want to know if the population will avoid outright extinction, or it could be a larger number if we want to have warning of impending disaster in time to take aggressive action to try to save the population (in fact, we will later argue that we should always use a number greater than zero). More indirect indices of future status, such as whether the population is expected to grow or decline in the coming years, are also valuable to conservation planners. Assigning concrete numbers to these measures of future status is the aim of population viability analysis. As we use the term in this book, PVA refers to a wide range of methods to analyze data and link the results to models of population growth and decline.

Having read the foregoing description of PVA, your first impression may be similar to the one that most people have of the field of accounting—that it sounds rather dry and esoteric. But as with accounting, a little more thought should lead you to acknowledge that PVA can also be incredibly useful.[1] To help convince you of this utility, before we say anything about how PVA is actually performed, we begin by listing specific conservation goals that a viability assessment can help to achieve. That is, we first address the important

---

[1]And, we believe, more interesting and valuable than accounting!

question: For what reasons might we wish to make quantitative predictions about future population status?

Potential Products and Uses of PVA

There are many uses for predictions of future population status. In Table 1.1, we group eight of these uses under the headings of assessment (in which our goal is simply to ask how well the population is doing) and management (in which our goal is to determine what interventions will reduce the population's likelihood of extinction). A recurring theme of this book is the reciprocal interplay between these two major uses of PVA. To illustrate these uses, we briefly describe specific examples in which PVA has been applied to populations of threatened and endangered animals and plants in the real world.

PVA Use 1: Assessing the Extinction Risk of a Single Population

If there is only a single remaining population of a threatened or endangered taxon, we would obviously want to know whether that population is likely to decline, and thus whether we should intervene to save it. We'd also like to know how soon extinction is likely to occur, as it would help us to determine the necessity of immediate intervention. Even if there are many extant populations of a rare species, conservation biologists charged with preserving one of those populations might profit from a quantitative assessment of its future state. A good example of such a population is that of the grizzly bears (*Ursus arctos*) in the Greater Yellowstone ecosystem (Yellowstone National Park and the surrounding National Forest lands). Grizzlies are common in Alaska and Canada, but because of its unique status as one of the few remaining populations in the lower 48 United States, the Yellowstone population is currently listed as threatened under the U.S. Endangered Species Act (ESA).

In one of the first quantitative assessments of population viability, Mark Shaffer (1978, 1981) and Shaffer and Sampson (1985) used computer simulations to ask whether the Yellowstone grizzly bear population was large enough to have at least a 95% chance of surviving for different periods of time into the future. He found that, given the number of bears in the population at the time of his analysis, the probability of persistence for 100 years was fairly high, but the chance of persisting for 300 years was far more bleak. A larger population, and hence a larger area, would be needed to achieve a reasonable likelihood of long-term persistence. This analysis, and a bevy of subsequent PVAs of the same population (Knight and Eberhardt 1985, Eberhardt et al. 1994, Doak 1995, Pease and Mattson 1999) have, via law suits as well as less confrontational means, had a significant impact on management policy. (For example, they resulted in reduced clear-cutting and mining development on the Forest Service lands that comprise the majority of the Greater Yellowstone ecosystem.) Moreover, they have helped to prevent the premature removal of the Yellowstone grizzly from the ESA's

Table 1.1 Potential uses of PVA "products"

Category of Use	*Specific Use*	*Examples*
Assessment of extinction risk	Assessing the extinction risk of a single population	Shaffer 1981, Shaffer and Samson 1985, Lande 1988a
	Comparing relative risks of two or more populations	Menges 1990, Forsman et al. 1996, Allendorf et al. 1997
	Analyzing and synthesizing monitoring data	Menges and Gordon 1996, Gerber et al. 1999,
Guiding management	Identifying key life stages or demographic processes as management targets	Crouse et al. 1987
	Determining how large a reserve needs to be to gain a desired level of protection from extinction	Shaffer 1981, Armbruster and Lande 1993
	Determining how many individuals to release to establish a new population	Bustamante 1996, Howells and Edwards Jones 1997, Marshall and Edwards Jones 1998, South et al. 2000
	Setting limits on the harvest (or take) from a population that are compatible with its continued existence	Nantel et al. 1996, Ratsirarson et al. 1996, Caswell et al. 1998, Tufto et al. 1999
	Deciding how many populations are needed to protect a species from regional or global extinction	Menges 1990, Lindenmayer and Possingham 1996

"threatened" list. Shaffer's quantitative approach to evaluating whether a single population was large enough to make near-term extinction unlikely also inspired many similar analyses for other taxa and spurred development of the new field of population viability analysis.

Another highly influential analysis of a single population was Russell Lande's (1988a) assessment of the northern spotted owl (*Strix occidentalis caurina*), then under consideration for listing as threatened under the U.S. Endangered Species Act due to logging of the old-growth forests on which the owl depends. Treating the entire subspecies as a single population, Lande used a relatively simple model with age structure and implicit spatial dynamics (as described in Chapters 6–9 of this book) to ask whether the population was declining in the face of logging pressure. Although Lande's best estimate of the rate of population growth suggested that the population was in fact in decline, the data were insufficient to rule out the possibility of a static or even a slowly growing population. Lande's finding that the population was most likely to be declining played a role in the eventual listing of the subspecies as threatened, with a consequent reduction in logging activity on federal lands within the owl's range. The uncertainty around Lande's estimate of the population growth rate also served a useful function in indicating that more data collection was needed to more accurately assess the population's viability. In fact, given the typical scarcity of data for the rare species that are most in need of PVAs, one of the benefits of population models is that they can help us to determine which types of data would be most useful to collect in the future.

PVA Use 2: Comparing Relative Risks of Two or More Populations

Perhaps of even greater utility than a single measure of risk for one population is the ability to compare the risk of extinction among multiple populations of the same taxon. Such comparative estimates are useful in at least three ways. First, if we want to know if the entire taxon is at risk of extinction, we can ask whether all or most of its populations are declining. For example, Forsman et al. (1996), motivated by Lande's (1988a) analysis of the entire subspecies, estimated that 10 out of 11 local populations of the northern spotted owl they studied were declining significantly, strongly supporting the need to address the impact of logging on the subspecies. Second, if we seek to save all existing populations of a taxon (for example, because each harbors unique alleles) but have limited funding, measures of relative risk would allow us to prioritize which populations should receive management attention first. Third, if we have determined that we need not or cannot save all populations, metrics of relative viability could form the basis for "triage" to select the most promising populations to preserve. Allendorf et al. (1997) advocated this use of PVA to determine which populations of Pacific salmon are most worthy of preservation. Having a tool to identify the healthiest populations may be especially useful to organizations such as The Nature Conservancy, that has the goal of preserving all native species within each of its large "ecoregions" (TNC 1997) but cannot afford to purchase lands harboring every extant population of every rare species.

PVA Use 3: Analyzing and Synthesizing Monitoring Data

Monitoring information is frequently gathered on rare or potentially threatened species. For example, Recovery Plans for species listed under the U.S. Endangered Species Act often specify that important population processes, such as survival, growth, and reproduction, be monitored to assess population health (Morris et al. 2002). Similarly, Habitat Conservation Plans, also implemented under the ESA, are required to contain a monitoring component, which ranges from estimates of relative population numbers to gathering of full demographic data (Harding et al. 2001a). Thus calls to collect monitoring data are common; unfortunately, carefully considered plans to analyze these data in such as way as to inform management decisions are far less frequent. For example, how will we decide whether evidence of improvement or decline in monitoring data is sufficient to indicate a real change in imperilment? Moreover, what if monitoring indicates that the survival rates in a focal population are high relative to other known populations or species but that rates of reproduction are low? The population models that form the basis for PVA provide a way to combine such seemingly contradictory monitoring data into a single assessment of population health and to gauge how a change in a monitored population translates into a change in our assessment of its viability.

In a simple but very useful application of a PVA model to monitoring data, Leah Gerber and her co-authors (1999) asked how many years of population census data were needed to conclude with reasonable certainty that the eastern North Pacific gray whale population met criteria for delisting under the ESA. This formerly very rare population had increased rapidly from 1965 to the mid 1990s. But while delisting occurred in 1994, at that point there had been no quantitative assessment that the population was truly out of danger of near-term extinction. Gerber et al. used the simplest of PVA methods (presented in Chapter 3) to show not only that the decision to delist was warranted, but also that it could have been made safely several years earlier.

PVA Use 4: Identifying Key Life Stages or Demographic Processes as Management Targets

In addition to facilitating assessment of likely future population status, PVA is also a potentially powerful tool to identify effective management actions. For example, when different population processes or different life stages contribute differentially to population growth, conservationists will want to know which processes or stages would have the greatest impact on viability if targeted for management. Similarly, we need a means to assess whether a management strategy that affects multiple processes and stages (sometimes in opposing ways) is having the intended effect. Once again, population models can provide answers to these questions.

A now-classic example of the use of PVA to guide management decisions is the work of Deborah Crouse, Larry Crowder, and their associates on the loggerhead sea turtle (*Caretta caretta*), a threatened species that nests on beaches along the southeastern coast of the United States (Crouse et al. 1987, Crowder et al. 1994). Two major threats, trampling of eggs and hatchlings on beaches and drowning of older-aged turtles in fishing nets, were hypothesized to underlie declining numbers of loggerheads. PVA models showed that management efforts aimed solely at protecting eggs and hatchlings would not reverse declines, even if they were 100% effective. In contrast, installing TEDs (turtle excluder devices) into fishing nets to prevent drownings could result in a growing population even if the devices do not eliminate all fishing-related mortality. The ability to provide a quantitative projection of population growth under different management scenarios played an important supporting role in the decision to require the year-round use of TEDs by the shrimp-trawling fleet in the United States (Crowder et al. 1994).

PVA Use 5: Determining How Large a Reserve Needs to Be to Achieve a Desired Level of Protection from Extinction

As Shaffer found in his analysis of the Yellowstone grizzly bear population, and as we will see repeatedly throughout this book, the current size of a population has a strong effect on its risk of extinction—bigger is better. Of course, larger populations require more space. If we can estimate the amount of space each individual needs, we can use PVA models to ask how large a reserve needs to be to achieve a low probability of extinction for a particular population inhabiting the reserve.[2] Armbruster and Lande (1993) exploited this aspect of PVA to ask how large national parks in semi-arid regions of Africa must be to buffer populations of African elephants (*Loxodonta africana*) from extinction in the face of droughts of varying frequency and severity. After considering uncertainty in some of their parameters, they conclude that at least 2,500 square kilometers are needed for safe management of elephant populations.

PVA Use 6: Determining How Many Individuals to Release to Establish a New Population

Closely related to the preceding use of PVA is the problem of determining the optimal number of individuals to release when attempting to establish new populations. Translocation and release of wild-caught or captive-raised individuals to sites where the species was once present, but has since been extirpated, is often used to initiate new populations and hence to reduce species-wide extinction risk. Such translocation programs face an inevitable tradeoff: given limitations on the number of individuals that can be generated by captive breeding programs or be safely removed from existing populations, releasing more of those individuals at one site reduces the number of new populations that can be established. Thus it is advantageous to have a way to ask how small a new population "inoculum" can be and still have a reasonable chance of successful establishment (measured as the probability that the new population will still be in existence at a future time horizon). PVA models offer a tool to quantify how establishment success might scale with the number of individuals released, and they have been used to assess the likelihood of success for reintroductions of European beaver, wild boar, and capercaillie into Scotland (Howells and Edwards Jones 1997, Marshall

---

[2]Of course, it is important to preserve both the *quality* and quantity of habitat.

and Edwards Jones 1998, South et al. 2000) and release of bearded vultures in the Alps (Bustamante 1996).

PVA Use 7: Setting Limits on the Harvest or "Take" from a Population that Are Compatible with Its Continued Existence

Populations of some species (e.g., ginseng) become a concern for conservationists when direct harvesting by humans depresses their numbers to low levels. Other species suffer from indirect human-caused sources of mortality, such an inadvertent capture in fishing nets (loggerhead sea turtles, short-tailed albatross, harbor porpoise), collisions with motorized vehicles (grizzly bears, desert tortoise, Florida manatee), and destruction of required habitat due to resource extraction (northern spotted owl, red-cockaded woodpecker, Leadbeater's possum), agriculture (cheetah, San Joaquin kit fox, blunt-nosed leopard lizard), and urban development (cougars, Santa Cruz long-toed salamander). The question often arises as to how much human-added mortality is compatible with the continued existence of a species or population. For example, the U.S. Endangered Species Act terms all human-related causes of mortality for listed species as "take," and allows the taking of listed species under some circumstances. The quantitative models on which PVAs are based can be used to translate take into elevated extinction risk; this relationship is usually not straightforward. This approach has been used to identify acceptable levels for intentional harvest of both plants (Nantel et al. 1996, Ratsirarson et al. 1996) and animals (Powell et al. 1996, Heppell and Crowder 1996, Tufto et al. 1999), as well as inadvertent mortality caused by other human activities (Caswell et al. 1998).

PVA Use 8: Determining How Many (and Which) Populations Are Needed to Achieve a Desired Overall Likelihood of Species Persistence

Finally, estimates of risk for separate populations can often be combined to calculate the probability that at least one population will still be in existence at a given future time. In this way, we can gauge how much safety from extinction an entire species would gain if we preserved particular numbers and combinations of extant populations. Such analyses are important both to sets of isolated populations and for metapopulations (archipelagoes of habitat islands with populations linked by dispersal). For example, David Lindenmayer and Hugh Possingham (1996) analyzed the number and spatial arrangement of fragments of old growth montane ash forest in southeastern Australia that would have to be maintained uncut to ensure the continued existence of Leadbeater's possum. In a quite different circumstance, Eric Menges (1990) performed a simple version of a multi-population analysis for 15 populations of the endangered Furbish's lousewort (*Pedicularis furbishiae*), a plant that is restricted to the banks of a single river in Maine. Populations in moist sites close to the river have a high risk of extinction due to catastrophic events such as ice scouring in winter and burial beneath collapsing river banks, whereas those in drier sites are likely to be overcome by the inexorable encroachment of other, competitively superior plant species. Thus virtually all local populations of Furbish's lousewort are predicted to go extinct in relatively short order, and the species cannot be preserved by protecting extant populations only. Rather, management of Furbish's lousewort must focus on assuring that the establishment of new populations is sufficient to compensate for local extinctions.

Just as frequently as they have been used to suggest preservation strategies, analyses of this type have been used to design strategies for restoring or reconstructing networks of critical habitat patches. Again, the northern spotted owl provides an example. Several different PVA analyses indicated that the U.S. Forest Service's original plan to regrow patches of the owl's old-growth forest habitat wouldn't work (Doak 1989, Thomas et al. 1990, Lamberson et al. 1992, 1994). The plan was to allow small, scattered stands of younger trees to regrow to old growth forest. The models all agreed that much larger patches of habitat, even if they were farther from one another on average, would result in much higher chances of the owl's persistence, a finding that resulted in a dramatic reworking of the final plan to recreate useable habitat for the species.

Types of Population Viability Analysis

The foregoing examples illustrate some of the many ways that quantitative models of population growth, which lie at the heart of PVA, can help to illuminate important questions in conservation biology. Although the acronym PVA is commonly used as though it signified a single method or analytical tool, PVAs are in fact based upon a range of data analysis and modeling methods that vary widely both in their complexity and in the kinds and amount of data they require. We now describe the types of PVA models covered in this book.

At the simpler end of the spectrum lie models that predict only the total number of individuals in a single population. The data needed to fit these models to a particular population consist of either exhaustive counts or estimates of the total number of individuals in the population, the number of an easily recognized subset of the population (such as territory-holding males or mothers with dependent offspring), or simply a measure of relative abundance. These data are usually obtained from a series of (typically annual) censuses. Throughout the book, we refer to analyses based only on the numbers of individuals as *count-based PVAs*.

Count-based PVAs treat all individuals in the population as though they were identical. Yet we know that for species such as the loggerhead sea turtle, some individuals (e.g., mature breeders) are likely to contribute far more to future population growth than others (e.g., hatchlings, who must survive for many years in order to become reproductive), and that very different fractions of the population may be in categories with high versus low survival or reproduction (e.g., mature trees versus seedlings). Thus, we may be able to obtain a more accurate assessment of the viability of such a population if we account for these differences in individual contributions and for the fractions of individuals of each type currently in the population. Species in which individuals differ substantially in age, size, developmental stage, social status, or any other attribute that affects their contributions to population growth are said to have *structured populations*. Building a PVA for a structured population requires more information than can be obtained from simple counts of the total number of individuals in the population. Specifically, we must estimate the rates of important demographic processes (survival, growth, and reproduction) separately for each type of individual in the population. Estimating these rates typically requires us to conduct a demographic study in which we mark individuals of each type and follow them for several years, each year recording whether they survived, their type (e.g., age or size), and the number of offspring they produced. Thus the data needed to perform a structured PVA are more expensive and labor-intensive to collect, but they allow us to explore how well management techniques aimed at different types of individuals will fare, as we have seen for the loggerhead sea turtle. In this book, we refer to PVAs that require data from a demographic study as *demographic PVAs*. Even though demographic PVAs can yield more informative predictions for most species, better predictions occur only when we have enough data to estimate the many parameters these models require. With less information, the simpler count-based approach can give much more reliable results, even for species with highly structured populations (Ludwig and Walters 1985).

At their simplest, both count-based and structured PVAs describe single populations, and they do not keep track of the actual spatial locations of individuals or of suitable habitat. Yet for species such as Furbish's lousewort that exist as metapopulations, assessing the fate of a single population will tell us little about the species' vulnerability to regional or global extinction. Moreover, for organisms such as the northern spotted owl and Leadbeater's possum that are threatened by habitat fragmentation, the spatial arrangement of remaining fragments, and hence the ease with which individuals can disperse between them, can strongly dictate population viability. Thus there is often a need for PVA models to explicitly include more than one local population or patch of suitable habitat. In this book, we refer to such models as *multi-site PVAs*. This category encompasses a broad range of model types. Incidence function models, which are based on presence–absence data and predict the extinction and colonization processes of classical metapopulations, are perhaps the best-known multi-site modeling framework. "Economy-class" multi-site models simply represent a set of isolated populations of a single species, with no movement of individuals between them, and are useful for calculating the probability that at least one population will persist for a given length of time. At the extreme opposite end of the spectrum from the simplest count-based PVAs lie so-called spatially explicit, individually-based (or SEIB) models, which track the actual positions of all individuals as they are born, move, reproduce, and die on a detailed habitat landscape (sometimes constructed using geographical information systems [GIS] databases). Not surprisingly, these highly complex multi-site PVAs have the most demanding data requirements of all PVA models. Not only does one need to estimate the contributions of each type of individual to population growth, one also needs to quantify their detailed movement behaviors, as well as the locations of different types of habitat on the landscape. Consequently, highly detailed multi-site PVAs will be made possible only by extraordinary data collection efforts. As such efforts will not be a feasible option for most rare species, we do not devote much space in this book to SEIB models. We do, however, describe informative analyses that can be performed with a more easily attainable quantity of data using simpler multi-site models.

A "Roadmap" to This Book

This book follows a sequence from simpler to more complex PVA models. Chapter 2 sets the stage for all the chapters that follow by describing several metrics that all types of PVA models use to gauge population viability. Chapter 2 also reviews general factors, both intrinsic to a population and imposed upon it by environmental forces, that influence any population's risk of extinction. Each of the models we describe in subsequent chapters will attempt to account for some or all of these factors. Hence we recommend that readers unfamiliar with the causes of extinction and the ways that extinction risk is measured read Chapter 2 before proceeding to the more methodological chapters that follow. Because readers familiar with these concepts may chose to jump straight to the following chapter, we recapitulate important theoretical results at the beginning of Chapter 3.

In Chapter 3, we describe procedures to perform the simplest type of count-based PVA, specifically ones in which the change in the number of individuals in the population from one year to the next is assumed to be unaffected by the current population size (so-called density-independent models), and the impact of environmental conditions on the change in numbers is assumed to be relatively small. In Chapter 4, we broaden the scope of count-based PVAs by reviewing how several factors omitted in Chapter 3 (i.e., density dependence, correlations in environmental conditions between years, large environmental perturbations, and within-year variation among individuals in the contributions they make to population growth) can be included in the models. The material in Chapter 4 is somewhat more advanced than that in the preceding chapter, and the data needed to incorporate these additional factors into count-based PVAs will often be lacking. Indeed, of the count-based PVAs for 70 species reviewed by Eldered et al. 2002, all used the simpler approach presented in Chapter 3. Therefore, we suggest that readers mainly seeking to understand the basics of common PVA methods may wish to skip Chapter 4 on first reading the book.

Chapter 5 takes up the important issue of observation error and how to account for its effects on measures of population viability for count-based models. Whenever population size is determined by a sampling procedure (e.g., aerial surveys, mark-recapture methods, or extrapolation from quadrat samples) rather than by exhaustive enumeration, vagaries of the sampling process will introduce variation into the counts. That is, the counts will be only estimates of the true population size, and some of the variation in the counts from year to year will reflect variation in the sampling process rather than biologically meaningful variation that actually influences population viability. In addition to inflating estimates of variability, observation error introduces uncertainty into our estimated viability measures, and Chapter 5 reviews how to account for this uncertainty.

Chapters 6 through 9 review techniques for performing demographic PVAs. These analyses are based on models known as population projection matrices or demographic matrix models. Most published PVAs fall into this category, and many authors still use "PVA" to strictly mean this type of quantitative analysis. Chapter 6 shows how to use data from a study in which marked individuals were followed through time to construct a projection matrix model. In Chapters 7 and 8, we demonstrate how to calculate measures of growth and viability for a structured population using a projection matrix model. In Chapter 9, we examine ways to use projection matrices to explore management options for structured populations, as was done, for example, for the loggerhead sea turtle. We note that in a general book on PVA methods, we cannot possibly cover the topic of projection matrix models as thoroughly as does Caswell (2001) in his comprehensive book *Matrix Population Models: Construction, Analysis, and Interpretation*, to which we refer readers interested in learning more about these models. Nonetheless, our book gives a complete introduction on how to construct matrix models and perform the most useful analyses needed for a demographic PVA.

Chapters 10 and 11 focus on multi-site PVAs. In Chapter 10, we describe how to measure two additional factors, dispersal of individuals and spatial environmental correlations, with which we did not concern ourselves in the preceding chapters but that may need to be estimated and incorporated when the joint viability of more than one population is being assessed. Chapter 11 covers several techniques for actually performing multi-site PVAs, such as incidence function models and methods based on both count and demographic data. As noted above, we concentrate on analytical methods most likely to be useable in many circumstances, eschewing the many spatial models that require extraordinary amounts of data to be useful.

In Chapter 12, we discuss cautions and caveats that help us to decide when it is useful, and more importantly when it is *not* useful, to perform a PVA. All conservation biologists are familiar with the following conundrum. It is difficult to make an accurate prediction of the future status of a population with only a limited amount of data. Yet it is precisely the rarity of threatened and endangered species that makes it both difficult to obtain a large quantity of population data and especially desirable to have some sort of population assessment. As a result, practitioners of population viability analysis will be perennially forced to operate in an arena in which data scarcity is the rule rather than the exception. In recognition of this reality, we will repeatedly argue in this book that measures of population viability should be viewed not as iron-clad predictions of population fate but as works in progress subject to updating as more data become available. But when data on a particular species are truly scarce, performing a PVA may do more harm than good by engendering a false sense of rigor. In such cases, basing conservation decisions on other methods makes far better sense. Thus we emphasize from the outset that although we view PVA methods as a potentially useful set of conservation tools, we do not see PVA as a panacea for all conservation problems.

PVA is an inherently quantitative endeavor, so naturally we rely on mathematical symbolism to describe the underlying models in the clearest, most concise way. To make it more transparent when the same basic quantity is being referred to in different PVA approaches, we have tried to use the same set of mathematical symbols consistently throughout the book. We have placed definitions of commonly used symbols into an Appendix (see pages 455–458) to which you may refer as you read successive chapters.

Our Modeling Philosophy: Keep It Simple

Population biologists have developed a vast array of complex and mathematically sophisticated models, many of which can be adapted to predict the likelihood of population extinction. This book does not attempt to review all of those models. Instead, we focus on the subset of all available PVA methods that we deemed to be the most practical given the types of data typically available for species of conservation concern. In making the decision of which methods to include, we used two simple and related rules of thumb that we think could be said to capture our general philosophy regarding the use of population models in conservation biology.

The first rule is: "Let the available data tell you which type of PVA to perform." More specifically, we should not seek to build a PVA model that is more complex than the data warrant. It is our view that when data are limited (as they almost always will be when we are dealing with the rare, seldom-studied species that are the typical concern of conservation planners) the benefits of using complex models to perform population viability analyses will often be illusory. That is, although more complex models may promise to yield more accurate estimates of population viability because they include more biological detail (such as migration among semi-isolated populations, the effects of spatial arrangement of habitat patches, and the nuances of genetic processes such as inbreeding depression, gene flow and genetic drift), this gain in accuracy will be undermined if the use of a more complex model requires us to guess at critical components about which we have no data. Instead, our philosophy is that the choice of models and methods in PVA should be determined primarily by the type and quantity of data that are available and not by the desire to include all interesting and possibly important processes. It is better to use a simple approach (keeping the simplifications in mind) than to construct a complex house of cards that relies on numbers with no empirical justification.

The second rule is: "Make sure you know what your model is doing." This rule will be easier to follow if we also heed the first rule—that is, if we keep our models simple. This second rule is germane to the question of whether one should build one's own PVA model or use one of the software packages designed specifically for PVA that are now widely available, such as ALEX (Possingham and Davies 1995), GAPPS (Harris et al. 1986, Downer 1993), INMAT (Mills and Smouse 1994), RAMAS (Akçakaya and Ferson 1992, Ferson 1994, Akçakaya 1997), ULM (Ferrière et al. 1996), or VORTEX (Lacy et al. 1995).[3] Careful use of these programs can certainly lead to a defensible viability assessment. However, naïve users of the programs run the risk of making two errors.

First, without fully understanding the underlying models used by the programs, users may build into the program incorrect assumptions about the biology of the species or population under study, leading to incorrect estimates of population viability. For example, some of these programs incorporate density dependence in very specific ways that may not be appropriate for the organism under study (Mills et al. 1996).

Second, inexperienced users may be lured by the array of options proffered by the software into including risk factors about which no data exist, thus violating our first rule. We hold that, even if one ultimately plans to use one of the software packages to perform PVAs, it is essential that one begins by learning how to build one's own model from scratch, so as to fully understand what the packages are doing. In this book, we demonstrate through worked examples how to go from raw data to a fully parameterized population model and then how to use the model for assessment and management. We illustrate this process by showing actual statistical analysis of the raw data, and by providing computer code that actually runs the models. Our

[3]For comparisons involving some of these programs, see Lindenmayer et al. 1995, Mills et al. 1996, and Brook et al. 2000b.

computer programs are written in MATLAB,[4] a matrix language that is relatively easy to learn and that is particularly well-suited to demographic PVA models. Readers who are not familiar with MATLAB and would like a quick introduction to its use in an ecological context should consult the appendix in Roughgarden (1997). We strongly urge readers to try to understand the structure of the programs we present as a way of truly knowing how the underlying models work. To make the programs as accessible as possible, we pepper them with extensive comments that indicate what the code is doing at each point.[5] Readers who wish to write their own programs in another language can treat the comments as a form of pseudocode. These programs are not simply another form of "canned" model in which all one needs to do is plug in some numbers. Rather, each program performs only a subset of the tasks involved in a complete PVA. In most cases, you will need to combine them, and even modify them, to suit the needs of your particular PVA. Indeed, a strength of building your own PVA model from scratch is that it can be far more flexible than any pre-packaged software would allow. Nevertheless, the shards of MATLAB code presented in this book should allow you to rapidly do all the operations necessary for most PVAs. To make it easier for you to use the computer code we present, all of the programs and data files can be downloaded from the publisher's website (www.sinauer.com/PVA/).

We end by emphasizing that the worked examples in the subsequent chapters employ data from actual populations of rare or declining species. That is, we do not use common species for which data are abundant simply to "beautify" the outcome of our analyses. In this way, we aim to illustrate the constraints of assessing population status with limited data, constraints that are a realistic and unavoidable feature of population viability analysis.

---

[4]MATLAB is a widely available commercial product. It is relatively inexpensive for students and relatively expensive for others (see http://www.mathworks.com/ for current prices and licensing arrangements). We use it because of its strong advantages for matrix manipulations and, indeed, for compactly programming many other population models. It has also increasingly been adopted by many theoretical ecologists as a lingua franca. There are several less expensive options that still allow you to reap the advantages of MATLAB. For example, SCILAB, a shareware program that uses MATLAB-like syntax, can be downloaded at http://www-rocq.inria.fr/scilab/.

[5]The "help" feature of MATLAB also makes the program easy to learn. To use it, simply type "help" followed by a MATLAB keyword or the name of a MATLAB function (examples: "help for", "help eig") in the command window, and MATLAB will respond with a detailed explanation. If you are a MATLAB novice, you should make liberal use of the "help" command when trying to understand the programs in this book.

2

The Causes and Quantification of Population Vulnerability

Whatever approach is taken to the development and interpretation of a population viability analysis, every PVA is based on the same basic biological and mathematical principles, which taken together determine extinction risk. The specific endangering factors will differ from one species or population to another: one group may suffer increased mortality due to contact with humans in fragmented landscapes, whereas another may experience reduced fecundity due to competition with invasive exotics. However, the multitude of ecological and genetic factors that can threaten a population—and may possibly be mitigated through proper management—can all be combined into a more limited set of rates and processes that include:

- the life history of the species
- the average environmental conditions
- the extrinsic variability in the biotic and abiotic factors influencing a population (environmental stochasticity)
- the intrinsic variability caused by small population sizes (demographic stochasticity)

This small set of processes in turn influences the mean and variance of birth, death, and growth rates of a population, which themselves determine population growth and extinction risk (Figure 2.1). The recognition that this small set of processes and causal links together determine population viability underlies all PVA models. A clear understanding of the features of population dynamics that determine viability is important to understanding why particular processes must be measured in order to construct a PVA.

In this chapter, we first describe the most important factors that influence population viability. In particular, we discuss the importance of mean demographic rates and the different forms of temporal variability that affect these rates. We then give a brief overview of why temporal variability is so important in determining population growth and persistence. Following

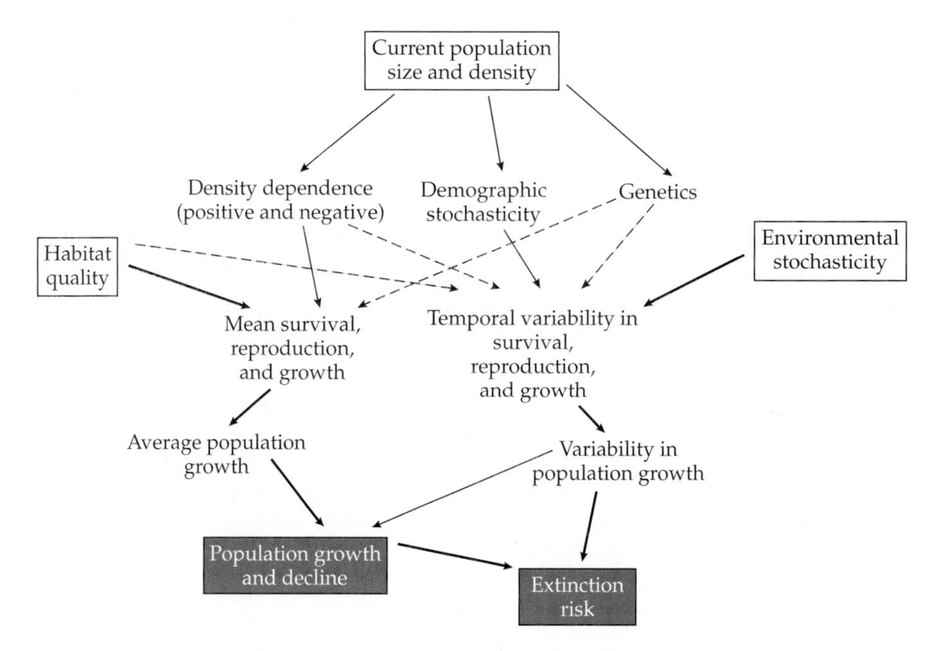

Figure 2.1 One way to represent the different factors influencing population viability. The construction of this boxes-and-arrows diagram emphasizes the role of variability in influencing population growth and the probability of extinction. The width of each arrow indicates the general strength of an effect.

this discussion of the most common factors influencing population viability, we discuss several other factors that are less frequently included in PVA models but should be generally understood and carefully considered when assessing population viability. These factors are spatial variation, density dependence, and genetic processes. Finally, we discuss several different metrics that PVAs of each of the types described in Chapter 1 can use to express a population's viability (or conversely, its risk of extinction).

Mean Vital Rates and Population Viability

A population's health is ultimately determined by the performance of individuals, summarized by their birth, death, and growth rates. Throughout this book, we will refer to these three components of individual performance as *vital rates*. The average values of vital rates are the bread and butter of demography and can easily be translated into measures of mean population performance, using either life tables or projection matrix models (Chapters 6–9). Two of the most common of these measures are R_0, the so-called *net reproductive rate*, which represents the average number of female offspring produced by a female over her entire life); and λ (the Greek letter "lambda"), the *annual population growth rate*, which is defined by the equation

$$N_{t+1} = \lambda_t N_t \qquad (2.1)$$

Here, N_t is the population size in year t and λ_t is the annual population multiplication rate, expressing population size in year $t + 1$ as a multiple of the preceding year's number, N_t. If growth rate is the same in all years, λ with no subscript is used, and we will use λ_A to symbolize the arithmetic mean of different λ_t values over many years. Each of these measures give a quick assessment of population viability. Obviously, if the average individual in a population does not replace itself over its lifespan ($R_0 < 1$), then the whole population will eventually decline. Similarly, if on average the size of a population in one year is only a fraction of its size in the previous year ($\lambda < 1$ or $\lambda_A < 1$), then the population is headed towards extinction.

Even though the interpretation of these basic measures of mean population performance is clear, how to reverse a population decline can be less so unless we understand the proximate causes of poor reproduction, survival or growth. Much of population viability analysis focuses on the relative importance of the different forces influencing different mean vital rates, in an effort to pick out the most effective interventions to reverse population declines (see Chapter 9).

For some high-profile species there are good, or at least adequate, estimates of most vital rates; but for many species we can only estimate the growth rate of the population as a whole from census data, using the relationship $\lambda_t = N_{t+1} / N_t$ to calculate the rate of population change (see Chapter 3). In this case, it can be hard to conduct a rigorous analysis of how best to intervene to improve poor population growth rates, because our data (and hence the models we construct) don't explicitly account for the underlying causes of population growth (i.e., the birth, death, and growth rates of individuals). Still, the mean rate of population change gives basic information on whether a population is or is not declining in most years, and thus whether further research is needed to understand and reverse any downward trends.

But viability, and thus PVA, involves more than just the mean rate of population growth. Although estimates of mean performance are the centerpiece of any PVA, the variability around mean rates is also key to determining viability. In particular, we will see that even if on average λ exceeds 1.0, a population can still be highly endangered if growth rates vary through time. Just as important, interpretation of data and estimation of model parameters are complicated by variability, and correct estimation of mean rates requires that variation be accounted for properly. We will now describe the different kinds of temporal variation that can impinge on vital rates and thus on overall population growth.

Temporal Variability in Vital Rates

In the real world, most vital rates undergo detectable temporal variation. For a demographic PVA, we think of variation in vital rates such as fecundity and survival, while for a count-based PVA, we think of the net result of these changes: the variation in overall population growth. Often this variability is not directional, which would result in changing mean rates of birth, survival and growth, but instead consists of "random" changes around the same mean rates. This random variability though time is known as *temporal stochasticity*. There can also be important spatial variability in vital rates, as well as interactions between spatial and temporal variation, leading to correlations among the temporal fluctuations observed in different places, or spatio-temporal variation. Here we outline the different causes and consequences of temporal variability and briefly touch on problems of measuring them.

Three kinds of temporal stochasticity are usually distinguished: environmental stochasticity, catastrophes, and demographic stochasticity. To these, we will add a fourth type that is rarely mentioned: bonanzas. Later in the chapter, after exploring the importance of temporal variability for population viability, we will return to the complications of spatial and spatio-temporal variability.

Sources of Temporal Variability: Environmental Stochasticity

The term *environmental stochasticity* describes temporal variation in vital rates driven by changes in the environment that are inherently erratic or unpredictable (unlike, for example, predictable seasonal changes). The causes of this variation are changes in the biotic and abiotic forces impinging on the population, such as changes in predator or pathogen numbers, prey quality or abundance, rainfall, winter temperatures, and so on. Changes in any of these forces may cause survival, reproduction, and growth rates to differ from year to year.

All vital rates vary over time, but variation is much higher for some rates than for others. For example, annual newborn survival is much more variable than is adult survival for most ungulate species (Galliard et al. 1998), and for many other species the survival of reproductive adults shows lower variation than do other, less critical vital rates (Pfister 1998). Basically, environmental stochasticity is "normal" variation in vital rates from year to year caused by changes in environmental factors.

An important point to reiterate is that, as the term is usually used, environmental stochasticity does *not* include consistent trends in the environment that cause parallel trends in vital rates. For example, through a successional sequence, a colonizing plant species such as Furbish's lousewort may have initially high rates of growth and reproduction that fall off as time passes (Menges 1990). Although annual fluctuations around the trend of declining performance would be considered environmental stochasticity, the more or less deterministic trend in performance through succession is not. Similar trends that often arise in conservation settings include increased dispersal mortality due to ongoing habitat fragmentation and declining survival due to steady increases in the density of invasive species. Changes in mean rates present different problems for both parameter estimation and viability assessment than does environmental stochasticity.

Two final aspects of environmental stochasticity involve correlations in vital rate values within and between years. Not surprisingly, the values of different vital rates are usually correlated with one another. For many species, good years for survival also tend to be good years for reproduction and growth, and good years for the survival of adults are also good years for the survival of juveniles, since the same environmental factors affect all these rates. These correlations can have substantial effects on predicted population viability. For the threatened desert tortoise, PVA models with and without the estimated correlations between different demographic rates differed little in mean growth rate (Doak et al. 1994). However, including the measured correlations resulted in much wider confidence limits of predicted future population sizes, including greater chances of near-term extinction (Figure 2.2). This is because the correlations tend to make a bad year bad for the whole population, and a good year good for the whole population. The net effect is to increase the variability in overall population growth rates, which will decrease population viability. (We explain more about why high variability drives down viability in a subsequent section of this chapter and in Chapter 3.). Conversely, for some species, the same events that kill adults will allow more successful recruitment of seeds or other propagules. In this case, including an estimate of correlated changes in different vital rates will raise estimates of population viability, because the correlations tend to reduce variation in λ_t values.

In spite of their strong effects, temporal correlations in vital rates are often not included in PVAs because they are difficult to estimate accurately with limited data (Ferson and Burgman 1995). However, such correlations should be incorporated into demographic PVAs whenever possible, and we discuss how to do so in Chapters 7 and 8. Even if the correlations cannot be estimated directly, one can run multiple models with different likely correlation patterns to try to bracket the real but unknown pattern (Ferson and Burgman 1995).

Correlations in the vital rates of a population can also occur between years, due either to persistent environmental conditions (such as multiyear droughts or warming events, including El Niño periods) or to persistent high or low population sizes of other species in the community, such as important predators and prey. Correlations in either the same or different vital rates between years result in "autocorrelation" in the values of λ_t—correlation between the value of λ_t one year with those in other years. Such

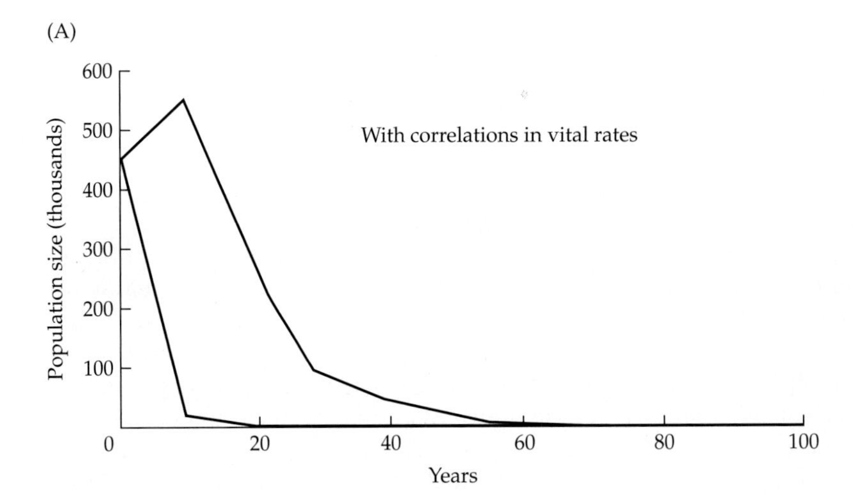

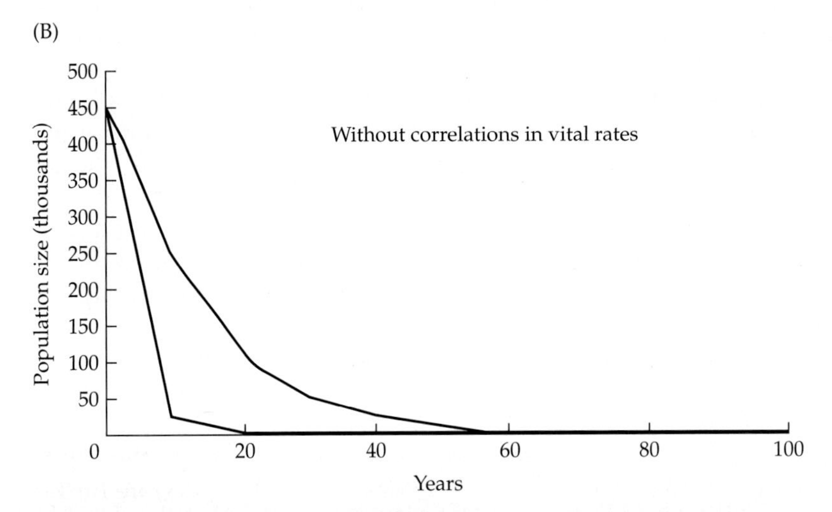

Figure 2.2 95% confidence limits around predicted future population sizes for the threatened desert tortoise of the western North American deserts. Top panel shows results from PVA simulations including the estimated correlations in the annual variation in vital rates, while the bottom panel is from simulations assuming uncorrelated variation in vital rates.

correlations can occur over different time scales. For example, a fire in many chaparral or forest communities decreases the chances of another fire in the short term, with increasing probabilities of another burn over time, as plants regrow and fuel accumulates. However, for the most part autocorrelation is strongest in the short-term, and most of these correlations are positive, with adjacent years being similar to one another (Steele 1985, Halley 1996). As we discuss briefly below, and in more detail in Chapter 4, this positive autocorrelation can alter extinction risk, sometimes in complex ways. Specifically, in the absence of density dependence, which we define and discuss later in the chapter, positive autocorrelation increases extinction risk, whereas it can increase or decrease the risk for density-dependent populations.

Sources of Temporal Variability: Catastrophes and Bonanzas

The conservation literature is filled with discussions about "catastrophes," extreme years with very low survival or reproduction, and their impact on population viability (e.g., Mangel and Tier 1993, 1994, Lande 1993, Young 1994, Ludwig 1999). However, equally or more important for many species are "bonanzas,"—extremely *good* years. For many species, these rare good years are when all or most reproduction or recruitment occur, often as a result of what might at first appear to be a catastrophe, such as a flood or fire (Burgman and Lamont 1992, Gross et al. 1998). In either case, bonanzas or catastrophes are generated by environmental variability and thus are part of the general phenomenon of environmental stochasticity. Indeed, the term *catastrophe* has been greatly overused in conservation biology, because "catastrophic" events are not always clearly separated from "normal" environmental stochasticity. Catastrophes and bonanzas will be clearly distinguished only when rare events lead to a bimodal distribution for vital rates or the population growth rate. If the frequencies of very high or very low survival years form a continuum with the frequencies of more average years, then there is little reason to peel these years off and treat them as distinctly different. Nevertheless, for some species there is clear evidence of catastrophes or bonanzas that are true outliers, often caused by epizootics, fires, truly severe droughts, or other infrequent events (Young 1994). In these cases, normal variability essentially never generates vital rates anywhere near as good or as bad as those found in the small number of truly exceptional years. For example, the normal annual mortality rates of the giant columnar saguaro cacti in southern Arizona is at most 5%, while rare freezing events, such as the one that occurred in 1961, can cause much higher mortality (Steenbergh and Lowe 1983). In this and similar situations, it is important to try to estimate the occurrence of catastrophes and bonanzas separately from normal environmental stochasticity.

Although infrequent, catastrophes can strongly influence extinction risk by increasing variability. Modeling work has shown that catastrophes can endanger populations that are otherwise nearly extinction-proof (Lande 1993). The importance of both catastrophes and bonanzas is that their frequency and magnitude can be difficult to quantify, and yet they may be critically important in determining long-term population viability. We are often in the situation in which we know that the occurrence of rare events is important, yet we cannot easily estimate the rates governing the frequency of these events or the magnitude of their effects. Aside from dispersal rates in multi-site PVAs, there is probably no other aspect of PVAs that more often forces us to rely on educated guesses, rather than data, than do catastrophe and bonanza years. For example, in a PVA for a small population of cougars in the Santa Ana Mountains of Southern California, Paul Beier (1993) wanted to include the effects of infrequent series of drought years. Without any empirical data on how such droughts affect cougars, he was forced to make educated guesses about the frequency, duration, and demographic effects (which he included as a reduction in carrying capacity) of droughts. Even though Armbruster and Lande (1993) were able to combine data on survival and reproduction of African elephants during mild and moderate droughts with long-term climate records on the frequency of such events, their approach towards quantifying drought year effects also had to rely on some guesswork and assumptions.

Sources of Temporal Variability: Demographic Stochasticity

The third type of temporal variation, demographic stochasticity, is distinctly different from the previous two. Both environmental stochasticity and catastrophes/bonanzas are due to alterations in environmental conditions that change, from one year to the next, the values of the mean vital rates across all individuals in the population. Because these changes are due to effects of the environment on individual performance, the magnitudes of such changes will often be unaffected by the size of a population. In contrast, demographic stochasticity is temporal variation in population growth driven by chance variation in the actual fates of different individuals within a year, and its magnitude is strongly dependent on population size. Demographic stochasticity is essentially the same as the randomness that causes variation in the numbers of heads and tails you get if you repeatedly flip a coin.

An example is probably the best way to explain demographic stochasticity. Consider the effort to reintroduce into the wild the critically endangered California condor, the objective of an intensive captive breeding effort. To date, over 89 releases have taken place in California and Arizona. The annual survival rate in the wild of the 28 birds released in Arizona was estimated to be 0.85 (Meretsky et al. 2000). This number is corrected for recaptures of ailing individuals and is quite low for a species with delayed maturity and a very low reproductive rate. Say that we have only a single pair of condors to release (one female and one male), and we want to know if both will survive their first year in the wild before they can breed. If each bird has an 85% chance of surviving, we would expect to see $2 \times 0.85 = 1.7$ live condors one year after the release. Although this is indeed the *expected* number of birds, it would be absurd to consider only this expectation (as, of course, we do not expect to see fractional condors). Rather, we are actually concerned with three possible outcomes: both birds survive, only one does, or neither do. It is straightforward to calculate the probabilities of these three events (Table 2.1). So we'd expect to see both birds survive about 72% of the time,

TABLE 2.1 Possible outcomes of releasing a pair of condors, each with a survival probability of 85%

Event	*Fate of female condor*	*Fate of male condor*	*Probability*
Both survive	Live ($p = 0.85$)	Live ($p = 0.85$)	$0.85 \times 0.85 = 0.7225$
One bird survives	Live ($p = 0.85$)	Die ($p = 0.15$)	$0.85 \times 0.15 = 0.1275$
	Die ($p = 0.15$)	Live ($p = 0.85$)	$0.15 \times 0.85 = 0.1275$
Neither survives	Die ($p = 0.15$)	Die ($p = 0.15$)	$0.15 \times 0.15 = 0.0225$

only one survive about 26% of the time ($12.75 + 12.75 \approx 26$), and neither survive about 2% of the time. [Notice that the expected number of surviving birds is $(2 \times 0.72) + (1 \times 0.26) + (0 \times 0.02) = 1.7$, just as we calculated above.]

Now consider what would happen in multiple releases, in each of which the "population" starts with only two birds. Assume that there is no environmental stochasticity—that is, the chance of surviving remains absolutely fixed at 85% for every individual in every population in every year. If we account for chance variation in the fates of individuals, we would *not* expect the number of surviving birds to always be the same, but rather to vary from 0 to 1 to 2 among populations (with the probabilities given in Table 2.1), *even though environmental conditions are identical.* Thus random variation among individual fates introduces variability into the amount of annual population change (and hence population numbers) that is distinct from the influence of environmental variation. In this example, even though the most likely outcome of releasing a single pair is for both birds to survive, 28% of all releases would be expected to fail in their first year, due to either to the loss of both birds, or the loss of only one. Thus, demographic stochasticity is a powerful force to consider when designing release programs. Although in this example we have considered chance variation in survival among individuals, demographic stochasticity can create variability in the number or sex ratio of offspring produced, or even in growth and maturation rates. Demographic stochasticity can cause outcomes to differ across separate populations and can also generate variability from year to year in the observed growth rate of a single population.

Just as it can influence the success of new populations, demographic stochasticity can also impact the fate of established populations because, just like environmental stochasticity, it creates variability around mean vital rates. However, demographic stochasticity creates substantial variability at only low population sizes, and thus is a critical factor only in fairly small populations. Table 2.2 shows the variability in survival rates that we would expect to observe for larger and larger populations of condors under the influence of demographic stochasticity. The larger the population, the less

Table 2.2 Percentage of populations expected to show different observed survival rates under demographic stochasticity[a]

	Population size						
Observed survival	*2*	*4*	*8*	*16*	*32*	*64*	*128*
0	2	0	0	0	0	0	0
0–0.1	0	0	0	0	0	0	0
0.1–0.2	0	1	0	0	0	0	0
0.2–0.3	0	0	0	0	0	0	0
0.3–0.4	0	0	0	0	0	0	0
0.4–0.5	26	10	2	1	0	0	0
0.5–0.6	0	0	8	7	2	0	0
0.7–0.8	0	37	24	13	18	15	6
0.8–0.9	**0**	**0**	**38**	**51**	**53**	**71**	**89**
0.9–1.0	0	0	0	21	27	14	4
1	72	52	27	7	1	0	0

[a]We assume that each individual has an annual survival rate of 0.85. The range of survival rates centered on the true mean (0.8–0.9) is in bold. With smaller population sizes, there is much more variability in observed survival, as each animal must either live or die.

likely we are to see large deviations from the expected survival rate of 0.85. Thus with even moderately large numbers, demographic stochasticity ceases to create very much variation. It has been argued that a good rule of thumb is to worry about demographic stochasticity only if a population is smaller than about 20 individuals (Goodman 1987, Lande 1993). But more recently, Lande (1998) has suggested that considerably larger numbers may be needed for the force of this inherent randomness to be safely ignored. Finally, Bruce Kendall and Gordon Fox (Fox and Kendall 2002, Kendall and Fox 2002) have recently shown that demographic stochasticity is likely to be *less* important than is usually predicted, due to inherent differences in survival rates among individuals. That is, due to genetic or environmental factors, even individuals of the same size, age, sex, and other characteristics do not have identical probabilities of surviving (as we assumed in the condor example and as is almost always assumed when including demographic stochasticity in PVAs). As a result, individuals with high probabilities of living will buffer the population mean, reducing the effects of demographic stochasticity, even at small population sizes. This same effect may also reduce demographic stochasticity in growth and reproduction, although for these rates individual differences could also have the reverse effect, *increasing* demographic stochasticity (Fox and Kendall 2002). Although not yet considered in any PVA analysis, these effects of individual variability are not, in theory, difficult to estimate and account for in reproduction or growth—

traits that can be repeatedly measured for the same individual. However, an individual's death can only be observed once, and thus how to practically measure individual variability in survival in wild populations remains unclear.

Given these conflicting results, how best to deal with demographic stochasticity in a PVA is a somewhat open question. It is not difficult to incorporate the simplest kind of demographic stochasticity into a PVA model by simulating the separate fate of each individual on a computer. However, the safer route is to use a PVA not to predict the probability of extinction, but rather the probability of falling to a population size at which there is a very high and immediate threat of extinction due to demographic stochasticity or other factors that come into play only at small numbers. If we set a so-called quasi-extinction threshold (which we will discuss in more detail later in the chapter) at the population size below which demographic stochasticity is likely to come into play, especially for the most important stages in the life cycle (often the reproductive individuals), it allows us to analyze viability without trying to fully account for the effects of demographic stochasticity. As a rough guideline, we suggest that if there are at least 100 individuals in the population for a count-based PVA, or at least 20 individuals in the most important life stages in a demographic PVA (often the reproductive adults), it is usually safe to ignore demographic stochasticity. This approach also makes sense because several other effects will begin to plague populations at these extremely small numbers (see the discussion that follows). In general, we deal with demographic stochasticity in this book in this way, thus avoiding the need to directly analyze its effects. However, in Chapters 4 and 8 we discuss how to include demographic stochasticity in a PVA when data on variation in reproductive success or survival among individuals are available.

The Effects of Temporal Variability on the Rate of Population Growth

In this section, we demonstrate why it is crucial to include the effects of temporal variability in any viability analysis. In the real world, population growth rates fluctuate over time as a result of all of the sources of variability we outlined in the preceding section. However, we saw that demographic stochasticity is of little importance unless population size is quite small, and that catastrophes and bonanzas are by definition rare events. Hence, to simplify this initial exploration of variability's effects, we will assume that variation in vital rates is caused solely by run-of-the-mill fluctuations in environmental conditions. How does this environmental stochasticity influence whether a population will grow or decline over the long term? The most important result is simple: Adding variation to population growth does *not* simply mean that growth is more variable; it means that populations mostly do worse than they would without variation. In other words, using a simple average to characterize the population growth rate in a variable environment is not just a simplification, it is actually wrong, consistently overestimating population performance.

To see why this is true, we need to quantify how much a population grows through time if its growth rate is variable. Again, to keep this explanation simple we will use a model for total population size, where environmental effects on vital rates are summarized into a single population growth rate, λ_t. First, we modify Equation 2.1 to allow the population growth rate to adopt one of two equally likely values at each time step. For example, assume that

$$N_{t+1} = \lambda_t N_t \quad \text{where } \lambda_t = \begin{cases} 0.86 & \textit{with probability } \frac{1}{2} \\ 1.16 & \textit{with probability } \frac{1}{2} \end{cases} \tag{2.2}$$

The arithmetic mean[1] of the two λ_t's equals 1.01. As above, we denote this arithmetic mean λ_A. If the population growth rate *always* had this arithmetic mean value, the population would increase in size by 1% per year. To compare the stochastic situation in Equation 2.2 with the results of using the arithmetic mean λ_A every year, we can "play out" both scenarios. Because $N_1 = \lambda_0 N_0$ and $N_2 = \lambda_1 N_1$, $N_2 = \lambda_1 \lambda_0 N_0$. More generally,

$$N_{t+1} = (\lambda_t \lambda_{t-1} \lambda_{t-2} \cdots \lambda_2 \lambda_1 \lambda_0) N_0 \tag{2.3}$$

For the deterministic case, if the starting population size, N_0, is 100, then after 500 years, $N_{500} = N_0 \cdot (1.01)^{500} = 100 \cdot 144.77 = 14{,}477$. Notice that since all the λ_t's are equal we can rearrange this equation to estimate λ, just from knowing the starting and ending population sizes and how much time has passed: $(N_{500}/N_0)^{(1/500)} = (14{,}477/100)^{(1/500)} = 1.01$.

In the stochastic case, because λ_t varies unpredictably from year to year, we could represent one possible outcome of population growth as $N_{500} = (1.16 \cdot 1.16 \cdot 0.86 \cdot 1.16 \cdot 0.86 \cdots 0.86) N_0$ where the term in parentheses abbreviates the product of 500 values of λ_t. We don't know the exact numbers of years with $\lambda_t = 1.16$ and with $\lambda_t = 0.86$, but they are likely to be about equal (250 each) over a period as long as 500 years. Therefore, the most likely outcome of the stochastic growth case is that $N_{500} \approx N_0 \cdot (1.16)^{250} \cdot (0.86)^{250} = 54.8$. Notice that the stochastic growth process is very likely to lead to population *decline*. In contrast, and as we calculated above, if the arithmetic mean value of λ

[1]The arithmetic mean is the simple average with which we are all familiar. That is, the arithmetic mean of the set of n equally likely numbers $x_1, x_2, x_3, \ldots, x_n$ equals $(x_1 + x_2 + \cdots + x_n)/n = \frac{1}{n}\sum_{i=1}^{n} x_i$. Or, if the numbers occur with the unequal probabilities $p_1, p_2, p_3, \ldots, p_n$, their arithmetic mean equals $\sum_{i=1}^{n} p_i x_i$.

(1.01) occurs every year, the population *grows*, ending up 264 times greater by year 500 than in the most-likely stochastic case.

To better measure what stochasticity does to population growth, we can rearrange the stochastic equation as we did in the deterministic case and use the starting and most-likely ending population sizes to estimate a most-likely stochastic growth rate: $(N_{500}/N_0)^{(1/500)} = (54.8/100)^{(1/500)} = 0.9988$. This value is the constant annual growth rate that would give the same final population size as does the most-likely outcome of the stochastic growth process. Because this value is less than one, it correctly predicts (unlike λ_A) that the population is likely to decline in a stochastic environment. It is also easy to show that this most-likely stochastic growth rate is the so-called geometric mean[2] of the λ_t values (which we will denote λ_G). In our hypothetical example in which λ_t can take on one of only two equally likely values, $\lambda_G = (0.86)^{1/2}(1.16)^{1/2} = 0.9988$. Importantly, the geometric mean of a set of numbers is *always* less than or equal to its arithmetic mean, with the difference between the two increasing with increasing variance in the numbers being averaged. In other words, with stochasticity in the annual growth rates, the population acts as though it were governed by a constant annual growth rate that is less than the arithmetic mean of the actual, variable, growth rates.

The key idea to get from this exercise is that adding variation to λ usually makes the population grow slower over the long haul than it would with the constant, arithmetic mean growth rate. This is a basic feature of a stochastic multiplicative growth processes: the "average" result is given by the *geometric* mean of the multiplier, *not* its arithmetic mean. This basic result occurs regardless of the distribution of λ_t values. Realistically, λ_t does not always equal one of only two values, as in our example, but can take on any value between 0 and some biologically possible maximum. If this distribution is approximately lognormal, the extent that increasing variance depresses the mostly likely stochastic growth rate can be easily calculated (Figure 2.3).

The dependence of stochastic population growth on temporal variability helps to explain many aspects of population viability. For example, the increase in extinction risk of a density-independent population caused by positive autocorrelation in the growth rates can be understood by thinking about population growth over two year time steps. Each two year growth rate is the multiple of two back-to-back λ_t values. If these values are positively correlated (with both typically being high or both being low), then the two year growth rates will have much higher variance than if there is no correlation (with many combinations of a high and a low λ_t occurring), and this

[2]In contrast to the arithmetic mean (defined in footnote 1), the geometric mean of a set of n possible outcomes $x_1, x_2, x_3, \ldots, x_n$, which occur with the respective probabilities $p_1, p_2, p_3, \ldots, p_n$, equals $x_1^{p_1} \cdot x_2^{p_2} \cdots x_n^{p_n} = \prod_{i=1}^{n} x_i^{p_i}$.

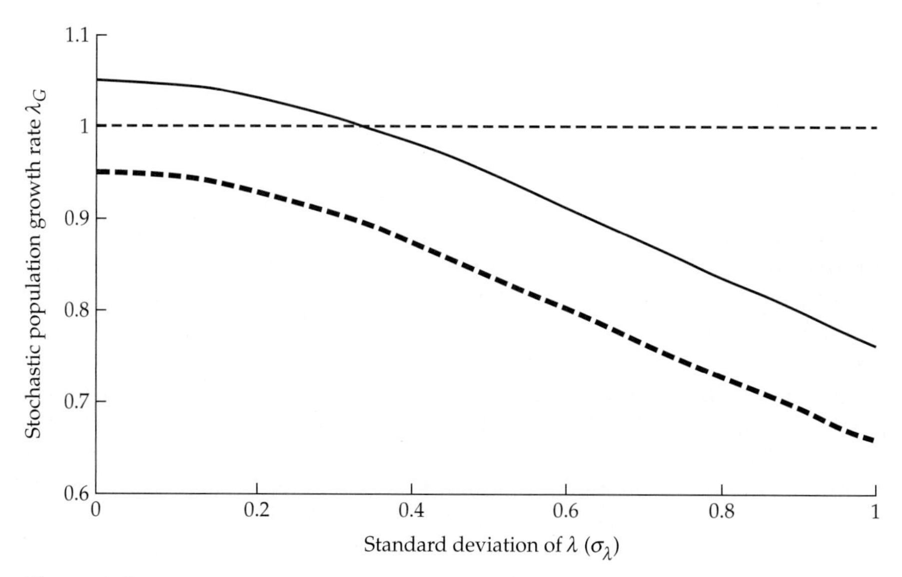

Figure 2.3 The most-likely stochastic growth rate (λ_G) is shown for different amounts of variation in annual population growth and for two arithmetic means: either λ_A = 1.05 (upper line) or λ_A = 0.95 (lower line). The calculations assume that λ_t is lognormally distributed, with mean λ_A and standard deviation σ_λ, so that the most-likely stochastic growth rate is $\lambda_G = \lambda_A \big/ \sqrt{1+\sigma_\lambda^2/\lambda_A^2}$. The horizontal dashed line indicates a stable population (λ_G = 1).

higher variance will in turn lead to a lower geometric mean growth rate, as we've just seen. This is just one example of how a basic understanding of stochastic growth rates can make sense of otherwise nonintuitive results.

Of course, if λ_t is variable, population growth is not just slower over the long term, but is also more variable. We treat this topic in considerable detail in Chapter 3, and here just illustrate the two main patterns in this variability. The first concerns our uncertainty about the fate of a population over time, which is key to the prediction of population size or extinction probabilities in the future. The likely future sizes of a stochastically growing population will rapidly diverge (how rapidly depends on the variance in growth rates). To illustrate this, we used data from a single population of the endangered California clapper rail population, which now exists as a set of small, largely isolated populations living in salt marshes around San Francisco Bay, California. For this population, there are five estimated λ_t values (Harding et al. 2001b), and we randomly chose between them each year to simulate the observed environmental variability in annual growth rates, using the MATLAB code from Box 2.1; this is the simplest, though often not the best, way to use count data to predict stochastic population growth (see Chapter 3 for

BOX 2.1 ***MATLAB code to simulate population trajectories using Equation 2.1, and drawing each year's annual growth rate from a list of observed rates.***

```
% PROGRAM RandDraw.m
% Does multiple simulations of discrete exponential growth
% trajectories with a set of observed lambda values.

%**************** SIMULATION PARAMETERS *********************
lams = [1.00, 1.98, 1.02, 0.92, 0.53];
        % The different lambda values to use. There can be as
        % many as you want. In the simulations each will be
        % drawn with equal probability.
numzero = 29; % starting population size
tmax = 100;      % length of simulations
numreps = 100;   % number of replicate trajectories to simulate
outname = 'temp24'; % name of file in which to store results
%***********************************************************


rand('state',sum(100*clock)); %this seeds the random number
   % generator, which must be done in all Matlab programs
numlams = length(lams); % how many lambda values are there?
intvalues = floor(numlams*rand(tmax,numreps)) + 1;
           % generates a matrix of random index numbers that
           % will be used to decide which lams value to use
           % for each year and each simulation
lambdas = lams(intvalues);
           % uses the indices to create a matrix of lambdas
           % to use for each year and simulation
initpopsize = numzero*ones(1,numreps);
           % sets initial population sizes
popsizes = [initpopsize;numzero*cumprod(lambdas)];
              % generates the population size every year
              % for all simulations, using the cumulative
              % products of the lambda values
% the next three lines find extinctions and set subsequent
% population sizes to zero
notextinct= popsizes > ones(tmax+1,numreps);
notextinct = cumprod(notextinct);
popsizes = popsizes.*notextinct;

maxN = 0.5*max(max(popsizes));
    % the biggest population size to graph (max popn size;
```

BOX 2.1 *(continued)*

```
    % will often be too high to see most other dynamics)
stochL = prod(lams)^(1/numlams)
        % geometric mean of the lambdas or stochastic lambda
expectedNt = numzero*(stochL^tmax)
                % this is the median population size at tmax
wk1write(outname,popsizes)  % write results to Lotus spreadsheet
% the next lines make a figure of the population sizes vs. time
plot([1:tmax+1], popsizes)
xlabel('Time');
ylabel('Population size');
axis([1 tmax 0 maxN]);
% the next lines make a histogram of final population sizes
figure % make a new figure window
Y=[0 1E-10 10:10:200 inf]; % set the category boundaries
N=histc(popsizes(tmax,:),Y); % count up the numbers of cases
N=100*N/numreps;     % convert numbers to percentages
Nmax=max(N);
X=(-5:10:215);       % values for x-axis of histogram
bar(X,N)              % plot a bar graph of the histogram
axis([-10 210 0 Nmax+5])
xlabel('Population size (<0=extinct, last bar >200)');
ylabel('Percent of populations');
```

another approach). To start the simulations, we used the estimated number of rails in the last year of the study (29 birds). While $\lambda_G = 0.9969$ for this population (predicting an almost stable population), with increasing time the probability that the population stays close to the starting population size rapidly diminishes (Figure 2.4). Indeed, the range of likely population sizes rapidly increases, and by 100 years, we see high probabilities of both extinction and very large numbers. This rapid increase in the uncertainty of future population sizes is a feature of all stochastic models, especially when there is no density-dependence.

The second basic aspect of stochastic growth is simply that more variation in λ_t will create more uncertainty in predicted population sizes at any point in the future. This increasing uncertainty about future population sizes can be quite dramatic, with modest increases in the variability of population growth leading to a much broader range of plausible population sizes and also more skew in the distribution of these sizes. Figure 2.5 shows this effect for simulated population sizes 50 years in the future, assuming a lognormal distribution for λ_t with differing variances but a constant $\lambda_A = 1.02$.

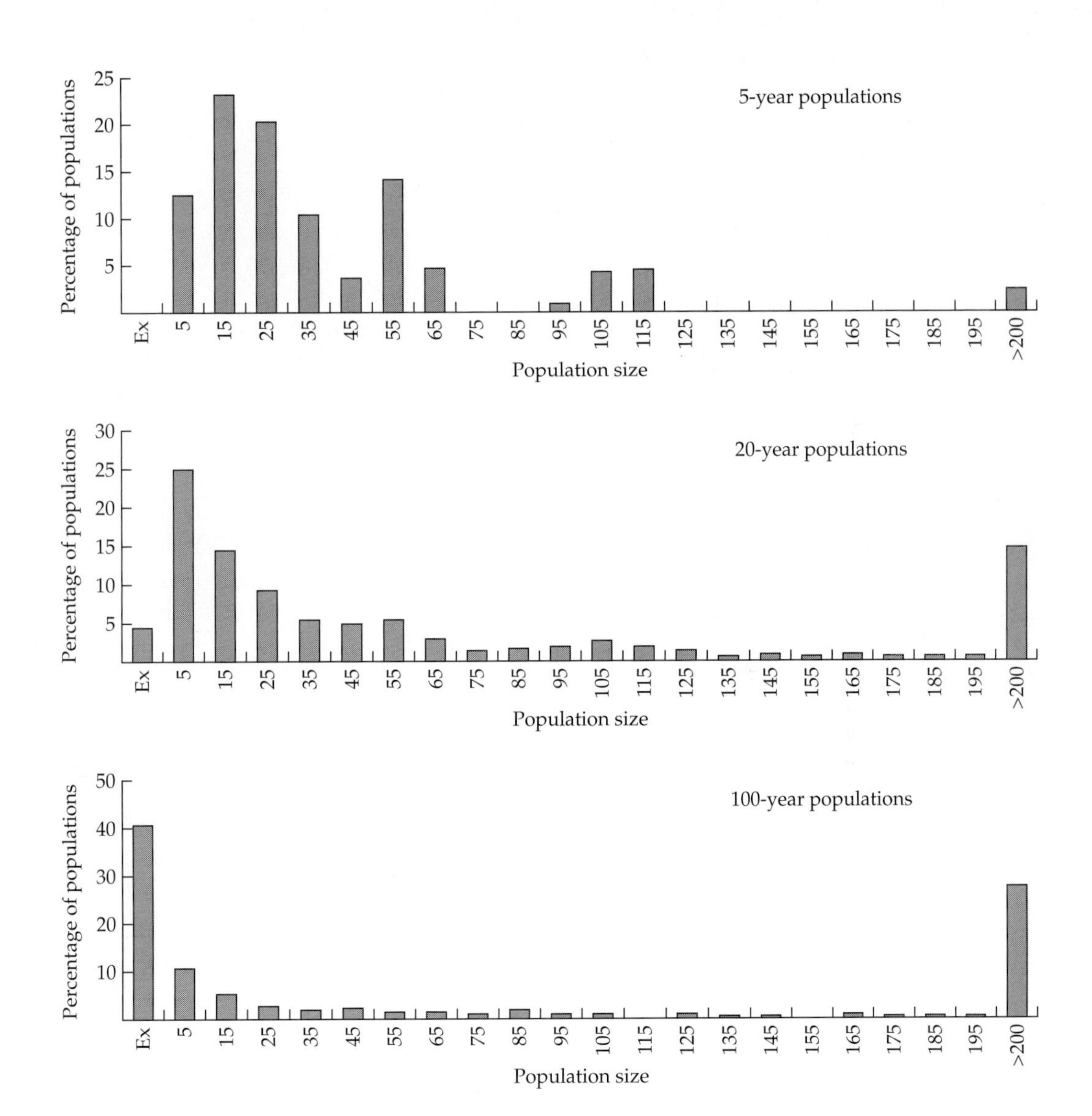

Figure 2.4 Histograms of population sizes for 1,000 simulations of the California clapper rail population in the Faber marsh area, San Francisco Bay, CA. From 1991 to 1996, yearly estimates of λ_t for this population were {1.00, 1.98, 1.02, 0.92, 0.53}. One of these values was randomly chosen to use in each of 100 years for each simulation. Notice the increasing divergence in population sizes from 5 to 20 to 100 years, by which point most trajectories were either extinct (bars labeled Ex) or very large.

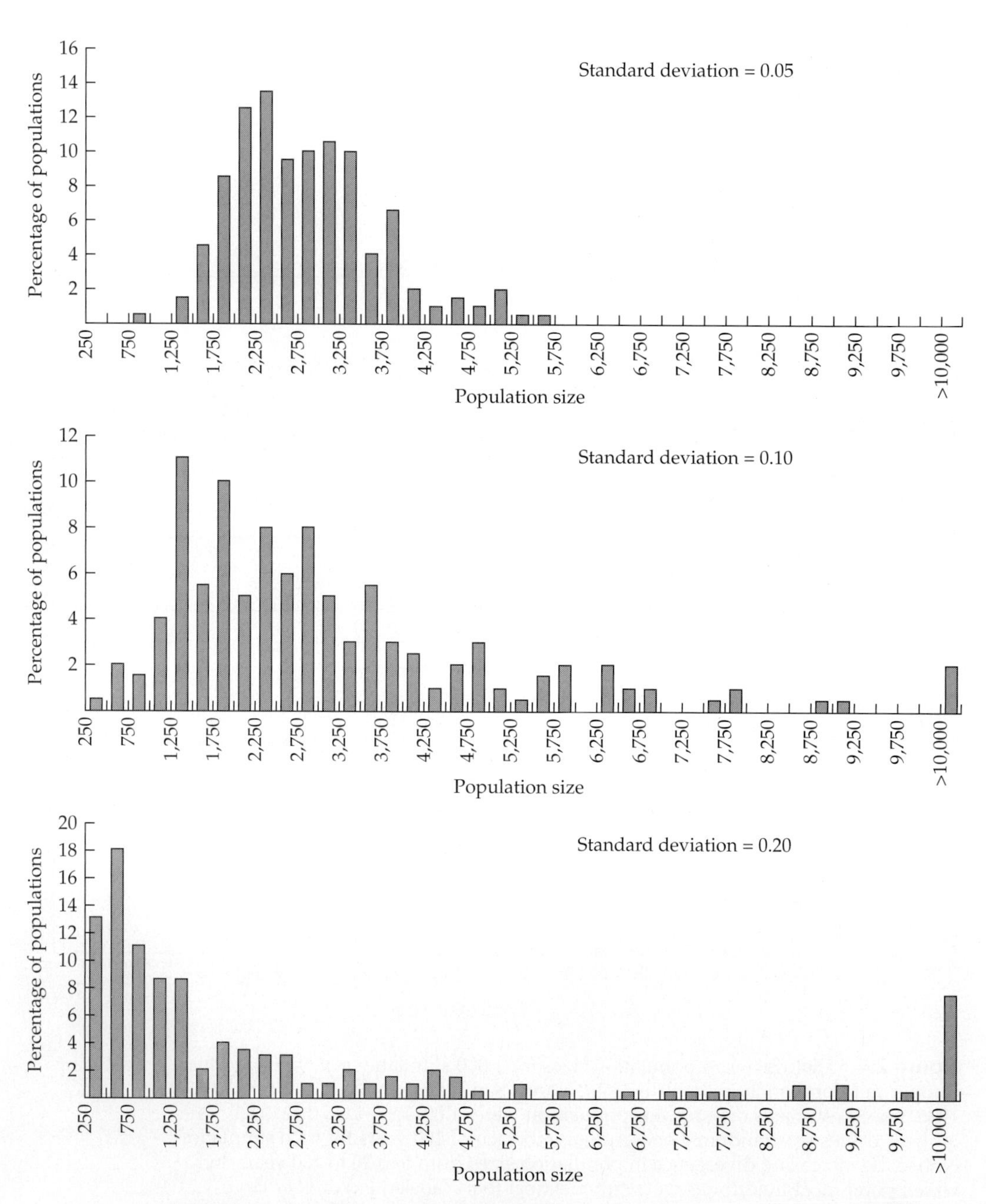

Figure 2.5 Histograms of population growth with increasing variability in λ_t. Population sizes for year 50 of simulated population growth with $\lambda_A = 1.02$ and standard deviations in λ_t ranging from 0.05 to 0.20. λ_t values were assumed to follow a lognormal distribution for these simulations, and initial population sizes were 1,000. Notice how much less predictable future population sizes become with more variable growth.

In summary, when we characterize population growth, use of a geometric mean to properly account for the variance in growth rates gives a more meaningful prediction about the most-likely outcome, whereas the standard arithmetic mean (which does not incorporate the variance in the growth rates) overestimates growth, and hence viability, most of the time. More intuitively, adding variation to population growth adds uncertainty to predictions of future population sizes, and this uncertainty grows over time. Both the general depression of population growth brought on by temporal variation and the increasing uncertainty with greater variability are key factors to bear in mind when considering how best to understand and predict population viability.

Two Other Aspects of Variability

SPATIAL VARIABILITY, SPATIO-TEMPORAL VARIABILITY, AND MOVEMENT While temporal variation of a single population is complicated enough, if multiple populations are being considered in a PVA, spatial variability in vital rates and population growth, as well as spatial correlation in temporal variation across sites, must also be considered. Here we just sketch the basic issues. We give a full explanation of these factors later on, when we discuss different approaches to multi-site PVAs (Chapter 10).

The most obvious complication that arises with multiple sites is that the mean and variance of vital rates, and hence of population growth rates, will usually not be equal across all sites and habitats. Spatial differences are handled in a straightforward manner by estimating the means and variances of vital rates or population growth rates individually for each site. For example, for the endangered Mt. Graham red squirrel, an endemic subspecies that occurs on one mountain in Arizona, annual population growth is somewhat higher on average but also more variable in one of the three primary habitats (spruce-fir forest) than in the other two (Figure 2.6A).

The more serious complication in a multi-site situation arises due to correlations in the temporal variation in rates across sites. For the Mt. Graham red squirrel, there is a high degree of correlation in annual fluctuations of squirrel numbers among two pairs of the three habitats, largely due to simultaneous masting of the conifers that are the squirrels' most important food source (Figure 2.6B). Positive correlations like these mean that poor years that drive one part of the population towards extinction are likely to do the same at other subpopulations. Conversely, if one population tends to do better in the years that another one does worse, the overall viability of a set of populations can be greatly improved (see Chapters 10 and 11). Thus, it is important to try to estimate this spatio-temporal correlation and include it in multi-site PVA models. A complication that frequently arises is that data from different populations for the same set of years is lacking; unless you measure the vital rates or population changes in the same years at different sites, there is no good way to estimate spatio-temporal correlation (see Chapter 10).

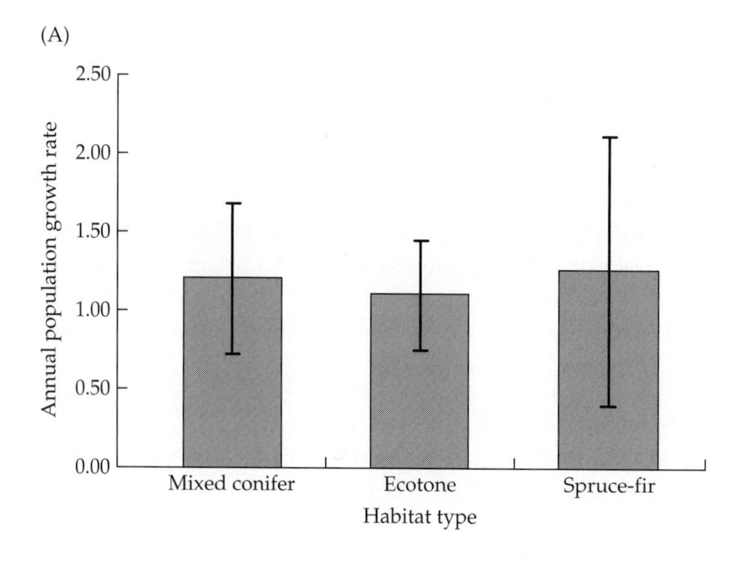

(B)

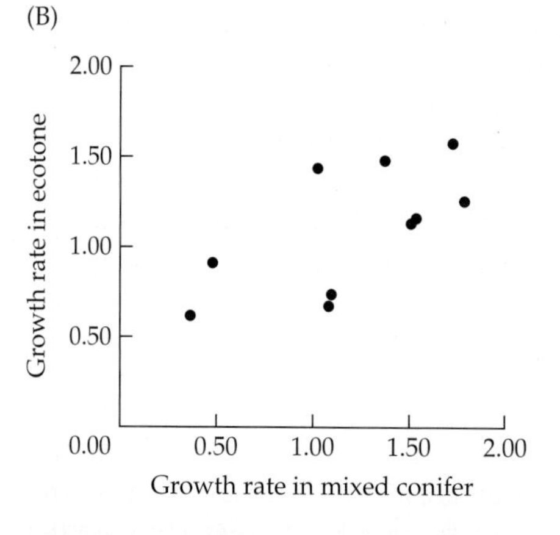

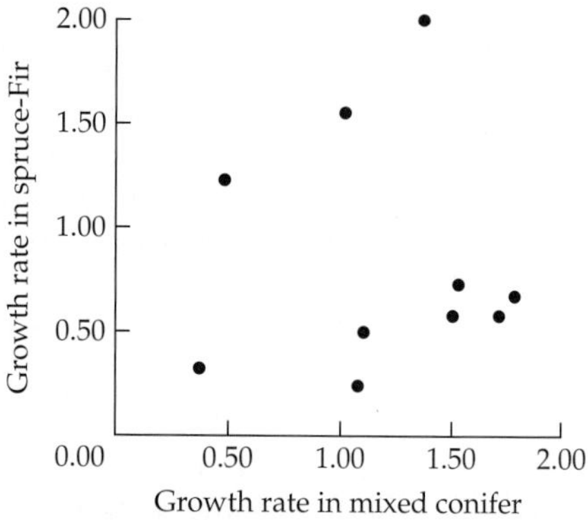

Figure 2.6 Spatial and spatio-temporal variation in population growth for the Mt. Graham red squirrel in three habitat types. (A) Arithmetic mean (±1 s.d.) annual population growth rates. (B) Correlation between population growth estimates across 10 different years. Population growth in the ecotone habitat is positively correlated with that in the spruce-fir ($r = 0.59$, $p = 0.07$) and mixed conifer ($r = 0.66$ $p = 0.04$) habitats, but squirrel population growth in the spruce-fir and mixed conifer habitats are uncorrelated ($r = -0.01$ $p = 0.98$).

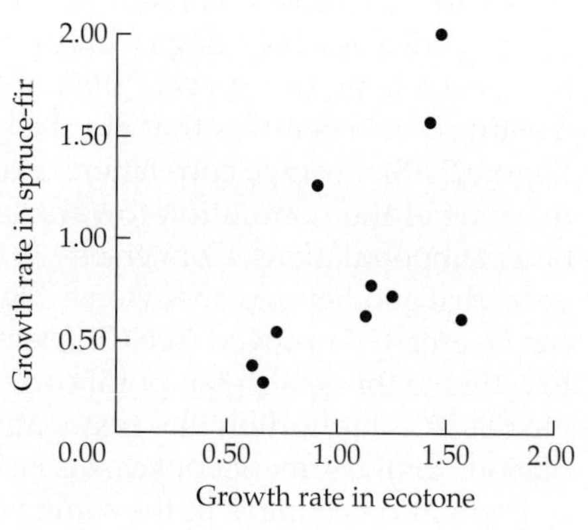

A final issue with spatial variability involves movement of individuals between habitats or populations. Understanding the effects of spatio-temporal correlations for population viability is inextricably tied to understanding movement, since movements are what link different populations. Even if each population is completely isolated from all others, there can be great value in having multiple populations with no correlation or with negative correlation in their temporal fluctuations. However, the benefits of having multiple populations can be much greater in a classic metapopulation, where populations are linked by dispersal events (and therefore extinctions can be balanced by colonizations). Still, unless successful movements are very common, without intensive effort it is typically quite difficult to estimate the parameters governing movement distances and frequencies, and we therefore focus much of our later discussion on how to perform multi-site PVAs with only minimal information on movement (Chapters 10 and 11).

A SPURIOUS SOURCE OF TEMPORAL VARIABILITY: OBSERVATION ERROR We end our discussion of variability with a factor that, although it does not actually affect population viability, is nevertheless an important concern for PVA. We have seen that variability contributes strongly to population fate. However, measured variation in both vital rates and population counts will usually reflect the influence of observation error. Unlike environmental stochasticity, catastrophes and bonanzas, and demographic stochasticity, all of which are real biological phenomena that influence viability, observation error merely reflects our inability to measure vital rates or population size with absolute precision, and so has no effect on viability. Nevertheless, observation error can introduce biases and uncertainty into our estimates of population viability, and we need to account for these effects as much as possible.

There are two basic sampling and estimation concerns that commonly arise when trying to parameterize a PVA. The first involves possible biases in the estimation of mean rates as a result of sampling that does not reflect the true distribution of individuals with differing performance. A common practice is to do most sampling in what we think is the best habitat. However, the resulting data need to be applied cautiously if, in fact, most individuals in the population live in suboptimal areas. Similarly, there can be subtle biases towards marking more robust plants (they are easier to find) or less fit animals (they are easier to catch). With the exception of clear habitat differences, most of these possible biases can only be dealt with by a properly randomized design for collecting the data in the first place—that is to say, there is little chance of understanding their effects after the fact. With habitat effects, however, if very clear differences are apparent, it may be worth constructing a PVA model that separates individuals on the basis of habitat (see Chapter 11). More simply, adjustments can be made in the estimates of parameters if there is clear evidence that they reflect either overly poor or overly good performance when considering the entire population.

The second, more difficult set of concerns deals with estimating variation in rates. Unless we can accurately observe every individual in the population, observation error will add an undesired, spurious component to our estimates of variability in vital rates and population growth, above and beyond the true variation caused by demographic and environmental stochasticity, bonanzas, and catastrophes. Observation error can be caused by a host of factors, such as complex background vegetation that renders a varying proportion of the individuals difficult to detect, multiple counts of the same individual for mobile organisms (or by different members of the census team), incorrect species identification, or the error always introduced when a partial census (e.g., quadrat or transect sampling) or an indirect measure of abundance (e.g., scat, tracks or hair snags) is used to infer total population size. Given that measured variation is caused by both true variation in the vital rates and by limited and imperfect observations, the trick is to obtain an estimate of the true variability that is uncontaminated by observation error. Several methods to do so have been proposed for different types of count and vital rate data (e.g., Ludwig and Walters 1981, Kendall 1998, Saether et al. 1998a, de Valpine and Hastings 2002, Holmes 2001, White 2000), and we discuss them in conjunction with particular PVA methods in subsequent chapters. In particular, Chapter 5 is devoted entirely to methods for dealing with observation error in count-based PVAs. However, even when these methods aren't feasible (as it often the case) we can still use estimates of total variability, *knowing that this will give a somewhat pessimistic answer about viability*. In other words, when we can't untangle real variability from observation error, we can at least know the direction of the bias that observation error will impart.

Other Processes Influencing Viability

Above, we discussed means and variances in vital rates, which are factors that are almost always going to be of critical importance in determining viability, and for which some type of data, even if rather indirect, often exists. Several other basic ecological and genetic processes are also likely to influence viability, but in the majority of cases there are few or no data available to estimate their effects. We consider the most important of these processes now, addressing how they influence population viability and also how they are often dealt with in the absence of good data.

Negative Density Dependence

Probably the single most important factor that biologists want to include in PVAs but lack the data to estimate properly is density dependence. Density dependence is a change in individual performance, and hence population growth rate, as the size or density of a population changes. The presence of density dependence can cause a density-independent PVA to either over- or underestimate the true viability of a population, depending on the specifics of how the growth rate responds to density. In discussing these effects, it is useful to distinguish two types of density dependence, so-called negative and positive density dependence. In this and the next section, we discuss these two kinds of density dependence, give examples of each, and discuss their impacts on population viability.

Negative density dependence is a decline in average vital rates as population size increases (that is, the growth rate depends *negatively* on population size). This is the most familiar form of density dependence, embodied in the logistic equation presented in most introductory ecology textbooks, and typically caused by intraspecific competition for limited resources or by interacting species (e.g., predators and pathogens) whose impacts increase disproportionately as the density of the focal organism increases. An example of negative density dependence is the change in the inter-birth interval for African elephants in response to density, used in a PVA by Armbruster and Lande (1993; Figure 2.7). Above a density of 0.325 elephants/mile^2, each female elephant gives birth at longer intervals as density increases, probably due to food limitation. Note that this relationship is not linear, and also that it is only one of several density-dependent effects for elephants, which will combine to create the overall effect of density on population growth rate. A remarkable amount is known about performance at different densities for elephants, but for most species we will have estimates of performance for only a very limited set of densities, making it difficult to detect density-dependent effects. Still, we know logically that for any species, higher numbers must eventually result in lower per-capita performance. That is, the growth of all populations will at some point become limited, although that

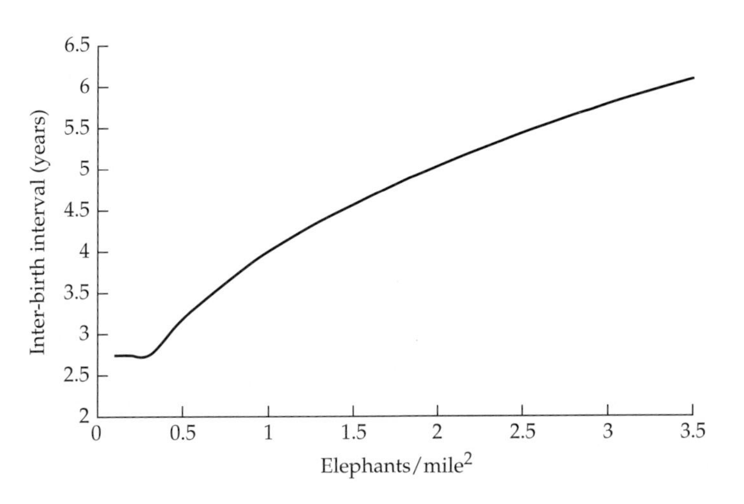

Figure 2.7 The pattern of density dependence in inter-birth intervals for African elephants (After Armbruster and Lande 1993).

point may be very much higher than current numbers, especially for declining populations.

Although the essence of negative density dependence is a decline in growth rate at high density, what happens at lower densities is not so clear. Over the entire range of densities, individual performance and hence population growth can exhibit a variety of qualitatively different responses. The standard logistic form of negative density dependence assumes that per-capita population growth declines linearly as numbers or densities increase; fecundity, for example, is assumed to be highest at very low numbers, and to decrease at a constant rate as numbers climb. Alternatively, negative density effects may be weak or nonexistent below some intermediate population size, so that fecundity (for example) remains quite flat as numbers initially increase, only falling off appreciably as density exceeds an intermediate threshold. This is more or less the pattern seen for the inter-birth intervals of elephants (Figure 2.7). Finally, it is conceivable that density effects can show the opposite pattern, with rapid declines at low densities and then an asymptoting response from moderate to high densities, although such responses appear to be much more rare.

The shape of negative density dependence can have important consequences for population viability. For example, linear density effects on fecundity (as assumed in logistic population growth) imply not only that fecundities will be lower at higher population sizes, but also that if the population dips to lower and lower numbers, fecundity will continue to increase as density effects are relaxed. These two density effects counter one another in a population that is being bounced up and down by environmental stochasticity. As the population gets larger—and therefore safer from extinction—fecundity will decline, slowing the population's escape from dangerously low numbers. In this way, negative density dependence increases the risk of extinction, holding a population at lower numbers where extinction is always a threat. Conversely, if the population dips very close to extinction, linear density dependence means that fecundity will rise, increasing population growth rate and thus allowing the population to jet upwards, away from immediate extinction. Because of these contrasting effects, a PVA accounting for linear density dependence may predict greater or lesser extinction risk than will a density-independent model, depending on the population sizes or densities at which data were gathered to estimate vital rates or population growth.

In contrast to linear density dependence, if negative density effects are strongly manifested only at relatively high densities, then there will be no added protection against extinction by increasing performance at low densities. In this case, assuming a linear density dependence function in a PVA model will dramatically and incorrectly increase our confidence that a population is safe from extinction, because its primary effect will be to unrealistically boost population growth at very low numbers. Again, the point is that the exact form and strength of negative density dependence on all the demographic rates will determine how density effects will influence extinction risk. We will explore ways to determine the shape of density dependence for count-based models in Chapter 4.

Although many methods exist to estimate and include negative density dependence in PVAs, we rarely know much about the features of negative density dependence for a rare species, and thus we must guess about its effects outside the narrow range of densities for which we do have data. Given the sensitivity of population viability to the details of density dependence, and the usual dearth of data to estimate them, many PVAs either omit density dependence altogether or include only the simplest kind—a cap or ceiling on population size beyond which a population cannot grow (but below which there is no density dependence; see Chapters 4, 8, and 11). A population ceiling can often be well-estimated from historical records or data for related species, and does not presume other, unknown effects. If more data are available, it is important to test for more realistic density dependence effects (Chapters 4 and 8). However, even for common species, data are often too scarce to confidently estimate the pattern of density-dependent effects, and this is even more true of rare species. Also, for currently threatened species, any data from a population when it was more common may also be from a time when current impacts on it were much less important, such that we might find that higher densities spuriously correlate with better, not worse, performance.

In sum, while it is worth looking for density dependent effects on vital rates or total population growth, considerable caution in interpreting the results and using them in a PVA is also warranted. If careful testing of your data show no significant signs of density dependence, it is usually best to proceed with a density independent model, as inclusion of a feature without any estimates of its rates throws doubt over the whole analysis. Still, in some cases with sparse data there are compelling reasons to think that density dependence will be important over the densities a population is likely to achieve in the near or medium term. In this case, a good approach is to construct a set of PVA models with differing strengths and forms of negative density-dependent effects. In any analysis of such a suite of models, the idea is to explore predictions that incorporate the range of likely patterns of density dependence in order to see how much our uncertainty over density effects will influence viability predictions.

Positive Density Dependence or Allee Effects

In contrast to negative density effects, positive density dependence is an *increase* in the population growth rate as population size increases (that is, the growth rate depends *positively* on population size). Positive density dependence is also known as the *Allee effect*, after Warder C. Allee, an ecologist at the University of Chicago who, in the 1950s, popularized its importance. Positive density effects may result from improvements in mating success, group defense, or group foraging as density increases. Evidence for Allee effects most often come from data on reproductive success or survival of individual organisms (see reviews by Courchamp et al. 1999 or Stephens and Sutherland 1999). Many such examples come from plants (Lamont et al. 1993, Widen 1993, Groom 1998), whose sedentary nature makes the estimation of reproductive success and survival easier. Usually, the negative effects of low numbers are only felt strongly when a population drops to fairly small sizes; at intermediate densities, there is no strong Allee effect. Two potential causes of Allee effects, group foraging and defense, will be irrelevant for many species. However, mate-finding problems at low densities (including problems plants face in getting the wind or insect pollinators to bring them their mates) are almost universal.

In contrast to the data showing Allee effects on individual vital rates, positive density dependence has only rarely been detected using census data. Both Fowler and Baker (1991) and Saether et al. (1996) used quadratic regression to look for a hump-shaped relationship between the per-capita population growth rate and population size, using data from mammals and birds, respectively. Such a relationship would be expected if an Allee effect is operating at low densities and negative density dependence is operating at high densities. Despite the fact that the authors specifically chose data sets that included years with low population size, when Allee effects would be most likely to appear, neither analysis found any evidence for them. Myers et al. (1995) used data from fish stocks (all with 15 or more years of data), some of which had been depressed to low levels by harvesting. They fit population models (Beverton and Holt 1957) with and without an Allee effect to see which model better predicted recruitment (that is, reproduction resulting in new fish being added to the adult population) as a function of the abundance of spawning adults. In only 3 of 128 stocks did the model with Allee effects fit significantly better than the model without Allee effects. In a successful effort to show Allee effects, Kuussaari et al. (1998) used census data collected from many small populations of a butterfly, the Glanville fritillary, over 4 years. Using logistic regression, they showed that the proportion of small populations that grew or remained constant (as opposed to decreasing) from year t to year $t + 1$ was positively related to population size in year t, as would be expected if an Allee effect is operating. In effect, Kuussaari and colleagues relied upon extensive replication of populations in space rather than on census data from a single population over a long period of time. Unfortunately, such extensive data will rarely be available for most threatened species.

Thus, while Allee effects are almost certain to operate in most populations, we do not have a good sense of the strength of such effects or the population sizes at which they will start to operate for most taxa. We can fairly easily incorporate Allee effects into either count-based PVAs (Chapter 4) or demographic PVAs (Beier 1993, Chapter 8). However, the limitations of our data can thwart any meaningful inclusion of positive density effects. As with demographic stochasticity, the best solution to this problem will often involve setting quasi-extinction thresholds high enough to avoid the low densities where Allee effects will strongly affect PVA results. Still, no overarching generalities exist as to what these thresholds should be, since they depend on searching behaviors (for mates or food) or social interactions that will be strikingly different depending on the species or population. It is widely speculated that for the passenger pigeon, Allee effects became operative at a population size in the tens of thousands, and contributed to the bird's demise (Halliday 1978, Ehrlich et al. 1988), whereas for some plants, Allee effects are only evident in populations of less than 20 individuals (Groom 1998). Such a broad range argues for caution in setting thresholds for Allee effects, using as much as is known about the natural history of the particular species with which you are dealing, and again, by building a suite of PVA models with contrasting assumptions about the onset and strength of these effects if they are to be included without clear supporting data.

Genetic Factors

The genetic problems that commonly arise at low population sizes are also a serious concern for viability analysis, and many PVAs have included the effects of genetic problems for population viability (e.g., Burgman and Lamont 1992). In general, the inclusion of genetic problems in a PVA requires estimates of two sets of processes. The first is the rate at which heterozygosity will be lost by a population of a certain size. (Heterozygosity is genetic diversity, measured by the probability that, for the average locus in the average individual in the population, there will be two different alleles.) The rate of loss of genetic diversity is usually estimated as the change in inbreeding level per generation (where inbreeding is the average probability that an individual's two copies of a gene are "identical by descent," which means they both came from the same ancestor). Second, after we establish how fast genetic diversity will be lost, the consequences of this loss must also be estimated. The decrease in individual performance with loss of genetic diversity is measured as inbreeding depression, which is commonly estimated as a certain percentage reduction in some vital rate (survival probability, litter size, growth rate, or another factor) with a given increase in inbreeding level (and hence, loss of heterozygosity).

The first step in making these calculations, estimating the rate of heterozygosity loss for small populations, is fairly straightforward, given some knowledge of the breeding system and demography (Falconer 1981, Hedrick and Miller 1992). However, for nondomesticated species, the strength of inbreeding depression, although usually substantial, is also highly variable (Schemske and Lande 1985, Ralls et al. 1988). Thus, for any particular rare species, for which there are likely to be no data at all on the magnitude of inbreeding depression, it is hard to confidently include inbreeding depression in a PVA based solely on generalities about average effects of inbreeding depression on fecundity and survivorship (Widén 1993, Lande 2002, Allendorf and Ryman 2002).

In the past, several authors have argued that if population sizes are low enough for significant loss of genetic diversity to occur, they are doomed due to demographic processes anyway (Lande 1988b). However, more recent modeling (Mills and Smouse 1994) and empirical work (Westemeier et al. 1998, Bouzat et al. 1998) has shown that, in fact, inbreeding can work synergistically with demography to impact population health. This interaction, which in part drives the "extinction vortex" envisioned by Gilpin and Soulé (1986), works as follows. Say that, due to environmental variation, a population dips to low numbers for a generation. This will result in a small increase in inbreeding levels, even if the population fairly rapidly increases again. However, the inbreeding depression resulting from this small loss of genetic diversity will also mean that the population growth rate will not be quite as high as it was before. If the population again dips to low numbers, it will tend to stay there longer, due to poorer per-capita performance, resulting in more genetic loss. Thus, each time a population dips to low numbers, it incurs a heavier genetic disadvantage and stays at lower numbers longer, increasing its risk of extinction due to environmental and demographic stochasticity. Thus, the interaction of environmental and demographic variability with genetic processes can strongly increase the chances of extinction (Mills and Smouse 1994). On the other hand, this effect will be countered to some extent by the elimination of deleterious alleles from the population by natural selection, a process known as purging, which can reduce the severity of subsequent bottlenecks. However, the speed with which deleterious alleles will be purged from a population depends on the rate of inbreeding and the genetic mechanism generating inbreeding depression (e.g., whether it is caused by a few genes of major effect or many genes of small effect), so that purging is certainly not a reliable antidote to the problems of inbreeding depression (Kalinowski et al. 2000).

Although the synergism between demographic problems and inbreeding depression is real and important, in general we do not advocate trying to put genetics into a PVA, simply because of the lack of data needed to do so. As before, we instead suggest setting a high enough quasi-extinction threshold to minimize the chances that genetic problems would dramatically change a PVA's conclusions. Here, "high enough" probably means an effective population size,[3] N_e, on the order of 50 individuals, resulting in only a

[3]Effective population size is a standard measure of relative population numbers that can be used to estimate the rate of inbreeding and genetic loss. One way to define N_e is as the number of equally contributing individuals in a idealized randomly mating population that would experience the same rate of genetic drift (loss of heterozygosity) as the population of interest (see Hartl and Clark 1997).

1% loss of genetic diversity per generation. The relationship between N_e and total population size is complicated by many factors (Falconer 1989). However, one fairly well-substantiated generality is that for many birds and mammals $N_e / N \approx$ one-half to two-thirds, where N is the total population size of *reproductive adults* (Nunney 1993; Nunney and Elam 1994), arguing for a quasi-extinction threshold of at least 100 breeding adults. This approach still basically ignores inbreeding problems and will always result in somewhat optimistic answers about population viability, since some significant inbreeding will occur even above this population size. However, as for density dependence, this solution seems better to us than including wild guesses about genetic effects that are open to endless argument and can weaken the credibility of the whole analysis.

Quantifying Population Viability

Before we leave this overview of factors contributing to population viability, it is important to think about how viability is quantified: What does viability really mean? Several metrics of viability are commonly used. All are related, but they are not identical in the answers they give about population safety. Because the similarities are pretty obvious and the differences often more subtle, we will emphasize how these metrics differ and indicate which of them we feel are the most useful gauges of viability. We will not present much of the math behind these different measures here, as this will come up in the following chapters.

Viability Metrics Related to the Probability of Quasi-Extinction

Viable populations are those that have a suitably low chance of going extinct before a specified future time. One key issue in the estimation of extinction risks is the setting of quasi-extinction thresholds. Up to now, we've often written in general terms about the risk of "extinction", but *no good PVA should attempt to evaluate the risk of utter population extinction*. This is because, as we've noted, many additional, difficult-to-evaluate population processes complicate the behavior of truly tiny populations. The basic goal of PVA is to predict the future with some reasonable degree of assurance. We can do this much better if we don't try to predict when the very last desert tortoise in Las Vegas County may die, but, rather, when the population will reach a small enough number that many additional genetic and ecological problems will further threaten it, making it perilously at risk. As we've mentioned above, the solution is to predict quasi-extinction, that is, the population falling below a quasi-extinction threshold set as the minimum number of individuals (or, often, females) below which the population is likely to be critically and immediately imperiled (Ginzburg et al. 1982). As you might guess, the size of the quasi-extinction threshold is often the subject of argument: A lower threshold will always yield a lower estimated risk of quasi-

Figure 2.8 Predicted extinction probabilities for a population of the California clapper rail. (A) Probability density function for extinction times. (B) Cumulative distribution function for extinction times. ▶

extinction over a set period of time than will a larger one. Concern about demographic stochasticity would argue for a threshold of 20 or more individuals, and preferably 20 or more *reproductive* individuals, whereas genetic arguments would favor much larger values—say, 100 breeders at a minimum. However, the reality is that many PVAs have used moderate to fairly low quasi-extinction thresholds, such as 50 individuals for the sentry milk vetch (Maschinski et al. 1997). At the extreme, PVAs for both the African elephant (Armbruster and Lande 1993) and the leopard darter (Williams et al. 1999) use thresholds of one, which is to say complete extinction (i.e., total loss of one sex). One likely reason for the use of low thresholds is that many threatened populations may already be below the levels that, in an ideal world, we would choose as minimum thresholds.

After defining a quasi-extinction threshold, the measurement of extinction risk can be done in several different ways, giving somewhat different answers about population health. To illustrate these differences, we used the count data for the California clapper rail in Faber marsh to estimate a range of extinction measures. (This kind of analysis is presented in Chapters 3 and 4.) For this analysis, we used a fairly low extinction threshold of 5, in part because the censuses used to estimate this population miss many birds. First, Figure 2.8A shows a probability density function (PDF) for the time required to first hit the quasi-extinction threshold, given the current population size. The probability density at time t is proportional to the probability that quasi-extinction actually occurs in a small interval of time centered at time t. Loosely speaking, the PDF represents a histogram of predicted quasi-extinction times. As the PDF shows, the risk of extinction rises rapidly until about 6 years and then falls off gradually. This fall-off is simply due to the fact that quasi-extinction is likely to have already occurred by these later times. An even more useful depiction of extinction probability is the cumulative distribution function[4] (CDF) of extinction times, shown in Figure 2.8B. The PDF is (roughly) the probability that quasi-extinction will occur during a small time window, whereas the CDF gives the probability that the population will have hit the quasi-extinction threshold *at or before* a given future time. For the clapper rail, there is a sharp rise in extinction risk from 5 to 15 years, and then a gradually slowing additional risk over many decades. The total risk of extinction over the next century is predicted to be about 70%. If we carried out the figure for several more centuries, we could see that it reaches an asymptote at 1.0, predicting certain extinction over the long haul. A plot of

[4]Mathematically, the CDF at time T is the integral of the PDF from $t = 0$ to $t = T$.

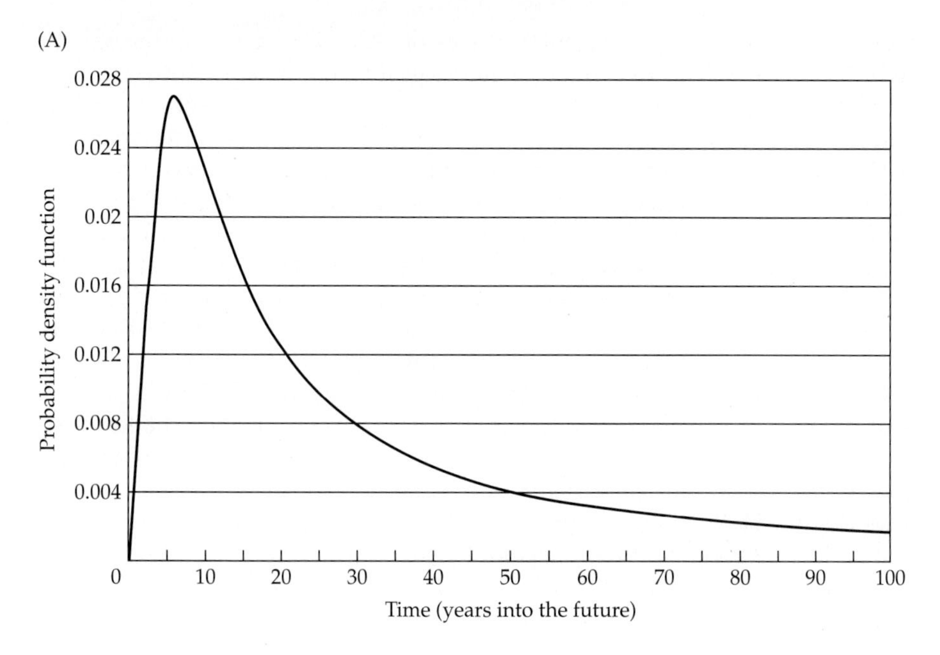

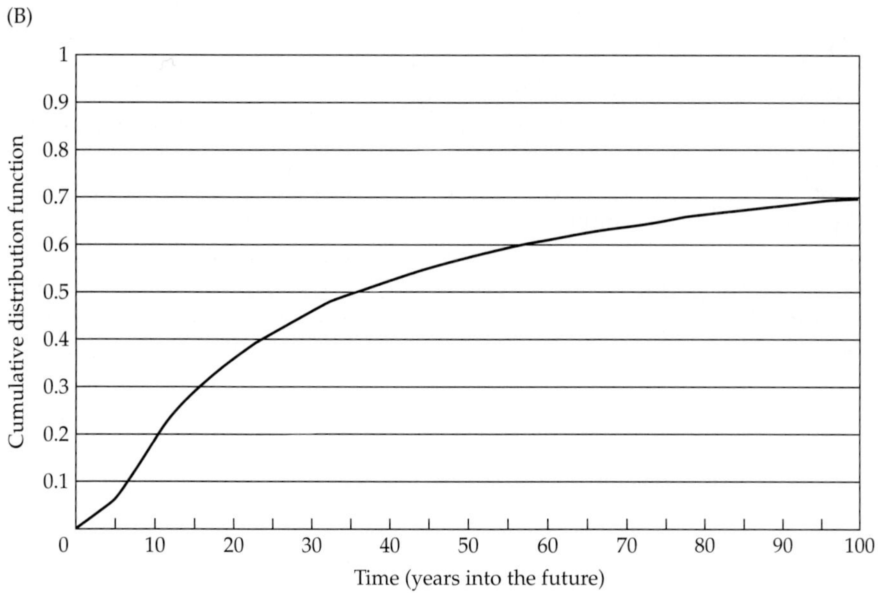

the extinction-time cumulative distribution function (CDF) versus years into the future (such as Figure 2.8B) is probably the single most useful way to present extinction risk information. Unlike simpler measures of risk (see below), it doesn't oversummarize the predictions, and it also provides an easy way to visualize different features of risk—in particular, the rate at which the overall risk of extinction increases through time.

Even though the whole distribution of extinction times is useful in understanding extinction risk, several specific numbers are often drawn from the CDF to more succinctly summarize population viability. The first is the ultimate probability of extinction, which represents the probability that the population will hit the threshold at *any* future time. Two other ways of summarizing risk are the mean and the median of the predicted extinction times, if the population in fact becomes extinct at all. For the clapper rail, the median time to extinction is about 35 years (the time at which the CDF line hits 0.50—a 50% chance of extinction before this time), while the mean time to extinction is 567 years. (You can't read this number from a plot of the CDF, but it is easy to calculate from the data.) Some authors also use the mode of the extinction time distribution (the most likely time to extinction) as a measure of risk. This number, 5.9 years for this example, is easily read off the PDF plot (Figure 2.8A). Finally, a good measure of viability is often the chance of extinction over some set time period. For the clapper rail, the probability of extinction within the next 20 years is about 35% (Figure 2.8B), suggesting that there is a substantial short-term threat to the population.

All told, then, we have five summary statistics for extinction risk: the probability of extinction by a given time, the probability of extinction ever occurring, and the median, mean, and modal times to extinction (given that it occurs eventually). Of these, the first three are the most useful, and the last two (which are among the most often used) are not recommended. First, the probability of extinction by some specific point in the future is different from all the other gauges of risk in being tied to clear management timelines and other real-world constraints. Deciding what a reasonable timeline is for management and risk evaluation is not just a biological decision, but also one that incorporates knowledge of future threats, funding deadlines, changes in political administration, and so forth. However, deciding on a good time horizon can be somewhat contentious or arbitrary, which has made this measure less used than some of the others.

Of the other four measures, the probability of ultimate extinction and the median time to extinction are, together, more meaningful. The median is a good description of the extinction time, because it is the time at which fully half of the possible paths the population might follow will have hit the threshold. The mean time to extinction is almost always an overestimate of population safety. This is because extinction times tend to be strongly right skewed (Figure 2.8A), with high extinction risks in the near term and a small chance of becoming extinct after a very long time. Thus, the mean typically overestimates the median, as the clapper rail predictions show quite dramatically (576 vs. 35 years). The modal time, or the most-likely instant at which extinction will actually occur, doesn't take into account the full distribution of extinction times and thus says little about when most paths go extinct.

Also, because of the skewed distribution of extinction times, the mode always predicts a shorter time to extinction than does the median (Figure 2.8A). In fact, the mean and modal extinction times are used simply because they are often easier to calculate than are other viability metrics.

Still, the median time to extinction alone is also not a sufficient metric of viability. What if the probability of extinction *ever* happening is very low? In this case, the median time to extinction is often quite low. This typically happens when a population is currently small and extinction prone, but the mostly likely scenario is for it to get quite large and extinction-proof. Thus, if extinctions occur, they will occur early. So, is the probably of ultimate extinction a better measure of risk? No, because the ultimate risk can sometimes be quite high, while the probability of extinction over a realistic management (and prediction) time-horizon is actually very small. Also, the ultimate risk is essentially a very long term prediction and as such can be highly sensitive to model assumptions and problems in parameter estimation (Fieberg and Ellner 2000). Because of these problems and the need to consider both the probability of extinction occurring at all and the speed with which it is likely to occur, probabilities of extinction over certain time horizons are the most useful and robust extinction risk estimates.

Given these considerations, we recommend that any PVA show the entire cumulative distribution of extinction times and also discuss the risk of extinction by certain time horizons that have biological and/or management significance. The cumulative distribution of extinction times provides so much more information than any of the summary statistics that it is well worth presenting it, as well as measures of uncertainty around it, when reporting PVA results. Furthermore, presenting extinction risk in this way emphasizes the time-dependent nature of viability assessment, forcing otherwise abstract discussions of risk to deal with the realities of management time-lines and the biology of the species.

Viability Metrics Related to the Population Growth Rate

Another way to summarize population viability is by an estimate of the population growth rate. In making an assessment of viability based on growth rates, you should always use an estimate of stochastic growth, as we discussed above under the section *Why variation is important*. Growth rate may be a more important indicator of possible future problems than is an extinction probability if the short-term risk of extinction is low, but the current population size is small and vulnerable. For example, the number of female Atlantic right whales is currently about 150, and the population is declining (Caswell et al. 1999). While the predicted risk of extinction over the next 100 years is close to zero, the population is almost certainly doomed over a 300-year horizon. Therefore, focusing our efforts on management to increase annual population growth above its estimated current value of $\lambda = 0.976$ is critical to achieving long-term viability. Another advantage of using population growth instead of extinction risk is that it may be much more reliably estimated with spotty or short-term data. In general, higher population growth will also decrease the probability of extinction, so these two ways of measuring viability or improvement in viability often give similar answers. However, they will not always yield identical answers about population health or risk, in large part because extinction risk is even more strongly influenced by temporal variability than is stochastic growth (Chapter 3).

Still, part of the appeal of using the population growth rate rather than the probability of extinction is to get away from the difficulties of predicting random events. Indeed, in 1994, Caughley suggested that there was a dichotomy of worldviews about rare species, with followers of the "small population paradigm" emphasizing (or really, *over*-emphasizing) the role of stochastic forces in determining extinction risk, while devotees of the "declining population paradigm" focused on the deterministic factors that lead to positive or negative population growth rates. In making this distinction, Caughley argued that randomness had little part to play in determining growth rates of populations and thus could be safely ignored much of the time. We frankly disagree with this assessment (as have others: Hedrick et al. 1996). Although a deterministic estimate of population growth rate can sometimes be a good gauge of population health, in most cases the effects of variability need to be considered. As our numerical example earlier in this chapter shows, variability can substantially change the answer about the magnitude and even the direction of population change. Thus, unfortunately, using population growth as a measure of viability does not exempt us from having to consider the complexities of variation in time and space.

Viability Metrics Related to Population Size and Number

Finally, population size and the number of populations are commonly used to measure population (or metapopulation) viability. Typically, some target population size or number is deemed safe from endangerment or "recovered." Implicit in any such measure of viability are assumptions about the population growth rate (presumably greater than one) needed to get to the target size, and the probability of extinction (presumably low) once that size is achieved. If these usually implicit aspects of a population size target are *explicitly* considered and estimated, then these numerical targets can be excellent, synthetic measures of viability. However, a target population size can also just be drawn from a hat, usually by the consensus of a committee of experts, but without clear supporting data. This may sound cynical, but several studies back up this claim. There are three situations in which target sizes for population safety are commonly used. One is in Recovery Plans produced under the ESA (United States Endangered Species Act). Tear et al. (1993) found that 28% of recovery plans set the population size for recovery at or below the number of individuals in existence at the time the plan was written, while 37% of plans set the target number of populations at or below the existing number at time of listing. Similarly, for ESA-listed plants, Schmeske et al. (1994) found that the number of populations set as recovery goals correlated very strongly with those in existence when the plans were written (with recovery numbers about twice the current numbers). This finding led Schemske and his co-authors to suggest that real biological criteria played little role in setting the recovery goals.

The second regulatory situation in which population size is extensively used to assess viability is conservation of marine mammals under the United States Marine Mammal Protection Act (MMPA). Historically, fisheries management has relied on the assumption that a carrying capacity exists for each population. Following this tradition, MMPA stipulates that a "depleted" population in need of protection has a population size less than 60% of its carrying capacity (Ralls et al. 1996). The advantage of such a criterion is that it provides a clear species-specific benchmark against which to measure current numbers. This criterion tends to be much more conservative than ESA recovery criteria. For example, The ESA recovery criterion for the California sea otter was set at 2,650 animals, while protection as a depleted population under MMPA would last until sea otter numbers are several times higher (Ralls et al. 1996). However, MMPA is also highly inflexible; if two species currently exist at the same numbers, but one was once very much more common, they are treated as having different risk, which may or may not be the case.

Finally, The World Conservation Union (IUCN) has adopted criteria for endangerment that rely on estimates of extinction risk, but these criteria have been roughly translated into population numbers, sizes, and trends (Mace and Lande 1991) so that they can be applied without exhaustive information. Because they were developed to be used with a variety of kinds and qualities of data, the criteria are rather complex. For example, critically endangered species (the category most severely at risk) are defined as showing any of the following: (1) reductions of at least 80% over the last 10 years or 3 generations (whichever is longer), or an expected reduction of at least 80% in the next 10 years or 3 generations; (2) an extent of occurrence of less than 100 km^2 or area of occupancy of less than 10 km^2; (3) a population of less than 250 mature individuals with evidence of decline; (4) a population of less than 50 mature individuals; or (5) a 50% probability of extinction in the wild in 10 years or 3 generations. As for the previous examples, if there is no possibility of doing even a basic PVA, the simple numerical criteria for endangerment under these guidelines are obviously useful, because the numbers are low enough to ensure that most species included in this categories will in fact be deeply imperiled—which is not to say that much larger populations cannot also be severely endangered.

As is probably clear, we firmly believe that the current population size and number can be no better than "second best" as a measure of viability. If we know nothing other than the current sizes of a set of populations, we should certainly use those sizes to perform triage, deciding which population(s) to manage first, but we must always do so with the knowledge that small but stable populations may be at lower risk than larger populations subjected to strong sources of variability. A far better approach, when the data are available, is to perform an analysis of population *dynamics* that can assess the stochastic population growth rate and the risk of extinction at specified future times. We will rely on these latter two viability measures for the rest of the book.

3

Count-Based PVA: Density-Independent Models

The type of population-level data that is most likely to be available to conservation planners and managers is count data, in which the number of individuals in either an entire population or a subset of the population is censused over multiple (not necessarily consecutive) years. Furthermore, to use such data for a PVA, it is not necessary to count the entire population; counts of breeding females, mated pairs, or plants actually in flower are all usable, as long as the segment of the population that is observed is a relatively constant fraction of the whole. Such data are relatively easy to collect, particularly in comparison with more detailed demographic information on individual organisms (see Chapters 6 through 9). In this chapter, we review a simple method for performing PVA using count data. The method's simplicity makes it applicable to a wide variety of data sets. However, several important simplifying assumptions underlie the method, and at the end of the chapter, we discuss how violations of these assumptions would introduce error into our estimates of population viability. In the next two chapters, we review methods to handle such violations, using more complex methods.

In a typical sequence of counts from a population, the observed number of individuals does not increase or decrease smoothly over time, but instead shows considerable variation around long-term trends (for an example, see Figure 3.6 later in the chapter). One factor that is likely to be an important contributor to these fluctuations in abundance is variation in the environment, which causes the rates of birth and death in the population to vary from year to year. As we mentioned in Chapter 2, the potential sources of environmentally driven variation are numerous, including temporal variability in climatic factors such as rainfall, temperature, and duration of the growing season. Most populations will be affected by such variation, either directly or indirectly through its effects on interacting species (e.g., prey, predators, competitors, diseases, and so on.). When we use a sequence of censuses to estimate measures of population viability, we must account for the pervasive effect of environmental variation that can be seen in most count data. To build a general understanding of how this is done, we first expand the overview of how environmental variation affects population dynamics that we gave in Chapter 2. We next review key theoretical results that underlie the simplest count-based methods in PVA. Then, with the necessary background in place, we delve into the details of using count data to assess population viability.

Population Dynamics in a Random Environment

Let us put more flesh on the bare-bones overview of how temporal variability influences population dynamics that we gave in Chapter 2. Specifically, let us return to the simple model for discrete-time geometric population growth in a randomly varying environment:

$$N_{t+1} = \lambda_t N_t \tag{3.1}$$

where N_t is the number of individuals in the population in year t and λ_t is the population growth rate, or the amount by which population size multiplies from year t to year $t + 1$. Recall that Equation 3.1 assumes that population growth is density independent (i.e., λ_t is not affected by population size, N_t). If there is no variation in the environment from year to year, then the population growth rate λ will be constant, and only three qualitative types of population growth are possible (Figure 3.1A): geometric increase (if $\lambda > 1$), geometric decline to extinction (if $\lambda < 1$), and stasis (if λ exactly equals 1). However, by causing survival and reproduction to vary from year to year, environmental variability will cause the population growth rate, λ_t, to vary as well, and unlike the simple example in Chapter 2 in which we assumed λ_t only took on two values, in reality we expect λ_t to vary over a continuous range of values. Moreover, if the environmental fluctuations driving changes in population growth include an element of unpredictability (as factors such as rainfall and temperature are likely to do), then we will not be able to predict with certainty what the exact sequence of future population growth rates will be. As a consequence, even if we know the current population size and both the average value and the degree of variability in the population growth rate, the best we can do is to make probabilistic statements about the number of individuals the population will include at some time in the future. That is, we must view change in population size over time as a stochastic process.

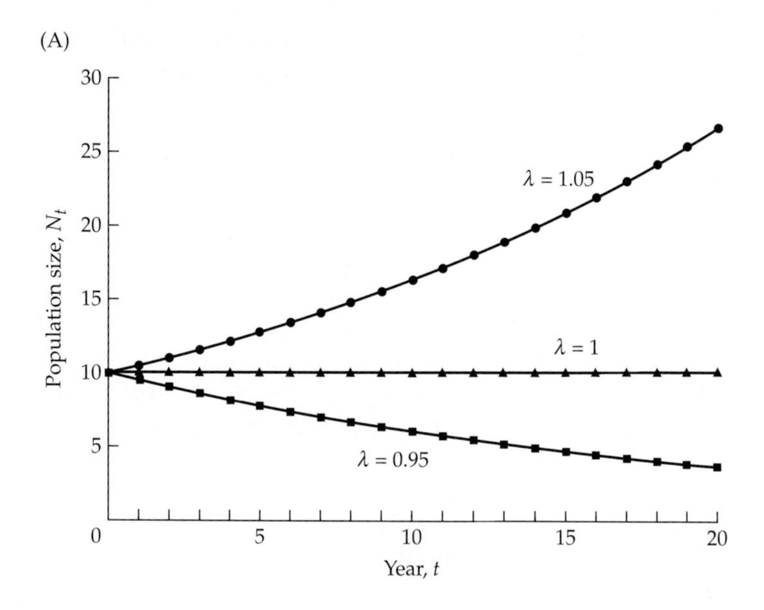

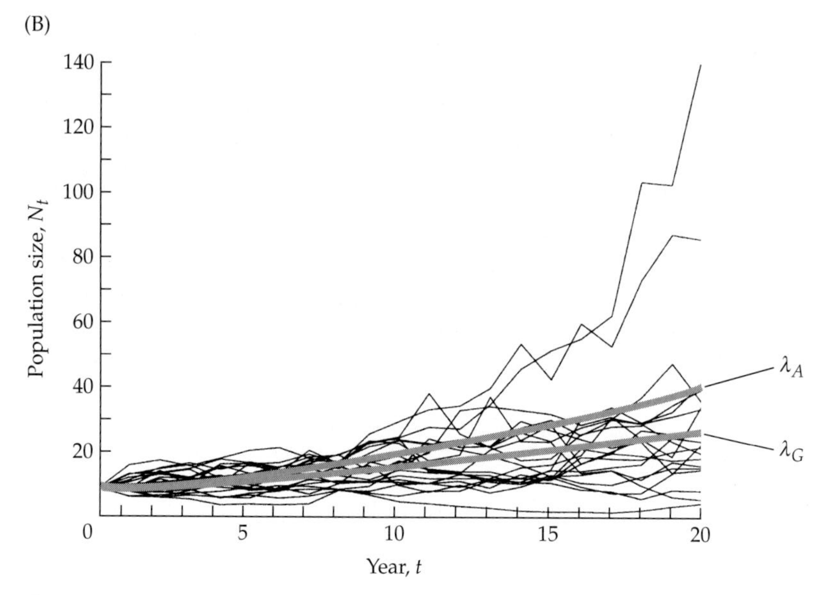

Figure 3.1 Patterns of deterministic versus stochastic growth. (A) Predicted population sizes under deterministic geometric growth with λ greater than, less than, or equal to 1. (B) Twenty realizations of stochastic population growth according to Equation 3.1. For all realizations, the population growth rate λ_t is log-normally distributed with an arithmetic mean of $\lambda_A = 1.0725$ and a variance of 0.047. The upper and lower broad gray lines show the deterministic predictions using the arithmetic and geometric mean growth rates, respectively.

To illustrate stochastic population growth, Figure 3.1B shows 20 realizations of Equation 3.1 in which the value of the population growth rate λ was drawn at random each year.[1] All realizations start at a population size of 10 individuals. Thus each realization can be viewed as a possible trajectory the population might follow given its initial size and the distribution of λ. Figure 3.1B illustrates three fundamental features of stochastic population growth. First, the realizations diverge over time, so that the farther into the future you go, the more variable the predictions about likely population sizes become. Second, the realizations do not follow very well the predicted trajectory based upon the arithmetic mean population growth rate, λ_A (this is the trajectory shown by the upper broad line in Figure 3.1B). In particular, even though in this case λ_A predicts that the population should slowly increase, a few realizations explode over the 20 years illustrated, whereas others decline. Thus extinction is possible even though the population size predicted by the arithmetic mean growth rate increases. Third, the endpoints of the 20 realizations shown are highly skewed, with a few trajectories winding up much higher than λ_A would suggest, but most ending well below this average prediction (in fact, three realizations ended lower than their starting population size). This skew is due in part to the multiplicative nature of population growth (see Equation 2.3). Because the size of the population after 20 years is proportional to the product of the population growth rates in each of those years, a long string of chance "good" years (i.e., those with high rates of population growth) can carry the population to a very high level of abundance, whereas strings of "bad" years tend to confine the population to the restricted zone between the average and zero abundance. We also saw in Chapter 2 that, because population growth is multiplicative, the geometric mean population growth rate λ_G is a better descriptor of the behavior of a typical realization. In fact, λ_G predicts the median population size at any point in the future; that is, for a large number of realizations, half will lie above and half below the population size predicted by λ_G (shown by the lower broad line in Figure 3.1B; in this particular sample of 20 realizations, more happened to fall below the median than above, but λ_G still does a better job of "splitting the difference" than does λ_A).

Skewness in the distribution of the likely future size of a population is a general feature of density-independent population growth in a stochastic environment. In fact, we can make an even more precise statement: the probability that the population will be of a certain size at a future time horizon will usually be well described by a particular skewed probability distribution,

---

[1]Specifically, λ is lognormally distributed (that is, the log of λ follows the familiar normal distribution); the lognormal distribution is one of several distributions that can appropriately describe random λ_t values because, unlike the normal, it never takes on negative values, and to be biologically realistic, a discrete-time population growth rate should never be negative.

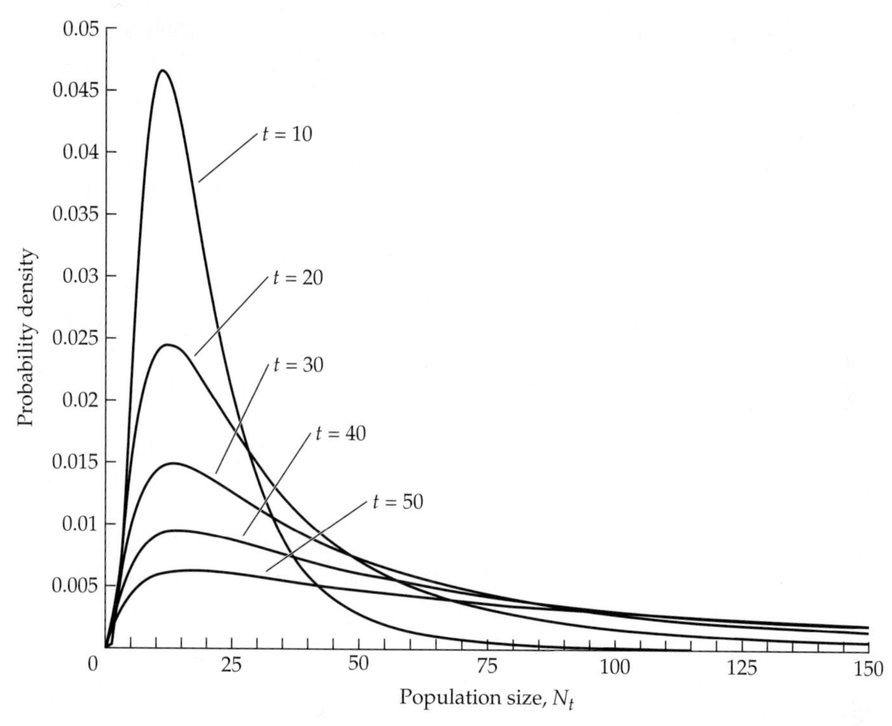

Figure 3.2 The probability that a population following the stochastic growth process illustrated in Figure 3.1B will lie within a certain size interval at a given time in the future is described by this lognormal probability density function.

the lognormal[2] (Lewontin and Cohen 1969). The particular lognormal distribution that would describe a large number of trajectories corresponding to the example in Figure 3.1B is shown in Figure 3.2 (compare this figure to Figures 2.4 and 2.5). The y-axis in Figure 3.2 is probability density; to obtain the actual probability that the population size lies between two particular values (the equivalent of the class boundaries in a histogram), we would calculate the integral of this probability density function between those two values of N_t. Thus population size is most likely to lie near the peak of the lognormal probability density function in Figure 3.2, but because the function is skewed to the right, the arithmetic mean population size will lie to the right of the peak. As Figure 3.2 illustrates, both the mean and the most likely population size increase over time, as does the probability density for larger population sizes. This makes sense given that most population trajectories tend to grow (Figure 3.1B).

[2]Thanks to the Central Limit theorem, this conclusion is true asymptotically regardless of the probability distribution from which the λ's are drawn. See Lewontin and Cohen 1969.

If population size itself follows a lognormal distribution, then the natural logarithm of population size will be normally distributed. This follows from the definition of a lognormal random variable; if N_t is lognormally distributed, then $\log N_t = X_t$, where X_t is *normally* distributed.[3] As does the lognormal distribution of N_t values (Figure 3.2), the normal distribution of log N_t values will change over time (Figure 3.3). Specifically, its mean will either increase or decrease, depending on whether separate realizations of the stochastic population growth process tend to grow (as in Figure 3.1) or decline, but its variance will strictly increase (as prediction becomes less certain over longer time intervals). In this chapter, we will use log N_t, rather than the untransformed population size, because a normal distribution arises naturally from the physical process of diffusion, which we will use below to approximate the extinction process.

Until now, we have talked about the growth of N_t. How do we characterize the growth of log N_t? Above and in Chapter 2, we argued that the best predictor of whether N_t will increase or decrease over the long term is λ_G, the geometric mean of the population growth rates each year. On the scale of log population size, the log of λ_G is the most natural way to express population growth. The geometric mean of λ is the value that would give the same average annual population growth rate as is observed over a long sequence of stochastically varying growth rates. That is, since $N_{t+1} = (\lambda_t \lambda_{t-1} \lambda_{t-2} \ldots \lambda_2 \lambda_1 \lambda_0) N_0$ (see Equation 2.3), λ_G is defined as

$$(\lambda_G)^t = \lambda_t \lambda_{t-1} \lambda_{t-2} \ldots \lambda_2 \lambda_1 \lambda_0 \quad \text{or} \quad \lambda_G = (\lambda_t \lambda_{t-1} \lambda_{t-2} \ldots \lambda_2 \lambda_1 \lambda_0)^{1/t} \qquad (3.2)$$

Converting this formula for λ_G to the log scale yields a new measure of "average" population growth:[4]

$$\mu = \log \lambda_G \approx \frac{\log \lambda_t + \log \lambda_{t-1} + \log \lambda_{t-2} + \cdots + \log \lambda_2 + \log \lambda_1 + \log \lambda_0}{t} \qquad (3.3)$$

As Equation 3.3 states, the correct measure of stochastic population growth on a log scale, μ, is equal to the log of λ_G or, equivalently, to the arithmetic mean of the log λ_t values. Thus if we can estimate the value of μ (which we will see how to do below), then we can immediately determine the geometric mean population growth rate, and thus whether the population will tend to grow or decline. In particular, Equation 3.3 predicts that if $\mu > 0$, then $\lambda_G > 1$, and most population trajectories will grow, whereas if $\mu < 0$, then $\lambda_G < 1$, and most trajectories will decline.

---

[3]In this book, when we write "log" we always mean the natural logarithm (following mathematical convention).

[4]Strictly speaking, the rightmost expression in Equation 3.3 approaches log λ_G only as t becomes large.

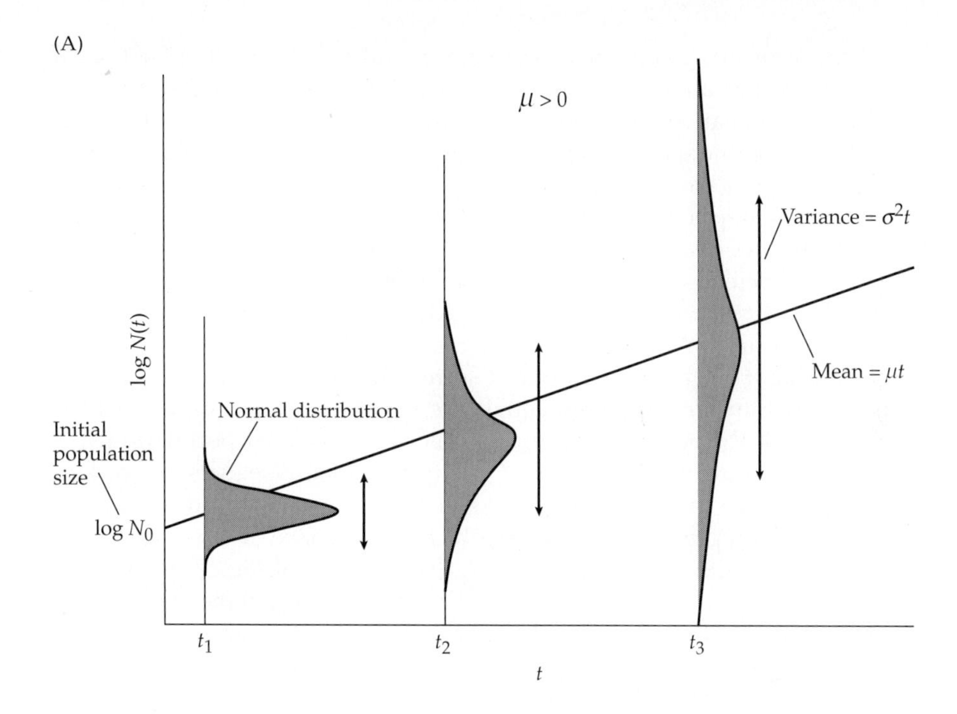

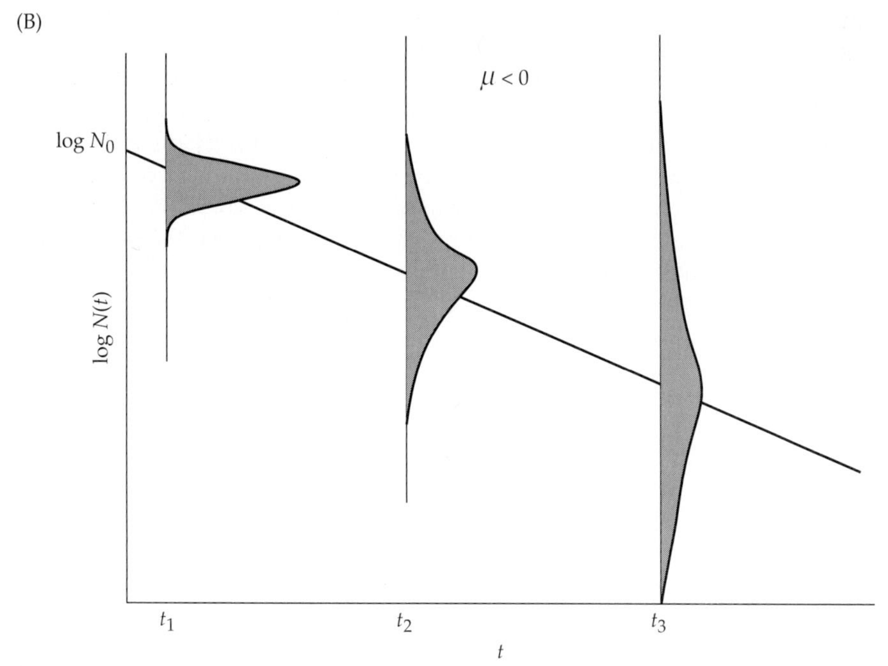

Figure 3.3 Normal distributions of the log population size, when the parameter μ is (A) positive or (B) negative.

Because of this relationship between μ and λ_G, the parameter μ is of direct interest as a metric of population viability, telling us the direction in which the population will tend to move over time. But remember that even if μ is positive, there is still some chance that the population will fall to low levels, even to zero, as Figures 3.1B and 3.3A show. Thus in addition to knowing the population's general tendency to grow or decline, we would also like to predict the full probability distribution for different population sizes, in particular so that we can calculate the probability that the population will fall below a specified quasi-extinction threshold at or before a specified future time horizon. To fully characterize the changing normal distribution of log population size (Figure 3.3), we will need two parameters. The parameter μ not only equals the log of λ_G ; it also represents the rate at which the mean of the normal distribution of log population size changes over time (that is, at time t, the mean of log N_t equals μt; see Figure 3.3). But we must also define a second parameter, called σ^2 (sigma-squared), that represents the rate at which the variance of the distribution increases over time (that is, at time t, the variance of log N_t equals $\sigma^2 t$; Figure 3.3). Whereas μ is approximated by the (arithmetic) mean of the log population growth rates (i.e., the mean of the log λ_t values; see Equation 3.3), σ^2 is approximated by the variance of the log λ_t's, as we will see below. Our use of the symbol σ^2 to represent the variance of log λ follows Lande and Orzack 1988 and Dennis et al. 1991, the two most important sources for the material presented in this chapter. Throughout the book, we use σ^2 *strictly* to represent the variance in the log population growth rate that is caused by environmental stochasticity.

Together, μ and σ^2 fully describe the normal probability distribution of future log population sizes. Specifically, a positive value of μ indicates an environment in which most realizations tend to grow, whereas a negative μ indicates that declining realizations predominate. The more the population growth rate λ varies from year to year as a result of environmental stochasticity, the greater will be the value of σ^2 and the greater the range of possible population sizes in the future. These two measures of the dynamics of log population size are used extensively in the calculation of extinction times. Because their verbal definitions are somewhat tortuous ("the mean and variance of the log population growth rate"), we will usually refer to them by their symbols. Nonetheless, to really understand how simple count-based PVAs work, you will need to fix the meanings of μ and σ^2 in your mind.

The Relationship Between the Probability of Extinction and the Parameters μ and σ^2

Because μ and σ^2 describe the changing probability that the log population size will lie within a given range (Figure 3.3), it makes intuitive sense that if we know their values (as well as the current population size and the quasi-extinction threshold), we can calculate the probability of quasi-extinction at any future time. In this section, we review the theoretical underpinnings behind the calculation of extinction probabilities using μ and σ^2. This section is more mathematically challenging than the rest of the chapter, and you may want to skip it. If you do, you will still be able to understand and use the following sections on how to do simple count-based PVAs, provided that you accept on faith the formulas relating μ and σ^2 to extinction risk. However, it is important to recognize the basic patterns in extinction times as well as the assumptions underlying the theory, so we urge you to read the last three paragraphs of this section even if you skip the rest.

For PVA purposes, we consider a population to be "extinct" (either literally so or at least in need of a radically different management approach) if it *ever* hits a quasi-extinction threshold. (In Chapter 2 we discussed several factors that determine the choice of a reasonable quasi-extinction threshold.) Thus to measure extinction risk, we want to calculate the probability that the population hits the threshold *at any point during the interval between the present time and a specified future time horizon*. We begin by calculating the probability that extinction first occurs in one small segment of that time interval, and then we sum the probabilities over all segments. This calculation is performed by viewing the changing normal distribution of log population size in Figure 3.3 as though it were a cloud of particles undergoing diffusion with drift. To envision this process, think of a large number of small beads released at a single point near the bottom of an infinitely deep river, with height above the river bottom standing in for log population size. The path of each bead represents a single realization of "population" growth. As time evolves, the cloud of beads will move downriver, but its vertical position will change as well, moving upward or downward depending on whether the individual beads tend to float or sink, and spreading out due to turbulent flow of the water. The rate of floating or sinking is represented by μ, and the rate of turbulent diffusion by σ^2. The river bottom represents the quasi-extinction threshold; we will assume that any bead that hits the bottom will adhere to it. (In mathematical terms, the quasi-extinction threshold is a so-called absorbing lower boundary.) Even if the individual beads tend to float, some of them will be pushed to the bottom by turbulence, just as some population trajectories will hit low levels even as most trajectories tend to grow (Figure 3.3A), and the more turbulent is the flow of the water (i.e., the larger is σ^2), the quicker some beads will be pushed to the bottom. Treating the change in log population size (or in the vertical position of a bead) as a process of diffusion with drift allows the rate at which realizations first hit the lower boundary to be calculated. (For full details on this diffusion approximation, see Cox and Miller 1965, pages 208–222, and Lande and Orzack 1988; for a clearly presented alternative derivation of Equation 3.4, see Whitmore and Seshadri 1987.) Using a diffusion approximation, the probability density for first hitting the quasi-extinction threshold (that is, going "extinct") in a small segment of time beginning at t is given by the so-called inverse Gaussian distribution:[5]

$$g(t|\mu,\sigma^2,d) = \frac{d}{\sqrt{2\pi\sigma^2 t^3}} \exp\left[\frac{-(d+\mu t)^2}{2\sigma^2 t}\right] \tag{3.4}$$

where $d = \log N_c - \log N_x$ is the difference between the log of the current population size N_c and the log of the extinction threshold, N_x. (The notation $g(t|\mu,\sigma^2,d)$ means that the value of g at time t depends on the values of μ, σ^2, and d.) Notice that the extinction probability depends only on the difference (on the log scale) between the current and threshold population sizes, not on their actual values.

Figure 3.4 illustrates how the shape of the inverse Gaussian distribution (Equation 3.4) is affected by the values of μ and σ^2. The most likely time to hit the threshold is the peak of the distribution. The position of the peak is shifted toward shorter times as the value of σ^2 increases. (Compare curves with the same value of μ but different values of σ^2.) This is because greater variation drives populations down to the threshold more rapidly, as we would expect from the analogy with diffusing beads. Figure 3.4A shows that more negative values of μ lead to shorter quasi-extinction times. (Compare the two curves with $\sigma^2 = 0.04$.) This also makes sense; if the mean of the normal distribution of log population size declines more rapidly, quasi-extinction should occur earlier. If μ is positive, the probability of quasi-extinction occurring in each time segment is smaller than if μ is negative. (Compare Figure 3.4 A and B, and note the different scales of the y-axes.) The more positive μ is, the smaller is the total area under the curve (which represents the ultimate probability of quasi-extinction, as we will see below). However, notice that as μ becomes more positive, the peak of the distribution (i.e., the most likely quasi-extinction time) actually *decreases.* (Compare curves for $\mu = 0.01$ and $\mu = 0.03$ with the same value of σ^2 in Figure 3.4B.) The reason for this somewhat counterintuitive result (as explained by Lande and Orzack 1988 and Dennis et al. 1991) is that when μ is strongly positive, most realizations tend to increase rapidly to high numbers, from which they

[5]Strictly speaking, Equation 3.4 is a "proper" probability distribution (i.e., a function that integrates to 1) only if $\mu < 0$. This merely reflects the fact that when $\mu > 0$, not all realizations hit the threshold, so the ultimate probability of extinction is less than 1 (see Equation 3.7). Equation 3.4 can be converted into a proper probability distribution for the time to hit the threshold *given that the threshold is reached eventually* by replacing the numerator of the exponential term by $-(d - |\mu| t)^2$. However, the "improper" distribution is actually more useful to us, as it reflects the full, unconditional probabilities of extinction. See Lande and Orzack 1988.

(A)

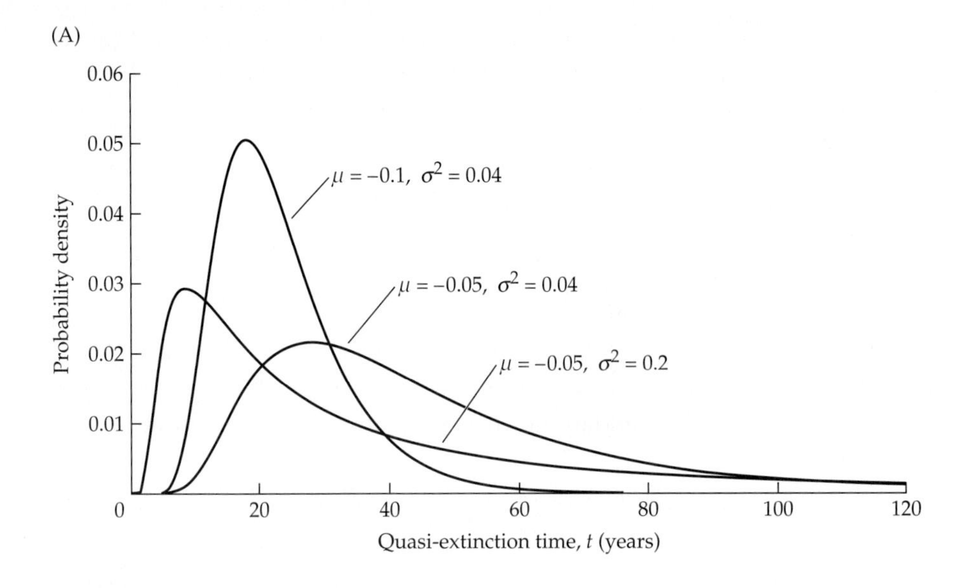

(B)

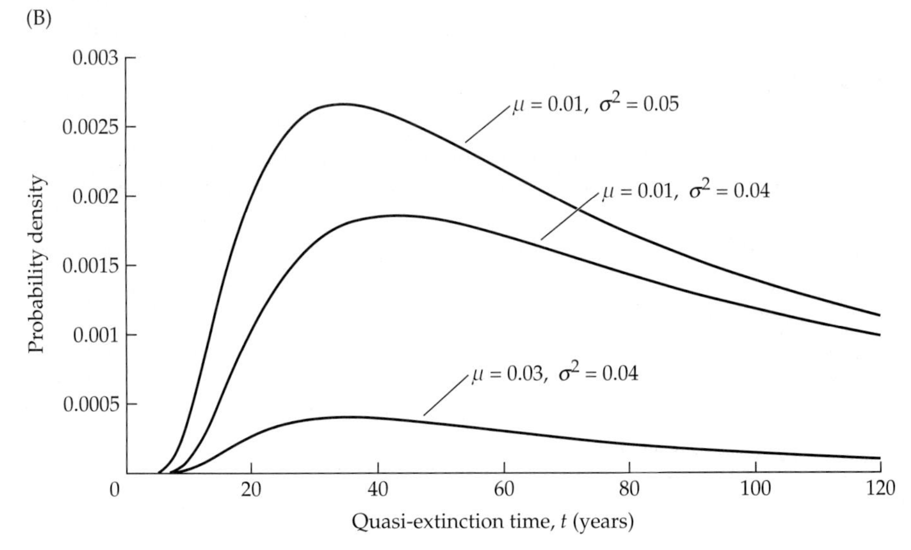

Figure 3.4 The inverse Gaussian distribution (Equation 3.3) with (A) $\mu < 0$ and (B) $\mu > 0$. Here, the current population size and the quasi-extinction threshold are $N_c = 10$ and $N_x = 1$, respectively.

are unlikely to ever hit the quasi-extinction threshold. It is only early in the growth process, when population size is still low, that a chance string of bad years can push the population down to the threshold. So, extinction is unlikely to happen, but if it does, it will occur rapidly.

The inverse Gaussian distribution gives the probability that quasi-extinction occurs in a very small time segment. To calculate the probability that the threshold is reached at any time between the present ($t = 0$) and a future time of interest ($t = T$), we integrate Equation 3.4 from $t = 0$ to $t = T$ (that is, we sum up the probabilities for each small interval of time). The result is the cumulative distribution function (or CDF) for the time to quasi-extinction,

$$G(T \mid d, \mu, \sigma^2) = \Phi\left(\frac{-d - \mu T}{\sqrt{\sigma^2 T}}\right) + \exp\left(-2\mu d/\sigma^2\right) \Phi\left(\frac{-d + \mu T}{\sqrt{\sigma^2 T}}\right) \tag{3.5}$$

where $\Phi(z)$ (phi) is the standard normal cumulative distribution function

$$\Phi(z) = \left(1/\sqrt{2\pi}\right) \int_{-\infty}^{z} \exp\left(-y^2/2\right) dy \tag{3.6}$$

$\Phi(z)$ is simply the area from $-\infty$ to z under a normal curve with a mean of zero and a variance of 1. Values of $\Phi(z)$ are tabulated in most basic probability reference books (e.g., Abramowitz and Stegun 1964), and can be calculated using built-in functions in mathematical software packages such as MATLAB and even spreadsheet programs such as Excel.

Figure 3.5 illustrates how the probability of hitting the threshold according to Equation 3.5 increases as the time horizon T is moved farther into the future. If μ is substantially negative, the probability of extinction increases rapidly with time, quickly reaching a value near 1. If the time horizon is not far into the future, increasing the value of σ^2 increases the probability that extinction will have occurred before the horizon is reached (once again, greater year-to-year variability leads to higher extinction risk). For positive μ, the probability of extinction increases slowly with time, and never reaches a value of 1. The probability of ultimate extinction (that is, at a time horizon of infinity) can be calculated by taking the integral of the inverse Gaussian distribution from $t = 0$ to $t = \infty$:

$$G(\infty \mid d, \mu, \sigma^2) = \begin{cases} 1 & \text{if } \mu \le 0 \\ \exp\left(-2\mu d/\sigma^2\right) & \text{if } \mu > 0 \end{cases} \tag{3.7}$$

Thus if μ is zero or negative, ultimate extinction is certain, and increasing σ^2 has no effect on the probability that the threshold is eventually reached. In contrast, if μ is positive, ultimate extinction is not a certainty, and increasing σ^2 increases the probability of ultimate extinction.

The rest of this chapter is devoted to the calculation of extinction risk estimates using count data, but even in the absence of any data, knowledge of

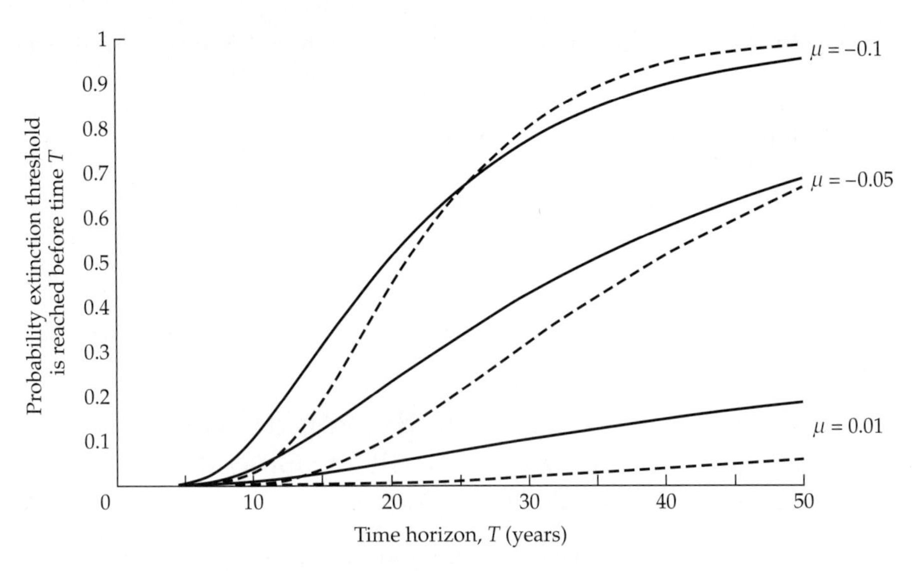

Figure 3.5 Cumulative distribution function (Equation 3.4) for the time to reach an extinction threshold of $N_x = 1$ from an initial population of $N_c = 10$; $\sigma^2 = 0.04$ (dashed lines) or 0.08 (solid lines).

how the CDF is affected by its underlying parameters can help us to make useful *qualitative* assessments of relative viability for a species of critical concern, especially if we can use natural history information to make inferences about the local environment or about the life history of the species in question. For example, we will frequently be able to make an educated guess that the environment one population experiences is likely to be more variable than another's in ways that will affect population growth. Such differences in σ^2 influence the extinction time CDF even when its other determinants (μ and the starting and threshold population sizes) are fixed (Figure 3.5). Similarly, some species will have life history features (e.g., long-lived adults) that buffer their populations against year-to-year environmental variation, while others do not. With a low σ^2, values of μ that are only slightly positive are sufficient to minimize extinction risks, whereas with high variance, much greater mean growth is needed. Thus we can state that, all else being equal, the greater the environmentally driven fluctuations in population growth rate, the greater will be the risk of extinction, especially for short-term time horizons, a statement that, though qualitative, nonetheless provides some useful guidance.

Before going on, it is also important to recognize that three key assumptions underlie the use of a diffusion approximation to derive the key Equations 3.4, 3.5, and 3.7. First, we assume that the environmental perturbations affecting the population growth rate are small to moderate, just as gas molecules typically diffuse through a series of small jumps.[6] In other words, we assume that large catastrophes or bonanzas (Chapter 2) do not occur frequently enough to be important. Second, we assume that the changes in population size are independent between one time interval and the next; that is, we assume that strings of good or bad years occur no more frequently than would be expected by chance. Third, we assume that the values of μ and σ^2 do not change over time, either in response to changes in population density or because of temporal trends in environmental conditions. We show how to test for violations of these assumptions in the following section, and how to deal with some of these violations (should they occur) in Chapter 4.

Using Count Data to Estimate the Population Growth Parameters μ and σ^2: An Illustration Using the Yellowstone Grizzly Bear Census

We argued in Chapter 2 that the extinction time cumulative distribution function or CDF, given by Equation 3.5 and illustrated in Figure 3.5, is perhaps the single most useful metric of a population's risk of extinction. To calculate it, we need only four quantities: the current population size N_c, the extinction threshold N_x, and the values of μ and σ^2. We turn now to methods for estimating μ and σ^2 from a series of census counts.

Let us assume that we have conducted a total of $q + 1$ annual censuses of a population at times $t_0, t_1, t_2, \ldots, t_q$, having obtained the census counts $N_0, N_1, N_2, \ldots, N_q$. The censuses need not have been conducted in consecutive years, but in a seasonal environment, the censuses should have been performed at the same time of year. Although μ and σ^2 can be used to describe future population size distributions (as shown in Figure 3.3), they are more directly measures of the mean and variance of the *change* in log population sizes. Thus it is not the raw census counts themselves that we want to use to estimate μ and σ^2, but the amount by which the natural logarithms of the counts change over each of the inter-census intervals. For example, over the time interval of length $(t_{i+1} - t_i)$ years between censuses i and $i + 1$, the logs of the counts change by an amount

$$\log N_{i+1} - \log N_i = \log\left(N_{i+1}/N_i\right) = \log \lambda_i \tag{3.8}$$

[6]Technically, diffusion models also assume that the state variable (in this case population size) is continuous, whereas in reality, a population can only contain an integer number of individuals. The assumption of small environmental perturbations is therefore tantamount to assuming that the *change* in numbers over a short time interval *relative to* the total number of individuals in the population is small, an assumption that is likely to break down at small population sizes (Ludwig 1996a). Using a moderate quasi-extinction threshold, as we advocated in Chapter 2, ameliorates to some extent the effects of this assumption.

In Equation 3.8, $\lambda_i = N_{i+1}/N_i$ is the population growth rate between census i and census $i+1$, emphasizing the fact that it is the logs of the population growth rates that we will use to estimate μ and σ^2. The counts from $q+1$ censuses allow us to calculate log population growth rates for q time intervals, although as noted above, these intervals need not be of the same length.

An example of count-based data that we will analyze extensively in this chapter comes from the ongoing census of adult female grizzly bears (*Ursus arctos*) in the Greater Yellowstone ecosystem. (For an earlier analysis of the same population, see Dennis et al. 1991.) This population is currently designated as threatened by the United States Fish and Wildlife Service, and is completely isolated from other grizzly bear populations. Each year, bear biologists count the number of unique female bears with cubs (offspring in their first year of life) in the entire Yellowstone population. Because the litters of one to three cubs remain closely associated with their mothers, females with cubs are the most conspicuous, and therefore most reliably censused, element of the population. Censuses were originally performed by observing bears at park garbage dumps, but they have been conducted by aerial survey since the closure of the dumps in 1970–1971. The counts of females with cubs are used to estimate the total number of adult females in the population. Specifically, the number of adult females in year t is estimated as the sum of the observed number of females with cubs in years t, $t+1$, and $t+2$ (Eberhardt et al. 1986). The logic underlying this estimate is that the interval between litters produced by the same mother is at least three years, so that females with cubs observed in years $t+1$ and $t+2$ could not have been the same individuals that were observed with cubs in year t. Yet, if they are observed in later years, they must have been alive in year t (although they may not have been *adults* in year t, which introduces some error). The estimated numbers of adult females from 39 annual censuses of the Yellowstone population, beginning in 1959, are shown in Figure 3.6 and listed in Table 3.1. (These data can also be obtained at the Interagency Grizzly Bear Study Team's website: www.nrmsc.usgs.gov/research/igbstpub.htm.)

We now review two methods to estimate μ and σ^2, illustrating both methods with the grizzly bear data in Table 3.1.

Estimating μ and σ^2 as the mean and variance of log (N_{i+1}/N_i)

If the censuses were conducted at regular yearly intervals, as was the Yellowstone grizzly bear census, the simplest method of estimating μ and σ^2 is to calculate the arithmetic mean and the sample variance of the log (N_{i+1}/N_i) values, given respectively by

$$\hat{\mu} = \frac{1}{q}\sum_{i=0}^{q-1} \log(N_{i+1}/N_i) \quad \text{and} \quad \hat{\sigma}^2 = \frac{1}{q-1}\sum_{i=0}^{q-1} \left(\log(N_{i+1}/N_i) - \hat{\mu}\right)^2 \qquad (3.9)$$

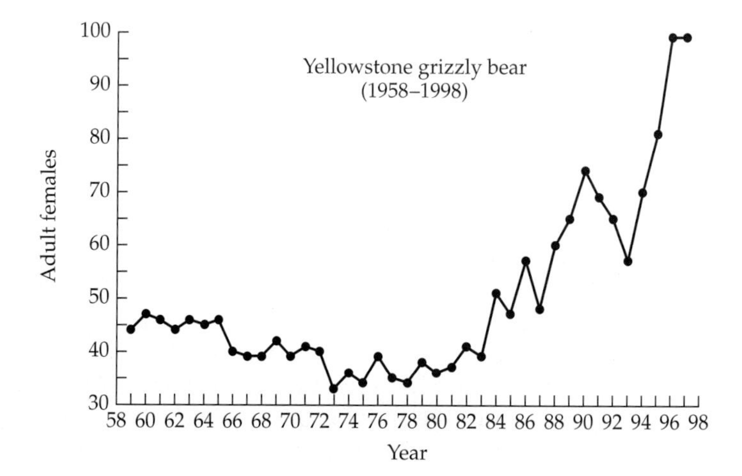

Figure 3.6 Census counts of adult female grizzly bears in the Greater Yellowstone population (data from Eberhardt et al. 1986 and Haroldson 1999).

(where the hats over $\hat{\mu}$ and $\hat{\sigma}^2$ indicate that they are estimates). Notice that in calculating $\hat{\sigma}^2$, we divide by $q-1$ instead of q, thus using the typical formula to obtain an unbiased estimate of variance from a sample of data (Dennis et al. 1991[7]). The values of $\hat{\mu}$ and $\hat{\sigma}^2$ are easily obtained by applying the functions AVERAGE and VAR (respectively) in Microsoft Excel to the log (N_{i+1}/N_i) values, or by using similar routines in any statistical package.

The estimates $\hat{\mu}$ and $\hat{\sigma}^2$ for the Yellowstone grizzly bear calculated using Equations 3.9 are given in Table 3.1. The value of $\hat{\mu}$ is positive, reflecting a general upward trend in the number of adult female bears over the sampling period (Figure 3.6). The value of $\hat{\sigma}^2$ is about 60% of the value of $\hat{\mu}$; recall that $\hat{\sigma}^2$ is a measure of the year-to-year variability in the census counts.

Estimating μ and σ^2 by Linear Regression

If the censuses were not taken at equal time intervals, we should not use Equations 3.9 to estimate μ and σ^2, because they do not account for the fact that both the mean and variance in population change should be larger for pairs of censuses that are separated by longer intervals of time.[8] In such cases, we can estimate the parameters using a linear regression method proposed by Dennis et al. (1991). In fact, the linear regression method, although somewhat more complicated to execute, has advantages over Equations 3.9 even when the censuses occurred every year. In particular, the output produced by most widely available regression packages allows us to easily

---

[7]We use the symbol $\hat{\sigma}^2$ to represent the unbiased estimate of σ^2. Dennis et al. (1991) use $\tilde{\sigma}^2$.

[8]An alternate formula that does account for the length of the time interval is presented in Dennis et al. 1991 (see their Equations 24 & 25), but as it does not have the advantages of their regression approach, we do not present it here.

TABLE 3.1 Estimated number of adult female grizzly bears in the Greater Yellowstone population[a]

Census, i	*Year, t_i*	*Adult females, N_i*	*log (N_{i+1}/N_i)*
1	1959	44	0.0660
2	1960	47	−0.0215
3	1961	46	−0.0445
4	1962	44	0.0445
5	1963	46	−0.0220
6	1964	45	0.0220
7	1965	46	−0.1398
8	1966	40	−0.0253
9	1967	39	0.0000
10	1968	39	0.0741
11	1969	42	−0.0741
12	1970	39	0.0500
13	1971	41	−0.0247
14	1972	40	−0.1924
15	1973	33	0.0870
16	1974	36	−0.0572
17	1975	34	0.1372
18	1976	39	−0.1082
19	1977	35	−0.0290
20	1978	34	0.1112
21	1979	38	−0.0541
22	1980	36	0.0274
23	1981	37	0.1027
24	1982	41	−0.0500
25	1983	39	0.2683
26	1984	51	−0.0817
27	1985	47	0.1929
28	1986	57	−0.1719
29	1987	48	0.2231
30	1988	60	0.0800
31	1989	65	0.1297
32	1990	74	−0.0700
33	1991	69	−0.0597
34	1992	65	−0.1313
35	1993	57	0.2054
36	1994	70	0.1460
37	1995	81	0.2007
38	1996	99	0.0000
39	1997	99	
		Mean:	$\hat{\mu} = 0.02134$
		Sample variance:	$\hat{\sigma}^2 = 0.01305$

[a]Data from Eberhardt et al. 1986 and Haroldson 1999.

place confidence limits on the parameters and to test assumptions of the underlying PVA model, as we now demonstrate.

REGRESSION PROCEDURE The basic idea of this method is to regress the log population growth rate over a time interval against the amount of time elapsed. However, the most straightforward way to do this does not conform to one of the assumptions of standard linear regression, namely that the variance in the dependent variable (log population growth, in this case) is constant over different values of the independent variable (time elapsed). As we noted above, the variance of log population size increases with time (see Figure 3.3), which means that the variance in population *change* over a time interval is also dependent on the time elapsed. In particular, if two censuses are separated by $t_{i+1} - t_i$ years, then $\log(N_{t+1}/N_i)$ has a variance of $\sigma^2(t_{i+1} - t_i)$. To make our regression conform to the assumption of equal variances, we need to transform the rate of population change (and also the time elapsed) to get rid of this time dependence. This is accomplished by dividing both the log population growth rate and the time elapsed by $\sqrt{t_{i+1} - t_i}$, which makes the variance in the transformed population change variable equal to σ^2 for any time interval.

Thus, to perform the linear regression, we first calculate a transformation of the length of each time interval to use as the independent variable:

$$x_i = \sqrt{t_{i+1} - t_i} \tag{3.10a}$$

We then calculate a transformed variable of population change, the dependent variable, as:

$$y_i = \log\left(N_{i+1}/N_i\right)/\sqrt{t_{i+1} - t_i} = \log\left(N_{i+1}/N_i\right)/x_i \tag{3.10b}$$

For example, if we have entered the values of t_i and N_i from Table 3.1 in the first 39 rows of Columns A and B in an Excel worksheet, we can perform the transformations in Equations 3.10a,b by entering the formulas =SQRT(A2-A1) in cell C1 and =LN(B2/B1)/C1 in cell D1 and then "filling down" the subsequent 38 rows of Columns C and D. Notice that when all adjacent censuses are one year apart, as in the grizzly bear data set, all the x values will equal 1, and y_i will equal $\log(N_{i+1}/N_i)$.

The final step is to perform a linear regression of the y_i's against the x_i's, *forcing the regression intercept to be zero.* (In some statistics packages, such as Excel's regression routine, this is accomplished by choosing "no constant" in the regression options.) By fixing the regression intercept at zero, we are enforcing the rule that there can be no change in population size if no time has elapsed. The slope of the regression is an estimate of μ and the regression's error mean square estimates σ^2 (Figure 3.7). For example, the following SAS command applied to the transformed grizzly bear data generates

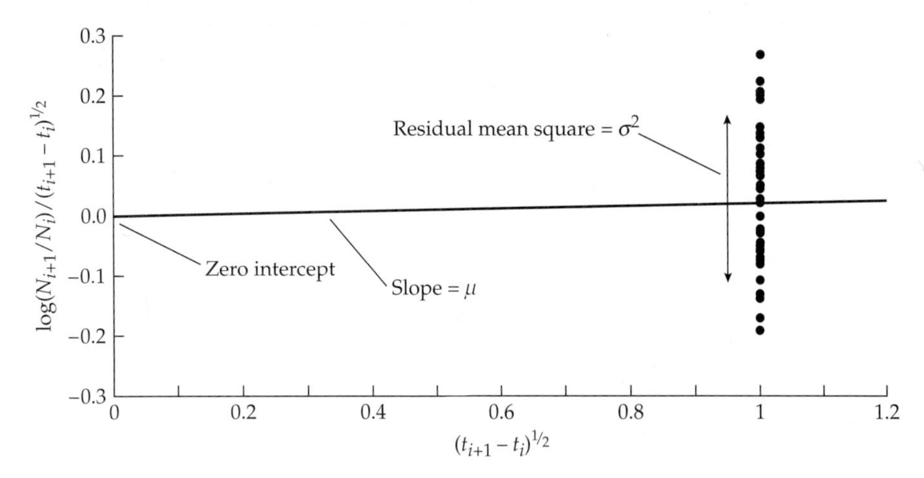

Figure 3.7 Estimating μ and σ^2 for the adult female population of Yellowstone grizzly bears using linear regression.

the output in Box 3.1:

```
proc reg;
  model y=x / noint dw influence;
run;
```
(3.11)

The command `noint` induces a regression with no intercept, and the commands `dw` and `influence` instruct SAS to print regression diagnostics of which we will make use below. In Box 3.1, the `Parameter Estimate for the Variable X` (0.021337) is the regression slope, which is our estimate of μ, and the second entry (0.01305) in the column `Mean Square` in the `Analysis of Variance` table is the error mean square, which is our estimate of σ^2. The output in Box 3.1 is equivalent to that produced by virtually all good statistics packages, although terminology may differ. (For example, the error mean square is sometimes labeled "residual mean square".) Notice that the regression yields the same estimates for μ and σ^2 that we obtained in Table 3.1. However, the regression output also provides additional useful information, as we now review.

USING REGRESSION OUTPUT TO CONSTRUCT CONFIDENCE INTERVALS FOR μ AND σ^2 Ideally, we want more than just the single best estimates of μ and σ^2. Because these will only be *estimates* based on the limited number of transitions in the data set, we would also like to know how much confidence we can place in them. Confidence intervals provide an upper and lower value between which the true value of each parameter is likely to lie. Confidence limits help us to place the best estimates of each parameter in context. For

BOX 3.1 *Output produced by SAS commands given in Equation 3.11[1].*

```
Model: MODEL1
NOTE: No intercept in model. R-square is redefined.
Dependent Variable: Y

                          Analysis of Variance

                              Sum of          Mean
        Source        DF     Squares        Square        F Value        Prob>F
        Model          1     0.01730       0.01730          1.325        0.2570
        Error         37     0.48295       0.01305
        U Total       38     0.50025

          Root MSE        0.11425       R-square         0.0346
          Dep Mean        0.02134       Adj R-sq         0.0085
          C.V.          535.45064

                          Parameter Estimates

                       Parameter        Standard     T for H0:
      Variable   DF     Estimate           Error    Parameter=0     Prob > |T|
      X           1     0.021337      0.01853351          1.151         0.2570

Durbin-Watson D                      2.570
(For Number of Obs.)                    38
1st Order Autocorrelation           -0.288
```

[1]Output referred to in the text is in bold.

example, if $\hat{\mu}$, the best estimate of μ, is positive, we would predict that most population trajectories will grow (as in Figure 3.3A). However, if the lower limit of the confidence interval for μ is negative, we cannot, based on the available data, rule out the possibility that the population will actually tend to decline over the long term.

In a proper linear regression, the estimate of the slope (i.e., $\hat{\mu}$, the estimate of μ) is normally distributed around its true value. Using this fact, many statistics packages will provide 95% confidence limits or standard errors for the regression slope. If yours does not, or if you want to calculate different confidence limits (say, 90 or 99% limits) it is easy to do so using the following formula:

$$\left(\hat{\mu} - t_{\alpha,q-1}\ \mathrm{SE}(\hat{\mu}),\quad \hat{\mu} + t_{\alpha,q-1}\ \mathrm{SE}(\hat{\mu})\right) \tag{3.12}$$

(Dennis et al. 1991). Here, $t_{\alpha,\, q-1}$ is the critical value of the two-tailed Student's t distribution with a significance level of α and $q-1$ degrees of freedom. For example, if we want to place a 95% confidence interval around μ for the Yellowstone grizzly bear, we would set α equal to 0.05 and q equal to 38 (the number of transitions on which the regression was performed; Table 3.1), and then obtain the value of $t_{0.05,\, 37}$. We can look up this value in

BOX 3.1 *(continued)*

Obs	Residual	Rstudent	Hat Diag H	Cov Ratio	Dffits	X Dfbetas
1	0.0447	0.3916	0.0263	1.0511	0.0644	0.0644
2	-0.0428	-0.3755	0.0263	1.0514	-0.0617	-0.0617
3	-0.0658	-0.5787	0.0263	1.0458	-0.0951	-0.0951
4	0.0232	0.2028	0.0263	1.0544	0.0333	0.0333
5	-0.0433	-0.3799	0.0263	1.0513	-0.0625	-0.0625
6	0.000663	0.0058	0.0263	1.0556	0.0010	0.0010
7	-0.1611	-1.4505	0.0263	0.9973	-0.2385	-0.2385
8	-0.0466	-0.4090	0.0263	1.0507	-0.0672	-0.0672
9	-0.0213	-0.1868	0.0263	1.0545	-0.0307	-0.0307
10	0.0528	0.4630	0.0263	1.0493	0.0761	0.0761
11	-0.0954	-0.8432	0.0263	1.0351	-0.1386	-0.1386
12	0.0287	0.2510	0.0263	1.0537	0.0413	0.0413
13	-0.0460	-0.4037	0.0263	1.0508	-0.0664	-0.0664
14	-0.2137	-1.9682	0.0263	0.9530	-0.3236	-0.3236
15	0.0657	0.5772	0.0263	1.0459	0.0949	0.0949
16	-0.0785	-0.6917	0.0263	1.0417	-0.1137	-0.1137
17	0.1159	1.0286	0.0263	1.0254	0.1691	0.1691
18	-0.1295	-1.1542	0.0263	1.0179	-0.1897	-0.1897
19	-0.0503	-0.4416	0.0263	1.0499	-0.0726	-0.0726
20	0.0899	0.7931	0.0263	1.0374	0.1304	0.1304
21	-0.0754	-0.6641	0.0263	1.0428	-0.1092	-0.1092
22	0.00606	0.0531	0.0263	1.0555	0.0087	0.0087
23	0.0814	0.7170	0.0263	1.0407	0.1179	0.1179
24	-0.0713	-0.6276	0.0263	1.0441	-0.1032	-0.1032
25	0.2470	**2.3163**	0.0263	0.9186	**0.3808**	0.3808
26	-0.1030	-0.9119	0.0263	1.0317	-0.1499	-0.1499
27	0.1716	1.5504	0.0263	0.9895	0.2549	0.2549
28	-0.1932	-1.7622	0.0263	0.9717	-0.2897	-0.2897
29	0.2018	1.8471	0.0263	0.9642	0.3037	0.3037
30	0.0587	0.5152	0.0263	1.0478	0.0847	0.0847
31	0.1084	0.9602	0.0263	1.0292	0.1579	0.1579
32	-0.0913	-0.8064	0.0263	1.0368	-0.1326	-0.1326
33	-0.0810	-0.7140	0.0263	1.0408	-0.1174	-0.1174
34	-0.1526	-1.3699	0.0263	1.0033	-0.2252	-0.2252
35	0.1841	1.6718	0.0263	0.9795	0.2748	0.2748
36	0.1247	1.1092	0.0263	1.0207	0.1824	0.1824
37	0.1794	1.6260	0.0263	0.9833	0.2673	0.2673
38	-0.0213	-0.1868	0.0263	1.0545	-0.0307	-0.0307

```
Sum of Residuals:                     0
Sum of Squared Residuals:        0.4829
Predicted Resid SS (Press):      0.5094
```

most standard statistics texts or compute it with the aid of mathematical software such as MATLAB[9]; even many spreadsheet programs can calculate it. For example, entering the formula =TINV(0.05, 37) into a cell in a Microsoft Excel worksheet yields the value $t_{0.05, 37} = 2.0262$. In Expression 3.12, SE($\hat{\mu}$) is the standard error of $\hat{\mu}$, the estimated regression slope. SAS supplies it in the regression output (Box 3.1) in the column labeled `Standard Error`; it is 0.0185. The standard error can also be calculated directly from the expression $\sqrt{\hat{\sigma}^2/t_q}$, where $\hat{\sigma}^2$ is the regression error mean square and t_q is the duration of the census in years (Dennis et al. 1991); for the Yellowstone grizzly bear data in Table 3.1, the census spans $t_q = 38$ years, from 1959 to 1997. Substituting the values of $\hat{\mu}$, $t_{0.05, 37}$, and SE($\hat{\mu}$) that we have just calculated into Expression 3.12 gives us a 95% confidence interval of (–0.01621, 0.05889) for the parameter μ. The way to view this confidence interval is to imagine repeatedly sampling at random 38 log growth rates from this population; the confidence intervals computed using Equation 3.12 would include the true value of μ 95% of the time. The fact that the confidence interval we have calculated ranges from negative to positive values indicates that, despite strong growth of the population in recent years (Figure 3.6) and the relatively long-term nature of the census, we still cannot rule out the possibility that the population will decline. Notice that the regression output in Box 3.1 also provides a test of the null hypothesis that the slope of the regression is zero. The relatively large probability associated with this test (0.2570) also tells us that we cannot say definitively that the population is growing, even though $\hat{\mu}$ is positive.

The results in the preceding paragraph highlight an important conservation lesson: We should be very skeptical about single estimates of viability metrics that are presented without an associated measure of uncertainty, such as a confidence interval. The case of the Yellowstone grizzly bear is particularly apropos. Recent years have witnessed intense pressure to remove the Yellowstone grizzly bear population from the Endangered Species list; much of that pressure is motivated by a desire to show that the Endangered Species Act functions effectively to bring about recovery of threatened and endangered species and populations. However, our analysis shows that it would be dangerous to conclude from the data in Table 3.1 that, because the best estimate of μ is slightly positive, the Yellowstone grizzly bear population is growing and can be safely delisted. The lower confidence limit on μ says that such a conclusion would be premature.

To know how much confidence we can place in extinction probabilities based on the estimated values of μ and σ^2, we also need to calculate a confidence interval for σ^2. Such an interval can be constructed using the chi-

[9]Specifically, the MATLAB function `tinv` will compute the inverse t statistic, but users must purchase the Statistics Toolbox to use `tinv`.

square distribution (Dennis et al. 1991). For example, if we want to place a 95% confidence interval around σ^2 for the Yellowstone grizzly bear, for which we have data on $q = 38$ transitions, we first obtain the 2.5th and 97.5th percentiles of the chi-square distribution with $q - 1 = 37$ degrees of freedom: call them $\chi^2_{0.025,37}$ and $\chi^2_{0.975,37}$. Once again, these can be looked up in a statistics reference or calculated with the aid of MATLAB[10] or a spreadsheet program; the Excel formulae =CHIINV(0.025, 37) and =CHIINV(0.975, 37) yield the values $\chi^2_{0.025,37} = 55.6680$ and $\chi^2_{0.975,37} = 22.1056$. We substitute these values into the following general expression for a 95% confidence interval for σ^2 based on $q - 1$ degrees of freedom:

$$\left((q-1)\,\hat{\sigma}^2 \big/ \chi^2_{0.025,q-1},\ \ (q-1)\,\hat{\sigma}^2 \big/ \chi^2_{0.975,q-1}\right) \tag{3.13}$$

(Dennis et al. 1991). Note that we can use this same approach to calculate confidence intervals with other degrees of coverage. For example, to calculate a 90% rather than a 95% confidence interval, we would substitute the values of $\chi^2_{0.05,\,q-1}$ and $\chi^2_{0.95,\,q-1}$ into the first and second terms in Equation 3.13, respectively.

For the Yellowstone grizzly bear, Equation 3.13 gives us the confidence interval (0.00867, 0.02184) for σ^2. Thus a true value of nearly 70% higher than the best estimate of $\sigma^2 = 0.01305$ is consistent with the available data (as is a much smaller value). We will return to Equations 3.12 and 3.13 when we discuss how to incorporate parameter uncertainty into estimates of extinction probabilities.

USING REGRESSION DIAGNOSTICS TO TEST FOR TEMPORAL AUTOCORRELATION IN THE POPULATION GROWTH RATE Another advantage of estimating the parameters μ and σ^2 by regression is that we can test the assumption of the diffusion approximation that the environmental conditions (and thus the log population growth rates) are uncorrelated from one inter-census interval to the next (that is, whether a particular interval was good or bad for birth or death is independent of whether preceding or succeeding intervals were good or bad). One test makes use of the Durbin-Watson *d* statistic (invoked by the dw option in SAS), which measures the strength of autocorrelation in the regression residuals (and we are interpreting the residuals as being the result of environmental perturbations). A residual is simply an observed value of y (i.e., the value of $\log\left(N_{i+1}/N_i\right)/\sqrt{t_{i+1}-t_i}$ for census *i*) minus the value predicted by the regression equation (which equals $\mu\sqrt{t_{i+1}-t_i}$). The two-tailed Durbin-Watson test evaluates the null hypothesis that all of the serial autocorrelations of the residuals are zero (Draper and Smith 1981). To

[10]Specifically, you can use the function `chi2inv` if you have purchased MATLAB's Statistics Toolbox.

perform the test, we compare both d and $4 - d$ to upper and lower critical values d_L and d_U. Specifically, if $d < d_L$ or $4 - d < d_L$, we conclude that the residuals show significant autocorrelation; if $d > d_U$ and $4 - d > d_U$, we conclude there is no significant autocorrelation; and otherwise, the test is inconclusive. Values of d_L and d_U for a significance level of $\alpha = 0.05$ and different numbers of data points in the regression, q, are given in Table 3.2; for other values of q, d_L and d_U can be calculated by linear interpolation, or by referring to Table 3.3. in Draper and Smith 1981. For the Yellowstone grizzly bear, $q = 38$, so that $d_L = 1.33$ and $d_U = 1.44$. The calculated value of d is 2.57 (Box 3.1). Because $d > d_U$ but $4 - d$ nearly equals d_U, this value of d leads to a test result that is right on the border between inconclusive and a finding of no significant autocorrelation. We conclude that there may be some autocorrelation in the residuals, but that it is probably weak. Another sign that the autocorrelation in the residuals is at best weak is the fact that the first-order autocorrelation of the residuals[11] ($r = -0.288$ with 37 degrees of freedom; see Box 3.1) has a small but nonsignificant p value ($p = 0.084$; this p value is calculated by most statistical packages and can be looked up in tables of two-tailed significance levels of the Pearson correlation coefficient, given in most basic statistics references). Given the possibility that successive log population growth rates may be slightly correlated (in this case negatively), a thorough analysis of the data should include an examination of the possible effect of the observed level of autocorrelation on the estimated viability of the grizzly bear population, using methods we describe in Chapter 4.

USING REGRESSION DIAGNOSTICS TO TEST FOR OUTLIERS We can also use the regression output to evaluate whether any particular transitions are outliers or are having a disproportionate influence on the parameter estimates. There are a variety of measures of the importance of a data point on the results of a regression. These include Cook's distance measure, leverage, and studentized residuals; different statistics packages provide different combinations of these measures. For example, the "influence" option in `proc reg` of SAS produces an output table with two useful measures of the effect of each data point on the regression results (Box 3.1). The column labeled `Rstudent` gives the studentized residual for each (x,y) pair in the regression. Data points with studentized residuals greater than 2 are suspected outliers (SAS 1990). `Dffits` is a statistic that measures the influence each data point has on the regression parameter estimates; for a linear regression with no intercept, a

[11]This autocorrelation can be computed easily as follows. Place the residuals for years 1 to $q - 1$ in one column of a spreadsheet, and the residuals for years 2 to q in the adjacent column. Then compute the Pearson correlation coefficient of the two columns of numbers. SAS performs this calculation for you when you specify the `dw` option in `proc reg` (see Box 3.1).

TABLE 3.2 Critical values for a two-tailed test of the Durbin-Watson statistic *d* at the $\alpha = 0.05$ significance level (from Table 3.3 in Draper and Smith 1981). *q* is the number of data points in a linear regression with one independent variable

q	d_U	d_L
15	0.95	1.23
20	1.08	1.28
25	1.18	1.34
30	1.25	1.38
35	1.30	1.42
40	1.35	1.45
45	1.39	1.48
50	1.42	1.50

value of `Dffits` greater than $2\sqrt{1/q}$ (where q is the number of data points, or transitions) suggests high influence (SAS 1990). For the grizzly bear data set, the critical value of `Dffits` is $2\sqrt{1/38} = 0.324$. By both the studentized residual and `Dffits` criteria, the 25th transition in the data set, corresponding to the 1983–1984 censuses, is flagged as an unusually large transition with a high influence on the estimate of μ. This transition represents a large *proportional* increase, in which the counts grew by more than 50% in a single year (Figure 3.6). Dennis et al. (1991) identified this same transition as an outlier in their analysis of the 1959–1987 census data.

If an outlier or influential observation is identified in your data set, you must decide whether to keep that data point or remove it from the regression. To make this decision, we recommend first examining why that point may be unusual. If the cause is "nonbiological," it might be sensible to delete the data point. For example, if unusual transitions are associated with years in which methodological problems with the census were known to have occurred, or if they correspond to a change from one method of censusing the population to another, they may represent errors in the counting process rather than periods of unusually high or low population growth. As we want our parameter estimates to include real environmental variation but to exclude observation error as much as possible (see Chapter 5), deleting such points would be justified. In the case of the Yellowstone grizzly bear, we are not aware of any such problems that occurred in the 1983 or 1984 censuses; hence we see no reason to delete the 25th transition as being unduly influenced by observation error.

On the other hand, there may be real biological reasons why a particular transition is an outlier. These reasons can be classified as one-time human impacts and extreme environmental conditions; the way we should treat these two types of outliers differs. If an outlier can be pegged to a one-time human impact that we do not expect to recur, that outlier should be omitted for the purposes of predicting the future state of the population. A hypothetical example of such an impact on the Yellowstone grizzly bear would be effects of closing the park garbage dumps (at which the bears had become accustomed to scavenge for food). Because this event is unlikely to occur again, any unusual transitions associated with it should not be allowed to influence our predictions for the future. The unusually high 1983–1984 transition, however, is not likely to be related to the closure of the dumps, which occurred in 1970–1971.

Finally, outliers could represent the consequences of extreme environmental conditions, either catastrophes or bonanzas (Chapter 2). Recall that the diffusion approximation used to derive an expression for the cumulative probability of extinction (Equation 3.4) assumed that such large fluctuations in the population growth rate do not occur. Strictly speaking, then, we should not use Equation 3.5 to calculate an extinction probability if our estimates of μ and σ^2 have been influenced by catastrophes or bonanzas. Instead, we must resort to computer simulation to calculate extinction probabilities, as we describe in Chapter 4.

In the case of the Yellowstone grizzly bear, the only suspected outlier is an unusually large *increase* in population size. Large increases inflate the estimated values of both μ and σ^2. As a higher value of μ reduces the probability of extinction but a higher value of σ^2 increases it (Figure 3.5), it is not clear how these two effects of the outlier on the extinction probability play out. Because of this, and because no environmental factors that could have caused the 1983–1984 transition to be a bonanza have been identified, we have chosen to retain this transition when estimating μ and σ^2. However, readers are encouraged to explore how the probability of extinction changes if we use parameter estimates ($\hat{\mu} = 0.01467$ and $\hat{\sigma}^2 = 0.01167$) obtained after deleting the 25th transition.

As the foregoing discussion indicates, there is no general rule regarding when to omit outliers, and the decision to do so must be decided on a case-by-case basis. Our point here is simply that the regression approach makes this decision possible by identifying outliers in the first place. When outliers are omitted, it is important that you state this fact explicitly in any reports you prepare on your PVA results, and that you carefully document why those points were omitted.

USING REGRESSION DIAGNOSTICS TO TEST FOR PARAMETER CHANGES A final advantage to the linear regression approach is that it provides a way to ask whether μ or σ^2 differ significantly in distinct segments of the time series. For example, for the Yellowstone grizzly bear, we might ask whether the parameter estimates differ before versus after the closure of the garbage dumps in 1970–1971, or before versus after the 1988 Yellowstone fires.

To test for changes in μ and σ^2 before versus after a pivotal census (e.g., 1988, the year of the fires), we recommend that you perform linear regressions in the following order. The first step is to test for differences in σ^2 before and after the pivotal census. To do this, divide the transitions into those leading up to and including the pivotal census versus those from the pivotal census onward. Transform each data set using Equations 3.10a,b. Next, perform separate linear regressions with zero intercept on these two transformed data sets (e.g., using two repetitions of SAS Command 3.9 above) to obtain two estimates of σ^2; call them $\hat{\sigma}_b^2$ and $\hat{\sigma}_a^2$ (the subscripts here stand for "before" and "after"). For example, if we divide the Yellowstone grizzly bear censuses into "before-fire" (1959–1988) and "after-fire" (1988–1997) periods, we obtain the estimates $\hat{\sigma}_b^2 = 0.01229$ and $\hat{\sigma}_a^2 = 0.01561$ from the error mean squares of the separate regressions. There are 29 and 9 transitions in these two data sets, respectively. We can perform a two-tailed test for a significant difference between two variances (in this case $\hat{\sigma}_b^2$ and $\hat{\sigma}_a^2$) by calculating their ratio with the larger variance in the numerator and then computing its probability from an *F* distribution with the appropriate numerator and denominator degrees of freedom (Snedecor and Cochran 1980). For the grizzly bear data, the ratio $\hat{\sigma}_a^2/\hat{\sigma}_b^2 = 1.270$ has $9 - 1 = 8$ numerator and $29 - 1 = 28$ denominator degrees of freedom. The probability of observing a ratio this large or larger if the two data sets have the same true value of σ^2 is 0.298 (obtained from an *F* table available in any statistics reference or by using the Excel formula =FDIST(1.270,8,28)). As this probability is larger than 0.025,[12] we conclude that there is no significant difference in the degree of variability in the grizzly bear population growth rate before versus after the 1988 Yellowstone fires.

Had the σ^2 values differed, we would use the two separate regressions to make different estimates of both μ and σ^2 and conduct any analysis of future trends using only the more recent set of estimates. (However, we would not if, for some reason, we thought that the earlier population dynamics might be a better indicator of what will happen in the future, for example if a threat to the population that was present only during the later time interval has now been removed.) In this case, however, we have determined that σ^2 does not differ significantly before versus after the fires, but we still need to test for differences in μ while allowing only a single estimate of σ^2 that applies to the entire census. We can do this by creating two x variables (call them x_1 and x_2) and performing a multiple linear regression of y on x_1 and x_2, once again forcing the regression to have an intercept of zero (Dennis et al. 1991). The new variable x_1 will have the values of x_i calculated from Equation 3.10a for all years before the fires and zero for all years after the fires. In contrast,

[12]A probability of 0.025 corresponds to an overall significance level of $\alpha = 0.05$ for a two-tailed *F* test (Snedecor and Cochran 1980).

variable x_2 will be zero for years before the fires and have values calculated from Equation 3.10a for all years after the fires. The y values should be computed from Equation 3.10b as before. The following SAS command performs this multiple regression of y on x_1 and x_2, yielding the output in Box 3.2:

```
proc reg;
     model y=x1 x2 / noint;          (3.14)
run;
```

The parameter estimates associated with variables x_1 and x_2 give us $\hat{\mu}_b$ = 0.01069 and $\hat{\mu}_a$ = 0.05564 for the periods before and after the fires, respectively. The single estimate of σ^2 for the entire census, once again obtained from the error mean square, is $\hat{\sigma}_m^2$ = 0.01303 (the subscript m here indicates that this estimate was obtained from a multiple regression; it is not necessarily the same as the estimate obtained from a single regression, as in Box 3.1). As the estimates of a regression slope are normally distributed, we can use a two-sample t test to ask if $\hat{\mu}_b$ and $\hat{\mu}_a$ are significantly different (Dennis et al. 1991). Specifically, we compute the statistic

$$T_{q-2} = (\hat{\mu}_a - \hat{\mu}_b) \Big/ \sqrt{\hat{\sigma}_m^2 [(1/j) + 1/(q-j)]} \quad (3.15)$$

where q is the total number of transitions in the data set and j is the number of transitions leading up to the pivotal census. For the Yellowstone

BOX 3.2 *Output produced by SAS commands given in Equation 3.14.*

```
Model: MODEL1
NOTE: No intercept in model. R-square is redefined.
Dependent Variable: Y

                    Analysis of Variance

                        Sum of         Mean
Source         DF      Squares       Square       F Value       Prob>F
Model           2      0.03118      0.01559         1.197       0.3140
Error          36      0.46907      0.01303
U Total        38      0.50025

      Root MSE       0.11415      R-square       0.0623
      Dep Mean       0.02134      Adj R-sq       0.0102
      C.V.         534.97870

                    Parameter Estimates

                  Parameter        Standard     T for H0:
Variable    DF     Estimate           Error   Parameter=0    Prob > |T|
X1           1     0.010690      0.02119667         0.504        0.6171
X2           1     0.055644      0.03804919         1.462        0.1523
```

grizzly bear, $q = 38$ and $j = 29$ (as above), and we substitute these and the values of $\hat{\mu}_b$, $\hat{\mu}_a$, and $\hat{\sigma}_m^2$ obtained above into Equation 3.15, yielding $T_{36} = 1.032$. The probability of obtaining this value from a t distribution with 36 degrees of freedom is 0.154 (which can be obtained using Excel formula =TDIST(1.032,36,1)). Once again, this probability does not support the hypothesis that μ differs before and after the fires. If it did, we would use the results of this regression to obtain separate estimates of $\hat{\mu}_b$ and $\hat{\mu}_a$ (with one estimate of $\hat{\sigma}_m^2$) and then decide which estimate of μ to use in viability calculations ($\hat{\mu}_a$ if we have reason to believe that post-fire conditions will continue, and $\hat{\mu}_b$ if we suspect that pre-fire conditions have returned).

Thus we have tested the assumption of the diffusion approximation that the parameters μ and σ^2 do not change, and found it to be reasonable (at least in reference to the periods before and after the 1988 fires). We now have justification for estimating μ and σ^2 with a single regression, as in Box 3.1. Once again, if any of the results *were* significant, we would be left in the position of having to decide, based on ancillary information, which set of parameter estimates to use. As an exercise, readers are encouraged to follow the procedure laid out above to test whether μ and σ^2 differ before versus after the dumps were closed in 1970–1971.

One caution about testing for parameter changes deserves mention. When we divide an already short data set into even smaller time intervals, we may be left with little statistical power to detect changes in parameter values. In fact, we may not even have sufficient power to detect simple trends in population size within each time interval. Hence it may be valuable to supplement the analysis described here with a power analysis for detecting population trends, using methods described in Gibbs (2000).

Using Estimates of μ and σ^2 to Calculate Probability of Extinction

Having estimated μ and σ^2 and made use of regression diagnostics to test some of the assumptions of the diffusion approximation, we are now ready to use Equation 3.5 to construct the extinction time cumulative distribution function (CDF), which we argued in Chapter 2 was the single most informative metric of a population's viability. Once we have obtained the estimates $\hat{\mu}$ and $\hat{\sigma}^2$ and determined a suitable quasi-extinction threshold, it is straightforward to use Equation 3.5 to calculate the probability that the population starting at the current size will have hit the threshold prior to each of a set of future times. The MATLAB code in Box 3.3 defines a function called `extcdf` that calculates the extinction time CDF. This function uses a second function listed in Box 3.3, `stdnormcdf`, which calculates the standard normal cumulative distribution function given by Equation 3.6. Later, we will use three other functions defined in Box 3.3 to compute confidence limits for the extinction time CDF. To make these functions accessible to MATLAB pro-

BOX 3.3 ***MATLAB code defining five functions used in calculating the extinction time cumulative distribution function and its confidence limits.***

File extcdf.m

```
function G=extcdf(mu,sig2,d,tmax);
% extcdf(mu,sig2,d,tmax) calculates the unconditional extinction
% time cumulative distribution function from t=0 to t=tmax for
% mean and variance parameters mu and sig2 and log distance from
% the quasi-extinction threshold d;
% Modified from Lande and Orzack, Proc. Nat. Acad. Sci. USA 85:
% 7418-7421 (1988), equation 11.
% REQUIRES THE FILE stdnormcdf.m

for t=1:tmax
    G(t)=stdnormcdf((-d-mu*t)/sqrt(sig2*t)) + ...
        exp(-2*mu*d/sig2)*stdnormcdf((-d+mu*t)/sqrt(sig2*t));
end;
```

File stdnormcdf.m

```
function phi=stdnormcdf(z)
% stdnormcdf(z) calculates the standard normal cumulative
% distribution function, using the built-in MATLAB error function
% erf;
% See Abramowitz and Stegun, 1964, Handbook of Mathematical
% Functions, Dover, New York, equation 7.1.22

phi=0.5*(1 + (erf(z/sqrt(2))));
```

File gammarv.m

```
function gamma=gammarv(alpha,beta,n);
% gammarv(alpha, beta,n) generates a vector of n random numbers
% from a Gamma(alpha,beta) distribution:
%    f(x)=exp(-x/beta)*x^(alpha-1)/[Gamma(alpha)*beta^alpha]
%      if x>=0;
%    f(x)=0 if x<0
% where Gamma(z) is the gamma function.
% Mean(x)=alpha*beta; Var(x)=alpha*beta^2
% See G.S. Fishman, 1973, Concepts and Methods in Discrete Event
% Digital Simulation, Wiley, New York, pp. 208-209.
```

BOX 3.3 *(continued)*

```
gamma=[];
for i=1:n
    X=0;
    k=floor(alpha);
    g=alpha-k;
    if k>0
        X=-log(prod(rand(1,k)));
    end;
    if g==0
        gamma=[gamma beta*X];
    else
        a=g;
        b=1-g;
        y=1;
        z=1;
        while y+z>1
           y=rand^(1/a);
           z=rand^(1/b);
        end;
        Y=y/(y+z);        % Y is a Beta(g,1-g) random variable
        Z=-log(rand);     % Z is a Gamma(1,1) random variable
        gamma=[gamma beta*(X+Y*Z)];
    end;
end;
```

File betarv.m

```
function beta=betarv(m,v,n);
% betarv(m,v,n) generates a row vector of length n, the elements
% of which are Beta random variables with mean m and variance v;
% See G.S. Fishman, 1973, Concepts and Methods in Discrete Event
% Digital Simulation, Wiley, New York, pp. 204-208.
% REQUIRES THE FILE gammarv.m

if v==0
   beta=m*ones(1,n);
elseif  v>=m*(1-m),
    fprintf(1,'Variance of Beta too large given the mean');
    pause;
```

BOX 3.3 *(continued)*

```
else
    a=m*(m*(1-m)/v - 1);
    b=(1-m)*(m*(1-m)/v - 1);
    beta=[];
    for i=1:n
        k1=floor(a);
        k2=floor(b);
        if k1==0 & k2==0
            Y=1;
            Z=1;
            while Y+Z>1
                Y=rand^(1/a);
                Z=rand^(1/b);
            end;
        else
            Y=gammarv(a,1,1);
            Z=gammarv(b,1,1);
        end;
        beta=[beta Y/(Y+Z)];
    end;
end;
```

File chi2rv.m

```
function x=chi2rv(df);
% chi2rv(df)generates a chi-squared random number with df
% degrees of freedom;
% See G.S. Fishman, 1973, Concepts and Methods in Discrete Event
% Digital Simulation, Wiley, New York, p. 213.
% REQUIRES THE FILE betarv.m

if mod(df,2)==0
    x=-2*log(prod(rand(1,df/2)));
else
    k=(df/2)-0.5;
    v=sum(-log(rand(1,k)));
    y=betarv(0.5,0.125,1);
    z=-log(rand);
    x=2*(v+y*z);
end;
```

grams, place them in separate files with a ".m" extension (e.g., `extcdf.m`) and place the name of the folder containing the files in MATLAB's path settings. The function `extcdf` returns a column array with `tmax` rows containing the probabilities that the extinction threshold will have been reached by each future time. In addition to `tmax`, calls to the function `extcdf` must provide three other arguments. These are the estimated values of μ and σ^2 and d, the difference between the log of the current population size and the log of the quasi-extinction threshold. Recall that it is only this difference, rather than the actual values of the population size and the threshold, that determines the extinction probability (Equations 3.4 and 3.5).

We argued previously that we should not place much faith in single estimates of μ and σ^2 because these parameters will not be estimated with high precision, given the limited amount of data typically available for threatened and endangered species and populations. For the same reason, we should not place much faith in the best estimate of the probability of extinction, which we will call $\hat{G}$ (see Equation 3.5), that is based only on our best parameter estimates μ and σ^2, without accounting for the uncertainty in these estimates. Thus we need a way to translate uncertainty in the estimates of μ and σ^2 into uncertainty in the probability of extinction, G. Although we cannot write down an expression for the confidence interval of G akin to Expressions 3.12 and 3.13, we can use a computer-based method known as a parametric bootstrap[13] to approximate the confidence interval for G. Specifically, because we know the probability distributions that govern the parameter estimates,[14] we can have a computer draw values of μ and σ^2 from the appropriate distributions. If both of those estimates lie within their respective confidence limits, given by Expressions 3.12 and 3.13, then we use them to calculate $\hat{G}$. By repeating this process many times, we obtain a range of values of $\hat{G}$, all of which will lie within the confidence interval for G. The extreme values of $\hat{G}$ define the upper and lower boundaries of the confidence interval. The number of bootstrap samples (i.e., the number of values of $\hat{G}$ we calculate) should be relatively large, so that we are reasonably sure to see the extreme values that define the confidence limits.

---

[13]An alternative bootstrapping approach, the nonparametric bootstrap, involves repeatedly sampling from the original data (i.e., constructing a set of q values randomly chosen (with replacement) from the observed log population growth rates, estimating μ and σ^2 and calculating G, repeating this process many times, and identifying (for example) the 2.5th and 97.5th percentiles of the resulting distribution of G values as the 95% confidence limits. For a discussion of the relative merits of the parametric and nonparametric bootstrap in a PVA context, see Ellner and Fieberg (2002).

[14]Specifically, $\hat{\mu}$ is normally distributed and $\hat{\sigma}^2$ has the distribution of a chi-square random variable multiplied by $\hat{\sigma}^2/q$. Furthermore, these estimates are independent, so their values can be obtained from separate random draws from the two distributions. For details, see Dennis et al. 1991.

The MATLAB code in Box 3.4 performs the bootstrap procedure described in the preceding paragraph. This program uses the function `extcdf` and the four other functions defined in Box 3.3. The user-defined parameters in Box 3.4 correspond to the values we have estimated for the Yellowstone grizzly bear population. Specifically, we use $N_c = 99$ (the number in the 1997 census; Table 3.1) as the current population size, a quasi-extinction threshold N_x of 20 reproductive females, following the guidelines we gave in Chapter 2 (remember that N_c and N_x together determine d), and estimates of μ and σ^2 from Box 3.1. The results from one run of the program with 500 bootstrap samples are shown in Figure 3.8. The best estimate of the probability that the population will decline from 99 to 20 female bears is quite low for short times into the future, reaching a value of 0.0018 at 50 years. (Note the logarithmic scale of the y-axis in Figure 3.8.) However, notice that the confidence interval for the probability of quasi-extinction widens rapidly as time increases; by 50 years, the 95% confidence interval

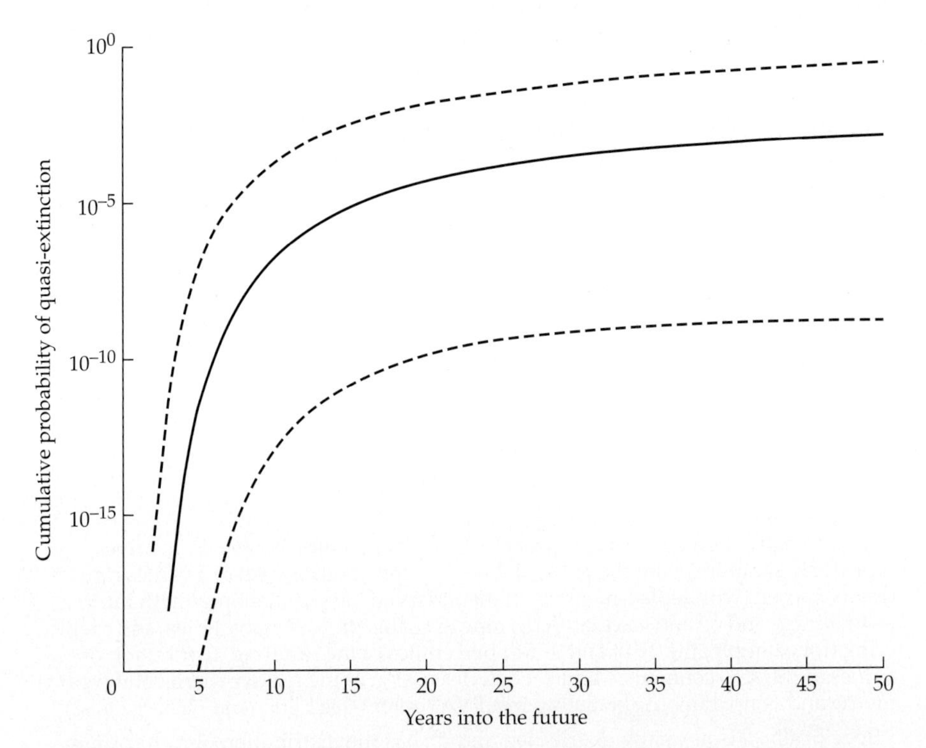

Figure 3.8 Extinction time cumulative distribution function for the Yellowstone grizzly bear population. Solid line is based on the best estimates of μ and σ^2; dashed lines delineate an approximate 95% confidence interval determined by a bootstrap (see Box 3.4). The x-axis shows the time required for a population of 99 female bears to decline to 20.

BOX 3.4 ***MATLAB code to calculate extinction probabilities and bootstrap confidence intervals[1].***

```
% Program ExtProb
%   ExtProb calculates the probability of extinction with
%   bootstrap confidence intervals for a density-independent
%   model using a diffusion approximation;
%   See Dennis et al. 1991, Ecological Monographs 61: 115-143;
%   REQUIRES THE FILES extcdf.m, chi2rv.m, stdnormcdf.m,
%   gammarv.m, and betarv.m

%****************************************************************
% Change the following user-defined parameters:
%****************************************************************
mu=0.02134;   % Enter the estimated value of mu
sig2=0.01305; % Enter the estimated value of sigma^2
CI_mu=[-.01621,.05889];  % Enter confidence interval for mu
CI_sig2=[.00867,.02184]; % Enter confidence interval for sigma^2
q=38;         % Enter the number of transitions in the data set
tq=38;        % Enter the length of the census (in years)
Nc=99;        % Enter the current population size
Ne=20;        % Enter the quasi-extinction threshold
tmax=50;      % Enter latest time to calculate extinction
              %   probability
Nboot=500;    % Number of bootstrap samples for calculating
              % confidence intervals for extinction probabilities
%****************************************************************

d=log(Nc/Ne);         % calculate log distance to quasi-extinction
                      %   threshold
SEmu=sqrt(sig2/tq);   % calculate standard error of mu
Glo=ones(1,tmax);     % initialize array to store lower bootstrap
                      %   confidence limits for extinction
                      %   probabilities
Gup=zeros(1,tmax);    % initialize array to store upper bootstrap
                      %   confidence limits for extinction
                      %   probabilities
rand('state',sum(100*clock));  % seed the random number generator

% Calculate the extinction time cdf for the best parameter
%   estimates
```

[1]The code in Box 3.4 uses functions defined in Box 3.3.

BOX 3.4 *(continued)*

```
Gbest=extcdf(mu,sig2,d,tmax);

% Calculate bootstrap confidence limits for extinction
%   probabilities

for i=1:Nboot,

    % Generate a random mu within its confidence interval
    murnd=inf;
    while murnd<CI_mu(1)|murnd>CI_mu(2)
        murnd=mu+SEmu*randn;
    end;

    % Generate a random sigma^2 within its confidence interval
    sig2rnd=inf;
    while sig2rnd<CI_sig2(1)|sig2rnd>CI_sig2(2)
        sig2rnd=sig2*chi2rv(q-1)/(q-1);
    end;

    % Calculate extintion probabilities given murnd and sig2rnd
    G=extcdf(murnd,sig2rnd,d,tmax);

    % Store extreme values
    for t=1:tmax
        if G(t)<Glo(t)
            Glo(t)=G(t);
        end;
        if G(t)>Gup(t)
            Gup(t)=G(t);
        end;
    end;
end;     %for i=1:Nboot

% Plot G (log-scale) vs t, using a solid line for the best
% estimate of G and dotted lines for the confidence limits
t=[1:tmax];
semilogy(t,Gup,'k:',t,Gbest,'k-',t,Glo,'k:')
axis([0,tmax,0,1])
xlabel('Years into the future')
ylabel('Cumulative probability of quasi-extinction')
```

for the probability of hitting the threshold ranges from 6.09×10^{-9} to 0.255. Thus even though the best estimate is that this population won't decline dramatically over the short term, we cannot say with certainty that the population has at least a 95% chance of remaining above 20 females over a period as short as the next five decades (and many authors have called for such high levels of safety over much longer periods). Thus due to the uncertainty in the parameter estimates $\hat{\mu}$ and $\hat{\sigma}^2$, we cannot predict the extinction probability very accurately for very far into the future, a caution that several authors have made (Ludwig 1999, Fieberg and Ellner 2000). The fact that, given the available information, we cannot make very precise statements about the extinction risk the Yellowstone grizzly bear population faces several decades from now further argues the need for caution when we consider delisting this population. Indeed, for a decision as momentous as delisting, it makes sense to use the more pessimistic bounds on our estimates of μ and σ^2 in order to act in a precautionary fashion. In this way, we can use uncertainty in extinction risk estimates to reach more valid conclusions about our understanding of population viability (Gerber et al. 1999). Wide confidence limits on extinction risk estimates realistically reflect our uncertainty given imperfect information, and quantifying this uncertainty is a useful product of a viability analysis.

Using Extinction Time Estimates for Conservation Analysis and Planning

We now give two examples that show how the extinction time CDF can be used to inform decisions about the viability of single populations or about which of several populations should receive the highest priority for acquisition or management.

Perhaps the most valuable use of the CDF is to make comparisons between the relative viabilities of two or more populations. Ideally, we would have a series of counts from each population. For example, Figure 3.9A,B shows the number of adult birds during the breeding season in two populations of the federally listed red-cockaded woodpecker, one in central Florida and the other in North Carolina. Applying the methods outlined above yields the CDFs in Figure 3.9C. Both because it has a more negative estimate for μ (−0.083 vs. −0.011) and a smaller current size, the central Florida population has a much greater probability of extinction at any future time than does the North Carolina population. This information could be very useful in deciding the population on which to focus management attention.

Often we will not have independent census data from each population about which we must make conservation decisions. However, if we have a single count of the number of individuals of a particular species in one population, we can use count data from multiple censuses of the same species at

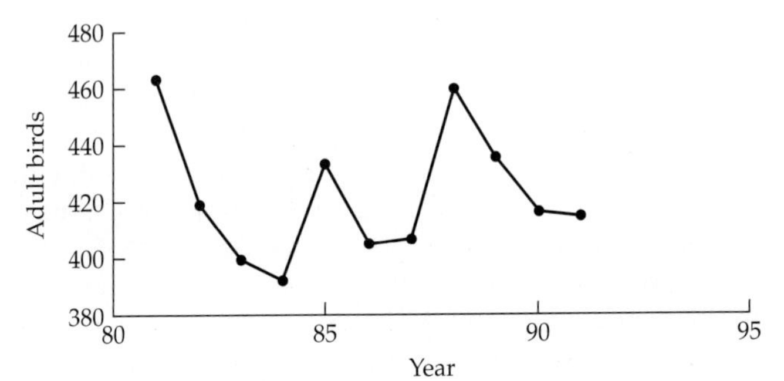

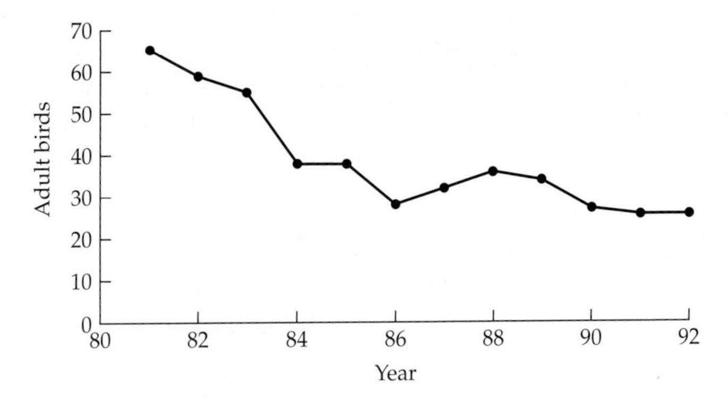

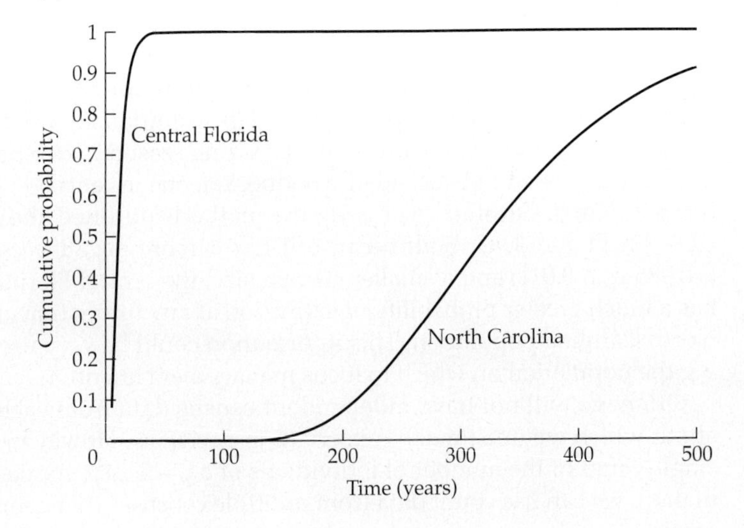

Figure 3.9 (A, B) Count data from two populations of the red-cockaded woodpecker (data courtesy of J. Walters [NC] and J. Hardesty [FL]); (C) A comparison of the best estimates of the extinction time CDFs for the two populations. (For both curves, initial population size equaled the last available count and the quasi-extinction threshold was 10 birds.)

a second location to make a *highly provisional* viability assessment for the first population when no other data are available. For example, we have good data on the Yellowstone grizzly population collected over a 39-year period, but there are no comparable records of population numbers over time for the other small and potentially threatened grizzly bear populations in western North America. One of these isolated populations of grizzly bears occupies the Selkirk Mountains of southern British Columbia and consists of about 25 adults, or roughly 12 adult females (United States Fish and Wildlife Service 1993). If we have no information about the Selkirk Mountains population other than its current size, we may as well use the CDF for the Yellowstone population to give us a *relative* sense of the viability of the Selkirk population. In so doing, we are assuming that the environments (including the magnitude of inter-annual variation) and the human impacts at the two locations are similar, an assumption that should be carefully evaluated using additional information on habitat quality, climatic variation, and land-use patterns. If we use the best estimates of μ and σ^2 from the Yellowstone population, the Selkirk population of 12 females has about a 4.2% chance of declining to only 5 females in 50 years, whereas this same chance is roughly four orders of magnitude less (about one in a million) for the larger Yellowstone population. (You should verify these numbers using Equation 3.5.) Note that the probability of extinction is a nonlinear function of current population size (Figure 3.10). For species of particular concern, it may be possible to improve upon this approach by compiling count data from multiple locations. We could then estimate *average* values for the parameters μ and σ^2 to provide ballpark assessments of viability for populations with only a single census, or we could choose the location with the most similar environment for comparison.

Key Assumptions of Simple Count-based PVAs

We have now seen how to go from census data on the number of individuals in a population (or well-defined subpopulation) to two metrics of population viability (μ and G) using a simple count-based PVA method. We have also seen how to calculate measures of uncertainty for both of these metrics. But as with any quantitative model of a complex biological process, the method described in this chapter relies upon simplifying assumptions. Before we leave this simplest and least data-demanding method of quantifying population viability, we return to the three assumptions discussed earlier in the chapter, and add one more key assumption underlying the method (Table 3.3). We briefly discuss how violations of each of these assumptions can cause the viability assessment we've been calculating to be in error. In Chapters 4 and 5, we describe methods that can be used to account for these complications, when the simpler approach described in this chapter is not appropriate.

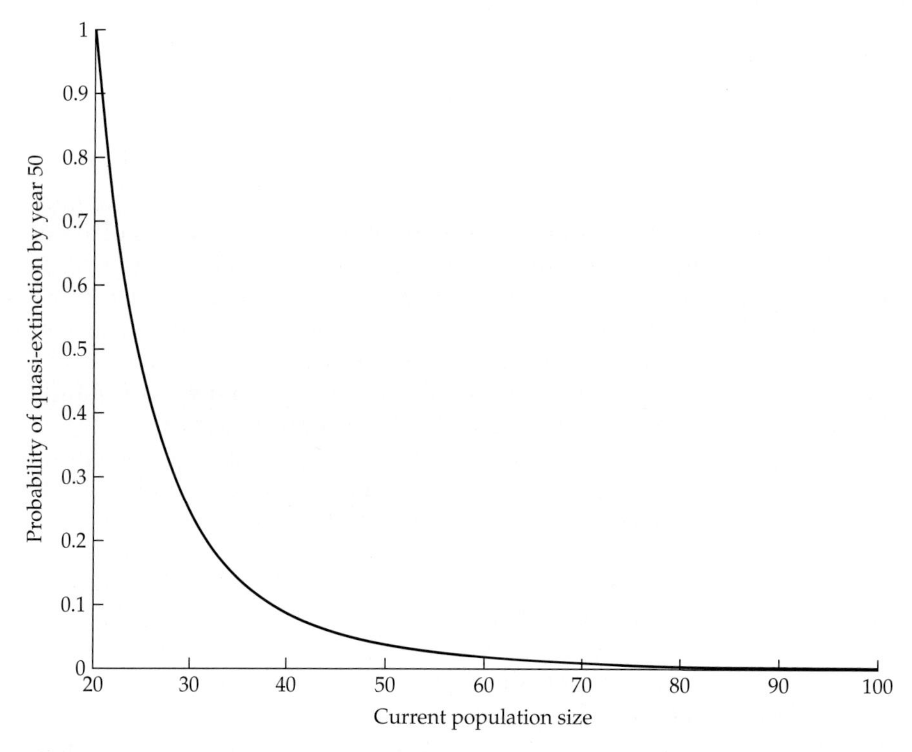

Figure 3.10 The cumulative probability of quasi-extinction G predicted by Equation 3.5 depends in a nonlinear way on the current population size. (This figure uses estimates of μ and σ^2 from the Yellowstone grizzly bear population, a quasi-extinction threshold of 20 adult females, and a time horizon of 50 years.)

Before reviewing the assumptions, we hasten to add that, rather than viewing these assumptions as a weakness, the fact that they are explicit is an advantage of a quantitative approach to evaluating viability, relative to an approach based upon general natural history knowledge or intuition. By evaluating whether the assumptions are met, we can determine whether our analysis is likely to give unreliable estimates of population viability, but more importantly, we can often determine whether violations of the assumptions are likely to render our estimates (e.g. of the probability of extinction) optimistic or pessimistic. By "optimistic" we mean that the true extinction probability is likely to be higher and the average log population growth rate lower than their estimated values. Conversely, by "pessimistic" we mean that the true extinction probability and growth rate are likely to be lower and higher (respectively) than estimated. If we know that the estimated metric is likely to be overly pessimistic, we should be more cautious in assigning that population a low viability rank, while a high but optimistic estimate should not inspire complacency.

TABLE 3.3 Key assumptions underlying simple count-based PVAs using Equation 3.5

Assumption	*Alternative methods that do not make this assumption are addressed in:*
Both the mean and variance of the population growth rate remain constant: • No density dependence • No demographic stochasticity • No temporal environmental trends	Chapter 4
Environmental conditions are uncorrelated from one year to the next.	Chapter 4
Environmental variation is small to moderate (i.e., there are no catastrophes or bonanzas).	Chapter 4
Census counts represent the true size of the entire population or of a constant proportion of the population (i.e., observation error is minor).	Chapter 5

Assumption 1: The Parameters μ and σ^2 Do Not Change Over Time

The diffusion approximation assumes that the mean and variance of the log population growth rate are both constant. However, there are three major factors that may cause this assumption to be incorrect: density dependence, demographic stochasticity, and temporal trends in environmental conditions. We now discuss each of these factors.

DENSITY DEPENDENCE Equation 3.1, the foundation on which the PVA method described in this chapter rests, assumes that the mean population growth rate is density-independent, and violations of this assumption can make the results obtained in this chapter extremely inaccurate. As discussed in Chapter 2, the ways in which density dependence (i.e., the tendency for population growth rate to change as density changes) may alter our estimates of extinction risk are complex. A decline in the population growth rate as density increases will tend to keep the population at or below a carrying capacity. Such regulated populations cannot grow indefinitely, and the probability of ultimate extinction is always 1 (although the time to extinction may be extremely long). On the other hand, declining populations may receive a boost as density decreases and resources become more abundant; because the models used in this chapter do not account for this effect, they may result in pessimistic estimates of extinction risk. Finally, the opposite effect may occur if a decline in density leads to difficulties in mate finding or predator defense or an increase in inbreeding, with a consequent *reduction* in population growth rate. The downward spiral created by these Allee effects results in extinction risks that become greater and greater as the population declines, and causes estimates of extinction risk made by ignoring these effects to be overly optimistic.

We can get a quick idea of whether the population growth rate is density-dependent by plotting $\log(N_{t+1}/N_t)$ versus N_t. A positive slope at low population sizes, or a negative one at larger sizes, indicates density dependence, and any indication of density dependence can be tested for significance using regression and other methods (Pollard et al. 1987, Dennis and Taper 1994). The growth rates for the Yellowstone grizzly bear suggests that there is no substantial density dependence over the range of population sizes found from 1959 to 1996 (Figure 3.11; Pearson correlation $r = 0.04$), so that the analytical methods we have applied to this data set in this chapter are not likely to be invalidated by density dependence, at least over time periods of moderate length.

Of course, no population is truly density-independent, so is the simple method ever useful? The key assumption behind Equation 3.1, and all of the results based on it, is actually not that density has *no* effect on the rate of population growth, but rather that its effect doesn't *change* over the range of

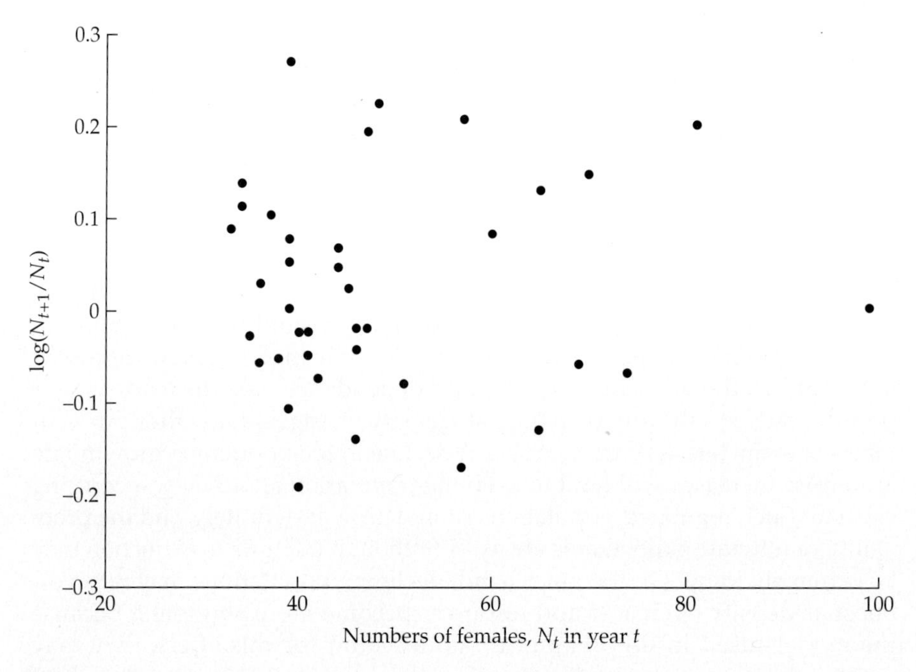

Figure 3.11 The annual log growth rates for the Yellowstone grizzly bear population do not show any obvious signs of density dependence.

population densities likely to occur in our predictions of the future. In other words, our viability estimates will be reasonably accurate as long as population numbers do not become so much lower or higher over the time horizons with which we are concerned that density effects on population growth will change dramatically. There are two factors characteristic of rare species that may minimize changes in density effects. First, because many endangered species are relatively long-lived and slowly-reproducing, they are likely to have only modest values of λ_t in most years. This means that population size, and thus density effects, will not change dramatically over relatively long time periods. Second, the very fact of endangerment implies that a population is not at high numbers, and so may not be subjected to the type of negative density dependence that is strongest at high densities. Certain decisions about how to conduct the analysis can also guard against the influence of changing density effects. In particular, analyzing extinction risk over a time horizon that is relatively short will prevent population realizations from reaching truly high densities (assuming that μ and σ^2 are such that this was even possible). With the best of intentions, many PVAs have attempted to make predictions over, say, 1,000-year time periods. Unfortunately, such predictions are not likely to be highly accurate, especially given the possibility of changing density effects. The second way to guard against density dependence is to set fairly high quasi-extinction thresholds. As discussed in Chapter 2, a high threshold means that we will be alerted to potential problems before the population can experience positive density dependence at very low numbers, which could strongly reduce population viability.

DEMOGRAPHIC STOCHASTICITY The assumption that σ^2 is constant made in deriving in Equation 3.5 is equivalent to assuming that demographic stochasticity is unimportant, and that environmental stochasticity is the only determinant of σ^2. Demographic stochasticity would cause the degree of variation in the population growth rate to be sensitive to population size (see Chapter 2). As we noted in Chapter 2, the simplest way to account for demographic stochasticity is to set the quasi-extinction threshold sufficiently high that its magnitude is negligible. In most cases, we will lack the data needed to do otherwise. However, in Chapter 4 we review methods to include demographic stochasticity when the appropriate data are available.

TEMPORAL ENVIRONMENTAL TRENDS Both μ and σ^2 may also change over time for reasons that are unrelated to population size. For example, environmental degradation (including alteration of natural disturbance regimes) and the effects of management could result in ongoing changes in μ or σ^2. While these effects can be either positive or negative, they will always compromise the reliability of a simple PVA that assumes constant parameters. An assumption behind all the extinction time calculations we've given above is that the estimates of μ and σ^2 are unbiased; that is, even if these estimates are uncertain, they are not systematically under- or overestimating the future parameter values. However, if the mean or variance in population growth is changing through time, data from the past may produce quite biased estimates of future rates and thus yield a false indication of extinction risk.

Two types of changes in μ and σ^2 may occur. The first is an abrupt change following some pivotal event, with μ or σ^2 showing one value for the entire period prior to the event and a second value afterward. We saw how to test for such changes in both μ and σ^2 and how to account for them in viability assessments when we examined the effects of the 1988 fires on the Yellowstone grizzly bear population. The second type of change is an ongoing trend in μ or σ^2 over successive years, and it presents greater challenges to PVA than does the former type of change. We can test for a linear change in μ using a regression of $\log(N_{t+1}/N_t)$ versus year; a significantly positive or negative slope indicates a temporal trend (and other regression models, such as polynomial regression, can be used to test for nonlinear trends). Similarly, if we calculate the squared deviation between each observed value of $\log(N_{t+1}/N_t)$ and either the constant value of μ or the value of μ in year t predicted by the preceding regression (if significant), and then regress these squared deviations against year, we can test for temporal trends in σ^2. If a trend is detected, we can either use the most current estimate of the parameter (if we can justify that it has now stabilized), or simulate on a computer the effects of an ongoing trend in μ or σ^2 (assuming that the past rate and pattern of change will continue into the future).

Assumption 2: No Environmental Autocorrelation

Both the diffusion approximation and the regression method used to calculate μ and σ^2 assume that the population growth rates over different time intervals are independent, having a common mean and variance but no tendency for adjacent λ_t's to be more similar to one another. The effects of such autocorrelation in population growth depends on the sign of the correlation. A positive autocorrelation means that one good year is likely to be followed by another, and, more importantly, that a bad year is frequently followed by a similarly bad year. When the growth rate is density-independent, the overall effect of a positive correlation is a higher extinction risk, since the events that drive populations to low numbers tend to come in series. Conversely, negative autocorrelation will delay extinction, because poor-growth years will tend to be followed by good ones that will buoy the population back up. Thus omitting positive and negative autocorrelation when it actually occurs will cause the resulting viability estimates to be optimistic and pessimistic, respectively. As we saw earlier in the chapter, the Durbin-Watson d statistic and the first-order autocorrelation coefficient of the residuals provide ways to ask whether autocorrelation is present in the data. If your data do show a significant autocorrelation, methods in Chapter 4 can be used to incorporate it into a count-based PVA.

Assumption 3: No Catastrophes or Bonanzas

Recall that the diffusion approximation assumes that population size changes by small amounts over short time intervals. One violation of this assumption that will cause viability estimates to be optimistic is the existence of intermittent catastrophes, such as rare ice storms, droughts, and severe fires, which introduce the possibility of sudden declines in abundance not accounted for in our estimate of σ^2. More detailed methods have been developed to include catastrophes in estimates of time to extinction (see the methods of Mangel and Tier 1993 and Ludwig 1996a, which also allow for density dependence). However, with most short-term count data, we will lack sufficient information to estimate the frequency and severity of rare catastrophes, information that more detailed methods require if they are to provide more accurate assessments of extinction time. Thus, in practice we may need to be content with the statement that if catastrophes do indeed occur, our assessments of extinction risk based upon short-term census data will likely underestimate the true risk. If catastrophes do occur but with similar intensity and frequency across multiple populations, the methods in this chapter may still give a reasonable picture of *relative* viability. Of course the converse, failure to account for rare good years (bonanzas), will have a pessimistic effect on the estimated extinction risk.

Assumption 4: No Observation Error

Observation error causes failure to count accurately the true number of individuals in a population (or a defined subpopulation) at any one time. We outlined its manifold potential causes in Chapter 2. Observation error will lead to a pessimistic measure of viability over the short term, because it will cause $\hat{\sigma}^2$ to overestimate the true environmentally driven component of variation in the counts, and a higher $\hat{\sigma}^2$ predicts a greater likelihood of extinction, especially over short time horizons (see Figure 3.5). Repeated sampling of the same area (see Chapter 5) and "ground-truthing" indirect measures of abundance (Gibbs 2000) are two ways to estimate the magnitude of observation error. We must also be aware of the fact that short sequences of counts will tend to misrepresent the true environmental component of variability because they will tend not to include extreme values. That is, a short time run of data is most likely to underestimate the true variability of population growth (Pimm and Redfearn 1988, Arino and Pimm 1995).

Another source of observation error is variability in the fraction of the population that is censused each year. Although count-based PVA can be reliably performed using counts of a subset of the population only, the subset that is censused should be a constant fraction of the total. If this fraction varies, it will inflate our estimate of σ^2, making for an overly pessimistic estimate of extinction risks. For example, it is quite possible that following poor food years, fewer female grizzlies reproduce, while in good years a greater fraction give birth. If so, then the variability in our estimates of grizzly numbers (and hence growth rates) would be due not only to changes in population numbers, but also in the fraction of adult females we observe (since only females with cubs are counted). One life history feature that may often cause this assumption to be violated is dormant or diapausing stages in the life cycle, such as seeds in a seed bank or diapausing eggs or larval stages of insects and freshwater crustaceans. Because they are difficult to census accurately, these stages are typically ignored in population counts, but as a result the counts may not represent a constant fraction of the total population. For example, when the number of above-ground individuals in a plant population is zero, total population size is not necessarily zero, as some individuals may remain in the seed bank. If the subpopulation in the seed bank is more buffered from environmentally driven fluctuations than is the above-ground population (as is likely to be the case in environments that favor the evolution of dormant life stages in the first place), then extinction risks estimated from the above-ground population alone may underestimate the true value for the entire population, and thus provide a (potentially highly) pessimistic measure of population viability. For organisms such as desert annual plants in which a large and persistent fraction of the population is likely to go uncensused, PVAs based on counts of a conspicuous subset of the population are probably not an appropriate way to estimate extinction risk.

When to Use This Method

The procedure we have described in this chapter provides a straightforward method to obtain quantitative measures of population viability using data from a series of population censuses. The principal advantage of the method is its simplicity in terms of both its data requirements and the ease of calculating viability measures. Other than simply recording the presence or absence of a species, population censuses are the simplest and most common way that field biologists collect data on rare species (Morris et al. 1999). Users of count-based PVA need to be aware of its limitations when the underlying assumptions (Table 3.2) are violated. Because of these limitations, the method we have presented here is not a panacea for making conservation decisions in a world of sparse data. However, because the assumptions are explicit, they can be carefully assessed before carrying out the PVA. If any of the assumptions are violated, there are other methods available to account for them when data are available (see the next two chapters). However, even if some assumptions (say, the absence of density dependence or of occasional catastrophes) are dubious, this simplest of PVA methods can still be very useful. We will often be able to gauge in which direction our estimate is in error (something which cannot always be said about viability measures obtained in other ways, such as through expert opinion). Moreover, if we know that a particular factor (such as density dependence in the population growth rate) has been omitted in a consistent way across multiple PVAs, these PVAs may still provide us with useful guidance about how risks are likely to differ among species or populations. In essence, we argue that in many cases (especially with limited data), count-based PVA is best viewed as a tool that provides us with *relative* measures of the "health" of two or more populations. That is, while we would not put much credence in a particular numerical value of a viability measure, we can be more comfortable accepting that estimated extinction probabilities differing by orders of magnitude for two populations warn us of potentially real differences in the viability of those populations.

Much more than violations of the underlying assumptions, limited amounts of data are likely to hinder the use of count-based PVAs. We obviously must have enough years of data to obtain good estimates of both μ and σ^2. However, how much data is "enough" is not clear. To apply the methods we've described to a single population to estimate extinction risk, we suggest that *ten censuses should be viewed as a minimum requirement.* In simulation trials, this amount of data provided reasonably reliable estimates at least of the sign of μ (i.e., if the population tend to grow or decline) as well as the relative rankings of μ across populations (Elderd et al. 2002). In practice, however, the degree of reliability will depend on the magnitudes of environmental variability and observation error. Accurate estimates of extinction probabilities will require even more data (Fieberg and Ellner 2000; see Chapter 12).

Indeed, assessing the reliability of viability measures when the underlying assumptions of the method are violated and when data are limited is an area of active research in population biology. Competing claims have recently been made both for and against the utility of count-based methods (Ludwig 1999, Fieberg and Ellner 2000, Meir and Fagan 2000, Holmes 2001, Elderd et al. 2002). While the outcome of this controversy is not yet clear, for the moment we believe it would be a shame not to make use both of available data and of this simplest PVA tool to address viability questions that can not be addressed quantitatively in any other way.

4

Count-Based PVA: Incorporating Density Dependence, Demographic Stochasticity, Correlated Environments, Catastrophes, and Bonanzas

The count-based population viability analysis that we outlined in Chapter 3 was based on the simplest possible stochastic population model (Equation 3.1) and on measures of extinction risk derived using a diffusion approximation. As we noted at the end of Chapter 3, the model and the diffusion approximation rest on the following assumptions:

1. The population growth rate is unaffected by population density.
2. The only source of variability in the population growth rate is environmental stochasticity (in particular, we neglected demographic stochasticity).
3. There are no trends in the mean and variance of the population growth rate over time.
4. Environmental conditions in successive years are uncorrelated; hence the population growth rate in a given year does not depend on whether the previous year was good or bad.
5. Environmental variability is moderate; that is, there are no catastrophes or bonanzas.
6. The measured variability in population growth rates is real, rather than a result of observation error.

In reality, virtually every population is certain to violate one or more of these assumptions. The population growth rate will surely change at some point as the size of the population changes, and its mean or variance may show a temporal trend due to a changing environment or the impact of management. The growth rate of very small populations will vary due to both environmental and demographic stochasticity, as described in Chapter 2. Cycles in environmental conditions that last longer than one year, such as El Niño oscillations, may cause the environmental effects on the population growth rate to be correlated in two or more successive years, resulting in strings of good or bad years that can impact the viability of a population. Large fluctuations that are outside the range of normal environmental variation may occur due to rare floods, hurricanes, droughts, or other infrequent occurrences. Finally, some variability in observed growth rates is certainly due to imperfect measurements.

Unfortunately, even if we know that such violations are likely to occur, it is an inescapable fact that incorporating these effects into PVA models requires more and better data, which we will often lack for rare species. Hence we may have little choice but to use simpler methods and models, with the consolation that we can often understand the direction in which violations of the above assumptions will cause our viability estimates to err (as we discussed at the end of Chapter 2 and also in Chapter 3). However, when a greater quantity and quality of data are available and support the importance of these other factors, alternative models that avoid one or more of the assumptions of the simplest count-based methods can be employed. In this chapter, we explore some of these models, including ways to incorporate density dependence, demographic stochasticity, correlated environments, and extreme events such as bonanzas and catastrophes. When discussing density-dependent models, we deal in depth with the important topic of choosing among alternative models, and the tools we use here will reappear throughout the book. In Chapter 5, we deal with the problem of observation error and how to disentangle it from true variability in growth rates. As a caveat, we reiterate that the material in these two chapters is both more mathematically complex and less likely to be generally applicable (because the appropriate data aren't available) than are the methods presented in Chapter 3. Hence readers may wish to reserve this chapter for more in-depth study during a second pass through the book.

Density Dependence

No population can continue to grow indefinitely. As a population approaches the limits of its resources, or as the impacts of its natural enemies intensify, the annual population growth rate (N_{t+1}/N_t) will approach 1.0, so that population growth ceases. As we noted in Chapter 2, such negative density dependence is an important factor to consider in assessing population viability, for two opposing reasons. On one hand, negative density dependence places a cap on how far a population can move away from the extinction threshold, even if the population tends to grow when its size is small. The closer to the threshold a population is held, the greater is the chance that a string of bad years will cause it to actually hit the threshold. On the other hand, if a population is experiencing negative density dependence, the growth rates we measure at the current population size might *underestimate* the growth rates the population would experience if it were reduced to a smaller size. That is, the population might be more buffered against extinction than growth rates measured at intermediate densities would imply.

However, we have also seen that the population growth rate could *decline* as population size decreases, due to positive density dependence (Allee effects). Lower population growth results from either a depressed birth rate or an elevated death rate at low density. Low birth rates may be caused by an inadequate number of potential mates within an individual's neighborhood, whereas high death rates can result from failure to achieve an adequate group size to fend off the attacks of predators. Both effects could result from reduced foraging success of small groups. Finally, inbreeding depression could underlie a reduction in growth rate at low population size. Clearly, Allee effects, if they occur, will be important to include in PVAs, because they will increase the extinction risk of small populations.

In the following sections, we discuss ways to account for density dependence in count-based PVAs. The bulk of our attention is devoted to negative density dependence because evidence for it is more widespread and hence it has been incorporated into count-based PVAs far more often than have Allee effects. Nevertheless, we briefly indicate how to include Allee effects as well.

Two Population Models with Negative Density Dependence

In this section, we describe two discrete-time population models that have played an important role in understanding and assessing the viability of populations exhibiting negative density dependence. In the following section, we review general theoretical insights that the first (and simpler) of these two models yields regarding the influence of negative density dependence on population viability. Following this, we present a case study of a PVA that makes use of the second (more realistic) model.

Perhaps the simplest way to incorporate negative density dependence is to introduce a population ceiling to the density-independent population growth model of Equation 3.1. Specifically, let

$$N_{t+1} = \begin{cases} \lambda_t N_t & \text{if } \lambda_t N_t \le K \\ K & \text{if } \lambda_t N_t > K \end{cases} \tag{4.1}$$

We shall refer to Equation 4.1 as the *ceiling model*. As in Chapter 3, N_t is the size of the population in year t, and the population growth rate λ_t is assumed to vary from year to year due to environmental stochasticity alone. The parameter K is the population ceiling. In Equation 4.1, the population's growth rate does not depend on its size as long as it remains below the ceiling; for any increment of growth that would take the population above the ceiling, we simply reset population size to K. Note that K is somewhat different from the carrying capacity in the familiar logistic equation, because in Equation 4.1 there is an abrupt rather than a gradual decline in the growth rate as population size increases. Nevertheless, many authors refer to K in Equation 4.1 as the "carrying capacity" to reflect the fact that population size is forced to remain within some bounds (in contrast to populations governed by Equation 3.1).

As we shall see in the next section, the mathematical simplicity of the ceiling model allows some clear insights into the viability of negatively density-dependent populations. Nevertheless, for many populations, the growth rate will change gradually rather than abruptly as population size increases. An alternative model that allows for a gradually changing growth rate is the *theta logistic model*:

$$N_{t+1} = N_t \exp\left\{ r\left[1 - \left(\frac{N_t}{K}\right)^{\theta}\right] + \varepsilon_t \right\} \tag{4.2}$$

(modified from Gilpin and Ayala 1973), variants of which have served as the basis for several published PVAs that incorporated negative density dependence (Saether et al. 1998a, 2000b). The mean population growth rate is e^r when N_t is very small relative to the carrying capacity K. Unlike the other models we've covered, environmental variability in the population growth rate is not represented by variance in the usual growth parameters themselves (e.g., λ or r) but by a separate variable, ε_t (the Greek letter epsilon), which has a mean of zero and a variance that reflects the degree of environmental stochasticity. The parameter θ (the Greek letter theta) determines how the population growth rate declines as population size increases. As shown in Figure 4.1, the advantage of the theta logistic model is its ability to describe several different patterns of negative density dependence depending upon the value of θ. In particular, if θ is large, the population growth rate remains high until population size gets very close to K, at which point the growth rate drops off precipitously. Thus the theta logistic model approaches the ceiling model as θ approaches infinity. The well-known Ricker model[1] (Ricker 1954), which has seen much use in fisheries biology, is a special case of the theta logistic model obtained by setting θ equal to 1. Note also that as the carrying capacity K goes to infinity, Equation 4.2 becomes $N_{t+1} = \lambda_t N_t$, where $\lambda_t = \exp\{r + \varepsilon_t\}$. That is, as density dependence gets weaker and weaker (as the carrying capacity gets higher and higher), Equation 4.2 reverts to the density-independent model of Equation 3.1.

---

[1]In some sense, the Ricker model can be thought of as a discrete-time version of the familiar continuous-time logistic model.

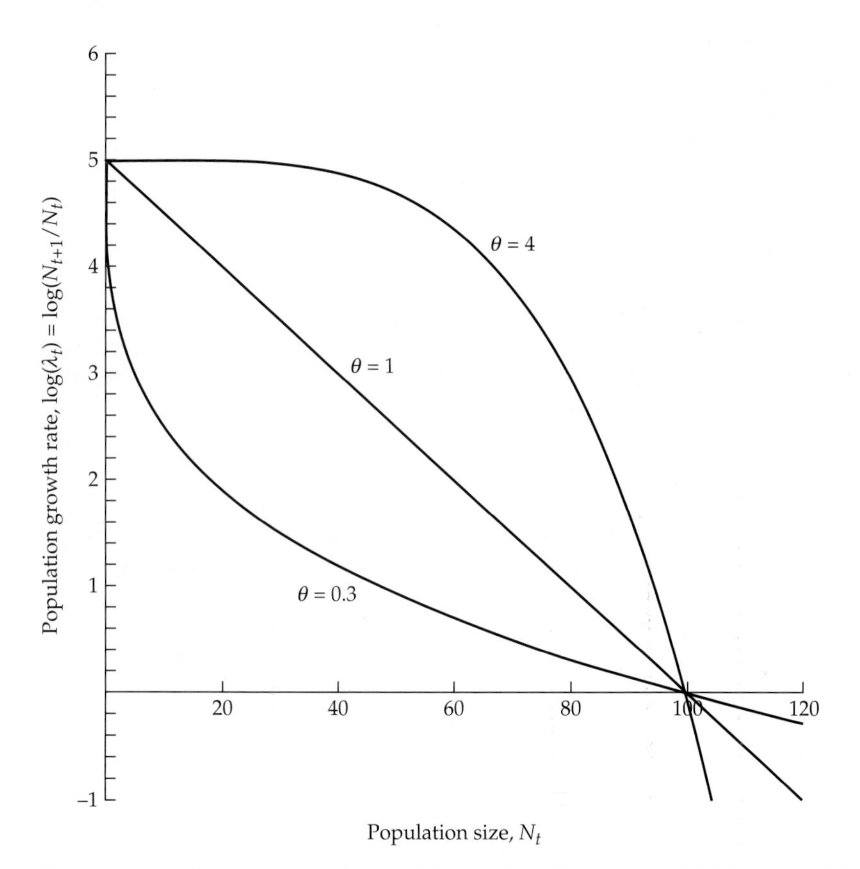

Figure 4.1 The parameter θ (theta) in the theta logistic model (Equation 4.2) governs the shape of negative density dependence. Specifically, the log of the population growth rate $\log(N_{t+1}/N_t)$ declines most steeply at low densities if $\theta < 1$, linearly with density if $\theta = 1$, or most steeply at high densities if $\theta > 1$. For all three values of θ, the carrying capacity, K, equals 100 and the log population growth rate at low density, r, equals 5.

Extinction Risk Predicted by the Ceiling Model

In Chapters 2 and 3, we discussed several metrics of extinction risk, including the geometric mean population growth rate, the mean time to reach the extinction threshold, and the probability of reaching the threshold by a given future time horizon. In the ceiling model, just as for the density-independent model (Equation 3.1), any population with a geometric mean growth rate less than 1.0 is doomed to eventual extinction. However, the two models differ in an important way when the geometric mean growth rate is greater than 1.0. Whereas the density-independent model predicts (unrealistically) that some population trajectories could grow to infinite size (so that the probability of ultimate extinction is less than 1; see Equation 3.6), the presence of an upper limit to population size in the ceiling model (or, for that matter, in any other model with negative density dependence) causes *all* trajectories to hit the extinction threshold *eventually*. (That is, the probability of ultimate extinction is *always* 1.) Sooner or later, a sufficient number of bad years will occur in succession to drive the population down to the extinction threshold, regardless of the height of the ceiling. However, such a sequence of years may take a *very* long time to occur by chance if the population tends to hover near the ceiling and the ceiling itself is high.

To see exactly *how* long, we can examine another viability metric, the mean time to extinction. The principal advantage that the ceiling model possesses over other models with negative density dependence is its mathematical simplicity, which makes it possible to calculate an analytic expression for the mean time to reach the extinction threshold from any initial density.[2] We caution that the same limitations of the mean time to extinction that we discussed in Chapter 2 also apply to the ceiling model, namely that the probability distribution of extinction times is highly skewed so that the mean is not very representative of the behavior of most possible outcomes. Nevertheless, other measures of extinction risk are not easy to calculate for this model, so we use the mean extinction time to gauge how the *relative* extinction risk depends on the height of the ceiling and the values of μ and σ^2, the mean and the variance of the log population growth rates, log λ_t. (Note that μ and σ^2 are exactly the same parameters as in Chapter 3). If the variation in the λ_t's is not too large, the ceiling model can be approximated by a diffusion model with a reflecting upper boundary at the ceiling (in addition to the absorbing lower boundary at the extinction threshold; see Chapter 3). From this diffusion approximation, an expression for the mean time to reach the quasi-extinction threshold N_x from a current population size of N_c is:

$$\overline{T} = \frac{1}{2\mu c}\left[e^{2ck}\left(1 - e^{-2cd}\right) - 2cd\right] \tag{4.3}$$

where $c = \mu/\sigma^2$, $d = \log(N_c/N_x)$, and $k = \log(K/N_x)$ (for details, see Lande 1993, Foley 1994, Middleton et al. 1995). If the population starts at the ceiling (i.e., $N_c = K$) and $N_x = 1$, then Equation 4.3 simplifies to

$$\overline{T} = \frac{1}{2\mu c}\left[K^{2c} - 1 - 2c\log K\right] \tag{4.4}$$

(Lande 1993). A MATLAB program to compute $\overline{T}$ using Equations 4.3 and 4.4 is given in Box 4.1.

---

[2]Mangel and Tier (1993) developed a numerical method to calculate the mean and variance of the time to extinction for any density-dependent model, but their method does not yield a simple expression for the mean extinction time, such as Equation 4.3.

BOX 4.1 *MATLAB code to plot the mean time to extinction for the ceiling model (Equation 4.1) as functions of the carrying capacity and initial population size.*

```
% PROGRAM tbar_ceiling
%    Plots mean time to extinction vs. K and starting population
%    size for the ceiling model, using expressions for the mean
%    time to extinction from Lande, Am. Nat. 142:911-927 (1993),
%    Foley, Conservation Biology 1:124-137 (1994), and
%    Middleton et al., Theor. Pop. Biol. 48:277-305 (1995).

mu=0.1;       % mean log population growth rate
s2=0.1;       % variance of log population growth rate
c=mu/s2;

% Plot mean time to extinction (N=1) vs. K for populations
% starting at K, according to Equation 4.4
K=[0:50:1000];        % vector of carrying capacities
K(1)=1;               % set smallest K to 1
k=log(K);
Tbar=(K^(2*c)-1-2*c*k)/(2*mu*c); % Eq. 4.4

subplot(2,1,1);       % upper of 2 plots on same page
plot(K,Tbar)          % plot Tbar vs K
xlabel('Carrying capacity, K','FontSize',12)
ylabel('Mean time to extinction, Tbar','FontSize',12)
title('Populations starting at K','FontSize',14)

% Plot mean time to extinction (N=1) vs. Nc for populations
% starting at various Nc<=K, according to Equation 4.3
K=200;        % carrying capacity
Nc=[0:10:K]; % vector of initial population sizes
Nc(1)=1;      % set smallest Nc to 1
k=log(K);
d=log(Nc);
Tbar=(exp(2*c*k)*(1-exp(-2*c*d))-2*c*d)/(2*mu*c); % Eq. 4.3

subplot(2,1,2);       % lower of 2 plots on same page
plot(Nc,Tbar)         % plot Tbar vs Nc
xlabel('Initial population size, Nc','FontSize',12)
ylabel('Mean time to extinction, Tbar','FontSize',12)
title('Populations starting below K','FontSize',14)
```

Equation 4.4 is useful for exploring how increasing the "height" of the ceiling (for example by increasing the size of a reserve and therefore increasing its carrying capacity) affects the mean time to reach the extinction threshold. The shape of the curve relating mean extinction time to the carrying capacity depends upon the parameters μ and σ^2 (Lande 1993). From a conservation standpoint, we would ideally like the mean time to extinction to increase faster than linearly, so that we get more and more protection from extinction for each increment in reserve size. However, when μ is equal to or less than zero, the mean time to extinction always increases in a slower-than-linear fashion with an increase in the carrying capacity, as shown in Figure 4.2A. This "law of diminishing returns" makes intuitive sense. As we saw in Chapter 3, a negative value of μ (equivalent to $\lambda_G < 1.0$) implies that the population will decline over the long term, and we cannot very effectively delay the inevitable crossing of the extinction threshold simply by raising the ceiling. In contrast, if μ is sufficiently positive, growth of the population will tend to quickly push it back up to the ceiling following a bad year, so that raising the ceiling widens the gap between the zone of sizes in which the population will typically be found (that is, close to the ceiling) and the extinction threshold, thus making extinction less likely in the short term.

However, even if μ is positive, increasing the environmental variance σ^2 can cause the mean extinction time curve as K becomes large to go from faster-than-linear to linear and finally to slower-than-linear (Figure 4.2B; note that when σ^2 exactly equals 2μ, the curve becomes perfectly linear as K increases). Again, this makes sense given what we know from Chapter 3 about the effects of environmental variation. The more severe are the extremes of the population growth rate, the more likely is a string of extremely bad years that could drive the population to extinction, even if the ceiling is high. Thus the expression for the mean time to extinction in the ceiling model tells us that management efforts aimed at increasing the carrying capacity (e.g., by expanding the size of a reserve) are destined to fail if the population is declining over the long term *or* if it exhibits very high inter-annual variation in growth. Identifying strategies to increase the growth rate or ameliorate causes of variation in the growth rate may be a more productive management goal for such populations. On the other hand, if long-term population growth is relatively high and environmental variation low, increasing the carrying capacity can dramatically enhance a population's protection from extinction (Figure 4.2B).[3]

In Chapters 2 and 3, we saw the value of viability metrics based on the extinction probability, particularly the cumulative distribution function (CDF)

---

[3]An important assumption of the ceiling model is that μ and σ^2 are independent not only of population size but also of the height of the ceiling. If increasing the size of a reserve both "raises" the ceiling and improves μ or σ^2 (e.g., by reducing edge effects or by buffering environmental stochasticity), then it may have substantial benefits even when μ is initially negative or σ^2 initially large.

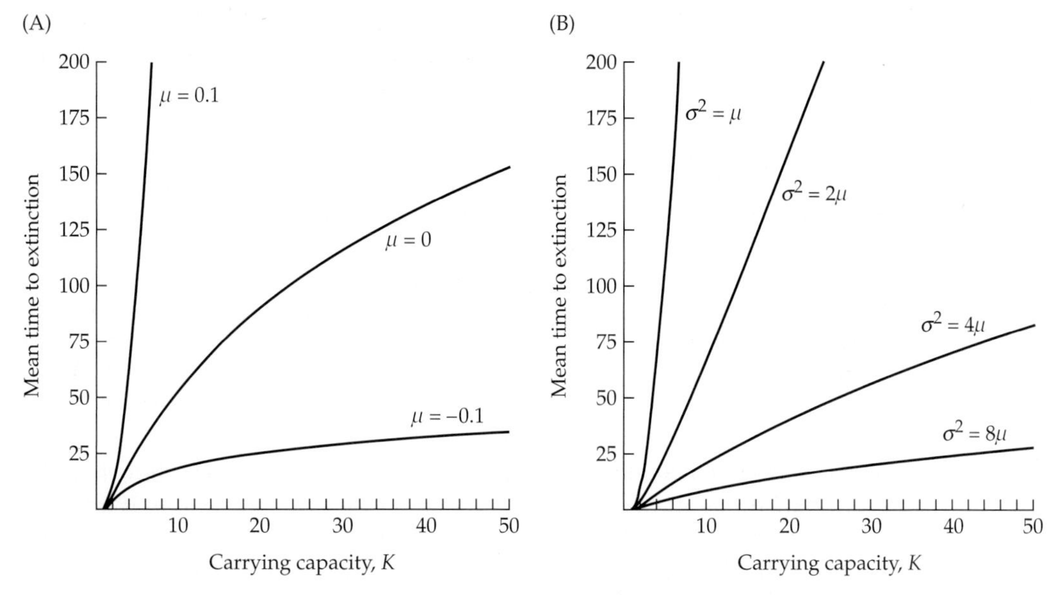

Figure 4.2 The mean time to extinction given by Equation 4.4 for populations starting at K in the ceiling model (Equation 4.1) with an extinction threshold $N_x = 1$. In panel A, σ^2 always equals 0.1, but μ changes from positive to zero to negative values. In panel B, μ is always 0.1, but σ^2 increases.

for the time to extinction (Equation 3.5). Unfortunately, there is no equivalent analytical expression for the extinction probabilities for any density-dependent model, including the relatively simple ceiling model.[4] Nevertheless, a straightforward alternative to an analytical formula is to estimate the extinction time CDF using computer simulations. The idea is to simulate many random trajectories that a population might take (as we did to create the 20 paths shown in Figure 3.1, but typically for tens or hundreds of thousands of trajectories), seeing when each becomes extinct (if ever), and using these results to build a distribution of predicted extinction times.

[4]Middleton et al. (1995) derived a solution to the diffusion approximation for the ceiling model using a Fourier series, which technically involves the sum of an infinite number of terms. In principle, one could (as did Middleton and colleagues for particular combinations of μ, σ^2, and K) use the Fourier series solution for the ceiling model to calculate a close approximation to the probability of extinction by time T, but the procedure required to do so is technically involved. (One needs to solve a set of implicit equations, substitute the results into the Fourier series, integrate the series term by term, and sum the resulting terms until the sum converges to a nearly constant value.) In practice, and given the peculiar way that density dependence is represented in the ceiling model, such a procedure is probably not worth the effort for specific PVA applications.

However, if we go to the work to make such a density-dependent simulation, we generally won't want to use the ceiling model. In the ceiling model, the assumption that population growth below the ceiling is density-independent was made primarily to allow the expressions for the mean time to extinction (Equations 4.3 and 4.4) to be calculated. Although this exercise helped give insight into how the existence of a ceiling affects population viability, the unusually sharp onset of negative density dependence in the ceiling model is probably not realistic for actual populations. Fortunately, with adequate data to estimate density-dependent effects, we can relax this confining stricture when we use computer simulations, as we demonstrate in the following section.

PVA for a Population with Negative Density Dependence: The Bay Checkerspot Butterfly

In this section, we demonstrate by means of a worked example how to use a more realistic negatively density-dependent model, in combination with computer simulations, to calculate the probability of extinction. The example we use is the Bay checkerspot butterfly, *Euphydryas editha bayensis*, which occurs in small populations restricted to serpentine outcrops in the San Francisco Bay area of California. We begin with the JRC population, one of three populations that were the subjects of a long-term population study conducted by Paul Ehrlich and his colleagues at Stanford University. The number of female butterflies in the population was estimated by capture-recapture methods each year from 1960 to 1986. The 27 consecutive estimates of population size are given in Table 4.1 (reproduced from Harrison et al. 1991) and shown in Figure 4.3. Note that the population size estimates undergo large fluctuations across the 27 years. There are two phases to conducting a density-dependent PVA with this kind of data. First we need to find the best model to describe the population's dynamics, and then we need to simulate the model to predict population viability.

TESTING FOR DENSITY DEPENDENCE AND IDENTIFYING THE BEST MODEL As we did for the Yellowstone grizzly bear population in Chapter 3, we can get a quick idea of whether the census data show evidence of negative density dependence by looking for a negative relationship between the log population growth rate and population size. Such a negative relationship does in fact appear likely for the JRC Bay checkerspot population (Figure 4.4), although there is clearly a lot of variability in the growth rate due to environmental variation, demographic stochasticity, and/or observation error. But is there *statistical* support for the presence of negative density dependence in the checkerspot data? And if so, what shape of the density dependence function (Figure 4.1) would best describe the data? We can answer both of these questions by fitting a suite of population models to the counts in Table 4.1 and determining which model best fits the data.

TABLE 4.1 Census data for two populations of the Bay checkerspot butterfly[a]

	JRC population		*JRH population*	
Year, t	*Estimated number of female butterflies, N_t*	*Log population growth rate, $\log(N_{t+1}/N_t)$*	*Estimated number of female butterflies, N_t*	*Log population growth rate, $\log(N_{t+1}/N_t)$*
1960	90	0.6650	70	1.6094
1961	175	–1.4759	350	0.7621
1962	40	0.1178	750	0.0000
1963	45	1.3581	750	0.6242
1964	175	0.1335	1400	0.3567
1965	200	0.7538	2000	–0.1335
1966	425	0.0000	1750	–0.6650
1967	425	0.6325	900	–0.4463
1968	800	–1.1394	576	0.4135
1969	256	1.0243	871	–0.0603
1970	713	–1.2812	820	–1.2497
1971	198	2.2178	235	1.5871
1972	1819	–1.1517	1149	–1.1331
1973	575	–0.0140	370	–0.7374
1974	567	1.1657	177	0.5828
1975	1819	1.3795	317	1.1499
1976	7227	–2.1380	1001	–1.6617
1977	852	–1.3723	190	0.5849
1978	216	0.1219	341	–0.9266
1979	244	0.0901	135	–0.0770
1980	267	1.8818	125	0.9274
1981	1753	–0.5623	316	–1.0644
1982	999	0.5821	109	0.1127
1983	1788	–2.5260	122	–1.3700
1984	143	–0.5934	31	0.4372
1985	79	0.1738	48	–0.9808
1986	94		18	

[a]Data from Harrison et al. 1991.

To identify the best model, we advocate the use of Information Criterion statistics, which are computed with the aid of maximum likelihood methods. In the Appendix, we give a thumbnail sketch of the method of maximum likelihood for identifying best-fit model parameters. (To obtain more information on these methods than we have space to provide here, see Edwards 1972, Hilborn and Mangel 1997, and Burnham and Anderson 1998.) *Likelihood* has a technical meaning here: it is the probability of obtaining the observed data given a particular set of parameter values for a particular

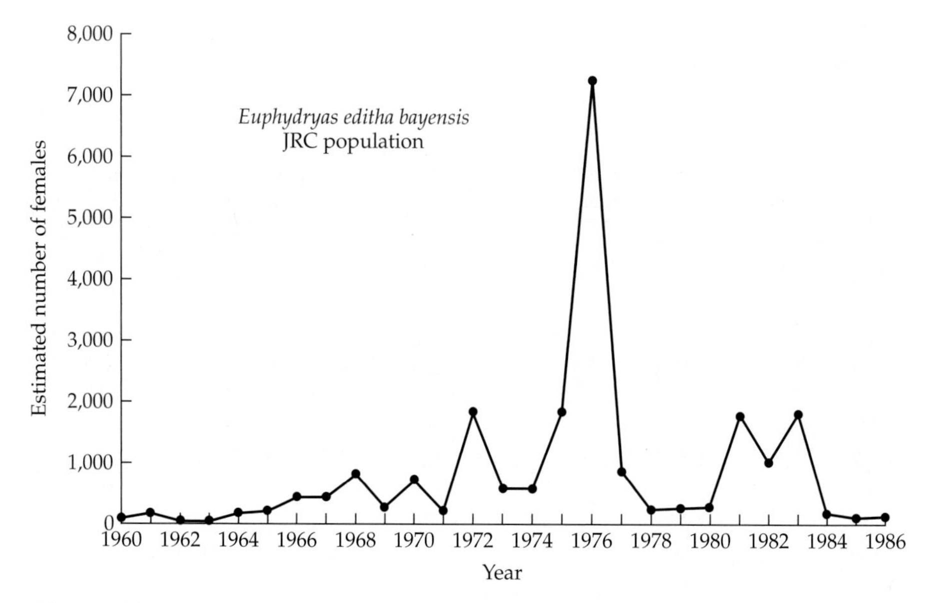

Figure 4.3 Census counts from the JRC population of the Bay Checkerspot butterfly, *Euphydryas editha bayensis* (data from Harrison et al. 1991).

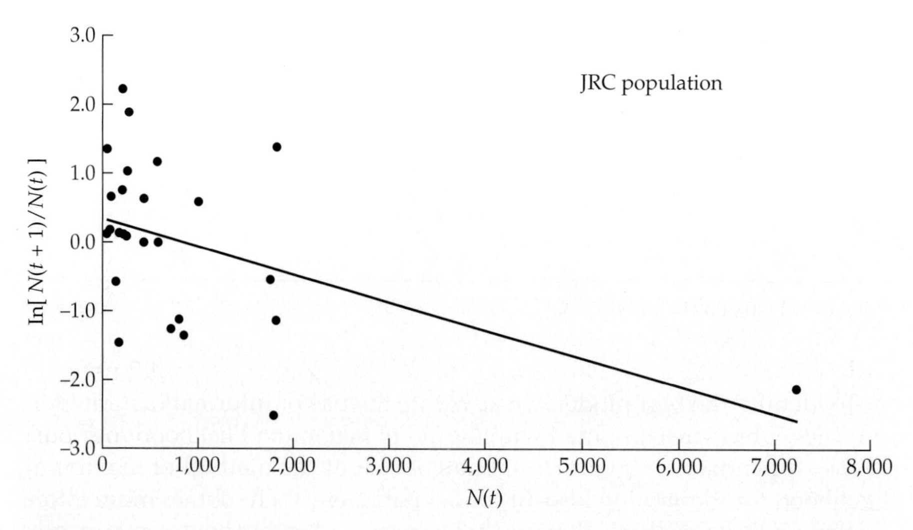

Figure 4.4 The log population growth rate for the JRC checkerspot population is negatively related to population size. The relationship remains significantly negative even if the outlying point at the far right is removed. Line is the prediction from the best-fit Ricker model.

model. For reasons of computational accuracy, it is usually better to calculate the log of the likelihood. The maximum log likelihood for a particular model is obtained when the values of the parameters that have the highest probability of generating the observed data are substituted into the model. Information Criterion statistics combine the maximum log likelihood for a model with the number of parameters it includes to provide a measure of the model's "support"—how good a description of the data it provides, given how complex it is. Support is higher for models with *higher* likelihoods and *fewer* parameters. More complex models are penalized because more parameters will almost always lead to a better fit to the data, but at the cost of less precision in the estimate of each parameter. Comparing the Information Criteria for two or more models will then help us to choose the model that provides the best description of the data.

Briefly, the procedure we will use to assess the presence and pattern of negative density dependence in the checkerspot data involves three steps:

Step 1: Fit three models to the data using nonlinear least-squares regression of $\log(N_{t+1}/N_t)$ against N_t, where N_t is the count obtained from population census t.[5] The regression equations for the three models are:

Model	*Regression Equation*
The density independent model	$\log(N_{t+1}/N_t) = r$
The Ricker model	$\log(N_{t+1}/N_t) = r(1 - N_t/K)$
The theta logistic model	$\log(N_{t+1}/N_t) = r[1 - (N_t/K)^\theta]$

As we noted above, the first two models are simplifications of the theta logistic model.

Step 2: Compute the maximum log likelihood of each model using the least-squares estimates of the parameters and the residual variance (defined below).

Step 3: Calculate Information Criterion statistics for the three models, and use them to decide which model provides the best description of the data.

We now provide more detail on each of these steps.

STEP 1: ESTIMATE MODEL PARAMETERS WITH NONLINEAR REGRESSION Any statistical software capable of performing nonlinear least-squares regression can be used in Step 1. For example, the SAS program in Box 4.2 uses the `nlin` procedure to perform the necessary regressions. The parts of the output produced by the SAS program in Box 4.2 that are important for our purposes are reproduced in Box 4.3. To calculate the maximum log likelihood for each model, we need the output labeled `Estimate` for each parameter

[5]Unlike in Chapter 3, in this chapter we will assume that censuses were taken at fixed time intervals.

BOX 4.2 *SAS program[1,2] to fit three models to the Bay checkerspot census data in Table 4.1*

```
data jrc;
infile "c:chkrspot.prn" firstobs=2;
input year nt yt;
run;

proc nlin data=jrc;
   title 'JRC Population - theta logistic model';
   parameters r=0.5 to 1.5 by 0.5 K=400 to 800 by 200 theta=0.2 to 1 by 0.4;
   bounds r>=0, K>=0, theta>=0;
   model yt=r*(1-(nt/K)**theta);
   der.r=1-(nt/K)**theta;
   der.K=( r*theta*nt*(nt/K)**(theta-1) )/(K*K);
   der.theta=-r*log(nt/K)*(nt/K)**theta;
   output out=fits p=ypred residual=yres;
run;

proc plot data=fits;
   plot yt*nt='*' ypred*nt='-'/overlay;
   plot yres*nt;
run;

proc nlin data=jrc;
   title 'JRC Population - Ricker model';
   parameters r=0.5 to 1.5 by 0.5 K=400 to 800 by 200;
   bounds r>=0, K>=0;
   model yt=r*( 1-(nt/K) );
   der.r=1-(nt/K);
   der.K=(r*nt)/(K*K);
```

[1]To use the program, the data in Table 4.1 must first be placed into a space-delimited data file named "chkrspot.prn"; in reading the data, the program skips the first row of the data file, which may containing column labels.

[2]Notes on nonlinear least-squares regression:

1. Users must supply SAS with the derivatives of the function to be fit to the data with respect to each of the parameters in the model (hence the `der.r`, `der.K`, and `der.theta` commands in `proc nlin`).
2. Parameter estimates are constrained to be non-negative using the "`bounds`" statement.
3. It is recommended that users repeat the NLIN procedure using different starting values, to be sure that the search converges to parameter estimates that correspond to the global minimum sum of squares.
4. It is a good idea to check the overlay plots of predicted and observed values produced by this program, to be sure that the best model fits the data reasonably well.

BOX 4.2 *(continued)*

```
   output out=fits p=ypred residual=yres;
run;

proc plot data=fits;
   plot yt*nt='*' ypred*nt='-'/overlay;
   plot yres*nt;
run;

proc nlin data=jrc;
   title 'JRC Population - density-independent model';
   parameters r=0.5 to 1.5 by 0.5;
   bounds r>=0;
   model yt=r;
   der.r=1;
   output out=fits p=ypred residual=yres;
run;

proc plot data=fits;
   plot yt*nt='*' ypred*nt='-'/overlay;
   plot yres*nt;
run;
```

(these are known as the "least-squares estimates"), as well as two pieces of output that can be used to estimate the so-called residual variance, which we will call V_r. The residual variance V_r is the mean squared deviation between the observed values of $\log(N_{t+1}/N_t)$ (i.e., the y values in the nonlinear regression) and the values predicted by the model. The residual variance, in addition to being a component of the maximum log likelihood, will also give us an estimate of the environmental variance in the log population growth rate, equivalent to $\hat{\sigma}^2$ from Chapter 3 (see the following steps). The residual variance is obtained by dividing the `Residual Sum of Squares` from the regression output (called error sum of squares by many statistics packages) by the number of data points in the regression (in this case, 26 for the number of estimates of the log population growth rate in Table 4.1). For example, for the density-independent model, the residual variance is 36.3976/26 = 1.3999. (Note that the residual variances are *not* the residual mean squares given in standard regression ANOVA tables such as the one in Box 4.3 (Burnham and Anderson 1998, pp. 15–17)). The least-squares

BOX 4.3 ***Key output from SAS program used to fit three models to data from the JRC Bay checkerspot population in Table 4.1. Key results are shown in bold.***

```
                JRC Population - density-independent model
Non-Linear Least Squares Summary Statistics     Dependent Variable YT

               Source                DF Sum of Squares      Mean Square

               Regression             1     0.000072745      0.000072745
               Residual              25    36.397556702      1.455902268
               Uncorrected Total     26    36.397629448

               (Corrected Total)     25    36.397556702


               Parameter      Estimate      Asymptotic               Asymptotic 95 %
                                            Std. Error            Confidence Interval
                                                                  Lower          Upper
               R           0.0016726923  0.23663524902  -.48568282538 0.48902821000

                         JRC Population - Ricker model
Non-Linear Least Squares Summary Statistics     Dependent Variable YT

               Source                DF Sum of Squares      Mean Square

               Regression             2     8.520152294      4.260076147
               Residual              24    27.877477153      1.161561548
               Uncorrected Total     26    36.397629448

               (Corrected Total)     25    36.397556702


               Parameter      Estimate      Asymptotic               Asymptotic 95 %
                                            Std. Error            Confidence Interval
                                                                  Lower          Upper
               R              0.3458095     0.24661973     -0.16318449      0.8548035
               K            846.0152803   517.10358675   -221.22547418   1913.2560348
```

parameter estimates and the residual variance for each model are listed in Table 4.2.

STEP 2: CALCULATE MAXIMUM LOG LIKELIHOOD VALUES FOR EACH MODEL With least-squares estimates of the model parameters and the residual variance in

BOX 4.3 *(continued)*

```
                   JRC Population - theta logistic model
Non-Linear Least Squares Summary Statistics      Dependent Variable YT
```

Source	DF	Sum of Squares	Mean Square
Regression	3	9.969340119	3.323113373
Residual	23	**26.428289328**	1.149056058
Uncorrected Total	**26**	36.397629448	
(Corrected Total)	25	36.397556702	

Parameter	Estimate	Asymptotic Std. Error	Asymptotic 95 % Confidence Interval Lower	Upper
R	**0.9940620**	1.15375841	-1.39264991	3.3807739
K	**551.3788238**	317.91269892	-106.26840990	1209.0260575
THETA	**0.4565906**	0.43302473	-0.43918206	1.3523633

hand, we can now calculate maximum log likelihood values for each model. In fitting the models using least squares, we implicitly assumed that the deviations between the observed and predicted log population growth rates followed a normal distribution with a mean of zero (and in so doing, we assumed that small deviations are more probable than large ones, and that positive and negative deviations of a given size are equally probable). The log likelihood of a model assuming normally distributed deviations (see Appendix) is

$$\log L = -\frac{q}{2}\log(2\pi V_r) - \frac{1}{2V_r}\sum_{t=1}^{q}\left[\log\left(\frac{N_{t+1}}{N_t}\right) - P_t\right]^2 \tag{4.5}$$

where P_t is the value of $\log(N_{t+1}/N_t)$ predicted by the model for census interval t, q is the number of measurements of the log population growth rate in the data set, and V_r is the residual variance we calculated in Step 1. The predicted values for the three models are:

Model	$\boldsymbol{P_t}$	
Density-independent	r	
Ricker	$r\,(1-N_t/K)$	(4.6)
Theta logistic	$r\,[1-(N_t/K)^{\theta}]$	

TABLE 4.2 Parameter estimates, residual variances, maximum log likelihood, AIC_c values, and Akaike weights for three models fit to data from the JRC Bay checkerspot population in Table 4.1 (number of data points q = 26)[a]

	Least-squares parameter estimates			
Model	***r***	***K***	***θ***	***Residual variance, V_r***
Density-independent	0.001673	—	—	1.3999
Ricker	0.3458	846.02	—	1.0722
Theta logistic	0.9941	551.38	0.4566	1.0165

[a]The most parsimonious model is indicated by asterisks.

Note that the right-hand side of Equation 4.5 includes a sum of squared deviations between the observed and predicted values of the log population growth rate. It was exactly this sum that we minimized by estimating the best parameter values using least-squares regression.

To calculate the maximum log likelihood for a given model, we could simply substitute into Equation 4.5 the appropriate form for P_i in Equation 4.6, the least-squares estimates of the parameters and the residual variance from Table 4.2, and the observed values of N_t and N_{t+1} from Table 4.1. However, as it turns out, when the least-squares estimates of the parameters are substituted into Equation 4.5, the sum on the right hand side equals q times V_r (in other words, the residual sum of squares that we got from the model-fitting output in Box 4.3), so that Equation 4.5 greatly simplifies to

$$\log L_{max} = -\frac{q}{2}\left[\log(2\pi V_r) + 1\right] \tag{4.7}$$

Here *max* signifies that the least-squares parameter estimates by definition produce the maximum value of the log likelihood. The maximum log likelihood values calculated using Equation 4.7 are given in the seventh column of Table 4.2. Note that the values are negative, because each represents the log of a small number (the product of the probabilities of each data point given the model, each of which is a number between zero and one). The model with the largest (that is, the least negative) maximum log likelihood is the theta logistic model. However, the theta logistic model achieves such a high likelihood by using the greatest number of parameters (r, K, θ, and V_r).

Number of parameters (including V_r), p	*Maximum log likelihood, log L_{max} (from equation 4.7)*	*AIC_c (from equation 4.8)*	*Akaike weights*
2	–41.266	87.054	0.07
3	–37.799	82.689***	0.62
4	–37.105	84.115	0.31

As noted above, we can usually improve the likelihood of a model simply by adding more parameters. But do the data really justify including additional parameters, each of which will be estimated with less and less precision? Information Criterion statistics provide a way to answer this question.

STEP 3: COMPARE THE SUPPORT FOR DIFFERENT MODELS USING INFORMATION CRITERION STATISTICS To find the simplest, best-fitting model, we need to account for the cost of using models that achieve high values of the maximum log likelihood by including many parameters. These costs include imprecision of parameter estimates and incorporation of spurious patterns from the data into future predictions. To measure and compare the support for different models, we recommend using the corrected Akaike Information Criterion (AIC_c; Hurvich and Tsai 1989, Burnham and Anderson 1998), which is more appropriate than other Information Criterion statistics when the number of data points used to fit the models is small relative to the number of parameters (as will usually be the case in PVA). Specifically, for each model we calculate

$$AIC_c = -2\log L_{max} + \frac{2pq}{q-p-1} \tag{4.8}$$

where p is the number of parameters estimated (*including* the residual variance, V_r) and q is (once again) the number of data points. The most parsimonious model is the one with the *lowest* value of AIC_c. The first term on the right hand side of Equation 4.8 gets smaller as the maximum log likelihood increases, so that AIC_c favors models with a high likelihood. However, the second term on the right hand side of Equation 4.8 gets *larger* as the number of parameters in the model increases, so that models with many parameters are less likely to have the lowest AIC_c value. In fact, the theta logistic model, which has the most parameters, does not have the lowest value of AIC_c (Table 4.2). Instead, the best model identified by the corrected Akaike Information Criterion is the Ricker model, in which the log population growth rate declines linearly with population size. To determine if this difference in AIC_c values is strong enough to firmly reject the more complicated theta logistic, we can also calculate Akaike weights, which quantify the probability that each of a suite of models is the best approximation to the truth (Burnham and Anderson 1998). For a set of R models, the weight of model i is:

$$w_i = \frac{\exp\left[-0.5\left(AIC_{c,i} - AIC_{c,best}\right)\right]}{\sum_{i=1}^{R} \exp\left[-0.5\left(AIC_{c,i} - AIC_{c,best}\right)\right]} \quad (4.9)$$

Here, $AIC_{c,i}$ is the corrected information criterion for model i and $AIC_{c,best}$ is the best (i.e., smallest) criterion of any model tested. The Akaike weights (Table 4.2) tell us that the Ricker model is twice as likely to be the best model as is the theta logistic. Unless we had some other compelling reason to use the more complicated model, this result supports use of the Ricker model. Also, note that the density-independent model has only a 7% chance of being the best description of the data, giving strong support to the inclusion of negative density dependence in estimation of extinction risks for this population. We now use maximum likelihood parameter estimates for the Ricker model from Table 4.2 to estimate extinction risk for the JRC Bay checkerspot butterfly population by means of simulations.

ESTIMATING EXTINCTION RISK FOR A DENSITY-DEPENDENT POPULATION BY COMPUTER SIMULATION To estimate extinction risk for a density-dependent population, we use a computer to predict the size of the population at one year intervals into the future, starting with the current population size. In making these predictions, we rely on the computer's ability to generate random numbers to introduce environmental variation into the population growth process. In this case, we are using Equation 4.2 with $\theta = 1$ (i.e., the Ricker model) as our population model, so we program the computer to generate a new value for ε_t each year and then substitute it, the parameter estimates, and the current population size N_t into the equation to yield a predicted value of N_{t+1}. Repeating this process over a number of years and for many replicate populations yields estimates of extinction risk.

To implement this algorithm, we must first decide what characteristics the environmentally driven deviations in the log population growth rate (that is, the ε_t's in Equation 4.2) should exhibit. For a number of reasons it makes sense for the ε_t's to be drawn from a normal distribution with a mean of zero.[6] By estimating V_r during the model-fitting step, we also have a way to estimate the variance of the ε_t's. The variance of the ε_t's is the environmentally driven variance in the log population growth rate, which we called σ^2 in Chapter 3. Hence to emphasize that these are the same quantity, we will use σ^2 to represent the true variance of the ε_t's, and $\hat{\sigma}^2$ to represent an estimate of σ^2. Whereas the maximum likelihood estimate of the residual variance, V_r, is the appropriate parameter to use when choosing among models with Information Criteria (Burnham and Anderson 1998), for finite data sets V_r gives a biased estimate of the environmental variance in the log population growth rate, σ^2 (Dennis et al. 1991). This is especially true when the number of annual transitions in the data set is small, as will often be the case for endangered populations. An unbiased estimate of σ^2 is

$$\hat{\sigma}^2 = \frac{qV_r}{q-1} \tag{4.10}$$

where q is (once again) the number of data points (Dennis 1989, Dennis et al. 1991, Dennis and Taper 1994). For the JRC Bay checkerspot population, the V_r estimate for the Ricker model from Table 4.2 yields $\hat{\sigma}^2 = 26 \times 1.0722/(26-1) = 1.1151$. Hence to produce appropriate values for the ε_t's, we multiply computer-generated normally distributed random numbers with a mean of zero and a variance of one (such as those produced by the `randn` function in MATLAB) by $\sqrt{\hat{\sigma}^2}$, to produce normally distributed random numbers with a mean of zero and a variance of $\hat{\sigma}^2$.

Box 4.4 provides a MATLAB program to compute the probability that a population described by the theta logistic model will fall below a specified quasi-extinction threshold at a suite of future times. In the section of the code labeled `SIMULATION PARAMETERS`, the appropriate values for the JRC Bay checkerspot population have been entered. There are several things to note about these parameters. First, we have set `theta` equal to 1, because the Ricker model was judged, using AIC_c, to provide a more parsimonious fit to the data than the theta logistic model (Table 4.2). However, the program allows for situations in which the best-fit value of θ in Equation 4.2 is not equal to 1.[7] Second, the starting population size (`Nc`) is the last census count in Table 4.1. Third, we do not attempt to predict very far into the future

[6]Because the only effect we want the ε_t values to have is to add variability to population growth, their mean should be zero. It is reasonable to assume that their distribution is normal if the additive impact of many environmental variables, each with small effect, determines the overall variability in log population growth, since by the Central Limit theorem, a random variable that is the sum of many other variables will be approximately normally distributed. Note that using least squares to estimate V_r assures that we will obtain the proper variance for a normal distribution of ε_t's using Equation 4.10.

[7]And the program can easily be modified to use other density-dependent population models.

BOX 4.4 *MATLAB code to predict the probability of extinction using the theta logistic model (Equation 4.2).*

```
% PROGRAM theta_logistic.m
%    Calculates by simulation the probability that a population
%    following the theta logistic model and starting at Nc
%    will fall below the extinction threshold Nx by time tmax

%****************** SIMULATION PARAMETERS ***********************
r=0.3458;            % intrinsic rate of increase
K=846.017;           % carrying capacity
theta=1;             % nonlinearity in density dependence
sigma2=1.1151;       % environmental variance
Nc=94;               % starting population size
Nx=20;               % quasi-extinction threshold
tmax=20;             % time horizon
NumReps=50000;       % number of replicate population trajectories
%*************************************************************

sigma=sqrt(sigma2);
randn('state',sum(100*clock)); % seed the random number generator


N=Nc*ones(1,NumReps); % all NumRep populations start at Nc
NumExtant=NumReps;    % all populations initially extant
Extant=[NumExtant];   % vector for number of extant pops. vs. time
for t=1:tmax,                   % For each future time,
                                %   compute new pop. sizes from
    N=N.*exp( r*( 1-(N/K).^theta )... % the theta logistic model
        + sigma*randn(1,NumExtant) ); % with random environmental
                                      % effects.
    for i=NumExtant:-1:1, % Looping over all extant populations,
        if N(i)<=Nx, % if at or below quasi-extinction threshold,
            N(i)=[]; % delete the population.
        end;
    end;
    NumExtant=length(N);      % Count remaining extant populations
    Extant=[Extant NumExtant]; % and store the result.
end;

% Compute quasi-extinction probability as the fraction of
% replicate populations that have hit the threshold by each
% future time, and plot quasi-extinction probability vs. time
ProbExtinct=(NumReps-Extant)/NumReps;
```

BOX 4.4 *(continued)*

```
plot([0:tmax],ProbExtinct)
xlabel('Years into the future');
ylabel('Cumulative probability of quasi-extinction');
axis([0 tmax 0 1]);
```

(tmax is small), for reasons we articulated in Chapter 3. Fourth, the number of replicate populations (NumReps) must be large to get accurate estimates of the probability of extinction.

Some output from the MATLAB code in Box 4.4 is shown in Figure 4.5. With 50,000 replicate populations, there is very close agreement among separate runs of the model regarding the fraction of populations hitting the quasi-extinction threshold by each future time. (Note, however, that these predictions do not account for uncertainty in the parameter estimates; see the following discussion.) Not surprisingly, the population is more likely to fall below a

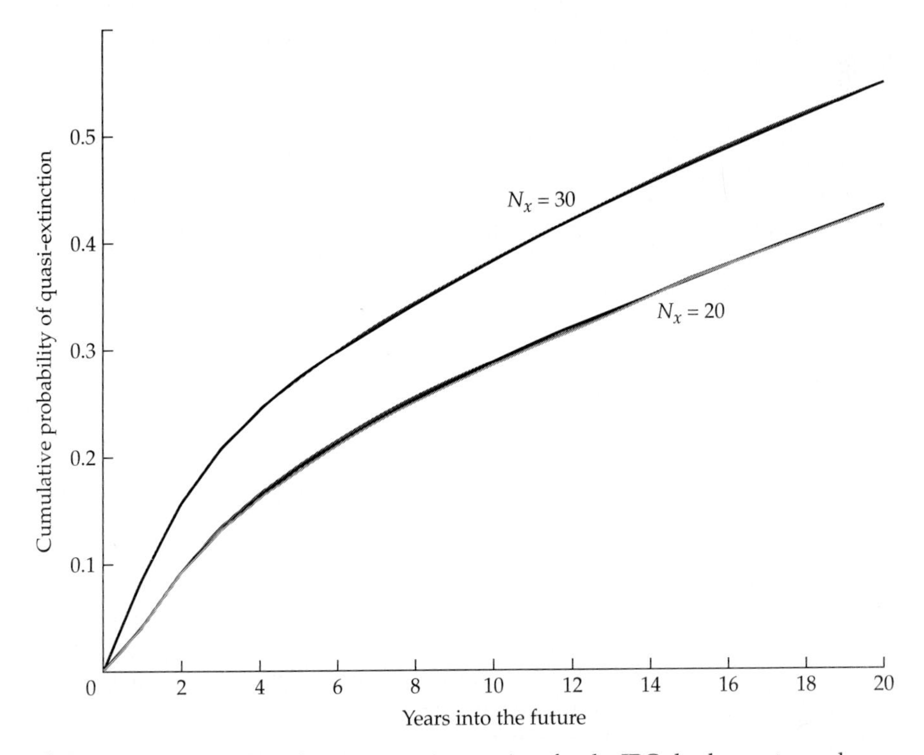

Figure 4.5 Probability of quasi-extinction vs. time for the JRC checkerspot population predicted by the simulation model in Box 4.4. Three runs of the program, each with 50,000 replicate trajectories, are shown for each of two quasi-extinction thresholds.

quasi-extinction threshold of 30 than a threshold of 20. Even with the lower threshold, the probability of extinction is rather high (> 0.4) at only 20 years. To see how including negative density dependence affected the outcome of our analysis, we can reanalyze the data from the JRC population using the methods in Chapter 3, which assume, inappropriately in this case, that the population growth rate is density-independent. This yields estimates for the parameters μ and σ^2 of 0.001673 and 1.4459, respectively. (Note that these same values are obtained from the least-squares estimates of r and σ^2 for the density-independent model in Table 4.2, after correcting the estimate of σ^2 for bias, using Equation 4.10.) Substituting these values into the CDF for the density-independent model (Equation 3.4) yields a predicted probability of 0.772 that the population will decline from 94 to 20 individuals in 20 years. Thus, including negative density dependence results in a predicted extinction probability at 20 years that is only about half the value predicted by a density-independent model. One reason for this difference is that the density-independent model fails to account for the increase in the average population growth rate as the population becomes very small, as we discussed in Chapter 2. Also notice that, by failing to account for the fact that some of the variation among the $\log(N_{t+1}/N_t)$ values is due to density dependence, we wind up with a higher estimate for the environmental variance (1.4559 for the density-independent model versus 1.1151 for the Ricker model), and, as we've seen repeatedly, higher variability increases extinction risk.

That being said, the JRC population actually went extinct in 1996, 10 years after the last census listed in Table 4.1 (McGarrahan 1997; fortunately, other, much larger populations of the subspecies are still extant). Of course, extinction of a single population is a unique event, whereas the probability of extinction reflects the outcomes of many possible realizations of population growth. Rather than ask whether an estimated probability of extinction correctly predicts the observed fate of a single population, it is more appropriate to ask whether the *relative* probabilities of extinction estimated for two or more populations correctly predict their *relative* extinction risk.

Consider the JRH Bay checkerspot population; census counts from this second population at Stanford University's Jasper Ridge Preserve are also listed in Table 4.1. The JRH population went extinct earlier than did the JRC population. As an exercise, readers should verify, using the same likelihood and Information Criterion approach we employed earlier, that the density-independent model provides the best fit to the counts from the JRH population. The estimates of μ and σ^2 for the JRH population are –0.05224 and 0.8410, respectively. Based on the density-independent model, these parameter estimates yield a predicted probability of 0.916 that the population would decline from 18 to 10 individuals in 20 years. This risk is much higher than the quasi-extinction probability predicted for the JRC population by the best-fit Ricker model, even though for the JRC population we used a higher threshold of 20 individuals. Thus the best estimates of the extinction probabilities successfully predict the order of extinction for the two populations. We should not make too much of this success, as with only two populations, it could easily have occurred by chance. Nevertheless, these results illustrate the value of quantitative estimates of extinction risk for comparing the relative viabilities of multiple populations.

Finally, we mention that individuals from both the JRC and JRH populations were periodically removed for studies performed by Paul Ehrlich and his associates at Stanford University. In our analysis, we have not explicitly considered the effects of this removal, which may have altered both the mean and variance of the population growth rate. For an assessment of how this removal may have affected population viability, see Harrison et al. 1991.

One important concern that we have ignored throughout this section is uncertainty in our parameter estimates. In Chapter 3 we suggested that extinction time predictions should take into account uncertainty in estimates of μ and σ^2 by generating a range of extinction predictions across the confidence intervals for these parameters. Although in this section we have only conducted simulations to predict the extinction time CDF for the best-fit parameter values, attempts to incorporate parameter uncertainty should be undertaken with density-dependent models as well. This is more time-consuming when each extinction probability must be generated by many thousands of simulation runs, and this entire process must be repeated many times for different parameter combinations.[8] Nevertheless, the increased speed of desktop computers makes this approach feasible for most PVA practitioners.

Positive Density Dependence, or Allee Effects

As we discussed in Chapter 2, Allee effects may be very important to include in PVAs, because they can greatly increase the extinction risk of small populations. Allee effects are well-documented at the individual level (e.g., Birkhead 1977, Widén 1993), and at some point become a logical certainty for all but asexual species. However, as we also discussed in Chapter 2, in spite of some excellent efforts to document positive density dependence at low population sizes, evidence of population-level Allee effects has remained extremely weak. These results leave us in something of a quandary as to

[8]Moreover, estimates of parameters in density-dependent models, such as r, K, and θ in Equation 4.2, will not in general be independent, as were estimates of μ and σ^2 for the density-independent model (Equation 3.1; Dennis et al. 1991). Thus we cannot simply draw values for each parameter independently. Instead, the most generally applicable approach to placing confidence limits on an extinction probability predicted by a density-dependent count-based model will be to conduct a nonparametric bootstrap. Briefly, sample with replacement q pairs of adjacent counts, N_i and N_{i+1}, from the data set, compute log population growth rates and estimate values for all model parameters (e.g., using least-squares regression as in the checkerspot example), and then calculate by simulation the probability of extinction. Now repeat this entire procedure many times. The limits of the central 95% of the resulting values provide a 95% confidence interval for the probability of extinction. Bootstrap confidence intervals may also need to be corrected for bias (see Dixon 2001 and Ellner and Fieberg 2002).

whether we should worry about including Allee effects in count-based PVAs. Information on how Allee effects influence the reproductive success or survival of individuals can be more explicitly incorporated into the demographic PVA models covered in Chapters 6 through 9 than in the count-based models described here. If census data do show evidence of Allee effects, we can include them in count-based PVAs in two ways. First, as we discussed in Chapter 2, if we can estimate a threshold population size below which Allee effects become strong enough to virtually guarantee population extinction, we can simply set the quasi-extinction threshold at or above the population size at which Allee effects become important. Above the threshold, we would not include positive density dependence in the model, although we might still include negative density dependence as described earlier. The second approach is to explicitly include Allee effects in the population model on which the PVA is based. A number of different models with Allee effects have been proposed, both in continuous time (Dennis 1989, Lewis and Kareiva 1993, Lande et al. 1994, Courchamp et al. 1999) and discrete time (Burgman et al. 1993, Stephan and Wissel 1994, Myers et al. 1995, Veit and Lewis 1996, McCarthy 1997, Amarasekare 1998a,b) formulations. As far as we know, none of these models have yet been applied to any rare species. Because discrete time models lend themselves most easily to census data, we briefly describe one discrete-time model with an Allee effect generated by mate-finding problems (Burgman et al. 1993, McCarthy 1997), and we use it to look for evidence of an Allee effect in the Bay checkerspot data (Table 4.1).

Let us assume that the larger a population becomes, the greater the chance that each individual will be able to find a mate. If so, then an appropriate function to describe the number of potential offspring produced per individual, O_t, in relation to population size, N_t, is $O_t = e^r N_t / (A+N_t)$, where A is a parameter controlling the population size at which Allee effects are felt. Justification for using this functional form comes from explicit consideration of the processes by which individuals search for mates (Dennis 1989, Veit and Lewis 1996, McCarthy 1997). When N_t is close to zero, the number of potential offspring per individual will be close to zero, because most individuals will fail to find a mate. But as N_t becomes large relative to A (so that $A + N_t \approx N_t$), and all individuals are able to find mates, the number of potential offspring per individual approaches a maximum of e^r. The parameter A represents the population density at which potential per-capita reproduction is one-half its maximum value. Thus A determines the range of sizes over which the population will "feel" the Allee effect.

We have used the term *potential offspring* above because negative density dependence will typically reduce the actual number of offspring each mated individual can produce, or the survival of those offspring, at high density. Assume that the fraction of the potential per-capita reproduction that is actually achieved is $e^{-\beta N_t}$, where the parameter β reflects the strength of negative density dependence (that is, high values of β imply that actual reproduction falls off steeply as population size increases). Multiplying population size in year t by the number of potential offspring per individual and by the fraction of potential reproduction that is actually achieved yields the population model[9]

$$N_{t+1} = O_t e^{-\beta N_t} N_t = \frac{N_t^2}{A+N_t} e^{r-\beta N_t} \tag{4.11}$$

We can add environmentally driven deviations to the population growth rate in the same way as in Equation 4.2 to yield a stochastic model

$$N_{t+1} = \frac{N_t^2}{A+N_t} e^{r-\beta N_t + \varepsilon_t} \tag{4.12}$$

where ε_t is the environmental perturbation in year t.

Equation 4.11 illustrates an important feature of models with Allee effects. Because reproduction falls as population size declines, the growth rate of small populations may be insufficient to allow them to avoid extinction (Figure 4.6). In a deterministic model, populations above the so-called Allee threshold will increase (at least initially), whereas populations below the threshold will always go extinct. If we add environmental stochasticity as in Equation 4.12, some populations below the threshold may be pushed above it by serendipitously favorable environmental conditions, and so avoid extinction for a time. On the other hand, environmental variability can push populations initially above the threshold below it, greatly increasing their extinction risk.

Just as we looked for evidence of negative density dependence by fitting density-independent and density-dependent models to census data, we can also compare the fits of a model with an Allee effect, such as Equation 4.12, and one without it (cf. Myers et al. 1995). For example, it is easy to modify the SAS code in Box 4.2 to fit Equation 4.12 instead of the theta logistic model to the census data from the JRC Bay checkerspot population.[10] One

---

[9]Equations 4.11 and 4.12 arise most naturally as a model for an organism that lives only one year or less, so that each individual contributes to next year's population only through reproduction, not survival. The model can easily be modified to allow overlapping generations, but at the cost of additional parameters that must be estimated.

[10]Specifically, change the call to the `nlin` procedure as follows:

```
proc nlin data=jrc best=1;
   title 'JRC Population - Allee effect model';
   parameters r=0.1 to 1.5 by 0.1 b=0.1 to 1 by 0.1 A=10 to 100 by 10;
   bounds r>=0, b>=0, A>=0;
   model yt=log(nt)-log(A+nt)+r-b*nt;
   der.r=1;
   der.b=-nt;
   der.A=-1/(A+nt);
   output out=fits p=ypred residual=yres;
run;
```

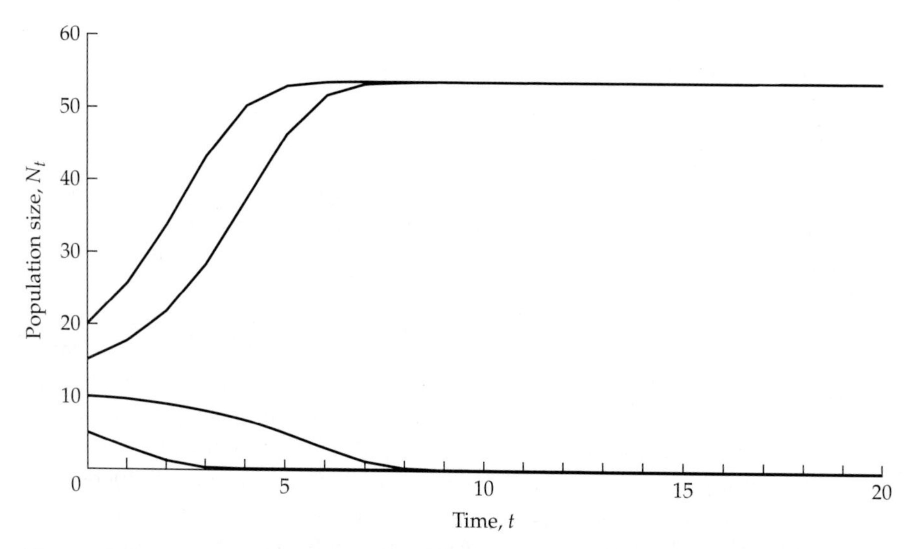

Figure 4.6 In a deterministic model with an Allee effect (Equation 4.11), populations starting below the Allee threshold (which occurs at a population size of approximately 12 in this graph) are doomed to extinction (parameter values: $r = 2$, $A = 50$, $\beta = 0.025$).

could then compute the AIC_c value for the Allee effect model using Equations 4.7 and 4.8, and compare it to the AIC_c value for the Ricker model (which, of course, lacks an Allee effect).[11] For the JRC Bay checkerspot data in Table 4.1, fitting the Allee effect model via least squares regression identifies a value of 0 as the maximum likelihood estimate of the parameter A, assuming that we constrain the estimate to be nonnegative[12] (try it!). When $A = 0$, the Allee effect model of Equation 4.12 reduces to the Ricker model (with $\beta = r/K$). Thus even without computing the likelihood or AIC_c values, we can see that the checkerspot data show no evidence of an Allee effect. Parsimony thus argues for the use of a model with negative density dependence only, as we have done. If you do find good evidence for Allee effects in a data set, the most practical way to derive extinction time estimates is to follow the same model fitting and simulation approach we detailed for negatively density-dependent dynamics in the preceding section.

[11] An advantage of using Information Criterion statistics to compare models is that they do not assume (as do likelihood ratio tests, for example) that the models are nested (i.e., that one model can be converted to another by setting one or more parameters equal to 0 or 1; for example, the theta logistic model (Equation 4.2) and the Allee effect model (Equation 4.12) are *not* nested). For details, see Burnham and Anderson 1998.

[12] As we did in the SAS code listed in Footnote 10.

Combined Effects of Demographic and Environmental Stochasticity

As we discussed in Chapter 2, very small populations may go extinct due to random variation among individuals in their survival or in the number or sex ratio of their offspring. By omitting demographic stochasticity in all we have discussed until now, we have ignored a factor that may often deliver the coup-de-grâce to populations that have been depressed to low levels by environmental stochasticity.

That being said, it not possible to estimate the magnitude of demographic stochasticity using count data alone, as demographic stochasticity is intimately connected to variation in the fates of individual organisms, not just the total number of individuals in a population. In essence, to estimate the amount of variability that demographic stochasticity introduces into the population growth rate, we need the type of data on survival and reproduction of individuals that is typically used to construct the projection matrix models we discuss in Chapters 6 through 9. However, when such individual-level data are available from a population that has also been the focus of a long-term census, the two types of data can be used in conjunction to incorporate both demographic and environmental stochasticity into the count-based PVA models we have described in this and the preceding chapter. Consequently, we give an overview here of how such a tandem approach might work in those rare cases in which census counts and individual-level information are available for the same endangered population. It is indicative of how rarely this situation will arise that the only published examples of this approach rely on intensive studies of small, isolated populations of otherwise common (and therefore not endangered) species. We cite most of these papers in describing how to approach this problem, and if you want to use this method you should consult them for more details.

Engen et al. (1998) proposed a method to estimate the separate contributions of demographic and environmental stochasticity to the variance in the population growth rate (also see Saether et al. 1998a,b, 2000a,b, Tufto et al. 2000, Saether and Engen 2002). We must first collect data on a set of marked individuals over several years, recording the number of offspring each marked individual produces each year and whether those offspring, as well as the marked parent, survive to the following year (for methods to conduct such a demographic study, see Chapter 6). The procedure of Engen and colleagues begins by calculating the contribution each marked individual in year t makes to the population in year $t + 1$. This contribution equals the number of that individual's offspring produced in year t that survive to year $t + 1$ plus 1 if the focal individual itself survives to the next year. For example, the contribution of an individual producing 5 surviving offspring would be 6 if that individual survives and 5 if not. Let us call the contribution of individual i in year t C_{it}, and the mean of those contributions

$$\bar{C}_t = \left(\sum_{i=1}^{m_t} C_{it} \right) / m_t$$

where m_t is the number of individuals whose contributions in year t are known. An estimate of the demographic variance in year t is the sample variance of the C_{it}'s, given by

$$V_d(t) = \frac{1}{m_t - 1} \sum_{i=1}^{m_t} \left(C_{it} - \bar{C}_t \right)^2 \tag{4.13}$$

Notice that Equation 4.13 is measuring *variation among individuals in the same year*, which is the driving force behind demographic stochasticity.

Data from each year in which the survival and reproduction of marked individuals were measured yield separate estimates of the demographic variance when they are substituted into Equation 4.13. These separate estimates should then be regressed against population size in each year, to see if the demographic variance is itself density dependent. (For example, high density may depress reproduction or survival of all individuals, reducing variation among them; Saether et al. 1998a.) If the regression is significant, the regression equation provides a way to update the demographic variance given the population size each year. If the regression is not significant, the average of the separate estimates of $V_d(t)$ across years may be used as an overall estimate of the demographic variance.[13]

Once we have calculated demographic variance estimates for each of n years using Equation 4.13, we can use the census data to obtain n estimates of the environmental variance from the expression

$$\sigma^2(t) = \left[\frac{N_{t+1}}{N_t} - f(N_t) \right]^2 - \frac{V_d(t)}{N_t} \tag{4.14}$$

where $f(N_t)$ is the (possibly density-dependent) population growth rate in year t predicted by the best-fit population model $N_{t+1} = N_t f(N_t)$. For example, if we were using the theta-logistic model (Equation 4.2), $f(N_t)$ would be $\exp[r\,(1 - (N_t/K)^\theta]$ with the values of r, K, and θ determined by nonlinear regression, as we did above. The term in square brackets in Equation 4.14 is the squared deviation between the observed population growth rate from year t to year $t + 1$ and the growth rate predicted by the model. This deviation is due to both environmental and demographic stochasticity. Subtracting from it our estimate of the contribution of demographic stochasticity to the variance in the population growth rate ($V_d(t)/N_t$) yields an estimate of

[13]Saether et al. 1998a, 2000b, and Saether and Engen 2002 advocate weighting the yearly estimates of $V_d(t)$ by m_t when computing the overall demographic variance. See White 2000 for an alternative weighting.

the variance in population growth due to environmental stochasticity alone. As with the $V_d(t)$ estimates, the $\sigma^2(t)$ estimates should be regressed against N_t to check for density dependence in the environmental variance. (For example, environmental perturbations to the population growth rate may be larger when population size is large and, consequently, refuges against extreme environmental conditions are scarce.) If they are density-independent, the separate estimates of $\sigma^2(t)$ should be averaged to obtain an overall estimate of the environmental variance; if density-dependent, the regression equation should be used to generate an appropriate value for $\sigma^2(t)$ each year of a computer simulation. Note that the estimate of σ^2 we obtain from Equation 4.14 is not identical to that used in Chapter 3 and earlier in this chapter. Those earlier calculations assumed that the observed variation in the log population growth rate was due solely to environmental stochasticity, even though in reality it always included some (hopefully very small) variability due to demographic stochasticity. Equation 4.14 removes this demographic variability, yielding an estimate of σ^2 that is a smaller, more accurate representation of true environmental stochasticity.

Notice that in Equation 4.14, the contribution of demographic stochasticity to variation in the population growth rate is the variation in an individual's contribution to population growth divided by the size of the population, $V_d(t)/N_t$, which becomes smaller and smaller as population size N_t increases. This makes sense, since as populations become larger, the variances in the contributions made by different individuals will tend to cancel each other out. In other words, Equation 4.14 predicts that the effects of demographic stochasticity will be important only for small populations, just as we would expect from our discussion of demographic stochasticity in Chapter 2 (also see Gabriel and Burger 1992 and Lande 1993).

If both demographic and count data are available so that V_d and σ^2 can be estimated separately, computer simulations can be used to predict extinction risk under both environmental and demographic stochasticity. For completeness, let us assume that linear regressions of $V_d(t)$ and $\sigma^2(t)$ against N_t are both statistically significant, with the regression equations given by $V_d(t) = a_d N_t + b_d$ and $\sigma^2(t) = a_e N_t + b_e$. We will also assume that the density-dependent population growth rate $\lambda(N_t)$ is lognormally distributed, with a mean given by the theta logistic model (that is, $M\{\lambda_t\} = \exp\{r[1 - (N_t/K)^\theta]\}$, where $M\{\}$ denotes the mean) and a variance given by $V\{\lambda_t\} = \sigma^2(t) + V_d(t)/N_t$, which is the sum of components due to environmental and demographic stochasticity. If $\lambda(N_t)$ is lognormally distributed with mean $M\{\lambda_t\}$ and variance $V\{\lambda_t\}$, then we can create random values for $\lambda(N_t)$ by computing values of $\exp\{X_t\}$, where X_t is normally distributed with mean and variance given by $M\{X_t\} = \log M\{\lambda_t\} - \frac{1}{2}V\{X_t\}$ and $V\{X_t\} = \log\left[1 + V\{\lambda_t\}/\left(M\{\lambda_t\}\right)^2\right]$ (see Equation 8.9). The MATLAB program `demstoch` listed in Box 4.5 uses the theta logistic model to simulate density-dependent population growth in the face of both demographic and

BOX 4.5: *A MATLAB program to simulate growth of a density-dependent population with both environmental and demographic stochasticity. It would be straightforward to hybridize this code with the program "theta_logistic" in Box 4.4 to compute the probability of extinction under both types of stochasticity.*

```
% PROGRAM demstoch;
%    Simulates the theta-logistic model with environmental and
%    demographic stochasticity.  Assumes the population growth
%    rate at time t is lognormally distributed with mean
%    exp(r(1-(Nt/K)^theta)) and variance Vd/N+sigma^2, where
%    both the demographic variance Vd and the environmental
%    variance sigma^2 may themselves be density-dependent -
%    i.e., Vd=ad*Nt+bd and sigma^2=ae*Nt+be

%**************** SIMULATION PARAMETERS **************
NumPops=10; % number of trajectories to simulate
Nc=10;      % initial population density
r=0.1;      % finite rate of increase
K=15;       % carrying capacity
theta=1;    % pattern of density dependence
ad=0;       % slope of demographic variance vs. Nt
bd=1;       % intercept of demographic variance vs. Nt
ae=0;       % slope of environmental variance vs. Nt
be=0.1;      % intercept of environmental variance vs. Nt
tmax=50;    % duration of simulation
%****************************************************

randn('state',sum(100*clock)); % seed the random number generator
N=Nc*ones(1,NumPops);          % start all trajectories at Nc

Ns=[N];                         % store initial densities
for t=1:tmax                    % for each future time,
    Mlam=exp(r*(1-(N/K).^theta)); % compute means of lambdas,
    Vlam=(ad*N+bd)./N + ae*N+be; % variances of lambdas,
    CV2=Vlam./(Mlam.^2);        % squared coefficients of
                                %    variation of lambdas,
    Mx=log(Mlam)-0.5*log(CV2+1); % means of Xt's, and
    SDx=sqrt(log(CV2+1));       % standard deviations of Xt's,
    N=N.*exp(Mx+SDx.*randn(1,NumPops));% then project populations
                                       % one year ahead using
                                       % lambda=exp(Xt),
    Ns=[Ns; N];                        % and store the results.
end;
```

BOX 4.5 *(continued)*

```
% For all trajectories, plot Nt vs t, with a log scale for the y
% axis
semilogy(Ns)
xlabel('Time (years)');
ylabel('Population density');
axis([1 tmax .01 100]);
```

environmental stochasticity, the magnitudes of which are determined by the simulation parameters a_d, b_d, a_e, and b_e estimated as we have just described. The variance in the population growth rate is computed in the line:

```
Vlam=(ad*N+bd)./N + ae*N+be;
```

At first glance, it may appear as though partitioning the observed variation in population growth into components due to demographic and environmental stochasticity was unnecessary, since the two parts of `Vlam` simply reconstitute the observed total variability with which we started. However, the importance of making this breakdown comes as population size drops. Although we have previously assumed the total variance in the population growth rate would remain fixed at the single observed value, the fact that the first part of `Vlam` has `N` in the denominator ensures that the total variance will increase due to the growing importance of demographic stochasticity at small population size (and conversely shrink with increasing numbers).

We can use the program `demstoch` to examine how environmental and demographic stochasticity combine to determine extinction risk (Figure 4.7). In Figure 4.7A, individuals vary little in their contributions to population growth (i.e., the demographic variance V_d is small), so demographic stochasticity is weak. Even so, the predominating effect of environmental variation can still cause some of the replicate trajectories (all of which start at a population density of 10) to fall low enough that we would probably label them "quasi-extinct" (note the log scale of the y-axis). However, if we ratchet up the variation in individual contributions to population growth, as in Figure 4.7B, we see qualitatively different behavior; any trajectories that happen to be pushed below a certain level by environmental variation embark on a precipitous, runaway decline that looks nearly deterministic in its predictability. The reason for this virtually inescapable slide to extinction is that, as population density falls, the variance in the population growth rate increases due to demographic stochasticity, and higher variance then leads to a lower geometric mean population growth rate (see Chapter 2), which causes the population to decline still more, further increasing the variance due to demographic stochasticity and further reducing the geometric mean, and so on, in an

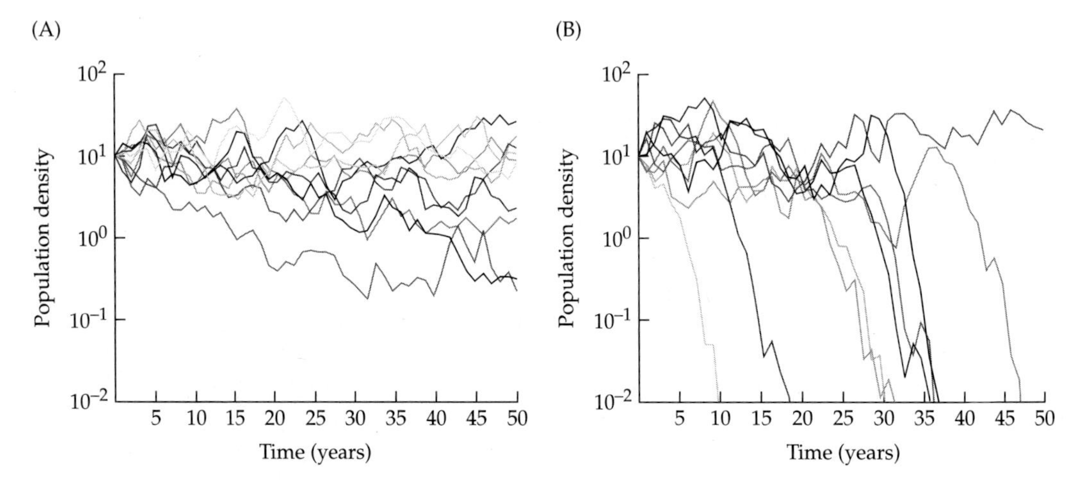

Figure 4.7 Population dynamics in the face of both environmental and demographic stochasticity, as simulated by the program `demstoch` (Box 4.5). Both panels show ten population trajectories simulated with the theta logistic model with parameters $(r, K, \theta) = (0.1, 15, 1)$ and a (density-independent) environmental variance of 0.1 (i.e., $a_e = 0$ and $b_e = 0.1$). In panel B (in which $a_d = 0$ and $b_d = 1$, so that $V_d = 1$ for all t), the (density-independent) demographic variance is ten times higher than in panel A (in which $a_d = 0$, $b_d = 0.1$, $V_d = 0.1$). When contributions to population growth vary greatly among individuals, populations that fall to low numbers are doomed to extinction (panel B).

inevitable downward spiral. Thus environmental and demographic stochasticity conspire in a "one-two punch," the former acting to push populations below some fuzzy threshold at which the later takes fatal hold. The height of this threshold is proportional to the demographic variance V_d (see Lande 1998, 2002). It is precisely this threshold-like behavior of demographic stochasticity that argues for the use of a quasi-extinction threshold, rather than measuring the risk of outright extinction, when lack of data on the variation in individual contributions (i.e., the C_{it}'s) forces us to assess extinction risk with a model that includes environmental stochasticity only. The challenge, of course, is to decide what that threshold should be without an estimate of the demographic variance. In such cases, the general guidelines for setting the quasi-extinction threshold that we gave in Chapter 2 are probably about the best we can do.

Although the method we have described in this section lets us estimate the effects of demographic stochasticity and to extrapolate its effects to low population sizes, one caveat needs to be mentioned about this procedure. It makes the implicit assumption that all variation among individuals in their contributions to population growth in a single year is, in fact, due to random chance, rather than to fixed differences between individuals related to their size, age, genotype, habitat, and so on. In other words, there is an assumption that all individuals are identical to each other, so that they have the same *expected* contribution to population growth in any one year. As we discussed in Chapter 2, fixed differences among individuals are most likely to inflate our estimate of the importance of demographic stochasticity, because variation due to fixed differences, unlike that caused by demographic stochasticity, will not increase as population size declines (Kendall and Fox 2002, Fox and Kendall 2002). Thus we should keep in mind that if a substantial amount of the variation in the C_{it}'s in Equation 4.13 is due to fixed inter-individual differences, we may overestimate the magnitude of demographic stochasticity at low population sizes, and thus overestimate the risk of extinction. In the terminology of Chapter 3, falsely attributing variation caused by fixed differences to demographic stochasticity will tend to make our viability assessments pessimistic.

Aside from this concern, this method does allow the simultaneous treatment of both environmental and demographic stochasticity in a count-based PVA. In Equation 4.13, the contribution each individual makes to the population in the next year accounts for both the surviving offspring it produces and its own survival. Thus two major factors contributing to demographic stochasticity have been accounted for (although variation in offspring sex ratio has not; for models that include sex ratio variation but no environmental stochasticity, see Legendre et al. 1999). Other approaches focus on either survival or reproduction. For example, Kendall (1998) proposed a maximum likelihood method to disentangle effects of demographic and environmental stochasticity when data on survival, but not reproduction, of individuals are available. However, as Kendall's method does not involve a combination of demographic and count data, we defer discussion of it until Chapter 8.

Environmental Autocorrelation

Up to now, we have assumed that the impacts of environmental factors on population growth are independent from one year to the next. For example, in the computer simulation in Box 4.4, we drew a new random number for the environmental deviation, ε_t, in each year, without regard to the value in the previous year. However, temporal environmental autocorrelation may cause the deviations in the population growth rate to be correlated among years, although this is not necessarily the case. For example, some correlated environmental conditions may have only a weak or insignificant effect on birth and death rates in a population, or changes that are caused by several environmental factors or that affect different population processes may cancel each other out. On the other hand, if environmental events in one year lead to long lasting changes in the age, size, or stage structure of a population, uncorrelated environmental events can still lead to autocorrelation in population growth rates. For example, bad weather conditions in one year could lead to depressed population growth rates for several years thereafter because of a reduction in the fraction of the population that is of prime reproductive age. Untangling autocorrelations in population growth rates that are due to direct autocorrelation in the environment versus interactions between the environment and the population's composition is not possible with census data alone, and in this chapter we will refer to both effects as environmental autocorrelation. In either case, if population growth is correlated in time, the risk of extinction may be poorly predicted by PVAs that assume environmental deviations are independent. In this section, we briefly review the impact of environmental autocorrelation on population viability. In addition, we summarize how to test for autocorrelation in census data, and how to incorporate it in viability assessments if it occurs.

Hypothetically, environmental factors could be positively or negatively correlated in time. Positive correlations occur when adjacent years in a sequence tend to be more similar than would be expected in a completely random sequence (for example, wet years tend to follow wet years, and dry years dry). Negative correlations occur if adjacent years differ more than would be expected by chance (dry years follow wet, and vice versa). In reality, positive correlations in environmental variables such as temperature are commonly encountered (Steele 1985, Halley 1996), but convincing examples of negative correlations are hard to find,[14] so we will only discuss positive correlations below.

In a density-independent model, positive autocorrelation elevates extinction risk relative to population dynamics with the same mean and variance in growth but no correlation (Figure 4.8). The reason is simple: strong positive correlation means that strings of bad years will be less likely to be broken up by the occasional good year that could pull the population back from the brink of extinction. A similar effect is seen in the ceiling model (Equation 4.1). Foley (1994) suggested that when the environmental effects on the population growth rate are correlated, the "effective" environmental variance in the log population growth rate is

$$[(1+\rho)/(1-\rho)]\,\sigma^2 \qquad (4.15)$$

where ρ (rho) is the correlation coefficient between the environmental effects on population growth in adjacent years and σ^2 is the environmental variance estimated as in Chapter 3. (See the following discussion for a way to estimate ρ.) Note that when ρ is positive, the effective environmental variance is greater than σ^2. To calculate the mean time to extinction in a correlated environment, Foley recommended using Expression 4.15 in place of σ^2 in Equation 4.4. We have already seen that a high degree of environmen-

---

[14]The grizzly bear data in Chapter 3 provide one possible example of (weak) negative autocorrelation.

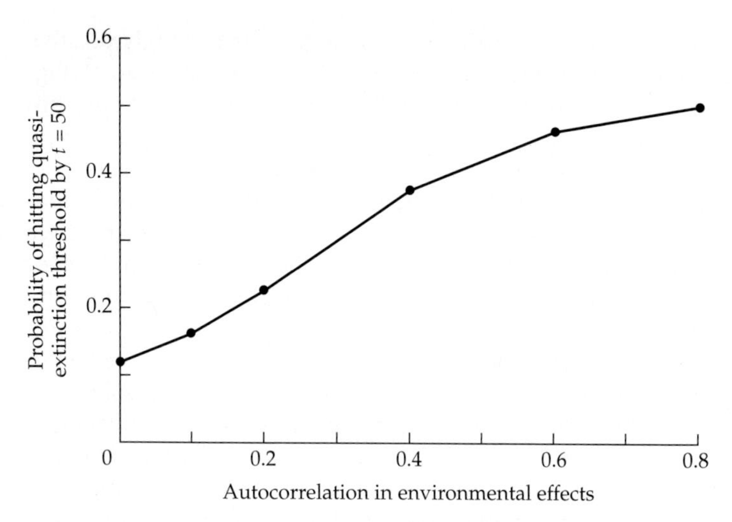

Figure 4.8 Positive autocorrelation in environmental effects increases extinction risk in a density-independent population. To make autocorrelation easier to add, we've modified the familiar density-independent model (Equation 3.1). The model is $N_{t+1} = e^{\mu+\varepsilon_t} N_t$, where μ and σ^2 are as in Chapter 3, and the environmental effect is $\varepsilon_t = \rho\varepsilon_{t-1} + \sqrt{\sigma^2}\sqrt{1-\rho^2}\, z_t$, where ρ is the correlation between environmental effects in adjacent years and z_t is a random number drawn from a normal distribution with a mean of zero and a variance of 1 (see Equation 4.16 for an explanation of ε_t). Starting population size and extinction threshold were 10 and 5, respectively. Ten-thousand replicate trajectories were computed for each level of environmental autocorrelation. Parameter values: $\mu = \sigma^2 = 0.01$.

tal variation reduces the mean time to extinction (Figure 4.2B), so by increasing the effective environmental variation, autocorrelation increases extinction risk in the ceiling model just as it does in the density-independent model (also see Johst and Wissel 1997).

The story is more complicated for fully density-dependent models such as the Ricker model (Ripa and Lundberg 1996, Petchey et al. 1997, Heino 1998, Ripa and Heino 1999). In that case, positive autocorrelation can either increase or decrease extinction risk, depending on the values of the other parameters in the model and the range over which the strength of autocorrelation is varied. A key determinant of how temporal correlation will affect extinction risk is whether the recruitment curve—that is, a curve plotting N_{t+1} against N_t—has a positive or negative slope at the equilibrium population size (Petchey et al. 1997, Ripa and Heino 1999). If the slope is positive, then positive autocorrelation will increase extinction risk; if the slope is negative, increasingly positive autocorrelation can decrease extinction risk over some range of values[15] (Figure 4.9). Two arguments can help to explain this divergent effect of environmental correlation. First, if the slope is positive, positive autocorrelation increases the variance in population size near the equilibrium, whereas the opposite is true if the slope is negative (Roughgarden 1975, Ripa and Heino 1999), and high variance increases extinction risk. This is the same effect we saw with the density-independent and ceiling models. The second argument involves the meaning of the negative slope of the recruitment curve in Figure 4.9B. The negative slope at the equilibrium indicates that the intrinsic dynamics of the population are "overcompensatory"; that is, the population will tend to repeatedly overshoot and then undershoot the equilibrium. Thus following a good year that pushes the population above the equilibrium, density dependence will push the population below the equilibrium in the next year (dramatically so if the slope is strongly negative). A second good year made more likely by positive environmental autocorrelation will tend to resist the population's inherent tendency to crash, thus keeping it close to the equilibrium and farther from the extinction threshold. So, when a population's own dynamics tend to cause "boom and bust" cycles, positive autocorrelation may help to stabilize dynamics and reduce extinction risk.

In summary, whether temporal autocorrelation will increase or decrease a population's risk of extinction depends on whether density dependence is operating, and if so, on the "shape" of density dependence near the population equilibrium (Figure 4.9). Thus we cannot make blanket statements about the impact of temporal environmental autocorrelation on population viability, and we must be careful to choose a model (or models) that incorporates density dependence in an appropriate way when including environmental correlations in a PVA.

How do we test for environmental correlations in a set of population counts? We advocate the following approach, which is slightly more elaborate

---

[15]The effect of positive autocorrelation on extinction risk is sensitive to the way environmental variation is entered in the model. In Figure 4.8, environmental variation is incorporated by multiplying the new population size predicted by the deterministic part of the model (i.e., $N_{t+1} = N_t \exp\{r(1-N_t / K)\}$) by a lognormal random deviate (i.e., $\exp\{\varepsilon_t\}$). In contrast, Ripa and Lundberg (1996) entered environmental variation into the Ricker model either by letting K vary randomly (that is, the strength of density dependence varies over time) or by introducing additive noise that is proportional to population size. In both cases, they found that, rather than decreasing and then increasing the extinction risk (as in Figure 4.8), increasingly positive environmental autocorrelation strictly decreased the probability of extinction. Unfortunately, it may be very difficult to determine which stochastic population model best fits a set of census data, and when exploring the consequences of environmental autocorrelation, it may be wise to use multiple models to see if their predictions agree (see Chapter 12).

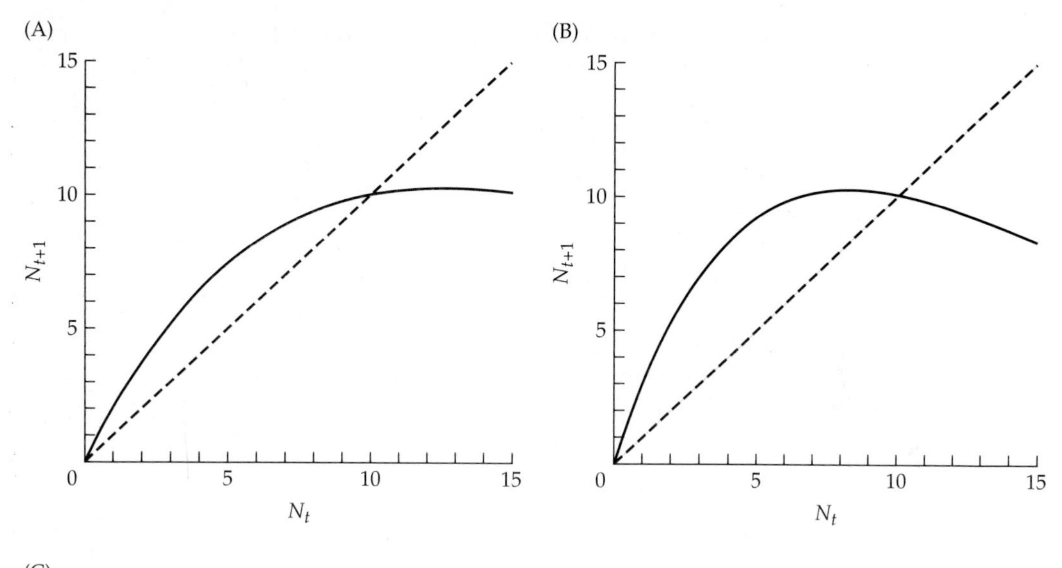

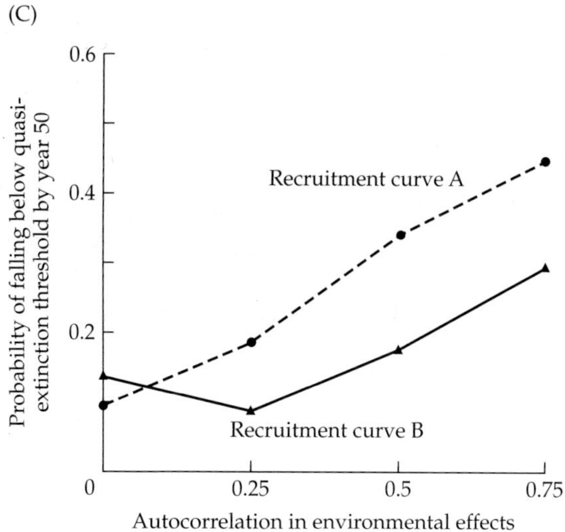

Figure 4.9 In a model with negative density dependence, temporal correlation in the environmental effects can increase or decrease extinction risk, depending in part on the slope of the recruitment curve (N_{t+1} vs. N_t) at the equilibrium. In graphs A and B, the solid line is the recruitment curve for the Ricker model — $N_{t+1} = N_t \exp\{r(1 - N_t/K)\}$ — and the dotted line connects all points where $N_{t+1} = N_t$; the equilibrium population size K lies at the intersection of the two lines (here, $K = 10$). When $r < 1$ as in graph A (where $r = 0.8$), the slope of the recruitment curve at the equilibrium is positive; when $r > 1$ as in graph B (where $r = 1.2$), the slope of the recruitment curve at the equilibrium is negative. Graph C shows the result of simulating the stochastic Ricker model $N_{t+1} = N_t \exp\{r(1 - N_t/K) + \varepsilon_t\}$ with the (autocorrelated) environmental effect ε_t defined in Equation 4.16 ($\sigma^2 = 0.05$; initial population size is K). Positive environmental autocorrelation increases extinction risk for recruitment curve A, but initially decreases it for recruitment curve B.

than that discussed at the end of Chapter 3, because of the need to incorporate density-dependent effects into the test.

1. Calculate the observed log population growth rates—that is, $\log(N_{t+1}/N_t)$—using all of the population censuses. (As usual, there will be one less log growth rate than the number of censuses; for example, see the third column of Table 4.1.)
2. Fit density-independent and density-dependent models to the data and use AIC_c criteria to choose the best model to describe population growth.
3. Using the best-supported model, calculate the predicted log population growth rate for each inter-census interval. First, substitute each observed value of N_t into the best-fit population model to calculate its associated predicted value of N_{t+1}. Second, divide the predicted value of N_{t+1} by the observed value of N_t. The log of this ratio is the predicted log population growth rate for year t. Repeat for all values of N_t.
4. Subtract the predicted log growth rates from the observed log growth rates; these differences represent the environmental deviations. If the best model is density independent, then these are simply the differences between each observed log growth rate and μ (see *Using regression diagnostics to test for temporal autocorrelation in the population growth rate* in Chapter 3).
5. Make a spreadsheet with two columns, the first containing the deviations for time intervals 1 through $q-1$, and the second containing deviations for intervals 2 through q (where q is the number of estimates of the population growth rate in the data set; note that both columns will have $q-1$ rows). Finally, calculate the Pearson correlation coefficient between the environmental deviations in successive years (e.g., using the function `corrcoef` in MATLAB or the function CORREL in Excel); this coefficient estimates the strength of "first-order" environmental autocorrelation (i.e., the correlation between adjacent years).

For example, Figure 4.10 shows the environmental deviations for the JRC Bay checkerspot population in year t versus the deviations in the previous year, using the parameter values for the Ricker model in Table 4.2 to calculate the predicted log population growth rates. The calculated correlation coefficient ($r = -0.189$, $n = 26$) does not differ significantly from zero. Thus there is no evidence that the impact of environmental factors on the JRC population is correlated from one year to the next, and we are justified in omitting such correlations from our computer simulation (Box 4.4).

In theory, one could test for correlations between environmental deviations that are separated by more than one year using time series analysis. (See Royama 1992 for an introduction to the application of time series analysis to population ecology.) These methods account for the fact that, if deviations in adjacent years are correlated, the correlation between years t and $t+1$ and between years $t+1$ and $t+2$ will generate an apparent correlation between years t and $t+2$, even if there is no direct correlation between them.

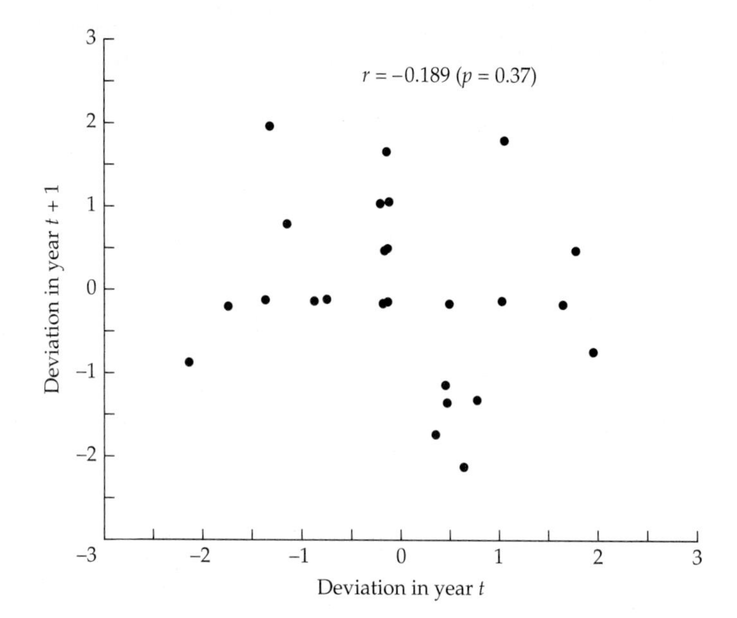

Figure 4.10 Deviations between observed and predicted log population growth rates for the JRC checkerspot population are not significantly correlated between adjacent years.

However, as the sample size decreases with an increase in the number of years between two environmental deviations we wish to compare, testing for long-term environmental correlations will usually require censuses that are longer than those available for most endangered populations.[16]

If we do find a significant correlation between the environmental deviations in successive years, we can incorporate environmental correlation into our viability assessment in the following way. Basically, we modify the computer simulation to make the environmental deviation in one year dependent upon the value of the deviation in the previous year. We must do this in such a way that, despite the fact that adjacent deviations are not independent, the expected variance of the environmental deviations still equals σ^2 as estimated from the data. We can accomplish this by generating the environmental deviations (for example, the ε_t terms in Equation 4.2) using the equation

$$\varepsilon_t = \rho\varepsilon_{t-1} + \sqrt{\sigma^2}\sqrt{1-\rho^2}\,z_t \tag{4.16}$$

where ρ is the correlation coefficient between successive environmental deviations and z_t is a random number drawn from a normal distribution with a mean of zero and a variance of 1 (Foley 1994, Ripa and Lundberg

[16]Halley 1996 has argued that environmental variation may be better represented by "1/f noise," which differs from autoregressive noise and may have different effects on population viability (Halley and Kunin 1999; also see Morales 1999). As it remains unclear whether 1/f noise provides a better description of environmental impacts on population growth, we follow the more traditional autoregressive approach here.

BOX 4.6 *MATLAB code to calculate the probability of quasi-extinction for the Ricker model with temporally autocorrelated environmental effects.*

```
% PROGRAM ricker_corr
%    Uses the stochastic Ricker model
%            N(t+1)=N(t)exp{r(1-N(t)/K)+e(t)}
%    with correlation rho between environmental deviations e(t)
%    in adjacent years to calculate (by simulation) the
%    probability that a population starting at Nc will fall below
%    the quasi-extinction threshold Nq by time tmax

%****************** SIMULATION PARAMETERS ***********************
tmax=50;              % time horizon
NumReps=10000;        % number of replicate populations to simulate
Nc=10;                % current population size
Nx=5;                 % quasi-extinction threshold
r=0.8;                % intrinsic rate of increase
K=10;                 % equilibrium population size
sigma2=0.05;           % total environmental variance
rho=0.1;              % first-order environmental autocorrelation
%****************************************************************

sigma=sqrt(sigma2);
randn('state',sum(100*clock)); % Seed the random number generator
beta=sqrt(1-rho^2); % Beta scales new random numbers so that
                    % total environmental variance=sigma2.
NumExt=0;           % No populations are extinct initially.
for i=1:NumReps,    % For each replicate population,
    N=Nc;           % start at current population size,
    eold=randn;     % generate an initial environmental deviation
    for j=1:tmax,   % then looping over time,
        enew=rho*eold + ...        % make new (auto-correlated)
            sigma*beta*randn;      % environmental deviate and
        N=N*exp(r*(1-N/K)+enew); % use the Ricker model to
                                   % project one year ahead.
         eold=enew; % Save old environmental deviation.
        if N<Nx     % Stop projecting if threshold is hit.
            break;
        end;
    end;
    if N<Nx
        NumExt=NumExt+1; % Store number of extinct populations.
    end;
end;
```

BOX 4.6 *(continued)*

```
NumExt/NumReps    % Compute probability of hitting Nq at or before
                  % tmax
```

1996). Each ε_t is the sum of a term due to correlation with the previous environmental deviation ($\rho\varepsilon_{t\text{-}1}$) and a new random term, scaled by the factor $\sqrt{\sigma^2}\sqrt{1-\rho^2}$ to assure that the variance of a long string of ε_t's generated in this way equals σ^2. Box 4.6 provides MATLAB code that incorporates environmental correlations into the Ricker model. This program was used to generate the results shown in Figure 4.9, and can also be easily modified to simulate correlations for a density-independent model of similar form (see the legend to Figure 4.8).

Catastrophes, Bonanzas, and Other Highly Variable Environmental Effects

Of the factors discussed in this chapter, catastrophes and bonanzas are perhaps the most difficult to include in a PVA in a defensible way, even though catastrophes may also be the greatest cause of concern for many species. The reason for this is that catastrophes and bonanzas are by their very nature infrequent events; consequently, we will usually have even less information about their frequencies and magnitudes than we will about density dependence, environmental correlations, and even demographic stochasticity. In practice, we may be forced to use indirect measures, such as the frequency of extreme climatic conditions, to infer the impact of catastrophes/bonanzas on population growth.

In reality, both catastrophes and bonanzas are likely to exert a range of effects on population growth. That is, not all severe droughts (or favorably wet years) are equally severe (or favorable). If we were omniscient, we could envision incorporating the full range of effects of catastrophes or bonanzas in the following way. In projecting the population forward by one year in a computer simulation, we could first decide whether the next year is to be a "typical" or an "extreme" year, using data on the long-term frequency of extreme environmental conditions (where we would use some criterion to determine the cutoff between "typical" and "extreme," such as whether a transition is an outlier in a regression analysis—see Chapter 3). If the year is to be typical, we would simply draw a population growth rate from the typical distribution of environmental variation. If the year is extreme, we would draw a value from a second distribution of conditions observed in extreme years. The problem is that censuses of threatened and endangered populations are almost always too short in duration to see more than one or two extreme events (if any), so that we typically have too little information to be able to construct a distribution for the severity of bonanzas or catastrophes or to estimate their true frequencies. Given this limitation, we argue that, if we aim to include catastrophes or bonanzas at all, we should use the actual severities observed in the data (or empirical estimates from other populations or related species), rather than incorporating a distribution of severities that will have poor empirical justification. However, unless your data on these rare events is very good, we also suggest that you run multiple models, ranging across reasonable values for the frequency and severity of both catastrophes and bonanzas (based upon your data). This approach will give a better sense of how much these rare events actually matter for extinction risk, and it will guard against results that depend critically upon very uncertain estimates.

Another problem is to decide whether to include catastrophes only, bonanzas only, or both. For example, in the analysis of the Yellowstone grizzly bear population in Chapter 3, only a single transition (the 1983–1984 transition) out of the 38 transitions in the data set was identified as an outlier. This transition represented an unusually large increase in population size, and so we might view it as a bonanza. But what should we do when constructing a simulation to account for extreme events? We could only include bonanzas occurring at a frequency of about once every 38 years in the simulation, because we never actually observed a catastrophe in the census (where a catastrophe would be an unusually *low* outlier identified by the same regression diagnostics used to identify the bonanza). On the other hand, if we assume that *either* bonanzas or catastrophes occur about once every 38 years, we would not be surprised to see only a single bonanza and no catastrophes in a 38-year period, *even if catastrophes actually do occur.* Because there is no clear answer to this puzzle, we advocate trying both possibilities. That is, run one set of simulations in which only events actually observed are included. For example, if the data show one unusually high outlier (i.e., a bonanza) and two unusually low ones (i.e., catastrophes), draw the magnitudes of extreme events from among these three possibilities. Then run a second set of simulations in which bonanzas and catastrophes are equally likely over the long run (e.g., by duplicating the single bonanza in the set of extreme values from which the simulation chooses). These two analyses should provide reasonable bounds on the possible outcomes, and both can be defended on the basis of the argument presented at the start of this paragraph.

The strategy we just articulated demands that we estimate the magnitudes of both catastrophes and bonanzas. How can we decide how severe a catastrophe might be for the Yellowstone grizzly bear population, when we never actually observed one? The simplest way to answer this question is to assume that a catastrophe would be just as bad relative to a typical year as the single observed bonanza was good. In the Yellowstone grizzly bear data (Table 3.1), the log population growth rate for the outlying 1983–1984 transition was 0.2683 (which translates into a population growth rate of 1.3077, or a 31% increase over 1 year). If we exclude this year and call the remaining log growth rates typical, their mean is 0.01467. Then, calculating the deviation between the log growth rate during the bonanza year and the mean log growth rate for typical years, and subtracting this deviation from the mean, we arrive at a catastrophic log growth rate (–0.2390) that is just as far below the mean as the bonanza growth rate is above it. Note that we have done this calculation using the log population growth rates, rather than the growth rates themselves, because it was on the log scale that the bonanza year was identified as an outlier. As exp{–0.2390} equals 0.7874, our hypothetical catastrophe represents a 21% decline in population size over one year. To reiterate the logic, we are simply saying here that it would not be implausible for the bear's population growth rate to occasionally drop as far below the average *as it has actually been observed to climb above it*, and to be cautious we would like to assess the consequences of such infrequent but unfavorable years in case they actually do occur.

We now explore the impact of these estimated bonanzas and catastrophes on the Yellowstone grizzly bear population, using a simple modification of the MATLAB programs we've been using (Box 4.7). The program allows for two different types of years. In "typical" years, the population growth rate is drawn from a lognormal distribution; that is, $\lambda_t = \exp\{x_t\}$ where x_t follows a normal distribution with a mean of μ and a variance of σ^2. Each year has a probability of 1 minus `probout` of being a typical year, and a probability `probout` of being an outlying (i.e., extreme) year. Thus to mimic the frequency of extreme growth rates for the Yellowstone grizzly bear, we set `probout` to 1/38. In extreme years, the program draws the growth rate at random from the values stored in the vector `outliers`. To explore the effect of bonanzas only, we place the single value 1.3077 in `outliers`. To account for both bonanzas and catastrophes, we add the value 0.7875 for the hypothetical catastrophe calculated in the preceding paragraph.

Our best estimate for the probability that the Yellowstone grizzly bear population would decline from 99 to 20 adult females at some point in the next 20 years, assuming no catastrophes or bonanzas,[17] is only 8.5×10^{-5}. Adding in only bonanzas at an average frequency of once every 38 years reduces this probability to 4.2×10^{-5} (based on the average of five runs of the program in Box 4.7, with 100,000 realizations in each run; obviously, to calculate such a small probability with greater precision, we would need to

[17]That is, we deleted the 25th transition in Table 3.1, computed μ and σ^2, and substituted the results into Equation 3.5.

BOX 4.7. *MATLAB code to calculate extinction risk in the presence of catastrophes and bonanzas.*[1]

```
% PROGRAM extremes
%    Simulates effect of catastrophes and/or bonanzas on
%    extinction risk.  Calculates the probability that a
%    population starting at Nc will hit the extinction threshold
%    Nx at or before t=tmax, using a density-independent model.
%    With probability "probout" each year, the population growth
%    rate is drawn at random from the vector "outliers", which
%    contains growth rates during catastrophes and/or bonanzas

% *************** SIMULATION PARAMETERS ***********************
tmax=20;              % time horizon
numreps=1000000;      % number of realizations to simulate
Nc=99;                % initial population size
Nx=20;                % extinction threshold
mu=0.01467;           % mean log population growth rate
sigma2=0.01167;       % environmental variance in log pop. growth
                      %    rate
probout=0.0263;       % annual probability of an outlying growth
                      %    rate

% vector of observed extreme growth rates
outliers=[0.7875, 1.3077];
%**************************************************************

rand('state',sum(100*clock));   % Seed uniform random number
                                % generator
randn('state',sum(100*clock));  % Seed normal random number
                                % generator
sigma=sqrt(sigma2);
numout=length(outliers);        % number of outlying growth rates
```

[1]Note that this program provides an alternative to using results based on diffusion approximations in Chapter 3. That is, rather than assume the log population growth rate is normally distributed and that large environmental perturbations do not occur, users can easily modify this program to choose population growth rates *only* from among those actually observed. Specifically, set "probout" to 1 and put all the observed population growth rates (*not* log population growth rates) in "outliers". (Of course, other modifications such as eliminating the "`if rand>probout`" statement would make the program faster.) Also see the program "randdraw" in Box 2.1.

BOX 4.7 *(continued)*

```
% Fill the matrix "lambdas" with population growth rates
for i=1:tmax                        % Looping over future times
   for j=1:numreps                  % and all realizations,
      if rand>probout                           % if year is 'typical'
         lambdas(i,j)=exp(mu+sigma*randn); % draw a 'typical'
                                           % lambda, but
      else                                      % in an extreme year,
         lambdas(i,j)= ...                         % draw lambda
            outliers(floor(numout*rand)+1);    % from "outliers".
      end;
   end;
end;

N=Nc*ones(1,numreps);  % Start all realizations at Nc,
for t=1:tmax           % project all realizations to
   N=N.*lambdas(t,:);  % t=tmax,
   N=N.*(N>Nx);        % set to zero any realization at or
end;                   % below Nx,
ProbExt= sum( (N<=Nx) )/numreps      % and compute
                                     % probability of
                                     % quasi-
                                     % extinction.
```

simulate many more realizations per run). When we include the possibility that both bonanzas and catastrophes may occur, the average probability of hitting the quasi-extinction threshold increases to 1.7×10^{-4}, which is still a rather small risk. In retrospect, it is not very surprising that including catastrophes and bonanzas of the observed (or at least justifiable) severity has little effect on the estimated short-term extinction risk. The chance that one of these extreme events occurs even once in a 20 year period is low, given the observed frequency of occurrence of once in 38 years, and although they represent outliers compared to the bulk of the observed population growth rates, their magnitude is not really all that extreme. Nonetheless, other populations may very well experience more extreme or more frequent catastrophes or bonanzas, which could have a strong impact (negative or positive) on population viability. We have now seen how such events can be incorporated into a PVA if their frequency and severity can be estimated.

While we've been discussing extreme years that are uncommon, for some populations typical environmental variability may be large enough that the usual assumption of normally distributed log growth rates may not be a good description of the true variation. This may be especially true for short-lived species living in highly variable environments, where most years are either a boom or a bust. Desert annuals, rodents, and some insect species fall into this category, with the annual population growth rate showing peaks at high and low values with few years in between. In these cases, more realistic predictions of extinction probabilities can sometimes be generated by using the observed growth rates themselves, rather than trying to estimate a mean and variance for the growth rate and drawing the simulated growth rates from a normal distribution. This is relatively easy to do in the absence of density dependence, and more difficult if density dependence does operate. Here is a procedure to assess whether extreme variation is common enough to deal with in this way:

1. Test for density dependence, as discussed earlier in the chapter.
2. If the best model is density-independent, use the methods in Chapter 3 to remove points that are extreme outliers.
3. Use a Lilliefors test to determine if the remaining log population growth rates are normally distributed. The Lilliefors test is a version of the one-sample Kolmogorov-Smirnov test that evaluates whether a set of numbers is likely to have come from a normal distribution. This routine is part of most statistics packages, often as an option under a Kolmogorov-Smirnov procedure.
4. If the test indicates there is very little chance that the log growth rates are truly normal (say, $p < 0.05$), you should reject the usual approach of generating log growth rates from a normal distribution with mean and variance μ and σ^2. Instead, simply draw a growth rate at random from the sample of observed rates in each year of a stochastic simulation.[18]

If your data do show significant density dependence, you can follow the same general procedure just outlined, but instead test the deviations between observed and predicted log population growth rates for normality (again using the Lilliefors test). However, now your simulation must add to next year's log population size predicted by the density-dependent model a random draw from the observed *deviations*.

Concluding Remarks

This chapter has reviewed methods to incorporate density dependence, demographic stochasticity, environmental correlations, and catastrophes and bonanzas into count-based PVAs. Even though we've tried to indicate how

[18]See the footnote to Box 4.7 for suggestions on how to do this.

to include different combinations of these effects in the same PVAs, we haven't done all the possible combinations. However, given the models and methods we have covered, it should be easy for you to put together many other combinations. As usual, your ability to include combinations of these factors will be limited not so much by an inherent difficulty in doing so, but more by the lack of data needed to estimate the strength of each factor for a single population. If such information is available for your focal population, you can view the fragments of MATLAB code presented in this chapter as building blocks from which a more comprehensive, multi-factor viability analysis can be constructed.

There are two additional problems we have not discussed in this chapter. The first is temporal trends in population parameters, that is, deterministic changes in the mean and variance of the population growth rate or the strength or pattern of density dependence over time. If you have the data to estimate how these changes will occur, simple modifications of the simulation models we have discussed will let you predict population dynamics for these situations (see *Temporal environmental trends* in Chapter 3 for a discussion of this issue). The second problem is much more difficult: how to untangle the effects of observation error from true variability in population numbers and hence growth rates. The next chapter is devoted to this problem.

Appendix: An Overview of Maximum Likelihood Parameter Estimation

The likelihood is simply the probability of obtaining a particular set of data given specified values of the parameters in a particular model. It makes intuitive sense that to find the best parameter values, we should seek to maximize this probability (i.e., we maximize the likelihood). To calculate the probability of obtaining the entire data set, we must first specify the probability that a single data point differs by a given amount from the value the model predicts it should be, given the parameter values and the values of the independent variables. For example, consider the checkerspot data in Table 4.1. For an observed value of N_t, the value of the log population growth rate $\log \lambda_t = \log(N_{t+1}/N_t)$ predicted by the Ricker model is $r(1- N_t/K)$ (see Equation 4.6). The deviation between the observed and predicted log population growth rates is $D_t = \log \lambda_t - r(1 - N_t/K)$. Let us consider such deviations to follow a normal distribution with a mean of zero and a variance of V_r (the residual variance)[19]. The probability of obtaining a deviation of size D_t is then governed by the normal probability density:

[19]Although we use normally distributed deviations here, we use maximum likelihood methods based on other probability distributions in Chapters 7, 8, and 11.

$$p(D_t|r,K,V_r,N_t) = \frac{1}{\sqrt{2\pi V_r}} \exp\{-(D_t)^2/2V_r\} = \frac{1}{\sqrt{2\pi V_r}} \exp\{-[\log \lambda_t - r(1 - N_t / K)]^2/2V_r\} \quad \text{(A1)}$$

Here, we use the notation $p(D_t \mid r,K,V_r,N_t)$ to emphasize that the probability of obtaining a deviation of size D_t depends on the values of the parameters r and K, the residual variance, and the independent variable N_t. The values of $\log \lambda_t$ and N_t are fixed features of the data set; the goal of maximum likelihood estimation is to identify the values of r and K that maximize (A1) over *all the observed values* of $\log \lambda_t$ and N_t. Some values of r and K will cause the value of $r(1 - N_t/K)$ to fall closer to one particular data point than to others. We want to choose the values of r and K resulting in predicted values that fall as close as possible to the greatest number of observed data points. Doing so will minimize the residual variance.

To find the values of r and K that provide the best fit to *all* the data points, we first compute the probability of obtaining the *entire* data set by computing the product of the probabilities of obtaining each data point:

$$L = \prod_{t=1}^{q} p(D_t|r,K,V_r,N_t) \quad \text{(A2)}$$

where q is the number of observations of the population growth rate in the data set (one less than the number of censuses). L is the likelihood of obtaining the data given the model. Because L is a product of many, potentially small, probabilities, it will typically be a very small number, which can be difficult to calculate accurately on a computer. Therefore, because the log of a small number is larger (in absolute value) than the number itself, it is convenient to take the logs of both sides of Equation A2 to yield the log likelihood function:

$$\log L = \sum_{t=1}^{q} \log[p(D_t|r,K,V_r,N_t)] = \sum_{t=1}^{q} \left[\log(2\pi V_r)^{-1/2} - \frac{1}{2V_r}\left[\log \lambda_t - r\left(1 - \frac{N_t}{K}\right)\right]^2 \right] \quad \text{(A3)}$$

Because the first term in the right-hand sum is independent of t, and there are q such terms, Equation A3 is the same as Equation 4.5.

Thus the best values of r and K are those that maximize Equation A3. There are several ways to find these values. The simplest way is to make a contour plot of the log likelihood versus r and K, and look for "peaks" in the likelihood surface. With finer and finer grids of r and K values, one can hone in on the best parameter values (the so-called maximum likelihood estimates), which are those that correspond to the highest peak in the likelihood surface. This graphical method for identifying the maximum likelihood estimates works well only if there are only one or two parameters to be estimated. Or one could take the derivatives of Equation A3 with respect to each of the parameters (e.g., r and K), set them equal to zero, and solve the resulting equations simultaneously to give the maximum likelihood estimates. To be sure that these values correspond to a maximum, not a minimum, in the log likelihood function (i.e., Equation A3), one should check that the second derivatives of Equation A3 are negative at these values. In the many cases in which the system of equations that results from setting the derivatives to zero is too complex to solve analytically, one can use gradient-based methods, which rely on expressions for the derivatives of Equation A3 to always move "uphill" on the log likelihood surface towards a maximum. This is the method that the SAS `nlin` procedure uses (see Box 4.2). Finally, one could use nongradient-based methods such as the simplex algorithm (described well in Press et al. 1995), which is implemented in MATLAB's `fminsearch` function.[20] Because in general the log likelihood function may have more than one maximum, it is important to check, once a maximum is found, that it is the global, not a local, maximum. For example, one should start gradient-based and nongradient-based searches at numerous starting values for the parameters, and choose the resulting set of parameter values (if there is more than one) that has the highest log likelihood.

[20]Note that `fminsearch` is a *minimization* routine, and so one must use it to find the parameter values that minimize the *negative* of the log likelihood function.

5

Accounting for Observation Error in Count-Based PVAs

We have now seen repeatedly that the degree to which population growth rates fluctuate due to environmental stochasticity strongly affects population viability. Therefore, to produce a useful count-based PVA, we must accurately estimate how much of the year-to-year variation in the census[1] counts, and hence in the estimated log growth rates (i.e., the values of log (N_{t+1}/N_t), is actually caused by variation in the environment. Count-based PVAs would be simpler to perform if we could assume that all of the variation in the estimated growth rates is due to environmental stochasticity. However, we know that several factors other than the environment may cause the growth rate estimates to vary from year to year. Two such factors are demographic stochasticity and density dependence.[2] In the preceding chapter, we saw how to account for these two sources of variation, either by choosing an appropriate quasi-extinction threshold or by using more complex models if the data needed to parameterize them are available. Unfortunately, there is yet another factor that can introduce spurious variation into the estimated growth rates, an unwanted source of variation that we will refer to as "observation error." Observation error is variation in the census counts (and hence the estimated population growth rates) caused by our inability to count precisely the number of individuals in the population or in the segment of the population that our census targets (e.g., adult females). Whereas demographic stochasticity and density dependence are true population processes that can have a real impact on population viability, observation error is not. Therefore, regardless of whether or not we have managed to account for demographic stochasticity and density

[1]To some users, *census* refers to an exhaustive count of all members of a population. In this book, we use the term in a looser sense to mean any attempt to quantify, by any method, the size of a population. Whenever a census does not involve a completely accurate and exhaustive count, observation error is possible.

[2]The two other factors discussed in Chapter 4, temporal environmental autocorrelation and bonanzas/catastrophes, are truly features of the environment.

dependence, we must still consider the possible contaminating influence of observation error.

As we mentioned in Chapter 2, one approach is to use the measured variation in the population growth rate as a surrogate for the environmentally driven variation, recognizing that viability estimates produced in this way will be overly pessimistic, since observation error will inflate the measured variation. Although this route is the easiest one to follow, it would obviously be preferable to quantify the amount of observation error and remove its effects from the analysis to the greatest extent possible. The goal of this chapter is to review some of the simpler and more practical methods available to reduce the influence of observation error on viability assessments. We begin by describing some potential sources of observation error. We then describe how observation error can be reduced and how the effects on viability assessments of those errors that do occur can be lessened. We break these solutions to the problem of observation error down into those that can be employed before a census is actually conducted, during the course of the census, and after the census data have been collected. At the end of the chapter, we direct readers to references for some of the more sophisticated methods for estimating model parameters in the face of observation error, methods that lie beyond the scope of this book.

Potential Sources of Observation Error

Before describing ways to reduce its influence, we first devote a bit more attention to defining what observation error really is, and what it is not. First of all, *observation error* is not strictly synonymous with *sampling variation* (although, as we will see, sampling variation is one potential source of observation error). Sampling variation occurs whenever any statistic (such as the mean log growth rate) is estimated using data about a limited sample from a larger "population." Here, we are using the term *population* not in the ecological sense (i.e., the set of all individuals of a given species in a prescribed area) but in the statistical sense (i.e., the set of all objects of a given type). For the purposes of conducting a count-based PVA, the statistical population we are interested in is the set of all possible growth rates that might occur in the ecological population under study. Considering that the growth rate can vary continuously over a broad range of values, the population of all possible growth rates is technically of infinite size. However, a limited set of yearly censuses can yield only a small subsample of growth rates from this infinite population. Even if we could measure the size of the ecological population, and therefore its growth rate, with perfect precision each year, *observing and counting every individual with complete accuracy*, we would still not expect to get exactly the same set of growth rates if we were to census the population annually for, say, repeated 20-year periods. Thus statistics such as the mean and variance of the log growth rates (which we use to estimate μ and σ^2, respectively) will differ among sets of census data due to sampling variation. The confidence limits on μ and σ^2 that we calculated in Chapter 3 used the theoretical sampling distributions of the mean and variance of a normal random variable to account for this sampling variation.

However, we are essentially *never* able to observe and count every individual in an ecological population. The population may be too large to count exhaustively, and even if it contains relatively few individuals, as do most endangered populations, they may be spread out over too large an area to search in its entirety. In these situations, the typical protocol is to *estimate* total population size at each point in time by extrapolation from a sample of the larger population. For example, we may exhaustively count the number of individuals in a set of study plots to get an estimate of mean density, which we multiply by the total occupied area to estimate total population size. Even if the count within each plot is completely accurate, our calculated mean density in a single year will be based on a limited number of plots and thus will be subject to sampling variation, which will in turn lead to observation error in the estimated population size. The important distinction to understand is that sampling variation is a consequence of limited amounts of data (even if those data are measured accurately), whereas observation error is *any* inaccuracy in estimates of population size, and hence population growth rate, whatever the cause.

Consider what effect this observation error might have on a viability assessment. As an unrealistic but informative example, let's assume that the true population size is absolutely constant over time, but that we must estimate it by sampling a set of plots that represent a known fraction of the population's entire geographical range. Unless the members of the population are distributed across the range in a perfectly uniform fashion each time we conduct the census (an extremely unlikely event for *any* organism), the mean numbers across all the plots, and thus the *estimates* of total population size, will vary from year to year, even though the true population size does not. That is, in some years, a relatively large fraction of the individuals in the population will happen to be located in our sampling plots at the time of the census, while in other years, few individuals will. If we now use the variance in the *estimated* log growth rates as a measure of the environmental variance σ^2, we would obtain a positive value, and would thus conclude that the population faces some risk of quasi-extinction *even though its true size never changes*. Thus the artifactual component of the variance in the log growth rate introduced by observation error inflates our estimate of extinction risk.

The preceding paragraph assumed that the counts within sampling plots were completely accurate. However, these counts may suffer from another source of observation error: failure to determine correctly the number of individuals actually present in each plot. Survey methods will often leave some individuals undetected (e.g., bears in dense vegetation may be impossible to see in an aerial survey), leading to undercounts. Conversely, counting some individuals more than once can lead to overcounts. If a large proportion of the sample counts taken from a population are undercounts, or a large proportion are overcounts, the mean of the samples will be a *biased* estimate of the true population mean. However, bias per se is not the main problem for estimating the population growth rate. For example, let's say that our census *consistently* misses 10% of the population, year after year. If the true population sizes in two successive years are 140 and 160 individuals, our estimated population sizes based on the census counts would be 126 and 144 (i.e., 90% of the true counts). However, our estimate of the population growth rate based on these population size estimates (i.e., 144/126 = 1.14 = 160/140) would actually give us the correct answer, because a fixed proportional bias in the two population size estimates does not affect their ratio.[3] The problem arises because the magnitude of the under- or over-count, and thus the degree of bias in the estimates of population size, will always vary somewhat from year to year. Imagine another pair of successive years in which the true population sizes are 140 and 160, as before. But now assume that 8% of the individuals are missed in the first year and 12% in the second year, so that now the estimated population sizes are 128.8 and 140.8. The resulting estimate of the population growth rate (1.09) is an underestimate of the true growth rate. If we switched the percentages of individuals missed in the two years, the resulting growth rate estimate (1.19) would be an overestimate of the true rate. Thus *variation* among censuses in the amount of bias in the counts will artificially inflate the estimate of the variation in the population growth rate (i.e., σ^2). Also, bias can sometimes be nonproportional (i.e., an over- or undercount by approximately the same *number* of individuals in every year); such density-dependent bias in the sample counts will translate into biased estimates of both μ and σ^2. Variation in the number or fraction of individuals detected results from many factors, including changes from census to census in the proportion of individuals located in dense vegetation, differences in light levels or weather conditions at the times aerial surveys are conducted, or simply differences among observers in their levels of training or inherent skill at locating and distinguishing individuals of the focal species.

Thus observation errors arising either from counting mistakes or from variation among samples used to estimate population size will often introduce an unwanted source of variation in census counts. Unfortunately, describing the potential sources of observation error is far easier than designing census protocols to avoid it or developing ways to reduce its influence on parameter estimates. Still, in the remainder of this chapter, we discuss a few basic approaches that can help to reduce the unwanted influence of observation error on estimates of population viability. We use the

[3]We would, however, want to keep the uncensused individuals in mind when setting a quasi-extinction threshold.

word *reduce* advisedly, as we can never entirely eliminate the influence of observation error, only attempt to minimize its effects.

Before proceeding, we note that observation error can have two distinct effects on estimates of population parameters such as μ and σ^2. First, it can introduce *bias* into the estimated parameters. A procedure for estimating a particular parameter is biased if, when applied many times to replicate sets of data generated from a "true" model, it yields a distribution of estimates whose mean differs from the true value of the parameter. That is, biased estimators tend to under- or overestimate the true parameter value. For example, as we have just seen (and will explore in more detail below), observation error usually causes the variance of the observed log population growth rates to be upwardly biased relative to the true value of σ^2. Bias is worrisome for PVA because it will lead to overly optimistic or overly pessimistic assessments of viability, depending on which parameter estimates are biased and in which direction. The second effect of observation error is that it can reduce the *precision* of parameter estimates. An estimation procedure that, when applied to replicate data sets produced by the same true model, yields a distribution of parameter estimates with a high variance is less precise than one that yields a low variance among the estimates, even if both produce unbiased estimates. The most general way to increase the precision of parameter estimates is to increase the amount of data used in the estimation process, but this is easier said than done for rare species. In this chapter, we will see examples of both bias and imprecision caused by observation error.

Considerations for Reducing Observation Error before a Census Is Initiated

In designing a census, care should be taken to assure that the census counts are as accurate as possible, and that the degree of accuracy changes as little as possible over the many years of a long-term census. Methods to accurately census a population are idiosyncratic to the type of organism under study, and we cannot possibly review this large field here. Instead, readers should consult one of the many taxon-specific books on population estimation (e.g., Sutherland 1996 for a range of taxa, Southwood 1978 for insects, Elzinga et al. 1998 for plants, and Thompson et al. 1998 for vertebrates, with even more specific treatments in Krebs 1989 and Wilson 1996 for mammals and Heyer 1994 for amphibians). A wealth of more general papers on the thoughtful design of monitoring plans have also been published (e.g., Vos et al. 2000 and Gibbs et al. 1999). However, a few general guidelines are worth noting here:

1. Make the first year of data collection a trial run to generate information needed to intelligently set sampling methods and effort. When possible, try several sampling methods before initiating the census or during the

first census period to help identify the method that yields the most accurate counts. In addition, deliberately oversample (i.e., invest more search time, search more plots, and so forth) on this first census, as this is the only way to evaluate how quickly sampling variation will decrease as you increase sampling effort. Alternatively, adopt a sampling method that is known to be fairly accurate when applied to organisms and habitats that are similar to the census situation you face.

2. After fixing the census methods to be used, formalize and write-up a clear sampling protocol, and use that protocol consistently in the successive censuses. The protocol should specify that the census is to be taken in the same season each year and under similar weather conditions. In addition, make sure that all observers have received consistent training prior to the census in how to search for and recognize the study organism. To assure consistency, it helps if the same person is charged with training the observers and leading the census for as long as possible, and it helps if she or he carefully trains the succeeding leader. Furthermore, for census data to be comparable across multiple observers, it is best if highly specialized skills are not needed to accurately collect the information. These procedures should reduce variation in bias from census to census.
3. If possible, establish a stratified,[4] randomly selected set of census plots, transects, flight routes, or roadside sampling stations at the outset of the census, and use these same locations in subsequent censuses. This should reduce the potential for confounding spatial variation in density or in degree of detectability related to habitat complexity with temporal variability in true population size.[5]
4. If it is not possible to count all members of the population, choose the most easily detected subset of individuals to census, or record information separately for each subset of the population, allowing separate analyses to be tried in the future. Often the most easily detected members will be the largest individuals in the population, which usually make the biggest contribution to population growth anyway.
5. Design data collection methods and data recording protocols that allow sampling effort to be clearly understood and quantified.

---

[4]By *stratified*, we mean that when the focal organism inhabits two or more distinct types of habitat, we place sampling units (e.g., transects) at random within each type of habitat. We then estimate total population size as the mean density across all units in one habitat type multiplied by the area of that habitat type, and sum these products over all habitat types. A stratified sampling design can yield more accurate estimates of total population size than a strictly random design (see Greenwood 1996).

[5]One downside of using the same sampling units every year is that the counts from the same unit in successive years will not be statistically independent. Fully accounting for the covariance this introduces requires methods more sophisticated than the ones we present here.

Most of these guidelines are patently obvious, but attending to them before initiating a census might greatly reduce observation error problems later on.

Quantifying Observation Errors while a Census Is Being Conducted

The most important thing that can be done at the time of the census to help account for observation error is to quantify its magnitude. If the census is conducted by counting individuals in replicate sample units (e.g., multiple quadrats or line transects), then simply recording the separate counts in each unit, rather than only the total or the mean count, permits us to account for observation error caused by sampling variation, as we will see in the following section. If the census protocol does not call for replicate samples in every year of the census, then it would be wise to perform repeat sampling of at least a fraction of the study area in at least some years. As noted above, this repeated sampling is especially useful in the first year of a census, allowing one to determine the sampling effort needed to yield acceptably accurate data. As an example of repeated sampling, one could count the number of individuals in an area on one day, and then repeat the count in the same area at one or more times shortly thereafter. If these repeated counts are sufficiently close in time, it is probably safe to conclude that the differences among them are largely due to observation error rather than an actual change in the number of individuals in the area. (Of course for mobile species, care must be taken to account for individuals that may have been induced to leave the area due to the disturbance caused by earlier censuses.) Ideally, these repeated censuses of the same area should be performed both in years when the population size is relatively large and in years of relatively low population size, because the degree of observation error itself may depend on population size. (For example, it may be easier to miss individuals when they are scarce and hiding places are abundant.) Although this repeated sampling will add to the time and expense of performing the census, it may provide invaluable information for correcting parameter estimates for observation error. We will see how to do so in the following section.

One other thing that should be done at the time of the census deserves further mention. If the census involves simply making a total count of the number of individuals seen (i.e., if there are no obvious sampling units), the amount of effort (e.g., the number of person-days spent searching, the number of flight hours or road miles driven, or the fraction of the total area occupied by the population that is searched) should be recorded each year, and the count adjusted for effort. (For example, dividing the count by the fraction of the occupied area searched gives a measure of total population size.) If multiple observers are involved in a census it is also important to keep separate tallies of the search time and success of each participant, in order to distinguish the effort of efficient experts from that of novices. In general, without recording sampling effort, among-year differences in observed population sizes cannot be divided into real differences versus observation error.

Correcting for Observation Errors after the Census Data Have Been Collected

Worrying about observation errors after data have been collected probably seems like shutting the stable door after the horse has escaped. However, at least some observation error is present in any data set, and hence there will always be a need for methods to reduce its influence on parameter estimates. Disentangling the variance in the estimated population growth rate that is due to environmental variation from the variance due to observation error is a complex statistical problem, and many sophisticated methods to attack this problem lie far beyond the scope of this book (see the end of the chapter). However, in this section we describe three simple approaches that can easily be applied to count-based data. As an up-front warning, we caution that all three of these methods have their limitations, which we note as we describe them. In many cases, none will work and simply proceeding without explicitly discounting for observation error (using the methods of Chapters 3 and 4) will be our only alternative.

Correcting Estimates of σ^2 for Observation Error when Counts Represent Means of Replicate Samples

As we have discussed above, the census counts we intend to use to build a count-based PVA will frequently involve means of replicate samples taken from the population. For example, to estimate the population size of an endangered plant, we might count the number of individual plants encountered in a set of quadrats or line transects, or for an animal species, we might repeatedly fly over the population in an airplane and count the number of individuals seen in each flight. Typically, we would then compute the mean of these replicate counts as a measure of the average density per sampling unit (e.g., a quadrat or a flight of a fixed duration) in each year, and then use either the means from every year (for a density-based PVA) or the means multiplied by the area occupied by the population (for an analysis based on total population size). The key point to recognize is that there is sampling variation associated with each of those means, and as we noted above, this sampling variation creates observation error. If we simply apply the methods of Chapter 3 to estimate σ^2 (i.e., if we compute the variance of the log population growth rates calculated from the sample means), the resulting value will represent an amalgamation of the true environmental variance and the sampling variation that affects the sample means. The larger the number of samples used to calculate the means, the smaller the observation error will be, but realistic constraints of time and money will often dictate a modest number of samples (especially for technically complex sampling methods such as aerial surveys). Hence simple estimates of σ^2 will frequently be biased by observation error. In addition, observation error will reduce the precision of simple estimates of both μ and σ^2.

To demonstrate the potential magnitudes of the bias and imprecision imparted by observation error, we performed a simple computer simulation. We began by generating a "true" population trajectory using the density-independent model:

$$N_{t+1} = N_t \exp(\mu + \sigma x_t) \tag{5.1}$$

where x_t is drawn at random from a normal distribution with a mean of zero and a standard deviation of 1. In this model, the true mean and variance of the log population growth rate are μ and σ^2, respectively. For each census t, we then mimicked the observation process by drawing n_t samples $N_{t,i}$ ($i = 1\ldots n_t$) from a negative binomial distribution[6] such that the mean of a large number of samples would equal the true population density, N_t. The negative binomial distribution commonly provides a close fit to the frequency distribution of the number of individuals in replicate samples from a population (Southwood 1978). The negative binomial is a discrete, strictly nonnegative probability distribution, meaning that it can only take on integer values of zero or greater, which is appropriate given that the number of individuals observed in each sampling unit can only be a nonnegative integer. More importantly, the negative binomial distribution can fit a wide variety of sampling distributions, ranging from randomly dispersed to

[6]The negative binomial distribution arises most simply in the following way. If individuals are distributed completely at random, the counts from replicate samples will follow a Poisson probability distribution, in which the mean and variance of the samples will be equal. But if individuals are aggregated, we can think of some samples as having been drawn from Poisson distributions with high means and other samples as having been drawn from Poisson distributions with low means. More specifically, if the means of a series of Poisson random variables are themselves drawn from a gamma probability distribution, $G(x|k, k/m) = (k/m)^k x^{k-1} e^{-(k/m)x}/\Gamma(k)$ (where $\Gamma(k) = \int_0^\infty z^{k-1}e^{-z}dz$ is the gamma function), then the resulting random numbers will come from a negative binomial distribution with a mean of m and a variance of $m + m^2/k$. Hence, to generate a sample y from a negative binomial distribution, we first generate a value of x from the gamma distribution $G(x|k, k/m)$, and then generate a value of y from the Poisson distribution $P(y|x) = x^y e^{-x}/y!$ (see Fishman 1973). The MATLAB statistics toolbox also includes a function (`nbinrnd`) for directly generating random numbers from a negative binomial distribution.

highly aggregated (Figure 5.1). A sampling distribution in which the variance is approximately equal to the mean (Figure 5.1A) implies that the location of each individual was determined at random, without regard to the locations of other individuals in the population. In contrast, an aggregated distribution, such as the one illustrated in Figure 5.1E, arises when individuals are highly clumped in space, so that most samples include few or no individuals but a few samples contain many individuals. It makes intuitive sense that when the counts vary as much among individual sampling units as they do in a highly aggregated distribution, the mean of the samples $\overline{N}_t = \frac{1}{n_t}\sum_{i=1}^{n_t} N_{t,i}$, which we use as our estimate of population density, will also vary considerably from one set of samples to another, so there will be a high potential for observation error in the estimated population size.

For our example, we used the true values $\mu = 0.0296$ (i.e., $\lambda_G = e^{\mu} = 1.03$) and $\sigma^2 = 0.01$, and we generated $n_t = 10$ samples for each of 21 annual censuses. We then used the mean densities, the $\overline{N}_t$'s, to estimate the log population growth rates between each pair of censuses, and computed the mean and variance of those estimated log population growth rates as our estimates of μ and σ^2. We will label these estimates $\hat{\mu}$ and $\hat{\sigma}^2$, and refer to them as the *raw* estimates of μ and σ^2, as they are uncorrected for observation error. We repeated this entire process 1,000 times to get a distribution of likely values of $\hat{\mu}$ and $\hat{\sigma}^2$ for each of several different levels of sampling variation. Specifically, we set the variance of the negative binomial distribution from which we drew the 10 samples in year t to equal 1, 2, 4, 8, or 16 times the true population density in that year, N_t, which was the expected mean of the samples. That is, we used sampling distributions similar to those illustrated in Figure 5.1 to mimic spatial arrangements of individuals that ranged from random to highly aggregated. Because the variance was a constant multiple of the mean, the actual variance among the samples was higher in years when true population size, and hence the mean of the samples, was high. We might well expect to see a higher variance among the samples when the mean is high than when the mean is low (although the opposite is also possible).

The distributions of the raw estimates of μ, the $\hat{\mu}$'s, are shown in Figure 5.2A. Even with relatively low variance among sample units (variance = mean), the distribution of the $\hat{\mu}$'s is quite broad (ranging from negative to positive values), meaning that it is difficult to estimate μ with precision using only 21 years of data. Nevertheless, even a high amount of observation error (i.e., high variance among samples) does not cause the mean of the $\hat{\mu}$'s to deviate appreciably from the true value of μ. In other words, the mean of the log population growth rates is a nearly unbiased estimator of the true μ. However, notice that the spread of the estimates increases with increasing among-sample variance; that is, observation error decreases the precision of the raw estimates of μ. Unfortunately, observation error has an even more

(A)

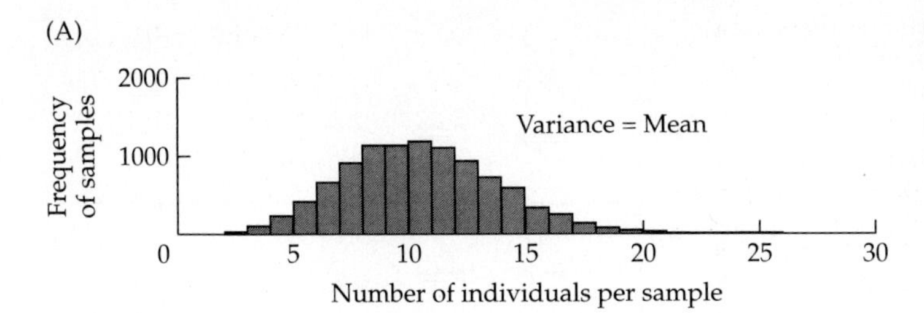

(B)

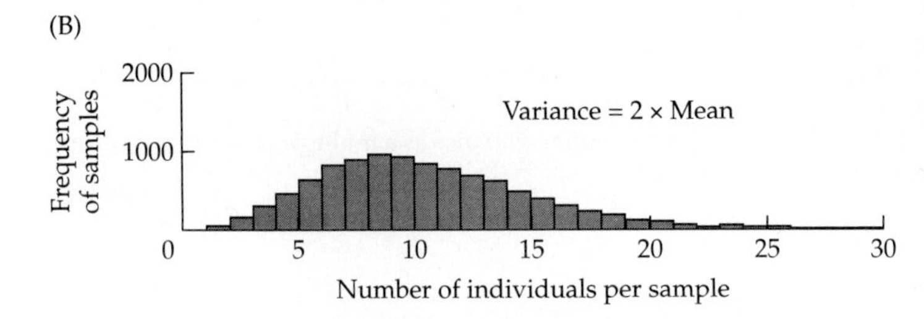

(C)

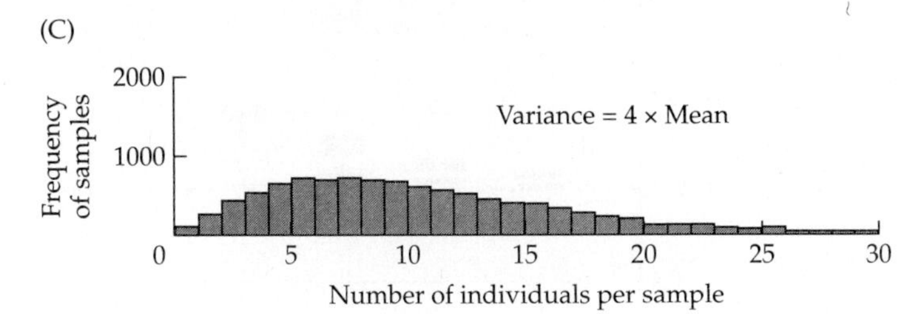

(D)

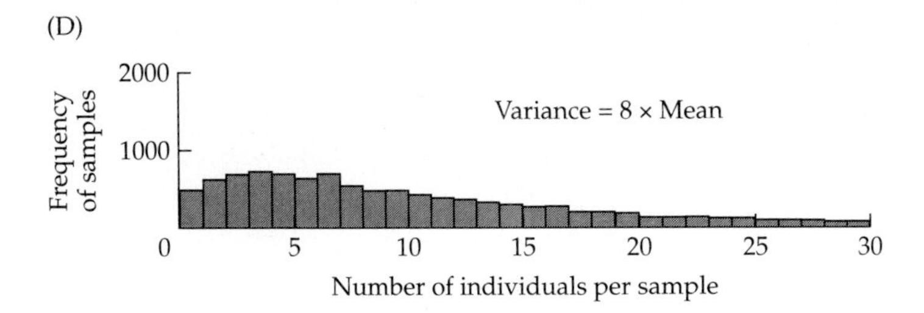

(E)

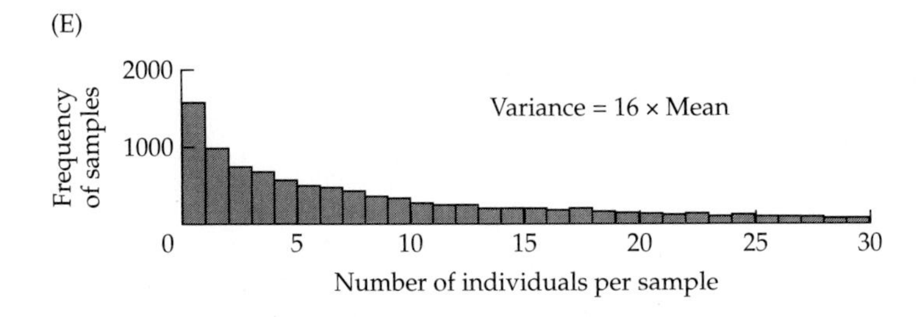

Figure 5.1 Negative binomial distributions describing the frequency of samples with differing numbers of individuals. All the distributions illustrated include 10,000 samples with a mean of 10 individuals per sample. Although the means are the same, the degree of aggregation of individuals among samples, which is reflected in the variance of the distribution as a multiple of the mean, increases from top to bottom. In Plot E (and to a lesser extent in Plots D and C), a few samples had substantially more than 30 individuals but are not shown.

Figure 5.2 (A) Sampling variation does not bias the raw estimates of μ, but it does reduce their precision. (B) Sampling variation in the means of replicate samples introduces bias into raw estimates of σ^2. (C) The corrected estimate of σ^2 calculated from Equation 5.4 reduces the observation error bias. Dashed lines indicates the true value of each parameter. The rectangular boxes contain the central 50% of the estimates, the heavy line in the center of each box is the median, the whiskers extend to the estimate closest to 1.5 times the interquartile range above and below the box, and more extreme estimates are indicated by separate lines. Estimates from 1,000 simulated data sets, each of 21 years' duration, are included.

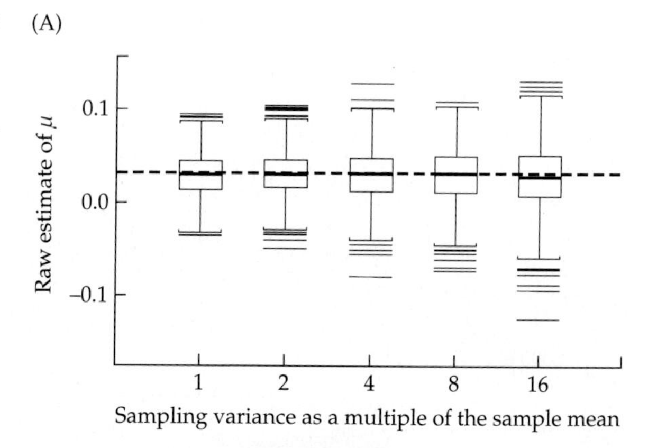

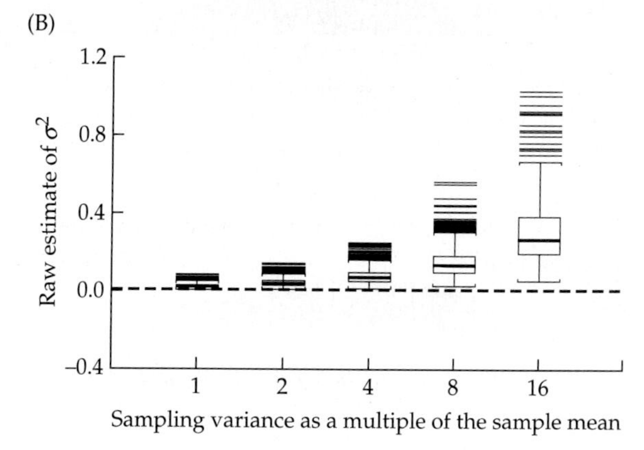

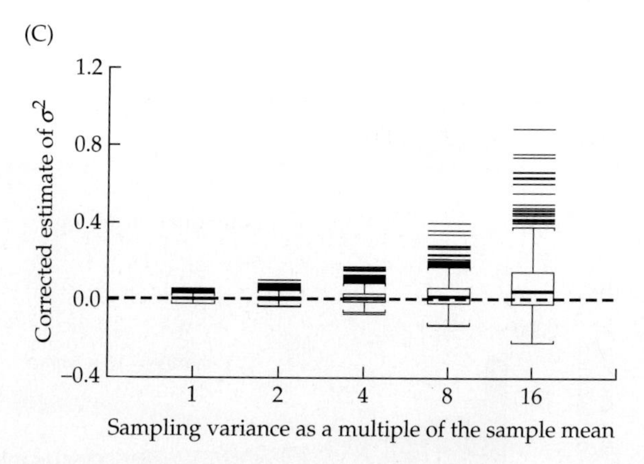

profound influence on raw estimates of σ^2. The distributions of those estimates are illustrated in Figure 5.2B; in all cases, the true value of σ^2 is 0.01. As for $\hat{\mu}$, $\hat{\sigma}^2$ based on only 21 years of data with 10 samples per year has relatively low precision (for example, the central 95% of the estimates lie between 0.011 and 0.056 at the lowest sampling variance shown in Figure 5.2B). But what is even more worrisome about the raw estimates $\hat{\sigma}^2$ is that they are biased upwards, and to an increasingly large degree as the sampling variance increases. For example, for the highest sampling variance shown in Figure 5.2B, the mean of the 1000 values of $\hat{\sigma}^2$ is 0.306, which overestimates the true value of σ^2 by a factor of 30. Thus if we do not correct our estimates of σ^2 for observation errors, we will obtain unrealistically high values that will lead to correspondingly pessimistic assessments of population viability. Also notice that increasing the amount of observation error progressively degrades the precision of $\hat{\sigma}^2$, just as it does for $\hat{\mu}$.

Fortunately, when counts are based on means of replicate samples, we can do better than to simply use the means alone to estimate σ^2, as we did in the preceding paragraph. Specifically, we can make use of information contained in the variances of each year's samples to partially correct $\hat{\sigma}^2$ for sampling variation. The logic underlying the correction is detailed in the following paragraphs.

If we consider the single interval between censuses t and t+1, our estimate of the log population growth rate over that interval will be $\log \lambda_t = \log(\overline{N}_{t+1} / \overline{N}_t)$. Because $\log \lambda_t$ is a function of the sample means $\overline{N}_t$ and $\overline{N}_{t+1}$, we can express the variance in $\log \lambda_t$ as a function of the sampling variances of these two means. A useful approximation[7] for the sampling variance in $\log \lambda_t$ is:

$$Var(\log \lambda_t) \approx \left(\frac{\partial \log \lambda_t}{\partial \overline{N}_t}\right)^2 Var(\overline{N}_t) + \left(\frac{\partial \log \lambda_t}{\partial \overline{N}_{t+1}}\right)^2 Var(\overline{N}_{t+1}) = \frac{Var(\overline{N}_t)}{\overline{N}_t^{\,2}} + \frac{Var(\overline{N}_{t+1})}{\overline{N}_{t+1}^{\,2}} \tag{5.2}$$

Because $\overline{N}_t$ and $\overline{N}_{t+1}$ are means across sampling units, then by the Central Limit Theorem their variances can be approximated by the squared standard error of the mean. That is,

[7]This approximation uses the so-called delta method, which gives an approximation for the variance of a function of independent random variables in terms of the variances of the individual random variables. Specifically, if $f(x,y)$ is a function of the random variables x and y, and if the variances of x and y are not too large, then

$$Var(f) \approx \left(\frac{\partial f}{\partial x}\right)^2 Var(x) + \left(\frac{\partial f}{\partial y}\right)^2 Var(y)$$

A slightly more accurate approximation is obtained by adding the following three terms:

$$\frac{1}{2}\left(\frac{\partial f}{\partial x}\right)^2\left(\frac{\partial f}{\partial y}\right)^2 Var(x)Var(y) - \frac{1}{4}\left(\frac{\partial f}{\partial x}\right)^4 (Var(x))^2 - \frac{1}{4}\left(\frac{\partial f}{\partial y}\right)^4 [Var(y)]^2$$

but in practice the contribution of these terms is often negligible.

$$Var(\overline{N}_t) = s_t^2 / n_t \tag{5.3}$$

where s_t is the standard deviation of the n_t sample counts used to compute $\overline{N}_t$. Substituting Equation 5.3 into 5.2 thus provides a way to estimate the variance in $\log \lambda_t$ that results from sampling variation, using all of the counts in each of the separate sampling units.

We now have to use this result for one inter-census interval to adjust our estimate of σ^2. A series of censuses performed in q+1 years, each comprised of replicate samples, yields q estimates of $\log \lambda$ and, using Equations 5.2 and 5.3, q estimates of the variance in $\log \lambda$. The simplest approach[8] is to use the average of the q variances as an overall measure of the mean variance in log λ that is due to sampling variation in the $\overline{N}_t$'s:

$$\overline{Var(\log \lambda)} = \frac{1}{q} \sum_{t=1}^{q} \left[\frac{s_t^2}{n_t \overline{N}_t^2} + \frac{s_{t+1}^2}{n_{t+1} \overline{N}_{t+1}^2} \right] \tag{5.4}$$

Subtracting this component due to sampling variation from the raw estimate $\hat{\sigma}^2$ obtained as in Chapter 3 [i.e., as the variance of the q values of $\log(\overline{N}_{t+1} / \overline{N}_t)$] produces an estimate of the environmental variance in $\log \lambda$ corrected for the sampling variation in the $\overline{N}_t$'s:

$$\sigma^2_{corr} = \hat{\sigma}^2 - \overline{Var(\log \lambda)} \tag{5.5}$$

The MATLAB code in Box 5.1 computes corrected estimates of σ^2 using both Equation 5.5 and an alternative approach that accounts for different amounts of sampling variation in different years. The sample data in Box 5.1 were generated using Equation 5.1 with $\mu = 0.0296$ and $\sigma^2 = 0.01$, and with the variance of the negative binomial samples equal to four times their expected mean (compare to Figure 5.1C). For this data set, the corrected estimate $\sigma^2_{corr} = 0.0101$ is much closer to the true value of σ^2 than is the raw estimate $\hat{\sigma}^2 = 0.0671$.[9]

How well does Equation 5.5 reduce the observation error bias across the many simulated data sets used in Figure 5.2? Most encouragingly, for small to moderate levels of sampling variation, the bias seen in Figure 5.2B is sharply reduced by using Equation 5.5 to estimate σ^2 (Figure 5.2C). Still,

[8]An alternative is to assign different weights to years with different levels of sampling variance; see Chapter 8 for a discussion of White's (2000) method to do this. Also see Box 5.1.

[9]However, for this data set, the estimate of μ, 0.073, is rather bad. Nevertheless, over many replicate data sets, estimating μ as the mean log population growth rate yields nearly unbiased estimates of the true μ even in the presence of sampling variation, as we saw in Figure 5.2A.

BOX 5.1 ***MATLAB code to correct a raw estimate of σ^2 for sampling variation when census counts represent means from replicate samples.***

```
% Program correct_sigma2
%   Corrects the estimate of sigma^2 for sampling variation when
%   census counts represent means of replicate samples


%************** USER-SUPPLIED INFORMATION ********************

% The matrix "samples" stores the counts from each sample;
% it has censuses (years) as its columns and individual samples
% as its rows; -1 is used as a place holder at the bottom of
% columns corresponding to censuses with fewer samples
samples=[
 10 11  7  0 14  0  6 26 22 12  6 10 22 20 39 31 23 21 14 14 19;
  8  4  6 13  1 14 23 10 11 11 10  8 11 27 19 18 16 15 32 38 21;
  1 10 11 27 36 16 18 13 29  7  5  2 16  7 25 15 21 12 15 18  9;
  7  8  1  9 17 14  9 11 24 13  4  8 12 28  8 15 21 11 30 28 15;
  0 16 10  5 16 17  4 19 13 14 21 10 14  6 21 12  8  8  6 14 32;
  7  6  5  4  2 15  3 10 13  9 15 22  6 33 14 23  8 14 27 28 17;
  5 11 10 21 10 23 12  7 14 12 11 11 22 10 44  5 12 17 21 42 40;
  2 16 15  8  6  6 12 14 15 18  9  5  4  7 20 22 14  5 16 13 30;
  6  5 12 11 10  6 18  8 17  7 22 13 -1  5 15  7  8 29 25 16 23;
  6  6 -1 25 10 14 19  4 16  7 19  6 -1 10 16 18 14 15 11 26 18];

% To calculate confidence limits using the method of White (see
% below), users must supply the critical values of the chi-
% squared distribution with degrees of freedom equal to the
% number of censuses minus one; see Chapter 3 of Morris and Doak,
% Quantitative Conservation Biology, for how to
% compute these values.  The following are critical chi-square
% values for p=0.025 and p=0.975 with 20 degrees of freedom:
chi2crit=[34.16958143, 9.590772474];

% Enter tolerance for computing corrected estimate and its
% confidence limits using White's method; smaller tolerance means
% more accuracy
tolerance=1E-8;

%************************************************************
```

BOX 5.1 *(continued)*

```
options=optimset('TolX',tolerance);         % set tolerance for the
                                            % function fzero below

% The vector n stores the number of samples for each census
n=sum(samples>=0);

% q=number of censuses minus 1
q=size(samples,2) - 1;

% Compute means and variances of the samples from each census
for t=1:q+1
     samplest=samples(1:n(t),t);
     Nbar(t)=mean(samplest);
     Vs(t)=var(samplest);
end;

% Compute raw estimates of mu and sigma^2 using the conventional % method
for t=1:q
    loglam(t)=log(Nbar(t+1)/Nbar(t));
end;
muest=mean(loglam);
disp('Raw estimate of mu:');
disp(muest);
s2raw=var(loglam);
disp('Raw estimate of sigma^2:');
disp(s2raw);

% Compute the component of total variance due to sampling
% variation
for t=1:q
     SampleVar(t)=Vs(t)/(n(t)*Nbar(t)^2) + ...
Vs(t+1)/(n(t+1)*Nbar(t+1)^2);
end;
MeanVar=sum(SampleVar)/q; % mean sampling variation across years

% Compute the simple corrected estimate of sigma^2
s2corr=s2raw-MeanVar;
disp('Sigma^2 with simple correction for sampling variation');
disp(s2corr);
```

BOX 5.1 *(continued)*

```
% Compute the corrected estimate of sigma^2 using the method of
% White, which weights by the sum of the environmental and
% sampling variance in each year; see White 2000.
Dev2=(loglam-muest).^2;
s2corrwhite=fzero(inline('sum(Dev2./(s2+SampleVar))-q+1',...
    's2','Dev2','SampleVar','q'),s2raw,options,Dev2,SampleVar,q);
disp('Sigma^2 with White''s correction for sampling variation');
disp(s2corrwhite);

% Compute confidence interval for White's version of the
% corrected sigma^2;
% WARNING: Users may need to try different values of 'start'
% below to get reasonable values for the confidence limits, as
% there may be more than one root to the equation being solved by
% the function fzero.

start=0.001;
s2lower=fzero(inline('sum(Dev2./(s2+SampleVar))-chi2', 's2', ...
    'Dev2','SampleVar', 'chi2'),...
    start,options,Dev2,SampleVar,chi2crit(1));

start=0.1;
s2upper=fzero(inline('sum(Dev2./(s2+SampleVar))-chi2', 's2', ...
    'Dev2','SampleVar', 'chi2'),...
    start,options,Dev2,SampleVar,chi2crit(2));

CI=[s2lower s2upper];
disp('Confidence interval for White''s correction:');
disp(CI);
```

[1] Note that this program provides an alternative to using results based on diffusion approximations in Chapter 3. That is, rather than assume the log population growth rate is normally distributed and that large environmental perturbations do not occur, users can easily modify this program to choose population growth rates *only* from among those actually observed. Specifically, set "probout" to 1 and put all the observed population growth rates (*not* log population growth rates) in "outliers". (Of course, other modifications such as eliminating the "`if rand>probout`" statement would make the program faster.) Also see the program "randdraw" in Box 2.1.

five caveats must be kept in mind. First, when the variance among samples is large, σ^2_{corr} reduces, but does not eliminate, the observation error bias. For example, when the sampling variance was 16 times greater than the mean, the average of the 1,000 corrected estimates was 0.076, substantially greater than the true value, although much closer to it than the mean of the uncorrected estimates (compare Figures 5.2B and C). As the sampling variance becomes large, the approximation used in Equation 5.2 breaks down. Second, even though it reduces the bias, the corrected estimate does not reduce the imprecision that observation error introduces into the raw estimates. (Note the similar ranges of the raw and corrected estimates in Figures 5.2 B and C). When the sampling variance is high, a large number of samples are needed to achieve precise estimates of σ^2 (Figure 5.3). Note that while increasing the sample size also reduces the bias in the raw estimates, the cor-

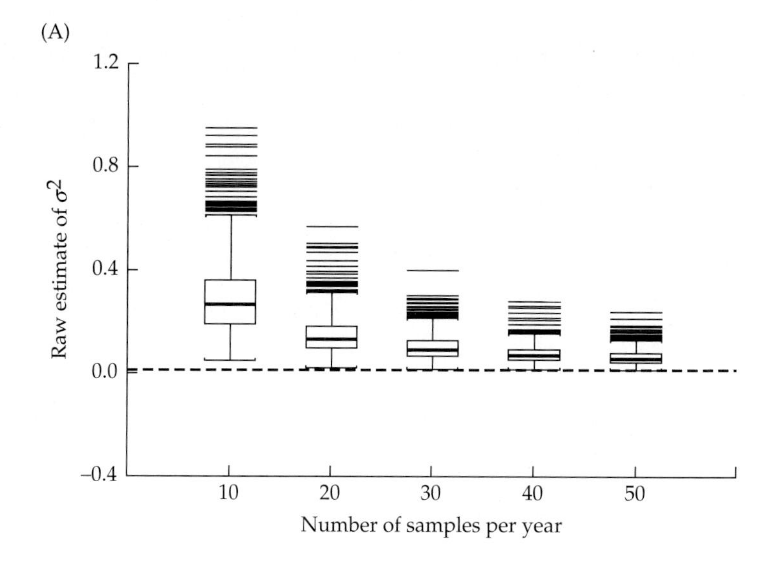

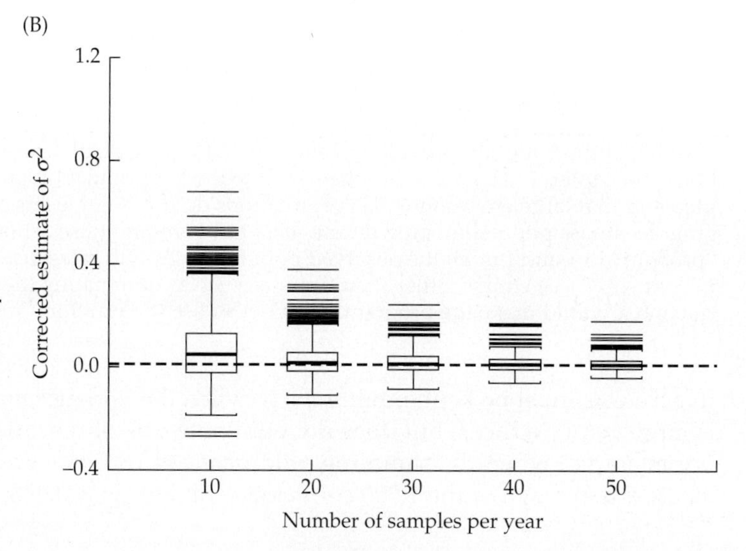

Figure 5.3 Increasing the number of samples taken per year decreases the imprecision in both the raw (A) and corrected (B) estimates of σ^2. Here, the true value of σ^2 is 0.01 (dashed lines), and the sampling variance is 16 times greater than the mean. Symbols as in Figure 5.2.

rected estimate is always less biased than the raw estimate (means of the estimates in Figure 5.3 for sample sizes of 10 through 50 were 0.0711 versus 0.2971, 0.0281 versus 0.1485, 0.0213 versus 0.1026, 0.0157 versus 0.0774, and 0.0146 versus 0.0643 for the corrected versus raw estimates, respectively). Because it will rarely be possible to estimate μ and σ^2 with as much precision as we would desire, it is important to compute confidence intervals for our estimates.[10] Third, the model used to generate the data sets in Figures 5.2 and 5.3 was density-independent (i.e., Equation 5.1). The method used in Box 5.1 may be less successful at correcting for sampling variation when the data reflect the influence of density dependence. Fourth, the approximation underlying Equation 5.2 assumes that the mean of the samples, $\overline{N}_t$, is an unbiased estimate of the true mean population density in year t. When undercounting and overcounting cause $\overline{N}_t$ to be a biased estimate of the true density, σ^2_{corr} will be less successful at removing the influence of observation error. For example, let's assume that counting errors cause the number of individuals observed in each plot to vary uniformly among years between 0% and 200% of the numbers actually present (that is, in some years all the counts are over-counts and in other years all are under-counts). Relative to the case when $\overline{N}_t$ is an unbiased estimate, this level of counting error increases σ^2_{corr} by approximately 20% when the variance of the samples equals their mean, but it increases σ^2_{corr} by only 3% when the variance is 16 times greater than the mean (results not shown). Taking steps to assure that the counts are as accurate as possible is the best way to reduce this bias. Finally, when the sampling variance is high, Equation 5.5 can produce negative values of σ^2_{corr} (Figures 5.2C and 5.3). Because a true variance cannot be negative, we would have little choice in such cases but to conclude that high sampling variation makes it impossible to rule out the possibility that the true value of σ^2 is actually zero. By running a pair of PVAs with σ^2 set both at zero and at the raw estimate, we could at least delimit the optimistic and pessimistic bounds of reality, respectively. Despite these caveats, when the variance among samples is modest, density dependence is weak, and the sample counts are reasonably accurate, Equation 5.5 presents a useful tool for obtaining better estimates of σ^2 in the face of sampling variation.

Using Repeated Censuses of the Same Area to Discount for Observation Error

We have to modify the approach taken in the preceding section to fit another commonly used method for censusing populations. Assume that in most

[10]An approximate confidence interval for σ^2_{corr} can be calculated using Equation 3.13 with σ^2_{corr} in place of $\hat{\sigma}^2$. However, we should recognize that the resulting interval is likely to be too narrow, as the range of possible values of σ^2_{corr} increases as the degree of sampling variation increases (see Figure 5.2C). For an alternative approach, see White (2000). Both Boxes 5.1 and 8.1 use White's method, which is further explained under *Discounting sampling variation* in Chapter 8.

years, we estimate population size by obtaining a single count of all individuals in an area that represents a known fraction of the total range of the population and by extrapolating that count to the population as a whole. (That is, if in year t we observe c_t individuals in an area that is a fraction b_t of the entire range, where $b_t \leq 1$, the estimate of total population size is $C_t = c_t/b_t$.) However, in some years, we repeatedly census a certain area (which may be smaller than the total area censused that year) over a short period of time in order to quantify the magnitude of observation error. We will assume that the variation among these repeated censuses is due to counting errors; it is not sampling variation because we are not censusing separate plots as we did in the preceding section. Because we do not have a set of replicate samples taken every year, we cannot simply use the sampling variances associated with the means of the samples (as we did in Equations 5.2 through 5.5) to correct C_t for observation error. However, we can try to extrapolate the observation error we saw in the repeated censuses to the whole-population scale, as follows. If V_t is the variance of the counts from the area repeatedly sampled in year t, a_t is the size of the area occupied by the entire population as a multiple of the area repeatedly sampled in year t (where $a_t \geq 1$), and C_t is the estimate of the total population size in year t, then the variance in C_t that is due to observation error is $Var(C_t) \approx a_t^2 V_t$. If $\log(C_{t+1}/C_t)$ is the estimate of the log population growth rate in year t, then replacing $\overline{N}_t$ with C_t in Equation 5.3 yields

$$Var(\log \lambda_t) \approx \frac{Var(C_t)}{C_t^2} + \frac{Var(C_{t+1})}{C_{t+1}^2} = \frac{a_t^2 V_t}{C_t^2} + \frac{a_{t+1}^2 V_{t+1}}{C_{t+1}^2} \tag{5.6}$$

Averaging the values of $Var(\log \lambda_i)$ across all years of the census and subtracting the average from the raw estimate $\hat{\sigma}^2$ computed using the total counts [i.e., the variance of the $\log(C_{t+1}/C_t)$'s] gives an estimate of σ^2 corrected for observation error. If repeated sampling was performed only once, the simplest option is to associate the single estimate of the observation error variance with each year's total count (i.e., use the single value of V_t for all t's in Equation 5.6).[11] However, if repeated sampling was performed in years with different population sizes, we can regress the separate variance estimates $Var(C_t)$ against the total counts in those years, and we can then use the resulting regression equation to predict the likely variances of the total counts from years in which repeated sampling was not performed.

As for the previous method, several caveats apply to Equation 5.6. First, it is based on the assumption that the $Var(C_t)$'s are not too large, and so it may not compensate completely for high levels of observation error. Second, as

[11]Alternatively, if we believe that counting errors are more likely in a larger population, we could set the observation error variance to be a constant multiple of the estimated total population size.

for N_t values in Equation 5.3, Equation 5.6 assumes that the C_t's are unbiased estimates of the true population size. This assumption may be correct when the observation error is due to sampling variation, the sampling is appropriately randomized, and the counts are fairly accurate, but it is not strictly true when the observation error results only from counting errors. Thus we can expect to be less successful at discounting for observation error in the present case than we were in the preceding section.

A Method to Reduce the Impact of Fluctuations in Population Structure and Observation Errors without Replicate Samples

In many cases, we will completely lack replicate counts taken during a single census. For example, every census may have taken only a total count, or the published source of the data may report only the means, not the separate counts in each sample. In such cases, we cannot use the methods described in the preceding sections. However, for a certain type of species and census, an alternative method to reduce the influence of observation error has recently been proposed by Holmes (2001). The method is most appropriate for species that have a short and well-known lifespan, with individuals that reproduce only once, and for which the census targets only a subset of the population (e.g., breeding females). For reasons we explain in detail below, we do not advocate the use of this method in other cases, although future development of this new technique may make it more widely applicable. The essence of the Holmes method is to use not the raw counts, but *running sums* of a predetermined number of contiguous counts, to estimate μ and σ^2. For example, if we are using a running sum with a length of three censuses, we would add the counts from the first, second, and third censuses, the second, third and fourth censuses, the third, fourth, and fifth censuses, and so on. With a total of n censuses and a running sum of length L, we would have $n - L + 1$ such running sums. These running sums are then used in a modification of the regression method of Dennis et al. (1991) to estimate μ and σ^2.

There are two distinct arguments for using running sums instead of single counts. First, when the counts include only a subset of a population, variation in the counts can be caused by fluctuations in the structure of the population (i.e., the relative proportions in different age, size, or developmental classes), as well as variation in environmental conditions. Ideally, we want our estimate of σ^2 to reflect only environmentally driven fluctuations in total population size, but the number of individuals in a particular class may vary more or less than does the number of individuals in the population as a whole, so that an estimate of σ^2 based on a subset of individuals may misrepresent the population-wide σ^2. Holmes (2001) argued that a running sum of the observed number of individuals in one class will more accurately represent the size of the population as a whole. To see why, consider a population of salmon in which individuals spawn at age 4 or 5 and then die. We census the population by counting adults as they return to spawning streams, but we know that the population also includes pre-breeding individuals that remain at sea and so are not counted. However, if we had additional demographic information, we could use the five most recent spawner counts to calculate the *total* population size at the present time. At the current census, the population will include, in addition to the observed spawners, one-year-olds that were produced by last year's spawners, two-year-olds that were produced by the spawners two years previously, and so on. Thus if S_t is the number of spawners in year t, then the total size of the population in year t is

$$T_t = S_t + f_{t-1}s_{1,t-1}S_{t-1} + f_{t-2}s_{2,t-2}S_{t-2} + f_{t-3}s_{3,t-3}S_{t-3} + f_{t-4}s_{4,t-4}S_{t-4} \quad (5.7)$$

where f_t is the number of offspring each spawner produces in year t and $s_{x,t}$ is the fraction of offspring born in year t that survive to age x and do not return to spawn. We know that the number of terms in the sum should be no larger than 5, because we know that individuals cannot live longer than 5 years. In addition, because we know that individual salmon spawn only once, we know that the spawners observed in a given year will not include any individuals observed to spawn in previous (or subsequent) years, so there is no double counting in Equation 5.7. According to Equation 5.7, the total size of the population at the current census is a *weighted* sum of the five most recent spawner counts, where the weights are determined by the values of the f_t's and the $s_{x,t}$'s, which change over time due to environmental variation. Unfortunately, in most cases, we do not know these values and so cannot compute the proper weights. Holmes proposed that in this situation, we use the *unweighted* sum of the five most recent counts as a surrogate for total population size, to at least partially reduce the potential for confounding fluctuations in population structure with environmental stochasticity affecting total population size.

The second reason to use running sums is that they can partially counteract observation errors. Counting errors will sometimes cause a single count to be an overestimate and sometimes an underestimate of the true population size. This is one reason why observation error artificially inflates the estimate of σ^2. However, the odds are small that over a set of contiguous censuses, *all* the counts will be overestimates or all of them underestimates. Rather, an overestimate will frequently be followed by an underestimate, and vice versa. That is, over a set of censuses, the positive and negative errors will tend to cancel each other out, so that the sum of the counts may be closer to the sum of the true population sizes than any one count is to the true population size in a single year. However, summing adjacent counts also removes some of the *real* year-to-year variation in population size, so that the variability of running sums may *underestimate* the annual variability of population size, especially if observation errors are actually small and the length of the running sum is large. The goal is to chose a running sum length that reduces observation error effects without leading to an underestimate of true population variability. The principal "rub" of the Holmes method is that there is no clear way to determine *a priori* what is the optimal running sum length, as we discuss in detail below. But first, we describe the method in its entirety.

As we noted above, the Holmes method employs a modified version of the regression method of Dennis et al. (1991) described in Chapter 3. In the Dennis et al. method, we regress the transformed log population growth rate between two successive censuses against the square root of the time elapsed between them. Instead, using the Holmes method we first match up all possible pairs of running sums separated by different elapsed times of $\tau = 1, 2, \ldots, \tau_{max}$ years. We then compute the log population growth rate over that time interval as $\log(R_{t+\tau} / R_t)$, where R_t is the running sum beginning in year t. Next, we compute the means and variances of the $\log(R_{t+\tau} / R_t)$ values for each value of τ.[12] Finally, we separately regress these means and variances against τ. The slope of a linear regression of the means against τ is an estimate of μ, and the slope of a linear regression of the variances against τ is an estimate of σ^2. A major difference between the regression method of Dennis et al. and that of Holmes is that in the former, the regression intercept should always be zero, whereas in the latter it is a fitted parameter. (In the case of the regression to estimate σ^2, the intercept represents the nonprocess error variance, that is, observation error plus "errors" due to fluctuations in population structure; see Holmes and Fagan, in press.) The reasoning behind these regressions is that the means and variances of $\log(R_{t+\tau} / R_t)$ are μ and σ^2 multiplied by the time elapsed between samples, τ (see Chapter 3 if you don't remember this idea). Thus, the slopes of these regressions are in fact measurements of μ and σ^2. Box 5.2 lists MATLAB code defining the function `dennisholmes` that estimates μ and σ^2 using Holmes' running-sum modification of the Dennis et al. regression method. Users must pass three things to the function: the name of a vector containing the census counts, the length of the running sum, and the maximum time lag in the linear regressions.[13]

To illustrate the ability of the Holmes method to reduce biases in σ^2 caused by observation error, we generated 1,000 simulated data sets from the stochastic density-independent model (Equation 5.1) with known values of μ and σ^2, as above. We then multiplied the "true" counts in each year by a

---

[12]Although the Dennis et al. method can easily handle them, missing censuses would cause the lengths of the running sums to differ when using the Holmes method. However, if there are relatively few missing census values, linear interpolation can be used to replace missing census counts.

[13]For example, if the vector `counts` contains the census counts, the command `[muest, s2est]=dennisholmes(counts, 3, 5)` will place the estimates of μ and σ^2 into the variables named `muest` and `s2est`, respectively, using a running sum of 3 and a maximum time lag of 5.

BOX 5.2 *MATLAB code defining the function "dennisholmes", which estimates μ and σ^2 using the method of Holmes (2001).*

```
function [mu,sigma2]=dennisholmes(RawCounts,L,taumax)
% dennisholmes(RawCounts,L,taumax) computes mu and sigma2
% using the Dennis-Holmes method;
% RawCounts=vector of census counts
% L=length of running sum
% taumax=maximum time lag in regressions;
% See Holmes, Proceeding of the National Academy of
% Sciences USA 98: 5072-5077 (2001).

NumCounts=length(RawCounts);

% First, compute running sums

RunSum=[];
for i=1:NumCounts-L+1
   RunSum=[RunSum sum(RawCounts(i:i+L-1))];
end;

% Second, compute means and variances of the log population
% growth rate at each time lag, tau

taus=[];      % array to store time lags
means=[];     % array to store means
vars=[];      % array to store variances
for tau=1:taumax                              % For each time lag,
   R=[];
   for i=1:NumCounts-L+1-tau                  % compute log growth
      R=[R log(RunSum(i+tau)/RunSum(i))];  % rates R using running
   end;                                         % sums, then compute
   taus=[taus; tau];
   means=[means; mean(R)];                    % their means and
   vars=[vars; var(R)];                       % variances
end;

% Third, compute regression slopes of means and vars. vs. tau

% Step 1: center tau, means, and vars by subtracting
% their respective means
taus=taus-mean(taus);
```

BOX 5.2 (continued)

```
means=means-mean(means);
vars=vars-mean(vars);

% Step 2: with centered dependent and independent variables,
% the right-hand sides of the following two lines
% give the slopes of linear regressions of the means and
% variances of the log growth rates on tau, which are
% our estimates of mu and sigma^2
mu=(taus'*means)/(taus'*taus);
sigma2=(taus'*vars)/(taus'*taus);
```

lognormally distributed random variable to mimic observation errors. The observation error distribution had a mean of 1 and a variance of either 0.01 or 0.05. For each corrupted data set, we then estimated μ and σ^2 by the standard method (i.e., as the mean and variance of the observed log population growth rates, using the counts contaminated with observation error) and by the Holmes method, using the function `dennisholmes`. Both methods produce unbiased estimates of μ, although with 20 years of data the estimates range widely around the true value (Figure 5.4A,C). However, the standard method yields estimates of σ^2 that are strongly upwardly biased, especially when the observation error variance is large (Figure 5.4B,D). For example, with an observation error variance of 0.05, the standard method gives a median estimate of σ^2 that is 10 times larger than its true value. Applying the Holmes method to the same data sets using a running sum length of 3 yields values for σ^2 that are again broadly distributed,[14] but that on average slightly underestimate the true value when the observation error variance is 0.01 and are nearly unbiased when the observation error variance is 0.05 (Figure 5.4B,D).

Thus the Holmes method does provide a way to reduce bias in estimates of σ^2 caused by observation error. However, as we noted above, its ability to do so without *underestimating* the true σ^2 hinges on making the correct choic-

[14]Because of this broad distribution, it is important to calculate a confidence interval for the running sum estimate of σ^2 so that we know how much faith to put in it. Holmes and Fagan (in press) show that an approximate $100(1-\alpha)\%$ confidence interval for an estimate $\hat{\sigma}^2$ obtained by the Holmes method is given by $[\,df\,\hat{\sigma}^2/\chi^2_{\alpha,df},\ df\,\hat{\sigma}^2/\chi^2_{1-\alpha,df}\,]$, where $\chi^2_{\alpha,df}$ is the 100αth percentile of the chi-square distribution with degrees of freedom given by $df = 0.333 + 0.212n - 0.387L$, where n is the number of census counts and L is the running sum length.

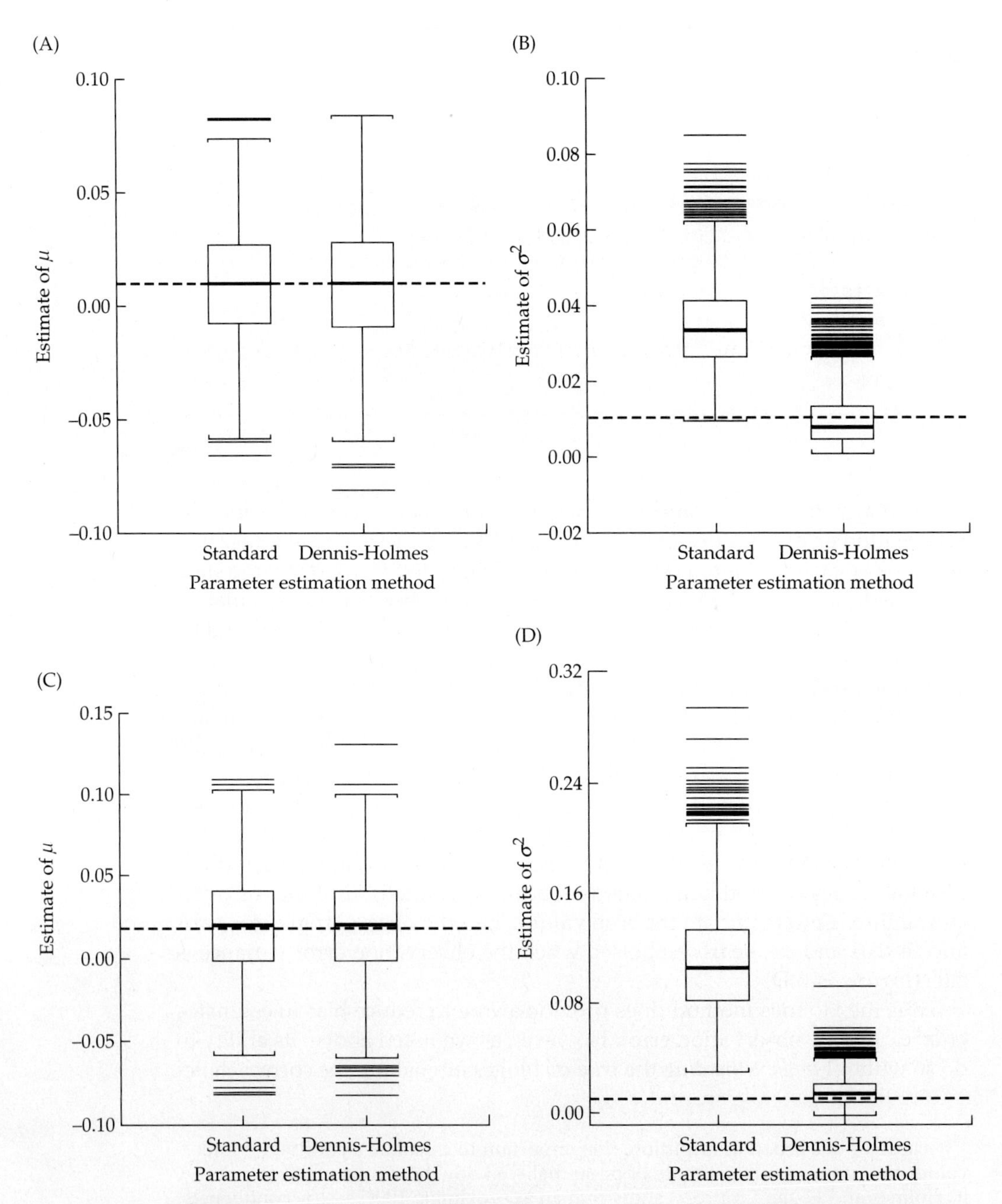

Figure 5.4 Demonstration of the effectiveness of the Dennis-Holmes method compared to the standard method of estimating μ and σ^2. In all cases the true value of both μ and σ^2 is 0.01 (dashed lines). The observation error variance was 0.01 (A,B) or 0.05 (C,D). Distributions of estimates for 1,000 simulated data sets, each of 21 years' duration, are shown. Symbols are as in Figure 5.2. The running sum length was 3, and the maximum time lag used in the regressions was 4.

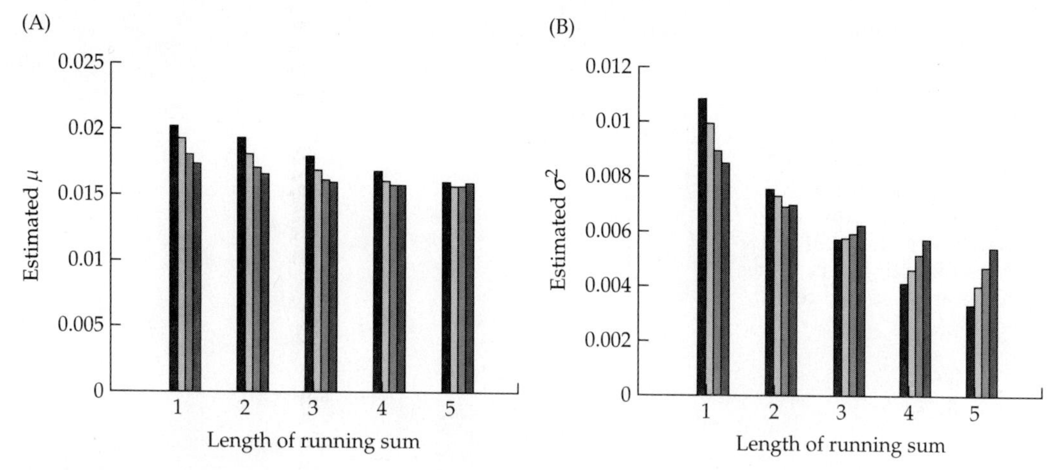

Figure 5.5 Estimates of μ and σ^2 for the Yellowstone grizzly bear using the Holmes method are sensitive to the choice of L (length of the running sum) and τ_{max} (maximum time lag in the regression). Within each group of bars with the same running sum length, τ_{max} increases from 3 (left) to 6 (right).

es for both τ_{max}, the maximum time lag in the regressions, and more importantly L, the length of the running sum. Both of these choices, but particularly the second, can strongly influence the resulting parameter estimates. For example, we applied the Holmes method to the census counts for the Yellowstone grizzly bear population (see the data in Table 3.1), varying the maximum time lag and running sum length (Figure 5.5). The estimate of μ is relatively insensitive to the choice of τ_{max} and L (Figure 5.5A), but there is a nearly threefold range of variation in the estimates of σ^2 across the values of τ_{max} and L (Figure 5.5B). Most notable is the fact that the longer the running sum, the smaller the estimate of σ^2 will be. Of course, we don't know what the "true" value of σ^2 is for the Yellowstone grizzly bear population (indeed, that is why we are trying to estimate it), so there is no clear way to determine which value of L yields the least-biased estimate. Using the simulated data sets above, where we do know the "true" value of σ^2, it becomes clear that choosing to compute too long a running sum will cause the resulting values to sharply underestimate the true value, thus producing overly optimistic viability assessments (Figure 5.6). Remember that this effect arises because calculating running sums averages out not only observation error, but also real variation in numbers due to environmental variability.

Thus we are left with a conundrum as to how to best determine the running sum length. It is clear from the logic underlying Equation 5.7 that if our goal is to minimize the effect of fluctuations in population structure on our estimate of σ^2, we should choose a running sum length that is close to the lifespan of the organism, hence approximating the total population size. Unfortunately, for long-lived organisms and with a census of limited dura-

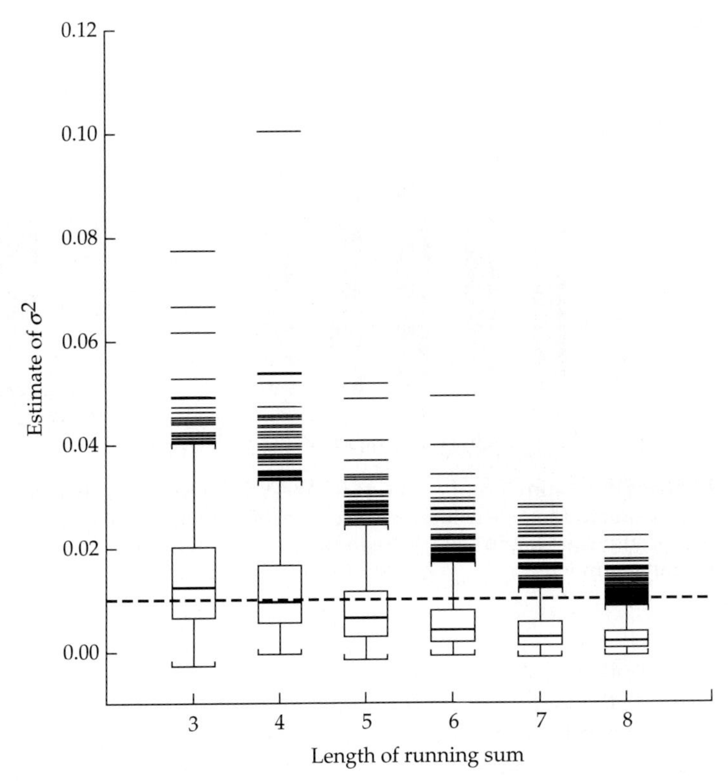

Figure 5.6 Using too long a running sum in the Holmes method can lead to severe underestimates of σ^2. The true value of σ^2 is indicated by the dashed line. Observation error variance = 0.05 and μ = 0.01. Each distribution shows estimates of σ^2 for 1,000 random data sets. Other symbols are as in Figure 5.2.

tion, it will be difficult to compute such long running sums and still have enough of them to estimate parameters accurately (for example, for grizzly bears, whose potential lifespan exceeds 30 years, even the 39 year census of the Yellowstone population would yield only 10 running sums of length 30 years). If our goal is to minimize observation error bias rather than the effects of fluctuations in population structure, then there are no clear guidelines as to how long the running sums should be. This question is currently the focus of active research. Until more precise guidelines emerge, we make two provisional recommendations:

1. Use the Holmes method only for organisms that live for only a few (but more than one) years and reproduce once (i.e., semelparous animals or monocarpic plants), and do not set the running sum length to be any longer than the known lifespan. If our goal in using the running sum is to approximate the total population size at the present time, adding more terms to the running sum is superfluous, and merely risks downwardly biasing the estimate of σ^2.

2. Repeat the analysis using a running sum length L that varies from 1 to the known lifespan, and see how much the predicted population viability is affected. The true value of σ^2 is likely to lie somewhere between the value obtained with $L = 1$ (which is likely to be an overestimate due to fluctuations in population structure and observation errors) and the value obtained with L equal to the lifespan (which is likely to be downwardly biased due to excessive averaging).

There is another reason to keep the running sums relatively short. Using short running sums, we are more likely to overestimate rather than underestimate the value of σ^2. The resulting somewhat pessimistic estimates of population viability might cause us to take action to manage a population sooner than we otherwise might, a sensible outcome if we are taking a precautionary approach. More generally, if observation errors or fluctuations in population structures are thought to be small relative to real variability, we recommend not using the Holmes method, as it is likely to underestimate σ^2 and thus overestimate viability. Finally, if your data seem appropriate for the Holmes method, we suggest that before proceeding with any analysis you look for emerging studies that explore the methodology and tradeoffs of this still very new approach (e.g., Holmes and Fagan, in press).

A Directory to More Advanced Methods for Estimating Parameters in the Face of Observation Error

In this chapter, we have described a few simple methods to reduce the influence of observation error and to discount it when it occurs. More powerful but also more complex methods exist for estimating population parameters in the face of observation error. In closing the chapter, we direct interested readers to references on two such methods.

The first approach we briefly describe is the so-called total least-squares method (Ludwig and Walters 1981, Ludwig et al. 1988, Ludwig 1999). To use it, one assumes that the observed log population size in year t equals the log of the true population size in year t plus a deviation due to observation error. In turn, the true log-population size in year $t + 1$ represents the size predicted by the deterministic part of the population model given the true population size in year t plus a deviation due to environmental stochasticity. One then writes a sum of squares term as a weighted sum of the squared deviations due to observation errors and the squared deviations due to environmental stochasticity. Then one uses numerical methods to find values for the model parameters as well as the observation error deviations (which are treated as unknown parameters) that minimize this sum of squares. To implement this method, one must first obtain an estimate of the observation error variance, for example as the variance of replicate samples (see Carpenter et al. 1994), or by repeatedly censusing the same area, as described above. Alternatively, one can use a prior estimate of the ratio of the observation error variance and the environmental variance (Ludwig and Walters 1981, Ludwig et al. 1988, Ludwig 1999). If one lacks an estimate of the observation error variance and simply guesses at the ratio of the two variances, biased estimates of the model parameters and associated extinction risk metrics will result if the guess is not reasonably accurate (see Figure 1 in Ludwig 1999). For the many count-based data sets for which no information about the observation error variance is available, the method of total least squares will be impossible to apply.

The second, more sophisticated approach, state-space modeling, does not require prior information about the observation error variance. A state-space model has two submodels, a population model that describes the dynamics of the "true" but unobserved population size in a stochastic environment and an observation model that describes how the observed population size is related to the true population size given possible sources of observation error. The data are a series of observed population sizes (i.e., census counts). To estimate parameters, one begins by writing a likelihood function to describe the probability of obtaining the observed population sizes given a particular set of values of the parameters in the population and observation submodels (including the environmental variance and the observation error variance). One then searches for parameter values that maximize the likelihood function.[15] In general, the likelihood function involves a complex product of integrals. Thus, calculating the likelihood for a set of parameter values requires either direct numerical integration (de Valpine and Hastings 2002) or stochastic integration schemes such those based on Markov Chain Monte Carlo techniques (Carlin et al. 1992, Bjørnstad et al. 1999). At present, implementing these methods is a highly technical endeavor. However, as they become more generally used in ecology over the coming years, we expect they will become widely applied tools for accounting for observation error in population viability analyses.

[15]See Chapter 4 for a general description of likelihood approaches to model fitting.

6

Demographic PVAs: Using Demographic Data to Build Stochastic Projection Matrix Models

The methods we have discussed in previous chapters assume that all individuals are identical, so that the models only need to consider the total number of individuals. But for many organisms, particularly the long-lived ones that are often of special conservation concern, individuals differ in important ways that affect their current and future contributions to population growth. For example, larger individuals often have a greater chance of surviving and a higher reproductive rate than smaller individuals. As a result, two populations that differ in the proportion of larger versus smaller individuals may have very different viabilities, even if they are of the same total size and experience the same sequence of environmental conditions. Populations in which individuals differ in their contributions to population growth are known as *structured populations*. The principal tool for assessing the viability of structured populations is the *population projection matrix* model, which divides the population into discrete classes and tracks the contribution of individuals in each class at one census to all classes in the following census. Different variables describing the "state" of an individual, such as size, age, or stage (e.g., seeds, seedlings, and adults for plants, or larvae, juveniles, and adults for fish or invertebrates), are most appropriate for different organisms as ways to classify individuals. As was true of the models for unstructured populations in Chapters 3, 4, and 5, projection matrix models must account not only for mean contributions of one stage to each of the others, but also for variability in these rates if they are to yield accurate assessments of population fate.

Projection matrix models have advantages and disadvantages relative to the simpler count-based models covered in the previous chapters. As we just indicated, one advantage is that structured models provide a more accurate portrayal of populations in which individuals differ in their contributions to population growth. A second advantage is that structured models help us to make more targeted management decisions. For example, we can use a structured model to ask whether, say, increasing the survival of younger or smaller individuals will have a greater impact on a population's growth rate than will enhancing the reproductive rate of adults. Such questions are simply impossible to answer with an unstructured model. The principal disadvantage of structured models is that they contain more parameters than do simpler models, and hence they require both more data and different kinds of data. Typically, the construction of a projection matrix model requires data from a demographic study of marked individuals, an endeavor that is both more time-consuming and more expensive than conducting a simple census of an unmarked population. Nevertheless, the advantages of incorporating population structure into viability assessments and management decisions will sometimes make more extensive data collection efforts worth the price, especially for keystone, indicator, or umbrella species, or for high-profile species for which it is easier to mount the necessary logistical and financial support.

In this and the subsequent three chapters, we discuss in some detail techniques that use deterministic and stochastic matrix models to assess the viability of structured populations and to identify management strategies to enhance the health of such populations. In this chapter we focus on the estimation of each year's demographic rates, and in Chapters 7 and 8 we discuss how to put together these estimates to incorporate environmental stochasticity into matrix model predictions. As we did for count-based PVAs, we begin with the simplest models for demographic PVAs, which include the effects of inter-individual differences and environmental stochasticity, but we omit density dependence, demographic stochasticity, temporal environmental autocorrelation, and catastrophes and bonanzas. One or more of these factors are often included in matrix models, but frequently the data needed to quantify them are lacking, and the issues of how to use demographic data to construct a matrix model are complex enough without at first considering these added complications. However, once we have covered the basics of how to construct and analyze simpler demographic models, we return to methods for including these complicating factors at the end of Chapter 7 and in Chapter 8. Finally, in Chapter 9, we discuss the use of projection matrix models in the management of at-risk populations.

Overview of Procedures for Building Projection Matrices

Individuals in structured populations may differ in any of three general types of demographic processes, the so-called vital rates. First, the probability of survival may depend upon an individual's state. Second, given that an individual survives from one census to the next, the probability that it will be in a particular state (e.g., a certain size class) will usually depend upon its state at the beginning of the time period. (The exception is age because all who *sur-*

vive one year must age one year.) Third, an individual's state may influence the number of offspring it produces between one census and the next. Throughout this book, we will consistently refer to these three vital rates as the *survival rate*, *state transition rate* (or *growth rate* for a size-structured population), and *fertility rate*, respectively. The elements in a projection matrix represent different combinations of these vital rates.[1] For example, the probability that a small individual this year will be a medium-sized one next year equals the probability it survives times the probability it grows. The projection matrix itself allows us to integrate the contributions of individuals in different states and to combine the effects of different vital rates into an overall measure of population growth and viability. Even though matrix elements typically combine several vital rates, it is still valuable to estimate the vital rates separately because particular threats or management actions may affect one particular vital rate and not others, and that rate may contribute to multiple elements in the matrix. Thus understanding the underlying vital rates allows us to account for the full effects of changes in management throughout the matrix, as we describe in detail in Chapter 9.

The construction of a stochastic projection matrix model involves four general procedures:

1. Conduct a detailed demographic study of a representative set of marked individuals, measuring their survival, state, and rate of reproduction each year over several years.
2. Determine the best state variable (e.g., age, size, or stage) upon which to classify individuals, as well as the number and boundaries of classes.
3. Use the demographic data to estimate the vital rates for each class in each year.
4. Use the class-specific vital rate estimates to build a deterministic or stochastic projection matrix model.

In the remainder of this chapter, we flesh out each of these general procedures. If you are completely unfamiliar with projection matrices, you may wish to first read both the section on *Basic matrix structure* later in this chapter and Chapter 7, so that you understand up front the target toward which all the sections that precede it are aiming.

Step 1: Conducting a Demographic Study

Unlike building the models in Chapters 3 through 5, which could be parameterized by using data on simple counts of the total number of individuals in a population (or an identifiable subset of the population), in building a

[1]Some sources refer to actual projection matrix entries as vital rates. Here, we use the terms *matrix elements* to indicate the matrix entries and *vital rates* to represent underlying survival, growth, and reproductive rates.

projection matrix model one must typically follow the states and fates of a set of known individuals over several years. Although conceptually simple, the collection of such demographic data represents the greatest investment of effort and expense involved in performing a PVA for a structured population. Hence we devote attention here to the proper design of an informative demographic study.

At the outset of a demographic study, the individuals to be followed are "marked" in a way that allows them to be re-identified at subsequent censuses. Useful methods for marking individuals (e.g., numbered tags, colored leg bands, toe-clipping, notches in shells, unique morphological features, DNA markers, or even simple map coordinates) vary depending upon the morphology and vagility of the study organism. Ideally, the mark should be permanent but should not alter any of the organism's vital rates (for example by making it more or less vulnerable to predation, or by affecting its mating success). At the time of the initial marking, the state of each individual should also be determined, for example by measuring its size (weight, height, girth, number of leaves or stems, and so on) and by determining its age (for example by tooth wear or by annual growth rings) if possible. Although sometimes you can conduct an entire demographic study using multiple measures of state, more commonly the best state variable to use is determined at the outset and used exclusively thereafter (see *Step 2: Establishing classes*).

Care should be taken to ensure that the individuals included in the demographic study are representative of the population as a whole (e.g., by including all individuals encountered along randomly located transects extended through a plant population, or by capturing animals to be marked at a variety of sites within the geographical area encompassing the study population). Some classes of individuals (e.g., the largest adults) may be relatively rare, yet may contribute disproportionately to population growth. Because a random sample of moderate size may include only a few or none of these individuals, a stratified random sampling scheme may be needed to ensure adequate representation of such individuals in the study. (For example, one could mark only the largest individuals in every other transect through a plant population, or make special efforts to capture and mark old individuals in an animal population.)

The marked individuals should then be censused at regular intervals. This same interval will be used for calculating the vital rates (e.g., survival over a year) and for projecting the future size and structure of the population. Because seasonality is ubiquitous, for most species and places, the best choice is to census, and hence project, over one-year intervals. At each census, one simply records which of the marked individuals are still alive and the current state of survivors (i.e., their age, size, or stage). In general, the census should be conducted at the time of year when individuals are most easily detected (perhaps at a time when birds are making territorial displays, or when individuals are forming conspicuous groups prior to breeding or migration). However, answering some management questions will require censusing the population more than once a year (e.g., to obtain separate estimates of survival over winter and summer when the population may be facing different threats).

A final consideration about the timing of the census is the need to estimate the number of offspring to which each individual gave birth over the previous inter-census interval. Consequently, it is also a good idea to consider the timing of reproduction while deciding when to conduct the census (see the section on *Reproduction* under *Building the matrix from the underlying vital rates*). For many organisms, especially those in markedly seasonal environments, reproduction is concentrated in a small interval of time each year. These organisms have so-called birth pulse populations (Caswell 2001). In contrast, organisms with so-called birth flow populations reproduce continuously throughout the year. For animals with birth-pulse populations and short-term parental care, it may be best to conduct the census immediately after the birth pulse, while newborns are still associated with their mothers. For many plants, the release of seeds represents the birth pulse. For such populations, it makes sense to conduct the census just before the pulse, while the number of seeds produced by each parent plant can still be determined. For birth-flow populations, frequent checks of potentially reproductive individuals at time points within an inter-census intervals may be necessary to estimate annual per-capita offspring production, or more sophisticated methods (e.g., DNA fingerprinting) may be needed to identify the parents of newborn individuals observed at the next census.

For some organisms, special procedures other than a census of marked individuals may be needed to quantify some aspects of their life histories. Many plants produce seeds that lie dormant in a seed bank for years. To quantify all aspects of the life cycle of such organisms, the demographer may need to supplement a census of above-ground plants with additional experiments. For example, one could add known numbers of seeds to some plots and leave adjacent control plots unmanipulated. By quantifying seedling emergence in the addition plots over subsequent years, and correcting for background germination by subtracting the number of germinants in control plots, one could estimate the fraction of seeds that exhibit 0, 1, or more years of dormancy. Another common problem requiring special procedures is the estimation of survival during juvenile dispersal. Dispersal often occurs over a short time period, can result in very low survival, and makes it very difficult to locate surviving individuals. Thus, quantifying survivorship over the dispersal period can require radio-tracking, marking a large number of potential dispersers and searching extensively for them later, or other extraordinary efforts.

Once individuals have been marked and the census protocol established, data collection should be repeated as many times as researcher motivation, time, and financial constraints allow. If at all possible, the aim is to construct a stochastic model; hence enough censuses are needed to estimate the variability in the vital rates (see *Step 4: Building the projection matrix*). In order to estimate all vital rates in each year of the study, at each census a fresh group of newborns (or one-year-olds, depending on when the census is conducted) must be marked and followed for at least their first one or two years of life. In addition, because older individuals that were marked initially will eventually die out of the population, it may be necessary to add new marked individuals in other stages to maintain adequate sample sizes.

Step 2: Establishing Classes

As noted above, a projection matrix model categorizes individuals into discrete classes. Yet two state variables, age and size, that are often correlated with an individual's demographic contributions are continuous. Hence the first step in constructing the model is to use the demographic data to decide which state variable to use as the classifying variable (often done with preliminary data early in the study), and if it is continuous, how to break the state variable into a set of discrete classes (best done at the end of the study).

One possible way to begin classifying individuals is by sex, but most projection matrix models keep track of only a single gender, such as all individuals for hermaphroditic species or females in species with separate sexes. Tracking only females is usually justified by the assumption that (in promiscuously mating species) a single male can fertilize several females or that (in monogamous species) male and female numbers are similar. In either case, the number of females would closely correspond to the rate of offspring production. When these assumptions do not hold, structured population models that explicitly include more than one gender are sometimes used (e.g., for monogamous species with unequal sex ratios, or gynodioecious plants), but they require additional demographic information (such as how reproduction depends on the sex ratio, survival differences between males and females, or the rates of seed production by females and hermaphrodites). We will not discuss two-sex models further here, but for more details, consult Chapter 17 in Caswell (2001).

Usually, we must decide whether to classify individuals by age, size, or stage. One important basis for making this decision is simple practicality. For many species, there is no way to determine nondestructively the age of individuals, precluding age as the classifying variable, whereas size can often be quickly measured in the field. Obvious morphological or life history stages (e.g., larval stages, dormant seeds, juveniles, and so on) that are easily recognized in the field are likely to be used when collecting demographic data and should usually be retained in classifying individuals in the model. It is important to remember that it is possible to use combinations of classifying variables, such as stage and size (e.g., a dormant seed stage plus different sizes of above-ground plants) or age and stage (e.g., 1-, 2-, and 3-year-olds plus all older individuals combined into an adult stage). The only requirement is that an individual can only be a member of one class at one time.

Basic features of an organism's biology also help to determine how best to classify individuals. For example, for indeterminate growers (in which equal-aged individuals may differ dramatically in size, but size strongly determines rates of survival, growth, and fertility), size is almost always a better classifying variable than age. For determinate growers, it is often convenient to combine all full-sized individuals into a terminal adult stage, provided individuals in this stage do not show substantial changes in survival or reproduction with age (e.g., due to senescence), in which case it makes sense to break up the adult stage into several classes.

Apart from practicalities and biological rules-of-thumb, there are several other features to consider when choosing a state variable and deciding how to break it into discrete classes. An ideal state variable will be highly correlated with all vital rates for a population, allowing accurate prediction of an individual's reproductive rate, survival, and growth. A major problem with some measures of size is that they do not accurately predict vital rates. For example, the future survival and growth rates of some plants with belowground storage organs may be only marginally predicted by, say, the current height of the stem, because aboveground height is not well correlated with belowground stores of carbohydrate. A second desirable feature in a state variable is accuracy of measurement. If repeatability of measurements is low, there will be little accuracy in estimates of growth, shrinkage, or stasis. Some measures of size, such as total leaf area, that would otherwise be excellent state variables, may be so difficult to measure accurately (at least with the amount of time that can be invested in measuring each plant when following hundreds or more individuals for a demographic study) that they are poor choices for a state variable. The best state variable is one that achieves a balance between accuracy and practicality.

When insufficient biological information is available to decide on a suitable state variable prior to the onset of a demographic study, graphical and statistical analysis of the study results can help to identify state variables that have the best power to predict differences in vital rates among individuals. These same tools yield estimates of the vital rates that are used to construct the projection matrix (see *Step 3: Estimating vital rates*). The appropriate statistical tool to use depends upon: 1) whether the vital rate is a binary property of individuals (for example, individuals either survive or they do not), a discrete but non-binary process (e.g., birds may fledge, say, one to five offspring in a breeding season), or a nearly continuous variable (for example, offspring number for fecund species may vary in an effectively continuous manner among individuals); and 2) whether the prospective classifying variable is continuous (e.g., age or size) or discrete (stage). Table 6.1 lists appropriate statistical tools for various combinations of vital rates and classifying variables.

We now provide examples of three of the analyses in Table 6.1, using data from endangered populations.

TABLE 6.1 Appropriate statistical tools for testing associations between vital rates and potential classifying variables

	Vital rate		
Classifying variable	*Survival or reproduction (binary)*	*Reproduction (discrete but not binary)*	*Reproduction or growth (continuous or nearly so)*
Age or size (continuous)	Logistic regression (or hazard functions)[a]	Generalized linear models (e.g., Poisson regression)[b]	Linear, polynomial, or nonlinear regression
Stage (discrete)	Log-linear models (G tests)	Log-linear models (G tests)	ANOVA (or nonparametric analogs)

[a]Simple logistic regression models predict steadily increasing or decreasing survival probabilities as a function of size or age, but logistic models can also include more complex patterns, such as survival rates that rise as animals move from youth to middle-age and then fall in age as animals senesce. If age is A and a, b, and c are constants, a logistic equation that can include such effects is: survival = $\exp(a + bA + cA^2)/[1+ \exp(a + bA + cA^2)]$. If b is positive and c is negative, survival will reach a maximum at an intermediate age. In spite of the flexibility of adding higher order terms to a logistic function (A^2, A^3, etc.), for some groups, it may be more natural to use a different equation to predict age- or size-dependent survival. In particular, a "proportional hazards function" (Eberhardt and Siniff 1988) has been formulated to flexibly model the usual patterns of age-dependent survival (or fertility) of large mammals that include juvenile, mature, and senescent phases. If you are fitting this or other functions to survival data using a nonlinear fitting routine in a statistical package, be sure to require that the fitting procedure uses binomial probabilities to calculate log likelihoods. (This is exactly the way maximum likelihood fitting of logistic regression models is done and is well-described in Sokal and Rohlf 1995.)

[b]See Footnote 4 in the text.

Example 1: Logistic Regression of Survival Versus Size

To determine whether survival is associated with a continuous classifying variable such as size or age, logistic regression is usually the appropriate analytical method to use. An example of such an analysis is provided by mountain golden heather (*Hudsonia montana*), a federally listed threatened plant found only in western North Carolina. Mountain golden heather is threatened by increased competition with other plants caused by suppression of natural wildfires. Frost (1990) undertook a demographic study of tagged plants between 1985 and 1989. In an annual census conducted at the end of each summer, he determined which individuals were still alive, and their size. (The plant is a ground-hugging species, and Frost measured size as the two-dimensional area of each individual.) In this and the following three chapters, we will draw on Frost's study extensively to illustrate how to conduct a PVA for a structured population.

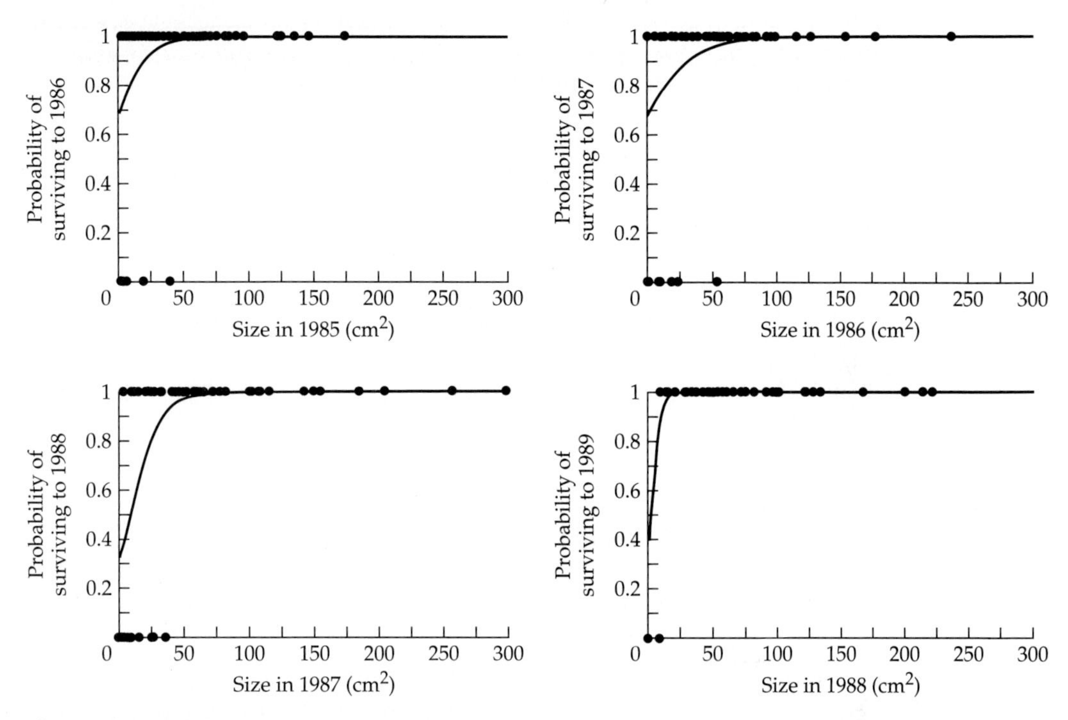

Figure 6.1 Survival versus size for mountain golden heather (data from Frost 1990). Points show fates of individual plants (0 = dead, 1 = alive). Best-fit lines were obtained by logistic regression (for regression coefficients, see Table 6.2).

To see if annual survival of mountain golden heather plants is related to size, we first assign to each individual that was alive at census i a binary survival variable ("1" if it lived to census $i + 1$ and "0" if it died). We then perform a logistic regression of the survival variable against size at census i (for example, using `proc logistic` in SAS[2]). Logistic regressions for survival versus size are shown in Figure 6.1. Despite relatively small sample sizes, there was a significant or marginally significant increase in the probability of survival with plant size in three of four years (Table 6.2). Therefore, there is statistical justification for classifying individuals by size in the projection matrix.

[2]If you are using SAS and are coding death as "0" and survival as "1", you must use the `descending` command to model the probability of survival, as the SAS default is to model the proportion of zeros. For example, if the variable `alive` contains the binary survival values, and `area` contains measures of plant size, the following SAS commands will fit a simple logistic regression model to the data:

```
proc logistic descending;
  model alive=area;
run;
```

TABLE 6.2 Results of logistic regressions of survival versus size using data for *Hudsonia montana* in Figure 6.1[a]

Independent variable	*Dependent variable*	b_0	b_1	P
Size in 1985	Survival from 1985 to 1986	0.7434	0.0734	0.0521
Size in 1986	Survival from 1986 to 1987	0.7336	0.0477	0.0602
Size in 1987	Survival from 1987 to 1988	–0.8144	0.0883	0.0076
Size in 1988	Survival from 1988 to 1989	–0.6981	0.2968	0.1847

[a]b_0 and b_1 are the coefficients in the logistic regression (Equation 6.2), and P is the probability of obtaining the observed value of b_1 if the true value is zero (i.e., if size has no effect on survival).

Example 2: Log-Linear Model of Survival Versus Stage

When the potential classifying variable is discrete (i.e., stages), statistical association between survival and the classifying variable should be assessed using log-linear models (G tests; Bishop et al. 1975). Walters and colleagues (Walters et al. 1988, Walters 1990) conducted a long-term study of the federally endangered red-cockaded woodpecker (*Picoides borealis*) in North Carolina. The red-cockaded woodpecker is a cooperative breeder, and Walters and colleagues classified male birds into stages according to their breeding status (fledglings, helpers-at-the-nest, floaters, solitary males (territorial birds without mates), and breeding males). The number of marked birds in each stage that survived or died over one year intervals in the period 1981–1986 are shown in Table 6.3 (data from Table 3.2 in Walters, 1990). The results of a G test using these data ($G = 182.86$ with four degrees of freedom, $p < 0.001$) confirm a strong influence of stage on survival.[3] On the basis of these and additional data, Heppell et al. (1994) developed a projection

[3]In SAS, this analysis can be performed by creating a data set with the variables `stage` (i.e., type of male), `fate` (1 if alive and 0 if dead), and `count` (the number of males of each type experiencing each fate, from Table 6.3), and by using the command:

```
proc freq;
  weight count;
  tables stage*fate/chisq;
run;
```

The output labeled `Likelihood ratio chi square` gives the desired result.

TABLE 6.3 Survival data for red-cockaded woodpeckers in different reproductive stages, from Walters (1990)

Stage	*Total number of bird-years*	*Fate at the end of a one-year interval*		*Proportion surviving one year*
		Dead	*Alive*	
Fledglings	616	345	271	0.44
Solitary males	131	50	81	0.62
Helpers-at-the-nest	273	60	213	0.78
Breeding males	838	201	637	0.76
Floaters	29	11	18	0.62

matrix model for red-cockaded woodpeckers that followed the number of males rather than females, as is more typically done. Two reasons for this choice are that populations appear to be limited by the availability of suitable territories (and male numbers more closely follow territory availability than do female numbers) and that only males help at the nest (and helpers do increase the number of young fledged per nest; Walters 1990).

Example 3: Polynomial Regression of Seed Production Versus Size

Unlike survival which, at the level of the individual organism, either occurs or does not, reproduction often takes on a more continuous range of possible values.[4] Exceptions are common. For example, large mammals may produce either zero or one offspring in a given year; these cases can be analyzed by logistic regression or *G* tests (Table 6.1), following the procedures used in Examples 1 and 2. Assessing relationships between more continuously varying reproductive output and continuous classifying variables such as size or age typically relies upon some form of regression analysis (linear, polynomial, or nonlinear regression). For example, the number of seeds produced in 1987 and 1988 by mountain golden heather plants of different sizes in Frost's (1990) study is shown in Figure 6.2. (Mountain golden heather produces all of its

[4]Of course, actual offspring production per parent is *always* a discrete (i.e., integer) number. However, when the number of offspring is large and approximately normally distributed, little error is introduced by using familiar statistical analyses such as linear or polynomial regression that assume the dependent variable is continuous. An alternative that makes sense when analyzing offspring numbers that are more modest is to use a generalized linear model (also known as analysis of deviance) that assumes the response variable is Poisson-distributed. This analysis is not available in all statistical packages, but can be performed, for example, using the `genmod` procedure in SAS.

fruits at the end of the growing season, so it is possible to quantify total annual seed production by counting fruits only once at summer's end, and then multiplying fruit number by the average number of viable seeds per fruit.) Because seed production is not necessarily a linear function of plant size, stepwise regressions of seed number against plant area, the square of area, and the cube of area were performed for the two years separately.[5] As plants with an area close to zero do not reproduce, and to prevent the regression equation from predicting negative seed numbers at small plant sizes, the regressions were forced to have a y-intercept of zero. The first-order terms are significant in the regression models chosen by the stepwise procedure for both years, indicating that there is a statistically significant (and positive) relationship between plant area and seed production. The squared and cubic terms in the regression were also significant in 1987, indicating that the relationship between seed production and size was nonlinear in that year[6] (Figure 6.2).

Choosing Among State Variables

When two or more potential classifying variables are statistically related to the vital rates, there are several ways to decide which of them to use. For example, a continuous measure of reproduction (e.g., seed production) might be significantly correlated with both age and size. If so, one could perform a stepwise regression, adding size to a regression model with age only, and age to a regression model with size only, to see which feature is a better predictor. (For example, if adding age to a size-only regression does not improve r^2 significantly, but adding size to an age-only regression does, size is the better predictor.) One can also test whether age or size better predicts a binary vital rate (e.g., survival) using log-linear models (for details, see Section 3.3 in Caswell 2001), although such analyses require that continuous state variables (age and size) be discretized first. Sometimes, age may continue to contribute explanatory power to a vital rate even after accounting for size, and vice versa. In such cases, matrix models that classify individuals by both size and age could be used (Law 1983). Although the initial classification of such a model is more complicated, subsequent analysis is the same as for a single classifying variable. In Chapter 11, we will discuss multiple classifications of this kind that involve different habitats or populations, but we don't pursue them further here.

---

[5]Specifically, a data set with the variables `year`, `seeds`, `area`, `area2`, and `area3` (the last two representing the square and cube of the area for each plant) was analyzed with the following SAS command:

```
proc reg;
   by year;
   model seeds=area area2 area3 / noint selection=stepwise;
run;
```

[6]If we remove the largest plant in 1987, which could be having an undue influence on the results, the stepwise regression identifies the best model as $y = 0.4828\,x + 1.38{\times}10^{-5}\,x^3$, which is still nonlinear.

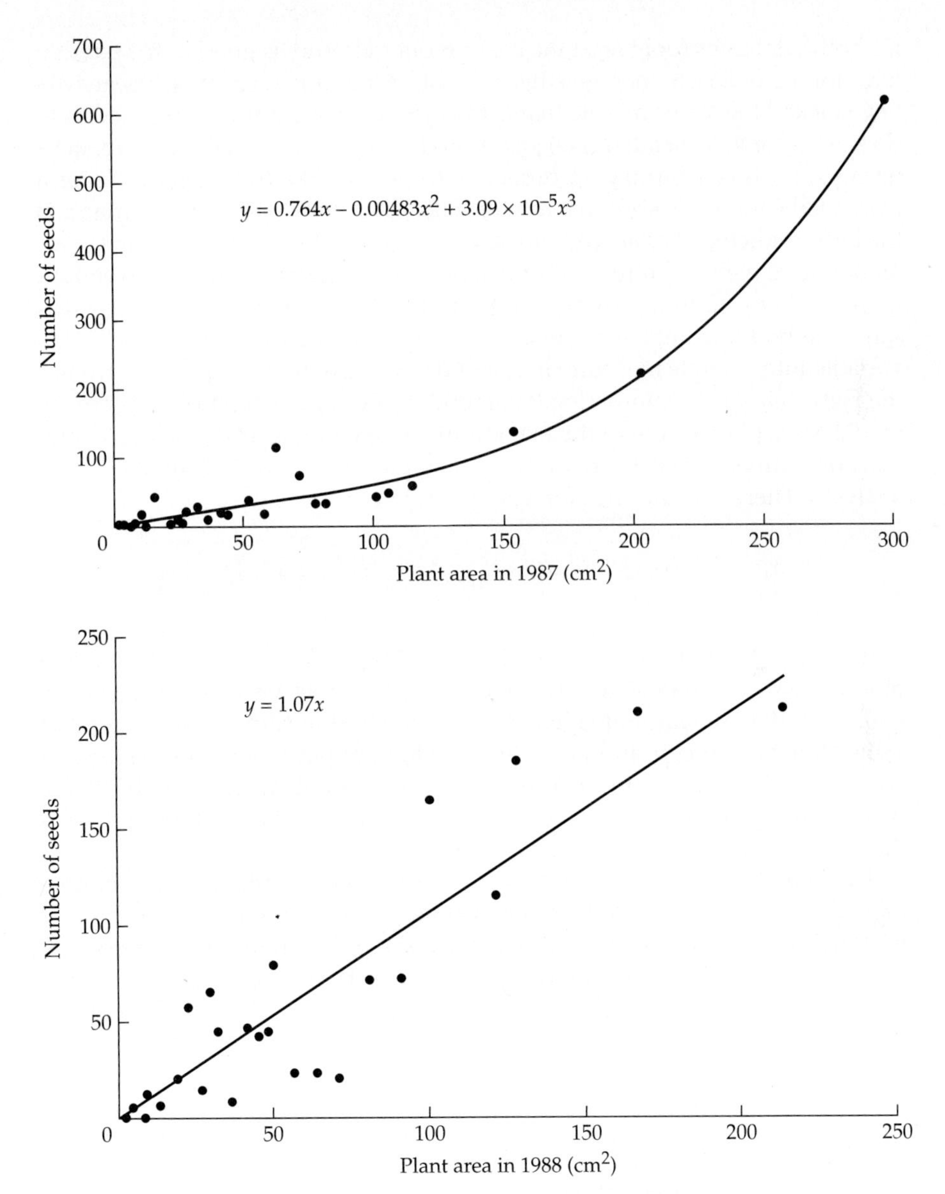

Figure 6.2 Seed production versus size for mountain golden heather in two years (data from Frost 1990). Seed number was estimated by counting fruits in photographic images of census plants and multiplying by the average number of viable seeds per fruit (for details, see Gross et al. 1998).

If you are trying to choose between two or more different state variables (e.g., age versus size, or several different ways of measuring size), then you should assess the ability of each state variable to predict *both* survival *and* fertility.[7] Most of the time the same state variable will prove the best predictor for both vital rates (or at least the best for one rate and as good as the alternatives for the other). Sometimes this will not be the case. If so, most researchers have picked the state variable that is the most convenient to use and seems to do a good (or at least adequate) job of predicting the two vital rates. An alternative (and less arbitrary) approach is to use Akaike Information Criteria (corrected for sample size) to more formally determine which state variable is the best predictor over all vital rates. We introduced the use of AIC_c in Chapter 4, where we used it to distinguish between different density-dependent models of population growth (see Equation 4.8). Now we wish to use this same method to combine the results of two different analyses for each state variable into a single measure of model fit. To make this comparison, we use the facts that the maximum log likelihood of the combined survival/fertility model is simply the sum of the maximum log likelihoods from the two analyses, and similarly that the total number of parameters is the sum over both analyses. Therefore, the AIC_c for a single state variable is

$$AIC_c = -2\left(\log L_{\max,s} + \log L_{\max,f}\right) + \frac{2p_s n_s}{n_s - p_s - 1} + \frac{2p_f n_f}{n_f - p_f - 1} \quad (6.1)$$

where $L_{max,\,s}$ is the maximum log likelihood of the best (i.e., smallest AIC_c) model from the analysis of survival using the state variable in question, $\log L_{max,f}$ is the maximum log likelihood of the best fertility model using that same state variable, p_s and p_f are the numbers of parameters in the best-fit models, and n_s and n_f are the numbers of data points in each analysis. We would simply compute AIC_c for each state variable and choose the one with the lowest value.

To illustrate this procedure, let's assume we have determined that a simple logistic regression best predicts survival versus age, and a linear regression best predicts fertility versus age. The former regression has two parameters (see *Step 3: Estimating vital rates*) and the latter three (slope, intercept, and residual variance). For a least-squares regression of fertility (and for analyses

---

[7]Here we have deliberately omitted the third type of vital rate, the rate of transition among states (e.g., growth), for two reasons. First, state transitions are idiosyncratic to the state variable used [e.g., age can only increase, stages typically follow one another in a fixed sequence, and size can increase, remain the same, or decrease—it even might simultaneously increase by one measure (e.g., aboveground biomass) and decrease by another (e.g., number of stems)]. In a sense, we are tracking these idiosyncratic changes in state only because state influences survival and fertility which (unlike state transitions) actually change directly the number of individuals in the population, the quantity we seek to predict. So we want to choose the state variable that best predicts survival and fertility, and then account for transitions in the chosen state variable in whatever way is appropriate. Second, we can only use AIC_c to compare models fit to the same data; that is, we cannot compare the ability of one model to predict changes in the number of stems to the ability of another model to predict changes in aboveground biomass.

of variance using continuous fertility data), $\log L_{max,f}$ can be calculated from the residual sum of squares and the sample size (see Equation 4.7 and the explanation that accompanies it), while for a logistic regression of survival (and log-linear models of survival for discrete stages), the estimated maximum log likelihood $\log L_{max,s}$ is a direct output on all standard statistical packages. We would thus have all the elements we would need to compute AIC_c for an age-based analysis, using Equation 6.1. We could then compare the AIC_c value for this age-based model to the AIC_c value for a size-based model in which, say, a quadratic regression for fertility (four parameters) and a hazard function (three parameters; see Footnote a to Table 6.1) for survival best fit the data. The state variable with the lowest AIC_c provides the best predictive power over both vital rates. We could also use Akaike weights to decide how much better the fit of one state variable is than that of another (see Equation 4.9)—a variable that provides only slightly inferior predictive power can be better to use if it is cheaper and faster to measure in the field. One requirement of this analysis, though, is that we must fit all models to the same set of data (that is, we cannot compare the AIC_c values of an age-based model fit to survival and fertility data for one set of individuals and a size-based model fit to data from a different set of individuals).

Setting Class Boundaries

Once we have determined an appropriate classifying variable, we must use it to divide the population into discrete classes. For an age-structured model, the widths of all classes must equal the inter-census (and therefore projection) interval (e.g., one-year age classes for an annual census period, or five-year age classes if we plan to project the population over five-year increments). For a stage-based model, classes are synonymous with stages, so that in effect the choice of stages determines class boundaries. Defining classes is more complicated for a size-based matrix because the width and number of classes are flexible. Thus we must decide both how many classes to use and the exact boundaries between classes. Two considerations are important in deciding the number of classes. On one hand, we want the number of classes to be large enough to reflect real differences in vital rates as a function of the classifying variable. If we use only a few, broad classes, we will be lumping individuals of very different sizes, and hence different vital rates, into a single class within which all individuals are assumed to be identical. Another reason to favor a larger number of narrower classes is that it more accurately reflects the time individuals require to advance from birth to reproduction. Such time delays are caused by the processes of growth and maturation in structured populations, and can strongly affect some results of a viability analysis. For example, assume that a certain organism takes at least 10 years to grow to a size at which it can begin to reproduce. If we divide the population into 10 pre-reproductive size classes, and we only allow individuals to advance by at most one size class per year, then we can ensure in the model that no individual can reproduce until it has survived for at least ten time intervals. If instead we use only two pre-reproductive size classes, then our model will allow some individuals (albeit with small probability) to begin reproducing after only two time intervals.

Size-dependent changes in vital rates and time lags thus argue for a large number of classes. On the other hand, if we divide the population into too many classes, some classes will include few or none of the marked individuals in the demographic study that we are using to estimate the vital rates. This can be especially problematic in estimating growth probabilities; with many narrow classes there will be very few individuals ever observed to make any particular transition, resulting in poor estimates of most growth probabilities. To some extent, this problem is reduced by statistical procedures for estimating some vital rates (i.e., survival and fertility) using the data for individuals in all states simultaneously (see *Step 3: Estimating vital rates*). Nevertheless, standard estimation of size transitions (discussed later) requires that there be some individuals in each class, and more generally, it is prudent to avoid creating size classes about which we have no empirical information.

To balance the opposing pressures to create many classes with few individuals versus few classes with many individuals, we advocate a practical approach to deciding the number and boundaries of size classes. Begin with a relatively large number of classes and set the boundaries at sizes that are convenient for biological or management reasons. For example, we might choose as the boundary between two classes the minimum size for reproduction, the maximum size at which individuals are vulnerable to a particular threat, or the size at which a particular management technique will begin to be effective. One should also examine scatter plots of survival, growth, and reproduction as functions of size to look for sharp discontinuities (if any); these values serve as natural break points between classes. Now check the data set to see that there are at least some marked individuals in each class. If not, reassign the class boundaries or combine adjacent classes until you achieve an adequate sample size in each class with which to estimate size transitions. More formal procedures to make these decisions exist (Vandermeer 1978, Moloney 1986), but they do not fully consider how particular choices of class boundaries affect all vital rates, and they are in practice rarely used (see Caswell 2001, p. 169–171, for a description and critique of these formal methods).

Step 3: Estimating Vital Rates

Once the number and boundaries of classes have been determined, we can use the demographic data to estimate the three types of class-specific vital rates.

Survival Rates

The survival rate is simply the expected proportion of individuals in class i at the last census that are still alive at the current census. How to best estimate survival depends on whether the classifying variable is inherently discrete (i.e., stage) or continuous (age or size). For stage-based estimates of survival, one simply sums the number of marked individuals in a given stage at the previous census, determines the number of those individuals that are still alive at the current census *regardless of their state*, and divides the number of survivors by the initial number of individuals.

One could use this same approach if the underlying state variable used to make the classes is continuous. Specifically, one could use the class boundaries to first divide the marked population at the previous census into size or age classes and then estimate survival separately for each class. An important limitation of this approach is that estimates of survival for classes into which few marked individuals happen to fall will be based on a small sample size. Small sample size means that our survival estimate will be especially sensitive to chance variation around the true value. For example, assume the true survival probability for a given class is 0.5. If only four individuals in the data set fall into that class, we would not consider it to be unusual if either all or none of those individuals happened to survive over a particular inter-census interval, whereas such outcomes are very unlikely if the sample size is 100. A second problem with small sample size is precision; with only four marked individuals we can only estimate survival to the nearest 25% (i.e., the estimate would be 0, 25, 50, 75, or 100 percent if 0, 1, 2, 3, or 4 individuals survive; this is the same problem—here in a sample of individuals rather than a whole population—that creates demographic stochasticity; see Table 2.2). An example of this problem is for estimates of age-specific survival of Yellowstone grizzly bears. Eberhardt et al. (1986) had data for so few animals in most older age classes that the estimated survival rates were usually either 0 or 1. These sample size limitations can seriously constrain our choice of the number and boundaries of classes if we plan to divide the marked population into classes before estimating survival. Unfortunately, this is the standard way most researchers have estimated survival rates (Caswell 2001).

To get around this problem, we advocate a two-step procedure for age- or size-based models that effectively uses all of the data to estimate survival for each class, thus reducing sample-size limitations. First, use the entire demographic data set to perform a logistic regression of survival against age or size. This is precisely the regression we performed earlier to determine whether survival was associated with age or size (see *Step 2: Establishing classes*), the only difference being that here we will use it to estimate vital rates rather than as a simple statistical test for association. The second step is to use the fitted regression equation to calculate survival for each class. Because the regression equation is estimated from the entire data set, problems of chance variation and precision are less important. This procedure also yields smooth changes in survival among classes, which in most cases is probably more biologically realistic than the large upward or downward jumps in survival among adjacent classes that can result from small within-class sample sizes. But note that frequently we *do* want to allow one or more discontinuities, estimating a smooth survival function for most classes but doing a separate, stage-based calculation for others. This is especially true for the youngest or smallest classes (weanling mammals, fledgling birds, seedling plants), which often have much lower survival than do the next older or next larger classes. In this case, we can estimate survival of the youngest or smallest class as for a stage-based model and then perform a logistic regression using the data for older or larger classes separately to estimate their survival rates.

There are at least three ways to perform the second step in the above procedure. The simplest is to take the midpoint of each size class, substitute it into the best-fit logistic regression equation as the independent variable, and use the resulting value of the dependent variable as the estimated survival for that class. This method assumes that the actual sizes of individuals within a class are tightly clustered around the class midpoint. In reality, the actual sizes might be broadly distributed within the class, and the distribution might even be skewed toward sizes above or below the midpoint. (For example, because on average individuals must survive longer to reach a larger size, there may be more individuals at the smaller end of a given size class than at the larger end.) If there is a distribution of sizes within the class, the survival at the class midpoint may not accurately reflect the survival of the average individual in the class, particularly if the logistic regression equation is strongly nonlinear across the range of sizes in the class. A second option that better accounts for skewness in the distribution of sizes within a class is to use the median size from a random sample of individuals from the population that fall within the class. A third option is to use the actual sizes of individuals in the sample in the regression equation to calculate an expected value of survival for each individual and then average those values to estimate survival for the entire class. The sample should reflect the population-wide distribution of sizes or ages of individuals that are in the class,

TABLE 6.4 Estimated survival of *Hudsonia montana* plants in four size classes in four years, using data from Frost (1990)[a]

				Survival probabilities estimated using:			
				Class midpoints			
Plant area (cm^2)	Class midpoint	Class median in 1985	Number of plants in 1985	1985	1986	1987	1988
0–25	12.5	10.95	30	0.8404	0.7908	0.5718	0.9531
25–50	37.5	34.85	16	0.9706	0.9257	0.9239	1.0000
50–100	75	65.85	24	0.9981	0.9868	0.9970	1.0000
>100	135[b]	134.8	5	1.0000	0.9992	1.0000	1.0000

[a]Survival probabilities were obtained by substituting values of the regression coefficients b_0 and b_1 from Table 6.2 into the logistic regression Equation 6.2, and calculating the predicted survival using the midpoint, median, or distribution of plant size.

[b]Because it is difficult to define an upper boundary for the largest size class, we use the median of all size measurements over 100 cm^2 in the first year of Frost's demographic study as the midpoint of the largest size class.

and might simply be the subset of marked individuals that fall within the class (if marked individuals were initially chosen at random). If there are few such individuals or the sampling was stratified so that this distribution doesn't reflect the distribution of the population as a whole, we might supplement the detailed demographic data with additional information on the frequency distribution of sizes or ages in the population, data that are easier to obtain than detailed demographic information because they do not require us to mark and follow individuals over time.

To illustrate the estimation of class-specific survival, we return to the data for mountain golden heather. We first chose class boundaries that yield a reasonable number of plants in each of four size classes (Table 6.4). The equation for a simple logistic regression (see Figure 6.1) is:

$$y = \frac{e^{b_0+b_1x}}{1+e^{b_0+b_1x}} \tag{6.2}$$

where in this case, y is the predicted survival probability for plants with area x. For each year, we substitute the appropriate estimates of the regression coefficients b_0 and b_1 from Table 6.2 into Equation 6.2, and then we calculate the survival probabilities using all three methods discussed in the preceding paragraph (Table 6.4). In this case, the three methods yield broadly similar estimates of survival probabilities, with the maximum difference in survival between methods averaging only 0.015 across years and classes (but note much larger differences for the smallest plants in 1988). Such similarity is not guaranteed, however, and we recommend that you use the class medians or the distribution of sizes if you suspect that sizes are strongly skewed within

Survival probabilities estimated using:							
Class medians in 1985				***Distribution of plant sizes within each class in 1985***			
1985	***1986***	***1987***	***1988***	***1985***	***1986***	***1987***	***1988***
0.8245	0.7783	0.5380	0.9277	0.8247	0.7827	0.5585	0.8456
0.9645	0.9165	0.9058	0.9999	0.9600	0.9125	0.8920	0.9997
0.9962	0.9797	0.9933	1.0000	0.9957	0.9798	0.9917	1.0000
1.0000	0.9992	1.0000	1.0000	1.0000	0.9992	1.0000	1.0000

classes or that survival is a highly nonlinear function of size. But for the purpose of illustrating methods in this book, we will use the estimates of survival for mountain golden heather based on the simple class midpoint method. Note that regardless of the estimation procedure used, large plants have a high probability of surviving every year, whereas survival of smaller plants is lower and more variable from year to year (Table 6.4).

State Transition Rates, or Growth Rates

If individuals are classified by size or stage, then in addition to estimating the chance that individuals in each class survive, we must also estimate the probability that a surviving individual undergoes a transition from its original class to each of the other potential classes. (For age-structure populations, survival and state transition are one and the same, as all survivors must age.) The simplest way to estimate state transition rates is to first place *survivors* from the marked population into a class transition table (Table 6.5). The columns of the table represent a survivor's class at the previous census, and the rows represent its class at the current census. Dividing each element in the table by the corresponding column total yields the estimated class transition probabilities. Note that this procedure also yields an estimate of the probability that a survivor remains in its original class from one census to the next.

This procedure suffers from the same sample size problems that affect the procedure for estimating survival by first dividing the population into size classes, especially for rarely seen transitions such as rapid growth or shrinkage. (For example, note the small sample sizes for some size classes in Table 6.5.) More sophisticated methods that make use of the entire data set to get around these problems are available, but they are even more complex than similar methods for estimating survival and have hardly been used to date. Details of these procedures are provided by Easterling et al. (2000). Until it is clearer how well these methods can deal with different patterns of growth, stasis, and shrinkage, we suggest sticking with the simpler procedure outlined above, which is unlikely to result in serious errors unless the numbers of observed individuals in some classes are extremely small.

A Note on Alternative Methods for Estimating Survival Rates and Class Transition Rates Using Capture-Recapture Data and Size Versus Age Relationships

In the preceding discussion, we have assumed that it is possible to find every marked individual at each census, so that its fate (survival and state) can be determined unambiguously. The ability to locate with certainty all marked individuals is routine in demographic studies of sessile organisms (e.g., plants, fungi, many aquatic invertebrates), and may also be achieved for mobile organisms if they are large, occupy structurally simple habitats, and do not move far between censuses. However, for smaller and more mobile species in complex habitats, we will often fail to observe some proportion of the marked individuals at a given census even if they are still alive. This same

TABLE 6.5 Size transitions for mountain golden heather over four years[a,b]

	Size in 1985 (cm^2)			
Size in 1986 (cm^2)	***0–25***	***25–50***	***50–100***	***>100***
0–25	21 (0.84)	2 (0.1333)	2 (0.0833)	0 (0)
25–50	4 (0.16)	7 (0.4667)	5 (0.2083)	1 (0.2)
50–100	0 (0)	5 (0.3333)	15 (0.625)	1 (0.2)
>100	0 (0)	1 (0.0667)	2 (0.0833)	3 (0.6)
TOTAL	25	15	24	5
	Size in 1986 (cm^2)			
Size in 1987 (cm^2)	***0–25***	***25–50***	***50–100***	***>100***
0–25	12 (0.6316)	4 (0.2353)	1 (0.05)	0 (0)
25–50	7 (0.3684)	8 (0.4706)	3 (0.15)	0 (0)
50–100	0 (0)	5 (0.2941)	9 (0.45)	0 (0)
>100	0 (0)	0 (0)	7 (0.35)	6 (1)
TOTAL	19	17	20	6
	Size in 1987 (cm^2)			
Size in 1988 (cm^2)	***0–25***	***25–50***	***50–100***	***>100***
0–25	8 (0.8)	5 (0.3333)	2 (0.1429)	0 (0)
25–50	2 (0.2)	9 (0.6)	2 (0.1429)	1 (0.0769)
50–100	0 (0)	1 (0.0667)	7 (0.5)	5 (0.3846)
>100	0 (0)	0 (0)	3 (0.2143)	7 (0.5385)
TOTAL	10	15	14	13
	Size in 1988 (cm^2)			
Size in 1989 (cm^2)	***0–25***	***25–50***	***50–100***	***>100***
0–25	10 (0.7692)	2 (0.1429)	0 (0)	0 (0)
25–50	3 (0.2308)	6 (0.4286)	0 (0)	0 (0)
50–100	0 (0)	6 (0.4286)	8 (0.6154)	1 (0.1)
>100	0 (0)	0 (0)	5 (0.3846)	9 (0.9)
TOTAL	13	14	13	10

[a]Data from Frost (1990).

[b]The table shows the number of plants making each size transition and in parentheses, the proportions across the cells in each column of the table; these are the estimated size transition probabilities.

problem introduces observation error into estimates of total population size that are based on capture-recapture studies, and so can complicate the interpretation of census data used in count-based PVAs (see Chapter 5). In the context of demographic PVAs, to accurately estimate both the survival rates and the rates of transition among classes, we must also estimate the probability that a live, marked individual failed to be resighted in a given census. Fortunately, maximum likelihood methods have been developed to estimate survival rates and rates of class transition, as well as resighting probabilities, using data from capture-recapture studies. The literature on parameter estimation using capture-recapture data is too extensive for us to review here. Methods for estimating survival are described by Lebreton et al. (1992), and Nichols et al. (1996) describe how to estimate class transition rates. An overview of the use of capture-recapture models in the specific context of PVA is presented by White et al. (2002). Examples of the use of these methods in demographic PVAs are in Hitchcock and Gatto-Trevor (1997) and Caswell et al. (1999). Another good source of information is Gary White's website (http://www.cnr.colostate.edu/~gwhite/mark/mark.htm), from which users can obtain the program MARK, which implements maximum likelihood parameter estimation using capture-recapture data. As any viability analyst who intends to estimate vital rates for a demographic PVA based on a capture-recapture study will need to become familiar with the use of MARK (or similar programs), we refer readers to the original sources rather than attempt to describe here how the programs work.

Another method to estimate class transition rates from capture-recapture data was used to parameterize the influential demographic PVA for the loggerhead sea turtle (Crouse et al. 1987; also see Chapter 1). Briefly, this method uses data on the average number of years individuals require to grow from one benchmark size to another (i.e., the boundaries of a size class), as well as the estimated probability of surviving over those years, to calculate the rate of transition from one size class to the next. A plot of size versus age for recaptured individuals gives the average length of time an individual will reside in a size class, and the fraction of marked individuals in that size class in one year observed alive in subsequent years provides an estimate of the survival rate. One then calculates the probability that a survivor advances to the next size class at the next census in such a way that the projection matrix predicts the same average residence time in a size class as is observed in the actual size versus age data. For more details on this method, see pages 159–165 in Caswell (2001). One limitation is that, as data from many years are often combined to produce size versus age plots, it is often difficult to estimate year-to-year variation in class transition rates using this method.

Fertility Rate

Finally, we must estimate the average number of offspring that individuals in each class produce during the interval from one census to the next. Note that the fertility rate is simply the number of new individuals to which an adult in a given class has given birth over the interval, *regardless of whether those offspring survive to the next census*. (We will account for offspring survival when calculating the actual matrix elements.) If properly designed, the demographic study should yield an estimate of total offspring production

TABLE 6.6 Estimated number of seeds produced annually by mountain golden heather plants in four size classes[a]

Size of parent (cm2)	*Class midpoint, x*	*Estimated seed production in1987*	*Estimated seed production in 1988*
0–25	12.5	8.86	9.48
25–50	37.5	23.49	25.13
50–100	75	43.17	46.19
>100	140	97.08	103.88

[a]Data from Figure 6.2 (Frost 1990).

for each marked individual, as either the sum of offspring observed in repeated checks of marked individuals throughout an inter-census interval (if reproduction is spread out over the interval) or as the number produced during a brief reproductive pulse.

If individuals are classified by stage, the stage-specific rate of fertility is simply the arithmetic mean of the number of offspring produced over the year by all individuals in a given stage at the previous census.[8] For age- or size-classified models, we advocate using all individuals in the data set to estimate reproduction simultaneously for all classes, as we did when estimating survival. For example, we can substitute either the class midpoints, the median, or the distribution of individual sizes or ages within the class into the equation for a regression of offspring number versus adult size. Again using mountain golden heather as the example, substituting the class midpoints from Table 6.4 into the regression equations for seed number versus plant size (Figure 6.2) yields the class-specific seed production values in Table 6.6. (As for survival, fertility estimates based on median sizes and the within-class distribution of sizes yielded similar estimates.)

Some organisms have multiple modes of reproduction. For example, many plants have both sexually produced seeds and vegetatively produced daughter ramets. Because seeds and daughter ramets may have quite different survival and growth rates, and hence make different contributions to population growth, it may be valuable to explicitly include each type of reproduction in the projection matrix. For such organisms, we can use the procedures we have just outlined to obtain class-specific estimates of reproduction via each mode. For example, matrix models for wild leeks (*Allium tricoccum*) were constructed to incorporate both sexual reproduction and asexual bulb division (Nault and Gagnon 1993, Nantel et al. 1996). (Note that mountain golden heather does not reproduce vegetatively.)

[8]Hence the demographic study must include enough individuals in each stage to estimate mean offspring number accurately.

Step 4: Building the Projection Matrix

Having conducted a demographic study, divided the population into classes, and estimated class-specific vital rates, we are finally ready to build the projection matrix.

Basic Matrix Structure

Here is a typical projection matrix:

$$\mathbf{A} = \begin{bmatrix} a_{11} & a_{12} & a_{13} \\ a_{21} & a_{22} & a_{23} \\ a_{31} & a_{32} & a_{33} \end{bmatrix} \tag{6.3}$$

Both the number of rows and the number of columns in the matrix equal the number of classes into which we have chosen to divide the population. Matrix elements are indexed first by row and then by column. For example, a_{ij} is the element in row i and column j of the matrix **A.** The element a_{ij} indicates the number of individuals in class i *at the next census* that will be contributed by each individual in class j *at the current census.* Typically, we index the classes in the order in which they occur in the organism's life cycle, so that class 1 individuals are the most recently born members of the population. As a result, a_{1j} is the number of new, live individuals at the next census produced by each individual in class j at the current census. The first row of the projection matrix therefore represents reproduction[9]. Elements along the diagonal of the matrix (e.g., a_{22}, a_{33}, etc.) represent survival of pre-existing individuals without a change in state; elements below the diagonal represent survival with an "advance" in state; elements above the diagonal represent survival with "reversion" to a "lower" state (e.g., a smaller size).

Population projection matrices have stereotyped structures depending on whether individuals are classified by age, size, or stage, as the following matrices indicate:

$$\mathbf{A}_1 = \begin{bmatrix} 0 & F_2 & F_3 & F_4 \\ P_1 & 0 & 0 & 0 \\ 0 & P_2 & 0 & 0 \\ 0 & 0 & P_3 & 0 \end{bmatrix}$$

[9]Additional rows in the matrix may also correspond to reproduction, as we will see in the matrices for mountain golden heather (see *Putting It All Together* later in the chapter.

$$\mathbf{A}_2 = \begin{bmatrix} P_{11} & F_2 + P_{12} & F_3 & F_4 \\ P_{21} & P_{22} & P_{23} & 0 \\ 0 & P_{32} & P_{33} & P_{34} \\ 0 & 0 & P_{43} & P_{44} \end{bmatrix}$$

$$\mathbf{A}_3 = \begin{bmatrix} P_{11} & 0 & 0 & F_4 \\ P_{21} & P_{22} & 0 & 0 \\ 0 & P_{32} & P_{33} & 0 \\ 0 & 0 & P_{43} & P_{44} \end{bmatrix}$$

The three matrices each include four classes, but they differ in the distribution of nonzero elements within the matrix.[10] Matrix $\mathbf{A}_1$ is a typical age-structured matrix[11]; F_j and P_j represent, respectively, reproduction and the probability of survival of individuals in class j. In age-structured matrices, an individual that survives from one year to the next must advance to the next age class; hence the only nonzero elements in the matrix are on the principal subdiagonal (survival plus aging) and in the first row (reproduction). In a size-structured matrix, such as $\mathbf{A}_2$, the F_j's again represent reproduction, but $P_{j,j}$, $P_{j+1,j}$, and $P_{j-1,j}$, represent, respectively, the probabilities that an individual in class j will survive without growing, survive and grow to the next-largest size class, and survive but revert to the next-smallest size class. Because an individual in size class j can wind up in any of three classes next year (if it survives), the P's must have two subscripts. Note that individuals in class 2 can contribute to class 1 at the next census both by reproducing and by shrinking. In matrix $\mathbf{A}_2$, individuals can advance or revert by at most one size class in a year, but matrices that allow larger changes in size are easily constructed. Matrix $\mathbf{A}_3$ represents an organism with three pre-reproductive stages (classes 1–3) and an adult stage (class 4). Each year, individuals in the pre-reproductive stages can stay in the same stage or advance to the next stage (or die if $P_{jj} + P_{j+1,j} < 1$). Surviving adults can only remain in the adult stage. For some stage-based models, it might be sensible to allow individuals to revert to an "earlier" stage (e.g., reproductive individuals that lose a breeding territory may rejoin a "pre-reproductive" class). As we will see

[10]The notation a_{ij} used in Equation 6.3 and elsewhere refers to any matrix element, even elements that must equal zero (e.g., transitions to younger age classes). In contrast, the F and P notation in the matrices A_1, A_2, and A_3 is used to refer to nonzero elements representing, respectively, the production of new members of the population (i.e., reproduction) and class transitions of existing members. Both of these notations are in common usage.

[11]Age-structured matrices are often called Leslie matrices, after the British ecologist P.H. Leslie who pioneered their use (Leslie 1945). Similarly, size- and stage-structured matrices are often called Lefkovitch matrices, after their originator L.P. Lefkovitch (Lefkovitch 1965).

below, it is also possible to construct matrices that combine age-based, size-based, and stage-based classes. Regardless of how the classes are defined, to put specific numbers into the matrix we must express the F's and P's in terms of the estimated vital rates, as we now describe.

Building the Matrix from the Underlying Vital Rates

We begin by describing how to use estimated vital rates from a single inter-census interval to build a single projection matrix. We then describe how data from multiple censuses are used to estimated means, variances, and covariances of vital rates and of matrix elements, the key components of a stochastic projection matrix model. We use these components to assess population growth and extinction risk in Chapters 7 and 8 and to explore management options in Chapter 9.

In describing both the construction of a matrix from the underlying vital rates and methods for analyzing matrix models, it is useful to represent the three types of vital rates by the following symbols:

- s_j is the survival rate for class j
- f_j is the fertility rate for class j
- g_{ij} is the probability that an individual in class j at one census makes the transition to class i at the next census, given that it survives

To keep clear the distinction between vital rates and matrix elements, we will consistently use the above symbols for the vital rates and either F's and P's (as in the matrices $\mathbf{A}_1$, $\mathbf{A}_2$, and $\mathbf{A}_3$ above) or a_{ij} for the matrix elements. Moreover, we will consistently refer to the F's as *reproduction* (*not fertility*, which we reserve for a vital rate[12]) and the P's as *class transition probabilities*.

CLASS TRANSITION PROBABILITIES IN AGE-STRUCTURED MODELS In an age-structured model, an individual that survives an inter-census interval must be one age class older than it was at the last census. Thus survival is all we need to estimate class transitions for individuals alive at the previous census. We simply put the class-specific survival estimates, s_j, along the principal subdiagonal of the matrix (i.e., $a_{j+1,j} = P_j = s_j$; see the matrix $\mathbf{A}_1$ above). Note that all individuals currently in the oldest age class are assumed to die before the next census.

CLASS TRANSITION PROBABILITIES IN SIZE- OR STAGE-STRUCTURED MODELS In contrast with age-classified populations, survivors in a size- or stage-classified model may or may not undergo a transition to another class. Individu-

[12]The terms *fertility*, *fecundity*, and *reproduction* have all been used for both matrix elements and vital rates, and human demographers tend to use these terms differently than do ecologists, who cannot seem to agree among themselves what these words mean. Rather than trying to sort out this semantic morass, we will choose one set of definitions and apply them consistently.

als making a transition can advance in state, but they can also regress. For example, some of the mountain golden heather plants in Frost's (1990) study became smaller from one year to the next (Table 6.5), perhaps due to dieback, trampling by hikers, or consumption by herbivores. In a size- or stage-structured matrix, the entry P_{ij} (see the matrix $\mathbf{A}_2$ above) represents the product of two vital rates: s_j, the probability that an individual in size class j at the current census survives to the next census, and g_{ij}, the probability that a survivor from class j makes a transition to class i. Hence, to estimate P_{ij}, we simply multiply these two vital rates (i.e., $a_{ij} = P_{ij} = s_j g_{ij}$).

REPRODUCTION As with class transitions in a size- or stage-based model, the reproduction terms in the matrix combine two or more underlying vital rates, the fertility rate plus the survival rate of adults and/or offspring, depending on the timing of reproduction relative to the census. The key idea here is that all matrix elements, including those for reproduction, must account for the passage of time from one census to the next, with the accompanying chance that some individuals will die over that time. Therefore, the reproductive elements of the matrix must include not only fertility (e.g., the number of female cubs born in an interval), but also survival (e.g., the fraction of newborn cubs that survive to the next census). Four scenarios for the timing of the census relative to reproduction (birth pulse with a pre-breeding, post-breeding, and an intermediate census, and birth flow) necessitate different procedures to calculate the reproduction terms in the matrix, as we now describe.

CASE 1: BIRTH-PULSE POPULATIONS WITH A PRE-BREEDING CENSUS If we census the population immediately before a birth pulse, the youngest (i.e., class 1) individuals in the population will be those that were produced in the previous birth pulse, and they will be nearly an entire census interval in age (Figure 6.3A). Thus to contribute to class 1 at the next census, an adult in the current census must produce newborns during the birth pulse immediately after the census, and those newborns must survive for an entire inter-census interval. That is, the reproduction term F_j in a birth-pulse population with a pre-breeding census is the number of newborn offspring an individual in class j produces during the pulse times the survival of newborns for one inter-census interval; i.e., $a_{1j} = F_j = f_j s_0$, where s_0 is the one-year survival rate of newborns. In practice, the design of the demographic study frequently does not allow the initial number of newborns and their survival over the first year of life to be estimated separately. That is, if the population is censused only before the birth pulse, the best we can do is to use the number of surviving offspring associated with a female at the next census to estimate her net reproduction (i.e., her total offspring number times their survival) over the previous interval. One disadvantage of such a procedure is that it does not allow us to investigate whether changing offspring production (e.g., by improving the condition of mothers) or newborn survival (e.g., by

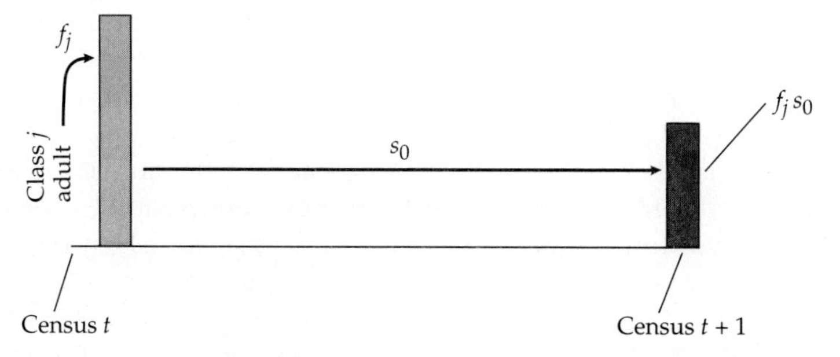

(B) Birth-pulse, post-breeding census

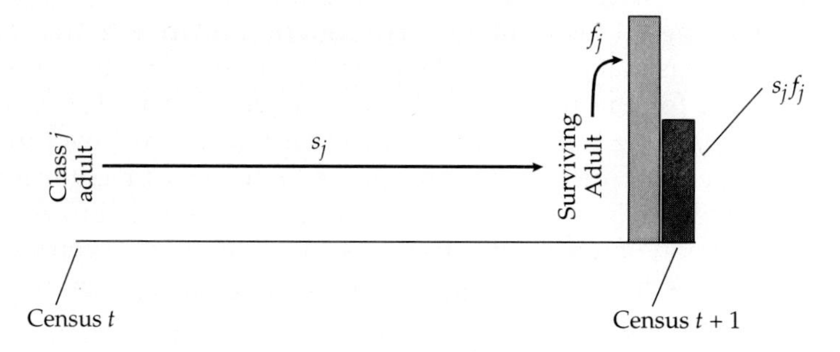

(C) Birth-flow

"Averaged" fertility

f_j

$\sqrt{s_j}\, f_j \sqrt{s_0}$

Class j adult

$\sqrt{s_j}$

Surviving Adult

$\sqrt{s_0}$

Actual fertility

Census t

Census $t + 1$

Figure 6.3 Graphical representation of how to calculate, for three temporal patterns of fertility, the matrix elements representing reproduction. light bars represent the number of newborns produced per live adult in class j; dark bars represent offspring alive at the next census. For birth-flow populations, fertility is represented by summing newborns produced throughout the inter-census interval and assuming their birth date is the same as that of the average newborn. See text for definitions of vital-rate symbols.

targeting management directly at newborns) would have a greater effect on population growth (see Chapter 9). If such an investigation is worth doing, we would need to census the population (or at least females in the reproductive classes) both before and immediately after the birth pulse, or if biological circumstances allow, use a post-breeding census design instead.

CASE 2: BIRTH-PULSE POPULATIONS WITH A POST-BREEDING CENSUS In contrast with a pre-breeding census, if we census a population immediately after a birth pulse, the youngest individuals we will see will have just been born (Figure 6.3B). The reproduction terms in the matrix now represent the number of such newborns that each adult in the current census will contribute to the next census. To make such a contribution, however, the adult itself must survive for nearly an entire inter-census interval and then reproduce. The reproduction term F_j in a birth pulse population with a post-breeding census is thus the survival of class j adults for one inter-census interval times the number of offspring an individual in class j produces during the pulse; i.e., $a_{1j} = F_j = s_j f_j$.

CASE 3: BIRTH-PULSE POPULATIONS WITH AN INTERMEDIATE CENSUS In some cases, it may be necessary to conduct the annual census at a point in time that lies between successive birth pulses. For example, if females go into hiding to give birth (as is true of bears, whose cubs are born in winter dens), a pre- or post-breeding census might be impractical. If so, offspring would first be observed at the next census. The reproduction element for class j would then be the average of the number of offspring at the next census attributable to each individual in class j at the previous census. Note that the number of offspring would be zero for individuals who died before the birth pulse, who produced no offspring, or whose offspring died before the next census. In principle, one could also parameterize a projection matrix model with a birth pulse occurring at an intermediate point between two censuses if one had the data to estimate both the probability that each adult survives from one census to the next pulse and the probability that an offspring survives from birth to the following census (see Equation 2.42 in Caswell 2001). In practice, such data can be obtained only by actually censusing the population near the time of the birth pulse, in which case it would be simpler to assume a pre- or post-breeding census when constructing the projection matrix.

Another useful approach in some cases is to "pretend" that the birth pulse occurs later than it actually does. For example, for demographic studies of songbird species, a sensible time to conduct the census might be just at the time of fledging, when offspring are still in (or near) the nest but are then large enough to band. Thus even though fledglings were actually born several weeks earlier, and some eggs or chicks may have died before they could fledge, we might treat fledglings as newborns (i.e., class 1) in constructing the matrix. If so, the reproduction terms in the matrix would represent the survival rate of an individual observed at the current census (including the fledglings, if it is possible for them to breed in the following summer) times the number of fledglings that individual can be expected to produce at the next census, given that it survives. That is, we might adapt the procedure for a post-breeding census of a birth-pulse population (see above), even though the time of fledging is not, strictly speaking, the birth pulse.

CASE 4: BIRTH FLOW POPULATIONS When reproduction occurs continually over an inter-census interval, the simplest approach to calculating the reproduction matrix elements is to pretend that all offspring are born at the midpoint of the interval. Reproduction for class j individuals is then the product of three things:

1. The probability that a class j individual survives from one census to the midpoint of the succeeding inter-census interval. If the survival rates are more or less constant over the course of a year, then six-month survival can be approximated as the square root of annual survival, $\sqrt{s_j}$.
2. The average number of offspring to which each surviving class j individual gives birth over an inter-census interval (summed over repeated checks performed throughout the interval), f_j.
3. The probability that an offspring produced at the midpoint of an interval survives to the next census. If the most recently born individuals at a census have an annual survival probability of s_0, the probability that an offspring born at the midpoint of an interval survives to the next census is approximately $\sqrt{s_0}$.

More complex approaches to estimating reproduction elements for birth-pulse and birth-flow populations are described by Caswell (Caswell 2001, pp. 171–173).

Putting It All Together: Estimating Projection Matrices for Mountain Golden Heather

We now illustrate the entire procedure of building a projection matrix using the data for mountain golden heather. This example highlights two important points about estimating matrix elements in a world of limited data. First, as noted above, we will often need to rely on information beyond that obtained in the annual census of marked individuals in order to estimate all the matrix entries. Second, data on certain vital rates will often be missing for one or more years; if so, we may have no other option than to use average values for those years.

In Tables 6.4 and 6.5, we divided the plants from Frost's census into four size classes. For our projection matrix to accurately reflect the plant's life cycle, we must add two additional classes. First, mountain golden heather has a persistent seed bank (as indicated by the appearance of seedlings after a fire in locations far from potential seed sources). Seeds in the seed bank would not be observed in the census of above-ground plants but could nevertheless contribute to population persistence, so we need to include them in our matrix. The second class we must add to the matrix is seedlings, which represent seeds that have germinated since the last census and which are smaller than any of the marked plants in Frost's census. It is important to include the seedling class because it reflects a realistic delay of at least one year between germination and the onset of flowering and because seedlings are unlikely to survive as well as already established plants. Our matrix will thus have six classes[13]: (1) seeds in the seed bank (really more of a stage than a size class); (2) seedlings (really an age class, since by definition seedlings do not remain so if they survive beyond one year after germination); (3) tiny plants (plants older than seedlings but between 0 and 25 cm^2 in area); (4) small plants (25–50 cm^2); (5) medium plants (50–100 cm^2); and (6) large plants (greater than 100 cm^2). We will use these class numbers (1–6) in presenting the construction of the matrix. Note that the widths of the larger size classes exceed those of the smaller classes; we did this to achieve better sample sizes in order to more accurately estimate class transitions for the larger classes. In Frost's study, plants were censused near the end of the growing season, when all of the fruits produced that season were present on the plants but had not yet released their seeds. Hence we will construct our matrix to reflect a pre-breeding census of a birth-pulse population.

Because all seeds do not germinate in the spring after they are produced, our matrix will need to have two types of reproduction terms: (1) contributions by above-ground plants to seeds in the seed bank (which appear in the first row of the matrix); and (2) contributions by above-ground plants to seeds that germinate the following spring to become seedlings (which appear in the second row of the matrix). Contributions to the seed bank (class 1) by individuals in class j, a_{1j}, are given by the product of : (1) f_j, the average number of seeds produced by plants in class j (from Table 6.6)[14]; (2) g_{11}, the fraction of those seeds that do not germinate the following spring; and, (3) s_1, the probability that a seed survives one year in the soil to be present in the population at the time of the next census. Knowledge of the plant's natural history suggests that only about 1% of the viable seeds dispersed at the end of the summer germinate the following spring (thus the fraction g_{11} = 0.99 do not germinate), and that viability of seeds in the seed bank declines by about 50% per year, so $s_1 = 0.50$. (For details on how these estimates were obtained, see Gross et al. 1998.) For example, in 1987 the estimated contribution of each large plant (class 6) to the seed bank, a_{16}, equals 97.08 seeds (Table 6.6) times 0.99 times 0.5, or 48.0546 seeds.

Contributions to the seedling class (class 2) by individuals in class j , a_{2j}, are products of four terms: (1) f_j, the average number of seeds produced by plants in class j; (2) the fraction of seeds that survive the winter in the soil;

---

[13]These class definitions differ slightly from those used by Gross et al. (1998), who combined seedlings and tiny plants into a single size class, and used different boundaries for the other classes.

[14]Since mountain golden heather is hermaphroditic, we do not correct seed numbers for the sex ratio.

(3) the fraction of surviving seeds that germinate the following spring, which equals $(1 - g_{11})$ or 0.01; and (4) the probability that a seedling survives its first summer. To estimate the second component, we assume (in the absence of data needed to do otherwise) that the rate of survival does not change from month to month, so that the fraction of seeds surviving the 7-month winter period from dispersal to spring germination is $(0.5)^{7/12} = 0.6674$, where 0.5 is the annual survival rate of seeds in the soil. In a set of plots separate from his census plots, Frost (1990) observed that newly emerged seedlings had a probability of 0.4681, 0.4444, and 0.9810 of surviving their first summer of life in 1987, 1988, and 1989, respectively. (Because Frost did not measure seedling survival in 1985 or 1986, we use the average of these three numbers for those years.) For example, in 1987 the estimated contribution of each large plant (class 6) to the seedling class a_{26} equals 97.08 seeds times 0.6674 times 0.01 times 0.4681, or 0.3033 seedlings. Repeating these steps for other size classes yields the reproduction terms in Table 6.7.

A seed remains in the seed bank at the next census if it survives (probability 0.5) and does not germinate (probability 0.99); hence a_{11} equals $0.5 \times 0.99 = 0.4950$. (Note that we have no way to estimate annual variability in this matrix element.) A seed in the seed bank will become a seedling at the next census if it survives the winter (probability 0.6674), germinates (probability 0.01), and survives the summer (probabilities given in the preceding paragraph). Thus, for example, in 1987, $a_{21} = 0.6674 \times 0.01 \times 0.4681 = 0.00031$.

Next we estimate the survival probabilities of seedlings, a_{32}. In his seedling plots, Frost (1990) found that seedlings still alive at the time of the census in 1987 and 1988 had a probability of 0.4545 and 0.5000, respectively, of surviving to the following census. In the absence of additional data, we use the average of these two numbers in the other years. Because a seedling cannot remain so for more than one year, all surviving seedlings automatically advance to the tiny plant class (class 3).

Finally, we estimate the size transition probabilities for plants older than seedlings by multiplying the size transitions in Table 6.5 by the survival probabilities in Table 6.4. For example, the estimated probabilities that individuals in size class 5 in 1987 are in size classes 3 through 6 in 1988 are 0.1425, 0.1425, 0.4985, and 0.2137, respectively. The results for all classes are given in Table 6.7.

The second through fifth columns in Table 6.7 give the estimates of the matrix entries *for particular years*. In Chapter 7, we will envision the projection matrix as changing from year to year due to changes in environmental conditions that affect the population's vital rates. To reflect this annual variation, we will represent the projection matrix by the shorthand $\mathbf{A}(t)$, which indicates that this matrix only applies in the interval between census t and census $t + 1$. Table 6.7 also shows the means and variances across years for each matrix element. We can also use the matrix element values in Table 6.7

TABLE 6.7 Estimated projection matrix elements for mountain golden heather for four years[a,b]

Matrix element	*1985*	*1986*	*1987*	*1988*	*Mean*	*Variance*
a_{11}	0.4995	0.4995	0.4995	0.4995	0.4995	0.0000
a_{13}	4.5782	4.5782	4.4234	4.7330	4.5782	0.0120
a_{14}	12.1425	12.1425	11.7319	12.5531	12.1425	0.0843
a_{15}	22.3167	22.3167	21.5620	23.0714	22.3167	0.2848
a_{16}	50.1895	50.1895	48.4923	51.8867	50.1895	1.4403
a_{21}	0.0004	0.0004	0.0003	0.0003	0.0004	0.0000
a_{23}	0.0039	0.0039	0.0028	0.0028	0.0033	0.0000
a_{24}	0.0102	0.0102	0.0073	0.0075	0.0088	0.0000
a_{25}	0.0188	0.0188	0.0135	0.0137	0.0162	0.0000
a_{26}	0.0423	0.0423	0.0303	0.0308	0.0364	0.0000
a_{32}	0.4773	0.4773	0.4545	0.5000	0.4773	0.0003
a_{33}	0.7059	0.4995	0.4575	0.7331	0.5990	0.0148
a_{34}	0.1294	0.2178	0.3079	0.1429	0.1995	0.0051
a_{35}	0.0831	0.0493	0.1425	0.0000	0.0687	0.0027
a_{43}	0.1345	0.2913	0.1144	0.2200	0.1900	0.0050
a_{44}	0.4530	0.4356	0.5544	0.4286	0.4679	0.0026
a_{45}	0.2079	0.1480	0.1425	0.0000	0.1246	0.0058
a_{46}	0.2000	0.0000	0.0769	0.0000	0.0692	0.0067
a_{54}	0.3235	0.2722	0.0616	0.4286	0.2715	0.0179
a_{55}	0.6238	0.4440	0.4985	0.6154	0.5454	0.0059
a_{56}	0.2000	0.0000	0.3846	0.1000	0.1711	0.0202
a_{64}	0.0647	0.0000	0.0000	0.0000	0.0162	0.0008
a_{65}	0.0831	0.3454	0.2137	0.3846	0.2567	0.0140
a_{66}	0.6000	0.9994	0.5385	0.9000	0.7595	0.0379

[a]Data from Frost (1990) and Gross et al. (1998).

[b]Matrix elements not listed in the table were zero in each of the four years.

to compute the covariance between each possible pair of matrix elements, which measures the similarity between the patterns of temporal fluctuation in the two elements. In Chapter 7, we will say more about how to calculate and interpret these covariances. More importantly, we will see that the means, variances, and covariances of matrix elements play an important role in analyzing the viability of structured populations in a variable environment. We could also compute means, variances, and covariances of the vital rates themselves, and we will discuss in Chapters 8 and 9 how to make use of them in a vital-rate-based PVA.

Summary

In this chapter we've gone through the basics of how to take field data from a demographic study and turn them into a set of vital rate estimates than can in turn be used to construct a projection matrix model. In doing so, we have emphasized how to construct a single matrix using the data from a single inter-census interval. We have done this in part because a single annual transition may be all the data you have. Even if data for only a single transition are all that are available, many useful analyses can still be done with a matrix model, as we will see when we discuss deterministic results in the beginning of Chapter 7 and in Chapter 9. However, with only a single matrix, we cannot assess the effect of environmental stochasticity on population viability, so ideally we would have access to data from a demographic study conducted long enough to allow the estimation of vital rates and matrix elements over multiple years. In either case, the task ahead is to use the estimated matrices to predict population growth and extinction probabilities. That is the topic of Chapters 7 and 8. We will also discuss how to treat several complications that we have ignored here, in particular demographic stochasticity, density dependence, and catastrophes and bonanzas. Then in Chapter 9, we will focus on the use of projection matrix models to identify the causes of low viability and to guide effective management of threatened populations.

7

Demographic PVAs: Using Projection Matrices to Assess Population Growth and Viability

In Chapter 6, we saw how to construct a series of projection matrices from the underlying vital rates (fertility, survival, and state transition rates) using data collected in a multiyear demographic study. Our goal in doing so was, of course, to assess population growth and the probability of extinction, accounting (at the very least) for the mean values of the vital rates as well as the effects of unpredictable environmental perturbations. To assess population viability, we can view these environmental perturbations at two levels. At the more fundamental level, changing environmental conditions affect the vital rates themselves. At a higher level, environmentally driven fluctuations in vital rates drive fluctuations in the matrix elements, which are themselves functions of the vital rates. (Remember from Chapter 6 that a single vital rate can influence multiple matrix elements, so fluctuations in a single vital rate and fluctuations in a single matrix element are not exactly equivalent.) Thus we can incorporate environmental stochasticity into demographic PVAs either at the level of the vital rates or at the level of the matrix elements. Although conceptually similar, these two ways of including environmental stochasticity use somewhat different methods and take advantage of somewhat different theoretical results. Consequently, we devote separate chapters to each approach: Chapter 7 covers ways to include stochastic environmental effects on matrix elements, and Chapter 8 deals with environmental influences on the underlying vital rates.

In describing methods for count-based PVAs in Chapter 3, we began with the simplest approach, which included environmental stochasticity but ignored the complications of demographic stochasticity, density dependence, extreme environmental perturbations (i.e., catastrophes and bonanzas), and sampling variation; we added these factors later (Chapters 4 and 5). By and large, we will repeat this pattern in describing demographic PVAs in Chapters 7 and 8. In part, our reason for doing so relates to the differences between the vital-rate and matrix-element viewpoints discussed in the preceding paragraph. Incorporating the complicating factors we listed above into demographic PVAs is more straightforward at the level of vital rates rather than matrix elements. Sampling variation is much easier to estimate and discount in vital rates than in matrix elements. Demographic stochasticity and density dependence will frequently affect certain vital rates and not others. Hence it is natural to estimate and model the effects of density dependence and demographic stochasticity on specific vital rates, which then may affect multiple elements in the projection matrix. Extreme environmental conditions can also exert effects on only a subset of vital rates, and the effects of a common environment will lead to correlations among vital rates. These are all topics we cover in Chapter 8.

In spite of these advantages of the vital-rate viewpoint, essentially all of the mathematical theory of population matrices rests on matrix elements, and thus it is easiest to understand the predictions of stochastic matrix models by initially focusing on the matrix-element viewpoint. In addition to covering the basic topics of stochastic matrix models in this chapter, we also consider a simple way to include extreme environmental effects expressed at the level of variation among entire projection matrices. We also briefly discuss ways to include another factor we touched on in Chapter 4, temporal environmental autocorrelation.

Regardless of the degree of complexity we build into the model and whether we are using it for assessment or management, a full understanding of demographic PVAs must begin with the fundamentals of how projection matrix models work. Thus we begin this chapter by describing how a projection matrix is used to "project" the future size and structure of a population in a deterministic (i.e., constant) environment and how to analyze the sensitivity of population growth to different changes in the matrix elements. Although we do not advocate the use of deterministic projection matrix models in PVA when estimates of environmental variation are available to construct stochastic models, the analytic results from deterministic models are also necessary components for approximating the behavior of stochastic models. Hence we explain these deterministic results before turning to fully stochastic projection matrix models. If you have no estimates of environmental variation, these analyses alone may form much of your PVA (along with methods described in the first half of Chapter 9). Readers who are already familiar with the concepts of population projection, convergence, eigenvalues and eigenvectors, and sensitivities can skip the following section.

Structured Populations in a Deterministic Environment

Population Projection

We saw in Chapter 6 that each entry $a_{ij}(t)$ in the projection matrix $\mathbf{A}(t)$ gives the number of individuals in class i at census $t + 1$ produced on average by a single individual in class j at census t. Considering all feasible values of i and j, the projection matrix as a whole thus summarizes the per-capita contributions of all classes at one census to all classes at the next census. Hence, if we know the numbers of individuals in all classes at one time, we can use the matrix to project the numbers in all classes, and hence the total population size, one census interval later. For the purposes of projection, it is convenient to represent the population structure by a population vector. A population vector is simply a column of numbers that indicates the densities[1] of individuals in each class in the population at one point in time. Each row in the vector represents the same class as in the corresponding row (or column) of the projection matrix. For example, if the average number of individuals of a particular bird species per hectare of habitat is 23.5, 14.2, and 7.3 in classes 1 through 3, respectively, we can concisely summarize the structure of the population at census t by the vector

$$\mathbf{n}(t) = \begin{bmatrix} 23.5 \\ 14.2 \\ 7.3 \end{bmatrix} \tag{7.1}$$

Note that we will follow the mathematical convention of using boldfaced, lowercase letters to represent vectors, and boldfaced, uppercase letters to represent matrices.

If we know the densities of individuals in all classes at census t, $\mathbf{n}(t)$, we can project the densities at the next census, $\mathbf{n}(t + 1)$, using the projection equation

$$\mathbf{n}(t + 1) = \mathbf{A}(t)\,\mathbf{n}(t) \tag{7.2}$$

Projection thus requires knowledge of how to multiply a matrix by a vector, which we now review.

Equation 7.2 is a concise version of the equation

$$\begin{bmatrix} n_1(t+1) \\ n_2(t+1) \\ n_3(t+1) \end{bmatrix} = \begin{bmatrix} a_{11}(t) & a_{12}(t) & a_{13}(t) \\ a_{21}(t) & a_{22}(t) & a_{23}(t) \\ a_{31}(t) & a_{32}(t) & a_{33}(t) \end{bmatrix} \begin{bmatrix} n_1(t) \\ n_2(t) \\ n_3(t) \end{bmatrix} \tag{7.3}$$

[1]Because the entries in a population vector need not be integers, it is natural to think of them not as numbers of individuals but as average densities (i.e., mean numbers of individuals per unit area). We can convert from densities to numbers by multiplying the average densities by the area occupied by the population.

where $n_i(t + 1)$ represents the density of individuals in class i at census $t + 1$, and the t in the notation $a_{11}(t)$ is to remind us that the elements in the projection matrix $\mathbf{A}(t)$ are assumed to vary from year to year. To calculate $\mathbf{n}(t + 1)$, use the following rules:

$$\mathbf{n}(t+1) = \begin{bmatrix} n_1(t+1) \\ n_2(t+1) \\ n_3(t+1) \end{bmatrix} = \begin{bmatrix} a_{11}(t)n_1(t) + a_{12}(t)n_2(t) + a_{13}(t)n_3(t) \\ a_{21}(t)n_1(t) + a_{22}(t)n_2(t) + a_{23}(t)n_3(t) \\ a_{31}(t)n_1(t) + a_{32}(t)n_2(t) + a_{33}(t)n_3(t) \end{bmatrix} \quad (7.4)$$

In effect, to calculate the entry in the first row of $\mathbf{n}(t + 1)$, we multiply the element in the first row of $\mathbf{n}(t)$ by the first-column element in the first row of $\mathbf{A}(t)$, add that product to the element in the *second* row of $\mathbf{n}(t)$ times the *second*-column element in the first row of $\mathbf{A}(t)$, and so on. For example, Hitchcock and Gatto-Trevor (1997) estimated an average projection matrix for a declining population of semipalmated sandpipers at La Pérouse Bay, Manitoba, Canada. They grouped birds that were one, two, or three or more years of age into separate classes, and built the following matrix to reflect a prebreeding census:

$$\mathbf{A} = \begin{bmatrix} 0.02115 & 0.074 & 0.0846 \\ 0.563 & 0 & 0 \\ 0 & 0.563 & 0.563 \end{bmatrix} \quad (7.5)$$

(Note that $\mathbf{A}$ represents a stage-structured population, not an age-structured one, and that, because it represents an average matrix, it is not a function of t.) If we start with the population vector in Equation 7.1 as $\mathbf{n}(0)$, then

$$\mathbf{n}(1) = \mathbf{A}\mathbf{n}(0) = \begin{bmatrix} (0.02115 \times 23.5) + (0.074 \times 14.2) + (0.0846 \times 7.3) \\ (0.563 \times 23.5) + (0 \times 14.2) + (0 \times 7.3) \\ (0 \times 23.5) + (0.563 \times 14.2) + (0.563 \times 7.3) \end{bmatrix} = \begin{bmatrix} 2.1654 \\ 13.2305 \\ 12.1045 \end{bmatrix} \quad (7.6)$$

Look carefully at each of the entries in this new vector to see that they simply sum up all the possible ways that an individual can arrive in a particular class at time $t + 1$. For example, the second vector element (the number of class 2 birds in the next census) is the probability that a class 1 bird survives to be a class 2 bird times the density of class 1 birds (0.563×23.5), plus the probability that a class 2 bird remains a class 2 bird (which is zero) times the density of class 2 birds (0×14.2), plus the probability that a class 3 bird regresses to be a class 2 bird (which is also zero) times the density of class 3 birds (0×7.3).

A simple shorthand for this multiplication process is the following set of three rules:

1. Write the original population vector in row form above the first row of the matrix and multiply the two rows:

$$\begin{array}{ccc} 23.5 & 14.2 & 7.3 \\ \underline{\times 0.02115} & \underline{\times 0.074} & \underline{\times 0.0846} \\ 0.4970 & 1.0508 & 0.6176 \end{array}$$

2. Add the resulting products ($0.4970 + 1.0508 + 0.6176 = 2.1654$) and write their sum in the first row of the new vector;
3. Repeat steps 1 and 2 using the subsequent rows of the matrix, writing the results in successive rows of the new vector until it is complete.

We reiterate that all matrix-vector multiplication does is automate the process of calculating the combined contributions of all classes in the population in one year to each class in the following year. With the knowledge of how to perform matrix-vector multiplication, the projection equation (Equation 7.2) allows us to calculate the population vector one census interval into the future. By repeatedly multiplying the projection matrix for each subsequent interval by the most recent population vector, we can calculate the population vector at any future time. As we have repeatedly emphasized, the projection matrix itself will change each interval due to environmental variation driving changes in the underlying vital rates. But methods of assessing viability of a structured population in a variable environment (our ultimate goal) make use of results obtained under a constant environment, so we review those basic results first.

Population Growth and Convergence

In a constant environment, the projection matrix $\mathbf{A}(t) = \mathbf{A}$ in Equation 7.2 will not vary from one time interval to the next, so projection is done by repeatedly multiplying the constant matrix $\mathbf{A}$ by the new population vector. Let's see what happens when we continue the process of projecting the semipalmated sandpiper population that we began in Equation 7.6. We get $\mathbf{n}(2)$ by multiplying $\mathbf{A}$ times $\mathbf{n}(1)$, $\mathbf{n}(3)$ by multiplying $\mathbf{A}$ times $\mathbf{n}(2)$, and so on, yielding the following sequence of population vectors (rounded to 4 decimal places):

$$\mathbf{n}(2) = \begin{bmatrix} 2.0489 \\ 1.2191 \\ 14.2636 \end{bmatrix}, \ \mathbf{n}(3) = \begin{bmatrix} 1.3403 \\ 1.1535 \\ 8.7168 \end{bmatrix}, \ \mathbf{n}(4) = \begin{bmatrix} 0.8511 \\ 0.7546 \\ 5.5570 \end{bmatrix}, \ \mathbf{n}(5) = \begin{bmatrix} 0.5440 \\ 0.4792 \\ 3.5534 \end{bmatrix},$$

$$\mathbf{n}(6) = \begin{bmatrix} 0.3476 \\ 0.3062 \\ 2.2704 \end{bmatrix}, \ \mathbf{n}(7) = \begin{bmatrix} 0.2221 \\ 0.1957 \\ 1.4506 \end{bmatrix}, \text{ and } \mathbf{n}(8) = \begin{bmatrix} 0.1419 \\ 0.1250 \\ 0.9269 \end{bmatrix} \qquad (7.7)$$

What do these numbers tell us? First, the sum of the elements in each vector represents the total density of individuals in the population, irrespective of their class. These sums, starting with $\mathbf{n}(0)$, are 45.0000, 27.5004, 17.5316, 11.2106, 7.1627, 4.5766, 2.9242, 1.8684, and 1.1938. So the total population declined by a factor of 27.5004/45.0000 = 0.6111 between years 0 and 1, but by somewhat less dramatic factors of 0.6375, 0.6394, 0.6389, 0.6389, 0.6389, 0.6389, and 0.6389 over the subsequent time intervals. Notice that despite initial fluctuations in the rate of decline, the proportional change in the population from one census to the next quickly approaches a steady value of 0.6389. This steady approach to a constant rate of population growth (or, in this case, decline) is known as convergence. The particular value toward which the population growth rate converges is traditionally denoted λ_1 (the reason for the subscript "1" will become clear momentarily). λ_1 is very similar to λ in the simple deterministic, density-independent model of a nonstructured population (i.e., Equation 3.1 with constant λ_t), the only difference being that, as we have just seen for the sandpiper population, the structure of the population may cause its growth rate to differ initially from λ_1, even though the environment is constant. For this reason, it is best to think of λ_1 as the *ultimate* or *long-term* growth rate of a structured population in a constant environment. Once convergence has occurred, the value $\lambda_1 = 0.6389$ predicts that the semipalmated sandpiper population will be 36.11% smaller each year than it was in the previous year, which closely matches the rate of decline observed by Hitchcock and Gatto-Trevor (1997). No population can long avoid extinction at such precipitous rates of decline. The key general point here is that simple matrix models predict that population growth will converge rather quickly to a constant, geometric rate of growth or decline. This convergence to the same growth pattern predicted by the simplest of nonstructured models (e.g., Equation 2.1) isn't too surprising, since here we are multiplying a population vector by a constant matrix. This is more complicated than, but obviously similar to, multiplying a total population size by a constant population growth rate.

What happens to the population structure over the same period of time? Because the density of each class is declining as total population density declines, it is easiest to see how the population structure changes by dividing each class-specific density in the population vector by the sum of the vector elements, in order to obtain a scaled population vector that gives the *proportion* of the total population in each class at a given census. Figure 7.1 shows these proportions for years 0 through 8. Just as the population growth rate converges to a constant value, so too does the fraction of the population in each class. The unique vector containing the ultimate proportions of the population in each class given the constant projection matrix $\mathbf{A}$ is known as the stable distribution,[2] traditionally denoted $\mathbf{w}$. (Note that $\mathbf{w}$ is often called the stable age, stable size, or stable stage distribution depending on whether

[2]Saying that a population is at the stable distribution does not imply that it is "stable" in the sense that it neither grows nor declines; rather, it implies that the *proportions* of individuals in each class are stable even as total population density grows or declines.

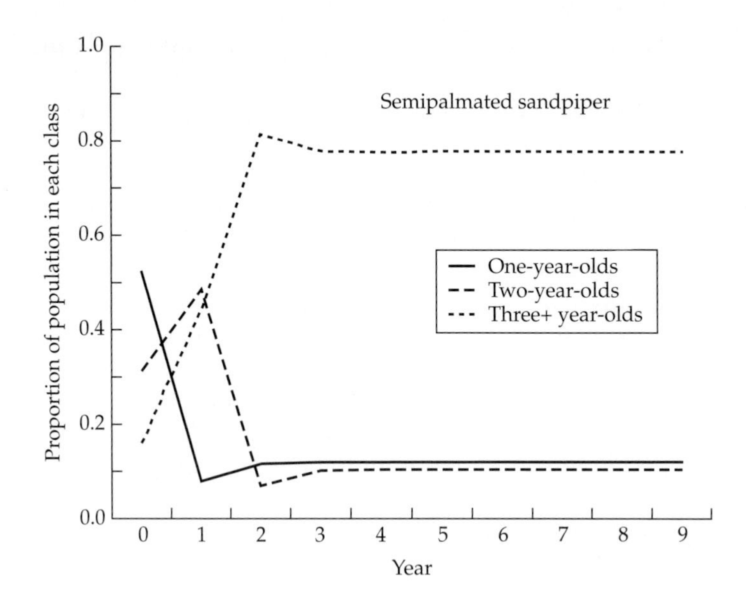

Figure 7.1 Convergence of population structure to the stable distribution in an unvarying environment.

individuals are classified by age, size, or stage). For the matrix **A** in Equation 7.5, the stable distribution is

$$\mathbf{w} = \begin{bmatrix} 0.1189 \\ 0.1047 \\ 0.7764 \end{bmatrix} \tag{7.8}$$

Note that the entries in **w** sum to 1, as true proportions should.

In a constant environment, as represented by an unvarying projection matrix, the population growth rate and the population structure will converge fairly rapidly to constant values in most cases. The potential exceptions are strictly semelparous (or monocarpic) organisms in which there is only a single reproductive class in the matrix and all individuals reproduce and die at exactly the same age. In this case, the rate of population growth can oscillate indefinitely (see Chapter 4 in Caswell 2001). However, it is difficult to point to real populations in which all individuals in a cohort become reproductively mature at precisely the same age or size and reproduce only once. For instance, Pacific salmon provide the textbook example of a semelparous organism, but in most salmon populations, some individuals return to their natal streams to spawn earlier than do other members of their cohort, so that reproduction would be spread over more than a single age class in the matrix (Ratner et al. 1997, Kareiva et al. 2000). Periodic cicadas and some bamboo species more closely approach this total synchrony of life cycles, but it is only for these extreme cases that sustained oscillations in population structure due to semelparity are likely to occur. However, populations of long-lived species with prolonged pre-reproductive periods will converge much more slowly than will populations of short-lived species.

Because of the phenomenon of convergence, assessing population viability in a constant environment would be trivial. We would only need to ask whether λ_1 is greater than, less than, or equal to 1 to know whether a population would ultimately grow, go extinct, or remain steady over time (just as for nonstructured models in a constant environment—see Figure 3.1A). However, as no environment is truly constant, assessing the viability of structured populations involves more work than simply calculating the value of λ_1 for a single projection matrix. Nevertheless, we will see that the value of λ_1 for an *average* matrix (a matrix composed of the arithmetic means of all matrix elements across years), as well as its corresponding stable distribution **w**, allow us to calculate an approximation for the population growth rate in a stochastic environment, so it is useful to have a more convenient method for calculating λ_1 and **w** than performing repeated projections until convergence occurs. This method involves the important topic of eigenvalues and eigenvectors.

Interpreting λ_1 and w as an Eigenvalue and Eigenvector

As seen above, we can calculate λ_1 and **w** by repeated projection until their values converge to any degree of precision we wish. These useful quantities can also be computed directly from the projection matrix. It is a standard result in demography (see Chapter 4 in Caswell 2001) that λ_1 and **w** represent the dominant eigenvalue and the dominant right eigenvector, respectively, of the projection matrix **A**. Mathematically, an eigenvalue of **A** and its associated right eigenvector are defined, respectively, as any single number (a so-called scalar) and a vector such that the product of **A** and the vector equals the product of the scalar and the vector[3] (the later product is simply the original vector with each element multiplied by the scalar). In other words, when we multiply a matrix by its right eigenvector, the resulting vector will simply be the original vector with each element multiplied by the eigenvalue. Note that this is exactly what happens when the population vector has converged to the stable distribution (see Equation 7.7 and Figure 7.1; the population *size* is changing but not the *proportions* of individuals in each class), so it makes sense that the stable distribution **w** is a right eigenvector of the projection matrix **A**.

Methods to calculate eigenvalues and eigenvectors are presented in any introductory linear algebra text. In practice, they are usually computed with the assistance of mathematical software. The MATLAB code in Box 7.1 defines a function `eigenall` that calculates eigenvalues and eigenvectors for the projection matrix **A**. If a projection matrix includes *s* classes, there will in general be *s* eigenvalues and *s* associated right eigenvectors. However, for

[3]That is, $\mathbf{Aw} = \lambda_1 \mathbf{w}$.

BOX 7.1 ***MATLAB code defining the function "eigenall", which calculates eigenvalues and eigenvectors of the matrix A.***

```
function [lambdas,lambda1,W,w,V,v]=eigenall(A);
%  eigenall(A) takes as input the projection matrix A and
%  returns eigenvalues, dominant eigenvalue, matrix with right
%  eigenvectors as columns, dominant right eigenvector (rescaled
%  to proportions), matrix with left eigenvectors as rows, and
%  dominant left eigenvector (rescaled relative to its first
%  element); eigenvalues and eigenvectors are sorted from largest
%  to smallest.

[W,lambdas]=eig(A);          % W=matrix with right eigenvectors of
                             %    A as columns
V=conj(inv(W));              % V=matrix with left eigenvectors of A
                             %       as rows
lambdas=diag(lambdas);       % lambdas=vector of eigenvalues
[lambdas,I]=sort(lambdas);   % sort eigenvalues from smallest to
                             %       largest
lambdas=flipud(lambdas);     % flip lambdas so that largest value
                             %       comes first
lambda1=lambdas(1);          % lambda1=dominant eigenvalue
I=flipud(I);
W=W(:,I);                    % sort right eigenvectors
V=V(I,:);                    % sort left eigenvectors
w=W(:,1);                    % w=stable distribution
                             %       (dominant right eigenvector)
w=w/sum(w);                  % rescale w to represent proportions
v=real(V(1,:))';             % v=vector of reproductive values
                             %       (dominant left eigenvector)
v=v/v(1);                    % rescale v relative to class 1
```

most projection matrices (the exception once again being those representing rigidly semelparous life histories), one of the eigenvalues will have a larger magnitude[4] than the others; that eigenvalue is λ_1, the dominant eigenvalue,

[4]Some of the subdominant eigenvalues of a projection matrix typically are complex numbers (i.e., $\lambda_n = a + b\sqrt{-1}$, where $n > 1$ and a and b are called the real and imaginary parts of λ_n). The magnitude of λ_n is $\sqrt{a^2 + b^2}$. The dominant eigenvalue of a projection matrix is purely real (i.e., $b = 0$), so that its magnitude is simply its absolute value. For any realistic population matrix, the dominant eigenvalue will also be positive. (For more details, see Chapter 4 in Caswell 2001.)

and its associated right eigenvector is the dominant right eigenvector. The subscript "1" simply denotes the fact that λ_1 is the eigenvalue with the largest magnitude.

Note that the function `eig` in MATLAB yields right eigenvectors in which the entries do not necessarily sum to 1 (and so cannot be interpreted directly as the proportion of the population in each class, as in Equation 7.8). However, an eigenvector divided by any nonzero constant is also an eigenvector. This means that we can rescale an eigenvector in any way we wish by dividing it by a constant, and it will still be an eigenvector. The most convenient way to scale the dominant right eigenvector produced by mathematical software is to divide it by the sum of its elements, which will convert its entries into proportions. The function `eigenall` in Box 7.1 rescales the dominant right eigenvector in exactly this way.

Reproductive Value

As we have seen above, the dominant right eigenvector, which is associated with the largest, or dominant, eigenvalue, is useful because it represents the stable structure toward which the population will converge in a constant environment. But each eigenvalue is also associated with another type of eigenvector, known (not surprisingly) as a left eigenvector. As we will see, the dominant left eigenvector, which is denoted **v** and is also associated with the largest eigenvalue, is also a very useful quantity.[5] It contains the so-called *reproductive values* of each class. Reproductive value is defined as the relative contribution to future population growth an individual currently in a particular class is expected to make. Reproductive value takes into account the number of offspring an individual might produce in each of the classes it passes through in the future, the likelihood of that individual reaching those classes, the time required to do so, and the population growth rate λ_1. (The population growth rate affects the reproductive values of individuals because, if the population is growing, offspring produced in the near future will contribute more to future population size than offspring produced in the more distant future, whereas the opposite is true in a declining population.) The left eigenvalues are easily calculated once the right eigenvectors have been obtained.[6] As was true of the stable distribution **w**, the vector of reproductive values **v** is often rescaled to make its entries more informative.

[5]Left eigenvectors are defined in an analogous way to right eigenvectors. Specifically, a projection matrix, when left-multiplied by a left eigenvector in row form, yields the same eigenvector multiplied by its associated eigenvalue: $\mathbf{v}'\mathbf{A} = \lambda_1\mathbf{v}'$ (where **v** is in column form as in Equation 7.9 and primes denote the transpose operation in which a column vector becomes a row vector, with elements ordered from left to right instead of from top to bottom).

[6]Specifically, they are the rows of the complex conjugate of the inverse of the matrix **W**, whose columns are the right eigenvalues of the projection matrix **A**; see MATLAB code in Box 7.1.

In particular, dividing each entry in the vector by the first entry gives us the reproductive values of each class relative to that of newborns (class 1), who will have a rescaled reproductive value of 1. For example, the MATLAB commands in Box 7.2 first define the projection matrix **A** for the semipalmated sandpiper population in Equation 7.5 and then use the function `eigenall` to calculate the vector of rescaled reproductive values:

$$\mathbf{v} = \begin{bmatrix} 1.0000 \\ 1.0973 \\ 1.1139 \end{bmatrix} \qquad (7.9)$$

This vector tells us that each sandpiper currently in classes 2 and 3 will contribute, respectively, approximately 10 and 11 percent more to future population growth than will an individual currently in class 1. The reason for these differences in reproductive value is that individuals in class 3, for example, have already reached the stage at which they have the highest rate of reproduction (as the matrix **A** shows), whereas an individual currently in class 1 has probability $1 - 0.563^2 = 0.683$ of dying before it reaches this most fertile stage. Because they measure potential contributions to future population growth, it makes intuitive sense that the reproductive values figure prominently in assessing the viability and management of structured populations, as we will soon see.

Eigenvalue Sensitivities: Measuring the Relationship between λ_1 and the Elements of the Projection Matrix

The ultimate rate of population growth in a constant environment, λ_1. depends on the magnitudes of all the elements in **A**, so changing any of them will change λ_1. However, changes in some matrix elements will have a

---

BOX 7.2 ***Fragment of MATLAB code that uses the function "eigenall" defined in Box 7.1 to generate Equations 7.8, 7.9, and 7.11 using the semipalmated sandpiper projection matrix in Equation 7.5.***

```
% Projection matrix for the semipalmated sandpiper population
A=[.02115 .074 .0846;.563 0 0;0 .563 .563];

[lambdas,lambda1,W,w,V,v]=eigenall(A);
lambda1
w
v
S=v*w'/(v'*w)
```

much larger effect on λ_1 than changes in others. It is useful to have a measure of how much changes in a particular matrix element will change λ_1. This is precisely what eigenvalue sensitivities do. The sensitivity of λ_1 to a_{ij}, represented by the symbol S_{ij}, is simply the partial derivative of λ_1 with respect to a_{ij}. The partial derivative measures the change in λ_1 that would result from a small change in a_{ij}, *keeping all other elements of the matrix* **A** *fixed at their present values.* S_{ij} can be expressed in terms of the elements of the stable distribution and the reproductive value vector as follows (Caswell 1978):

$$S_{ij} = \frac{\partial \lambda_1}{\partial a_{ij}} = \frac{v_i w_j}{\sum_{k=1}^{s} v_k w_k} \tag{7.10}$$

where v_i is the reproductive value of individuals in class i and w_j is the fraction of individuals in class j in the stable distribution vector $\mathbf{w}$. The denominator in Equation 7.10 is simply a constant[7] that appears in all the sensitivities, so the only thing that causes the sensitivities to differ among the matrix elements is the term $v_i w_j$ in the numerator. Thus the sensitivity of λ_1 to matrix element a_{ij} (remember that a_{ij} is the average contribution of each class j individual to class i in the following census) is directly proportional to the fraction of individuals in the population on which the element will act (the proportion of the population in class j when the population is at the stable distribution, w_j) times the future value of each individual that the element "creates" (contribution of each new individual in class i to future population growth, v_i). Relatively minor changes in λ_1 will result from small changes in matrix elements that represent either: (1) transitions from classes that represent a small proportion of the stable population structure; or (2) the production of individuals that are expected to make only minor contributions to future population growth. Also notice that because all the elements in $\mathbf{w}$ and $\mathbf{v}$ (including w_j and v_i in Equation 7.10) are positive, the sensitivity of λ_1 to changes in matrix elements will always be positive.[8]

A matrix containing the eigenvalue sensitivities for the semipalmated sandpiper projection matrix (Equation 7.5) can be calculated easily using MATLAB (see Box 7.2). The resulting matrix is

---

[7]Specifically, it is the so-called scalar product of $\mathbf{v}$ and $\mathbf{w}$, the summed products of the rows of the two column vectors. In vector notation, the scalar product is $\mathbf{v}' * \mathbf{w}$, where the prime denotes the transpose operation (see Footnote 5). $\mathbf{v}$ and $\mathbf{w}$ may be rescaled as in Equations 7.8 and 7.9.

[8]As we will see in Chapter 9, this is not true of eigenvalue sensitivities to changes in the underlying vital rates, which can be negative. For example, increasing the probability that an individual regresses into a less fertile class given that it survives may reduce λ_1.

$$\mathbf{S} = \begin{bmatrix} 0.1082 & 0.0953 & 0.7067 \\ 0.1187 & 0.1046 & 0.7755 \\ 0.1205 & 0.1062 & 0.7872 \end{bmatrix} \tag{7.11}$$

Each element in the sensitivity matrix **S** measures how a small change in the corresponding element in **A** would change λ_1, keeping all other elements in **A** constant. The sensitivities in **S** tell us that a small increase in matrix element a_{33}, which represents the fraction of class 3 birds that survive to next year, would result in a larger increase in the ultimate rate of population growth, λ_1, than would the same amount of increase in any other matrix element alone. In contrast, a small increase in the fecundity of class 2 birds would have the smallest effect on the ultimate rate of population growth. Notice how elements that are zero in the projection matrix (e.g., a_{23}) can nevertheless have large sensitivities. In the case of matrix element a_{23}, making its value larger than zero while keeping all other matrix elements constant would increase the proportion of class 3 birds that survive until next year (even though some would "regress" to class 2), and as class 3 represents the largest fraction of the population at the stable distribution (see Equation 7.8) and class 2 individuals have a relatively high reproductive value (see Equation 7.9), allowing some class 3 individuals to regress to class 2 would have a relatively large effect on future population size. Of course, in a matrix such as the one for the semipalmated sandpiper in which the classes are fundamentally age groups, it is not biologically possible for individuals to regress to a younger age class (a sad fact with which we are all personally familiar). Thus it is best to think of the eigenvalue sensitivities as representing hypothetical changes in λ_1 that would result if certain matrix elements *could* be changed, even if such changes are not really possible. Because eigenvalue sensitivities can be calculated for biologically impossible changes and because they take no account of linkages among matrix elements caused by shared vital rates (e.g., increasing the survival rate for a class should increase both the reproduction element and the class transition probabilities for that class, if the matrix reflects a post-breeding census), they must be used with care, an issue we return to in Chapter 9.

Another important point is that the sensitivity of λ_1 to a_{ij} measures the slope of a curve plotting λ_1 versus a_{ij} precisely at the value of a_{ij} in the matrix **A** (Figure 7.2). Using this slope to assess the effect of a change in a_{ij} will be valid if that change is small or if λ_1 changes *linearly* as a_{ij} changes. In reality, λ_1 may change in a nonlinear fashion as a_{ij} changes over a broader range of values (Figure 7.2). In Chapter 9, we present methods to assess the effects on λ_1 of large changes in matrix elements.

Eigenvalue sensitivities (and related quantities known as elasticities; see Chapter 9) calculated from a single, constant projection matrix have been used in a conservation context to identify matrix elements or their underlying

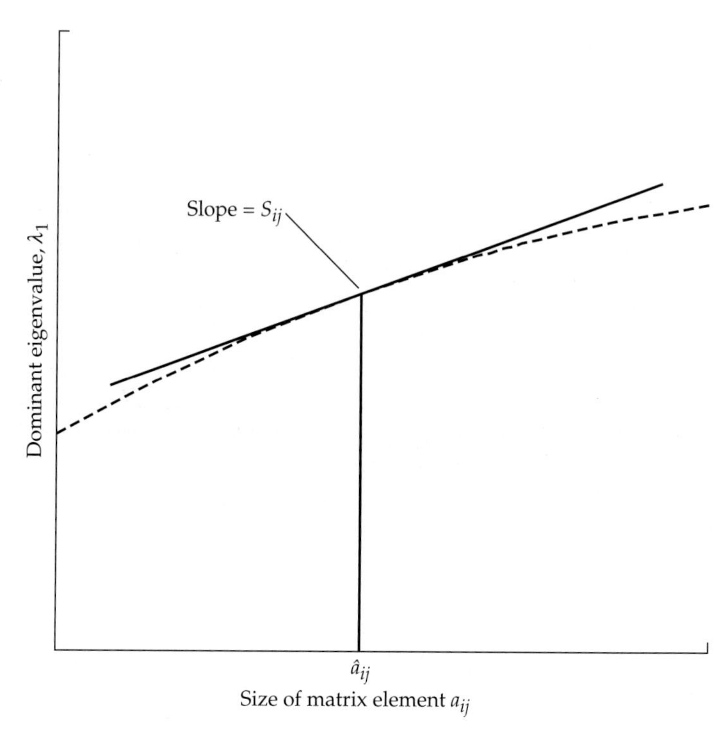

Figure 7.2 A graphical interpretation of eigenvalue sensitivity. $\hat{a}_{ij}$ is the current value of element a_{ij} in the matrix **A**. The dotted line shows how λ_1 would actually change if we were to change a_{ij} to a value other $\hat{a}_{ij}$, keeping all other matrix elements constant. The solid line shows an approximation for λ_1 using a linear extrapolation of the slope of the dotted line evaluated at $\hat{a}_{ij}$. This slope is S_{ij}, the sensitivity of λ_1 to a_{ij}. The approximation for λ_1 gets worse and worse the farther a_{ij} gets from $\hat{a}_{ij}$.

vital rates that, because they have a large effect on λ_1, should be the primary targets of management efforts aimed at enhancing the growth rate of a threatened population (Crouse et al. 1987, Heppell et al. 1994, 1996, 2000). In Chapter 9, we cover these methods in much more detail and also present other ways to ask such management questions when environmental conditions, and hence the projection matrix itself, are not assumed to be constant. However, as we will see in the next section, deterministic eigenvalue sensitivities are useful even if the projection matrix varies from year to year, because the eigenvalue sensitivities for the *average* matrix can be used to calculate useful approximations for the population growth rate and the probability of extinction in a stochastic environment.

Growth and Extinction Risk of Structured Populations in a Variable Environment

Having reviewed important deterministic results in the preceding section, we now turn to viability analysis for structured populations in a variable environment. Environments are not constant, and accounting for environmental variation in calculating extinction risk is just as important for structured populations as it is for unstructured ones (see Chapters 3 and 4). We now cover methods to analyze population growth and extinction risk in variable environments. In doing so, we will make use of the deterministic results we have just described. Just as we began our discussion of count-based PVA models in Chapters 2 and 3 by using computer simulations to build our intuition about stochastic population growth, we will use computer simulations to motivate viability analysis for stochastic structured populations.

Exploring Population Growth in a Variable Environment with Computer Simulations

The simplest way to model the growth of a structured population in a stochastic environment is to view the actual projection matrices estimated over a series of years as the demographic manifestations of possible states of the environment. We can then use a computer to choose one of the matrices at random each year and multiply it by the most recent population vector. Repeating this process yields the projected population vector, and hence the overall size of the population, at a specified future time. This process is analogous to the process of drawing population growth rates at random when simulating changes in total population size with a count-based model (e.g., Box 2.1), except that now we are drawing an entire matrix at random rather that drawing a population growth rate.[9]

Implementing the computer simulation we have just described allows us to include correlations among the values of different matrix elements within the same year. However, we must also decide whether the environmental conditions in one year are independent of conditions in the previous year (or years). That is, just as we did for count-based PVAs, we need to consider the possibility of temporal environmental autocorrelation. Conditions would be nonindependent if there is a cycle in the environment with a more or less regular period (i.e., a set sequence of environmental conditions that repeats itself over a fixed number of years). In this case, we would want our simulation to cycle through the estimated matrices in the order in which they actually occurred in the demographic study, and we would need to be sure that we have obtained an estimated matrix for all phases of the cycle. If the demographic study was conducted for a period long enough to estimate more than one matrix for each phase of the cycle, we could write our simulation to draw randomly among the appropriate matrices at each phase. Examples of cyclic environmental conditions that have been incorporated

---

[9]Because the dominant eigenvalue of a matrix represents the ultimate rate of population growth in a constant environment, one might think that we could simulate the growth of a structured population in a stochastic environment by calculating the dominant eigenvalues for each observed projection matrix and then drawing those values at random. *This procedure is not correct.* Due to the continual buffeting of a stochastic environment, the population will virtually never be at the stable distribution associated with each year's matrix, and it is only at that stable distribution that the population growth rate will exactly equal the dominant eigenvalue of the current matrix.

into demographic PVAs include water levels in the Florida Everglades (Beissinger 1995) and controlled burns recurring with a period determined by managers (Gross et al. 1998).

More commonly, environmental conditions are not rigidly periodic, but conditions in adjacent years may be correlated. For example, unusually wet or dry conditions tend to be repeated over several years in locations affected by El Niño events. If we have the data to estimate the correlation between environmental conditions in adjacent years, we can build into our simulation a certain probability that the same environmental conditions recur for two or more successive years. We can simulate the growth of a structured population in a correlated environment in three ways.

First, if we know how a particular vital rate depends on one or more continuously distributed environmental variables that are themselves autocorrelated, we can draw values for those variables in the same way that we simulated an autocorrelated population growth rate in Chapter 4 (see Box 4.6), use those values to calculate the vital rates, and then use the vital rates to construct a new matrix each year, using the methods in Chapter 6. Putting this procedure into practice requires simultaneous measurements of vital rates and environmental conditions over a sufficient number of years to uncover the relationship between them.

Second, we can modify this approach by directly estimating and simulating correlations in the vital rates themselves, a method we describe in detail in Chapter 8.

The third approach is to require that the environment can assume a limited number of discrete states and to draw a new environmental state each year, with each year's state dependent on the state in the preceding year, and each estimated matrix associated with a particular environmental state (for details, see Chapter 14 in Caswell 2001). Given the limited duration of most demographic studies of rare species, this approach will almost never be practical unless the number of environmental states is small (e.g., wet versus dry years, fire versus no-fire years).

If environmental conditions are aperiodic and uncorrelated, and moreover the probability of choosing a particular matrix does not change over time (even if some matrices are more likely than others[10]), then environmental conditions are said to be "independently and identically distributed" or "iid". A MATLAB program to simulate population growth in an iid environment is listed in Box 7.3. This program uses the four matrices estimated for mountain golden heather in Table 6.7. As shown in Box 7.3, the code assumes that each of the four matrices is equally likely to occur. However, the program is written in such a way that users can easily change the frequencies with which the different matrices are chosen, simply by changing

[10]Even if the frequencies of different matrices (i.e., environmental states) are not equal, so long as there is no autocorrelation and those frequencies do not change over time, the environment is still iid.

BOX 7.3 *MATLAB code to simulate growth of a structured population in an iid stochastic environment.*

```
% PROGRAM iidenv
%   Simulates growth of a structured population in an iid
%   stochastic environment

%************* USER-SPECIFIED PARAMETERS ***************
% Enter name of m file containing matrices; e.g. the file
% hudmats.m contains 4 matrices (A85, A86, A87, A88),
% one for each year of Frost's demographic study
hudmats;
% Change names of matrices below to correspond to the names in
%   the m file from the preceding command:
matrices=[A85(:) A86(:) A87(:) A88(:)];
% Enter probabilities of choosing each matrix;
% NOTE: sum of penv must equal 1
penv=[.25 .25 .25 .25];
% Enter time to predict future population size
tmax=50;
% Enter number of realizations of population growth to simulate
numreps=5000;
% Enter initial population vector
n0=[4264; 3; 30; 16; 25; 5];
%*****************************************************

rand('state',sum(100*clock));      % seed random number generator
s=sqrt(size(matrices,1));          % s=number of classes
cumdist=cumsum(penv);              % CDF for environmental states

Nend=[];
for i=1:numreps
   n=n0;
   for t=1:tmax
      x=sum(rand>=cumdist)+1;             % draw a matrix at random
      A=reshape(matrices(:,x),s,s);       % extract the chosen matrix
                                          %    from "matrices" and
      n=A*n;                              % project the population 1
                                          %    year ahead
   end
   Nend=[Nend sum(n)];       % store population size at tmax
end
hist(Nend,50)             % create histogram of population size at
                          %    tmax
```

BOX 7.3 (continued)

```
MeanN=mean(Nend)        % mean population size at tmax
MedianN=median(Nend)    % median population size at tmax
```

the entries in the vector penv. For example, if we wanted to draw the 1985 mountain golden heather matrix only 10% of the time, but draw the other three matrices with equal frequencies, we would change the elements of penv to 0.1, 0.3, 0.3, and 0.3. (Note that these frequencies must sum to 1.) We will return to this point when we discuss below how to include the effects of extreme environments on measures of population growth and extinction risk.

Each simulated realization of population growth begins with a population vector with entries equal to the numbers of marked plants in each size class over all study plots in the first year of Frost's demographic study, and runs for 50 years.[11] The initial total population density (the sum of the initial population vector) is 4,343. The distribution of total population density at year 50 for 5,000 independent realizations is shown in Figure 7.3. (The code can easily be modified to plot the final density in any one class, rather than the total density summed across all classes.) Notice that in all of the realizations, total population density has declined over the 50 years. Also notice that the final population density is lognormally distributed, just as was the case for a density-independent, unstructured population in a stochastic environment (see Figures 3.1B and 3.2). An asymptotically lognormal distribution for total population density is a general feature of density-independent growth of a structured population in a stochastic environment (Tuljapurkar and Orzack 1980). The lognormal distribution shows that the possible final population sizes are skewed, with a few realizations declining much less dramatically than the majority. Indicative of this skew, the mean of the realizations shown in Figure 7.3 (as calculated in Box 7.3) is 637.45, while the median is 576.94. This again shows the correspondence between matrix models and the nonstructured models we dealt with in earlier chapters, which respond similarly to the addition of environmental stochasticity. Note

[11]Because Frost did not measure the density of seeds or seedlings in his plots, we set their initial densities to match the proportions in the stable distribution for the average matrix. We used the numbers from the first year of Frost's study, rather than the last, because they better reflect the size distribution in the population at large, since no new marked plants were added to replace those that died over the course of the study.

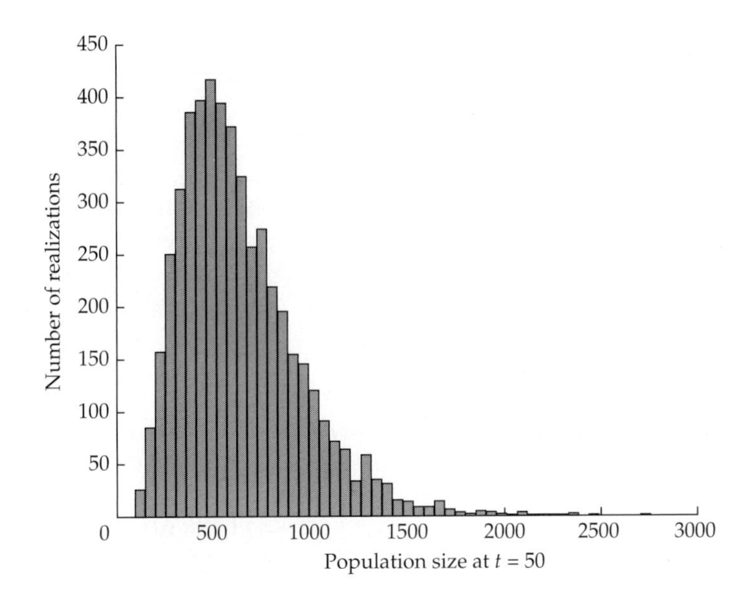

Figure 7.3 Results of simulating a population of mountain golden heather in an iid stochastic environment (see Box 7.3). The initial population size was 4,343.

that the mean matrix predicts deterministically that $N = 631.9$ at $t = 50$, an overly optimistic prediction for most realizations (again reminiscent of results for total counts, where we saw that the arithmetic mean gave an overly optimistic prediction of the populatin growth rate in a stochastic environment).

Estimating the Stochastic Log Growth Rate, Log λ_s

Our simulation results (Figure 7.3) predict that the mountain golden heather population will surely decline over a 50-year period if the four matrices we estimated from Frost's demographic data alternate at random. To measure a population's tendency to grow or decline in a variable environment over a longer term, it would be nice to have a measure of the stochastic growth rate for a structured population that is akin to the parameter μ, or equivalently $\lambda_G = \exp(\mu)$, for an unstructured population. The fact that population size (or density) is lognormally distributed in both unstructured and structured populations suggests that there may be parallels in the population growth rate as well. Indeed, just as the long-term growth of an unstructured population is poorly represented by the arithmetic mean growth rate, λ_A, we cannot predict very well the growth of a structured population by calculating the arithmetic mean of the yearly projection matrices[12] and computing its dominant

[12]See *An analytic approximation for log* λ_s later in the chapter for the calculation of the arithmetic mean of a set of projection matrices.

eigenvalue. Instead, long-term population growth is better predicted by the most-likely log population growth rate over a long sequence of years (that is, the arithmetic mean of the log ratios of population sizes in adjacent years). We will refer to this quantity as the stochastic log growth rate and represent it by log λ_s (following Caswell 2001). Just as λ_1 is similar to λ, log λ_s is similar to μ, but we give it a different name and symbol to emphasize the fact that log λ_s pertains to a structured population. The stochastic log growth rate can be estimated in two ways: by computer simulation, and by means of an analytical approximation. These methods each have advantages and disadvantages, as we now discuss.

CALCULATING THE STOCHASTIC LOG GROWTH RATE BY SIMULATION It is very straightforward to calculate the stochastic log growth rate by means of a stochastic computer simulation like the one in Box 7.3. Specifically, we project population growth over many successive time intervals, using a matrix drawn at random each time interval to calculate $\mathbf{n}(t + 1)$ from $\mathbf{n}(t)$. If $N(t)$ and $N(t + 1)$ are the total population densities in successive years (i.e., the sums of successive population vectors), then calculating the arithmetic mean of $\log[N(t + 1)/N(t)]$ over all pairs of adjacent years yields an estimate of log λ_s. For reasonable accuracy, the number of years of simulated population growth must be large, typically in the tens of thousands or more (Caswell 2001). The MATLAB code in Box 7.4 estimates log λ_s and an approximate 95% confidence interval[13] by using the mean and variance of 50,000 simulated population growth increments. We will discuss its output after we outline the second way to estimate log λ_s.

AN ANALYTIC APPROXIMATION FOR LOG λ_s A second method for estimating the stochastic log growth rate is due to Tuljapurkar (1982), who developed an approximation for log λ_s based on the assumption that the variation among the annual matrices is not large. Even though this assumption restricts the range of conditions in which the approximation will be reasonably accurate, Tuljapurkar's approximation allows us to gain far greater insight into how stochastic variation in the matrix elements affects a population's long-term stochastic growth rate than we could achieve by means of computer simulation alone. Specifically, it shows why variation in some matrix elements is more important than variation in others, and it illuminates the importance of covariation between the values of different matrix

[13]The confidence interval is based on the fact that, because of the Central Limit theorem, the arithmetic mean of $\log[N(t + 1)/N(t)]$ will be approximately normally distributed.

BOX 7.4 *MATLAB code to estimate log λ_s by simulation and by Tuljapurkar's approximation.*

```
% PROGRAM stoc_log_lam
%   Calculates stochastic log growth rate by simulation and by
%   Tuljapurkar's approximation; penv allows some matrices to be
%   chosen more often than others.
%   REQUIRES THE FUNCTION eigenall.m

%************* USER-SPECIFIED PARAMETERS ******************
% Enter name of m file containing matrices; e.g. the file
% hudmats.m contains 4 matrices (A85, A86, A87, A88),
% one for each year of Frost's demographic study
hudmats;
% Change names of matrices below to correspond to the names in
%   the m file from the preceding command:
matrices=[A85(:) A86(:) A87(:) A88(:)];
% Enter probabilities of choosing each matrix below;
% NOTE: sum of penv must equal 1
penv=[.25 .25 .25 .25];
% number of time intervals to simulate - should be large
maxt=50000;
%**********************************************************

rand('state',sum(100*clock));    % seed random number generator
s=sqrt(size(matrices,1));        % s=number of classes
numenvts=size(matrices,2);       % number of matrices
                                 %     (environmental states)
cumdist=cumsum(penv);            % CDF for environmental states

% Calculate mean matrix Abar and covariance matrix C, taking into
% account the fact that all matrices may not be equally likely;
% uses the fact that Cov(x,y)=E(x*y)-E(x)E(y), where E() denotes
% expectation
Abar=zeros(s^2,1);
Exy=zeros(s^4,1);
for i=1:numenvts
   A=matrices(:,i);
   Exy=Exy+penv(i)*kron(A,A);    % kron is Kronecker tensor
                                 %      product function
   Abar=Abar+penv(i)*A;
```

BOX 7.4 *(continued)*

```
end;                    % for i
C=(Exy-kron(Abar,Abar))*numenvts/(numenvts-1);
C=reshape(C,s^2,s^2);             % covariances of all matrix
                                  %      element pairs
Abar=reshape(Abar,s,s);           % mean matrix
% Calculate dominant eigenvalue for mean matrix, lambdabar,
% and S=matrix of eigenvalue sensitivities for mean matrix Abar;
[lambdas,lambdabar,W,w,V,v]=eigenall(Abar);
S=v*w'/(v'*w);

% Calculate stochastic log lambda by simulation
n=w;                    % start at stable distribution of Abar, with
                        %      total population size = 1
for t=1:maxt
   x=sum(rand>=cumdist)+1;         % choose a matrix at random
   A=reshape(matrices(:,x),s,s);   % extract the chosen matrix
   n=A*n;                          % project 1 year ahead
   N=sum(n);            % sum n to get new total pop. size
   r(t)=log(N);         % calculate log growth rate
   n=n/N;               % renormalize so sum(n)=1 to avoid pop.
                        %    sizes too large or small for computer
                        %    to handle
end

loglsim=mean(r)                 % simulated stochastic log growth
                                %    rate
dse=1.96*sqrt(var(r)/maxt);     % standard error of loglsim
CL1=[loglsim-dse loglsim+dse]   % approx. 95% confidence interval -
                                %    see Caswell 2001, eq. 14.62
lamsim=exp(loglsim)             % simulated stochastic growth rate
CL2=exp(CL1)                    % confidence limits on lamsim

% Calculate stochastic log lambda using Tuljapurkar's
% approximation
Svec=S(:);
tau2=Svec'*C*Svec;
loglams=log(lambdabar)-tau2/(2*lambdabar^2)
lams=exp(loglams)
```

elements with the same year. Moreover, it provides an important link to the count-based methods we covered in Chapter 3. Therefore we devote considerable attention here to explaining the approximation.

Tuljapurkar's approximation[14] is:

$$\log \lambda_s \approx \log \bar{\lambda}_1 - \frac{1}{2}\left(\frac{\tau^2}{\bar{\lambda}_1^{\,2}}\right) \tag{7.12}$$

where

$$\tau^2 = \sum_{i=1}^{s}\sum_{j=1}^{s}\sum_{k=1}^{s}\sum_{l=1}^{s} Cov(a_{ij}, a_{kl})\bar{S}_{ij}\bar{S}_{kl} \tag{7.13}$$

In Equation 7.12, $\bar{\lambda}_1$ is the dominant eigenvalue of the mean matrix $\bar{\mathbf{A}}$ obtained by averaging each element of the estimated annual matrices, weighting the values for each element by the frequency at which they are expected to occur. For example, if we assume the four mountain golden heather matrices are equally likely to occur, we calculate $\bar{\mathbf{A}}$ by averaging the four matrices in Table 6.7 element by element (i.e., we sum the four values of a_{11} and divide the sum by 4 to get $\bar{a}_{11}$, place the result in the first row, first column position of $\bar{\mathbf{A}}$, and repeat for all other elements). We then calculate the dominant eigenvalue $\bar{\lambda}_1$ of $\bar{\mathbf{A}}$ as in Box 7.1.

The quantity $\tau^2/\bar{\lambda}_1^2$ in Equation 7.12 approximates the temporal variance of the log population growth rate caused by environmental stochasticity; it is the structured-population equivalent of the parameter σ^2 for an unstructured population (see Chapter 3), and we will see that it plays a similar role in analyzing population viability. In the expression for τ^2 (Equation 7.13), $\bar{S}_{ij}$ is the sensitivity of $\bar{\lambda}_1$ to changes in $\bar{a}_{ij}$, which we calculate by substituting into Equation 7.10 the appropriate elements from the dominant right and left eigenvectors of $\bar{\mathbf{A}}$. The term $Cov(a_{ij},a_{kl})$ in Equation 7.13 is the covariance[15] between matrix elements a_{ij} and a_{kl}. If $i \neq k$ or $j \neq l$, then a_{ij} and a_{kl} represent

---

[14]The approximation in Equation 7.12 assumes that matrix elements are uncorrelated from one year to the next (i.e., it assumes an iid environment). Tuljapurkar also developed an approximation that takes temporal environmental autocorrelation into account, but it is far more difficult to compute, and do not add much insight into the basic effects of covariation on population growth.

[15]Covariance is the basic statistical measure of the tendency to vary in synchrony. For two variables x and y, $Cov(x,y) = \frac{1}{n-1}\sum(x_i - \bar{x})(y_i - \bar{y})$, where n is the sample size and $\bar{x},\bar{y}$ are the arithmetic means of the two variables. The correlation coefficient is simply the covariance of two variables divided by the product of their standard deviations.

(A)

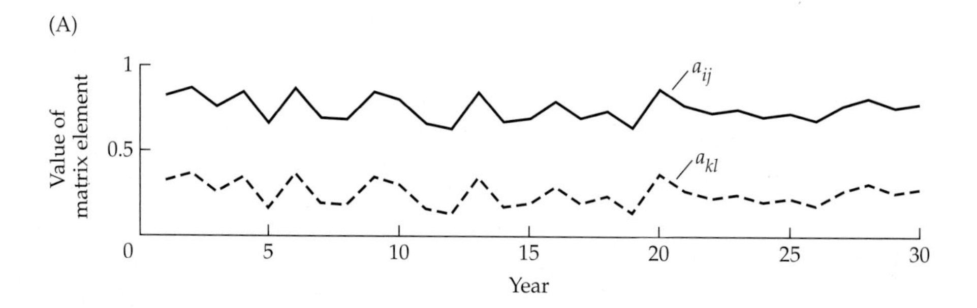

(B)

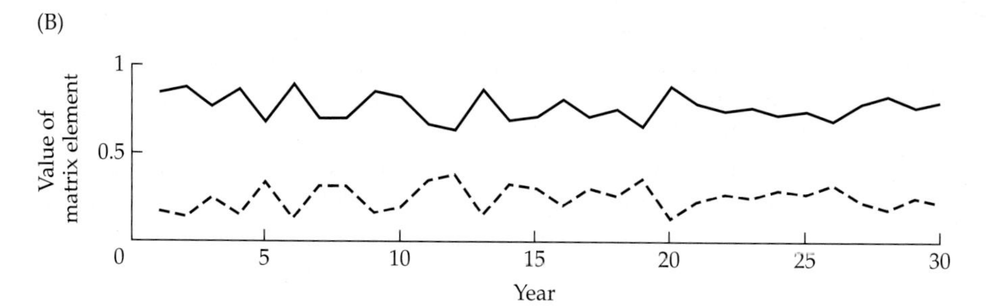

(C)

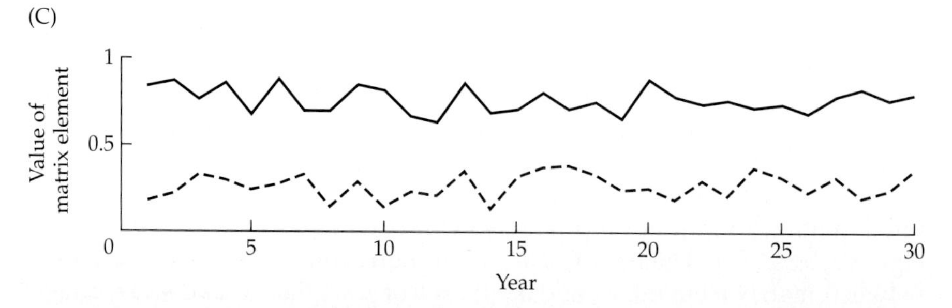

Figure 7.4 The covariance between two matrix elements a_{ij} and a_{kl} will be (A) positive if the two elements vary in synchrony, (B) negative if they vary in opposition, and (C) zero if they vary independently.

different matrix elements, and $Cov(a_{ij},a_{kl})$ is a measure of the tendency of these two elements to change in synchrony across years. As shown in Figure 7.4, covariances can be positive, zero, or negative. If $Cov(a_{ij},a_{kl})$ is large and positive, then a_{ij} and a_{kl} will show very similar patterns of variation over time. That is, years in which a_{ij} is near its maximum will also be years in which a_{kl} is near its maximum, and years of relatively low a_{ij} will also be years with relatively low a_{kl}. In contrast, if $Cov(a_{ij},a_{kl})$ is strongly negative, a_{ij} and a_{kl} will show opposing patterns of variation, with high a_{ij} always accompanied by low a_{kl} and vice versa. If $Cov(a_{ij},a_{kl}) = 0$, the two matrix elements vary independently of one another. If $i = k$ and $j = l$, then a_{ij} and a_{kl} represent the same matrix element; $Cov(a_{ij},a_{kl})$ is simply the variance of matrix element a_{ij} (such as the entries in the last column of Table 6.7), and the term on the right-hand side of Equation 7.13 becomes $\text{Var}(a_{ij})\bar{S}_{ij}^2$. To calculate τ^2 using Equation 7.13, we simply sum the products of the covariances and eigenvalue sensitivities for all possible combinations of the four indices *i*, *j*, *k*, and *l*, letting each index vary from 1 to the number of classes, *s*.

There are good reasons to expect both positive and negative covariances between different elements in the projection matrix. For example, good years for reproduction by individuals of one class will often be good years for reproduction by individuals in other classes, causing the first-row elements of the matrix to covary positively. Similarly, favorable environmental conditions may enhance both reproduction and survival of individuals. On the other hand, years in which individuals in a certain size class tend to grow may also be years in which shrinkage is less likely, causing the subdiagonal and supradiagonal elements in the same column of the matrix to covary negatively. Also, high reproduction may come at a cost of low growth or low survival.

Having defined the terms in Tuljapurkar's approximation, we can now discuss their biological interpretation. First, Equation 7.12 states that the more the log population growth rate varies from year to year (as measured by $\tau^2/\bar{\lambda}_1^2$), the smaller the stochastic log growth rate will be relative to the rate that would be expected in the absence of variation (i.e., $\log \bar{\lambda}_1$). This result is reminiscent of results from the unstructured case: environmental variation reduces the rate of population growth over the long term (see our discussion of the geometric mean growth rate in Chapter 2). However, we expect that variation in a particular matrix element, say a_{ij}, will contribute to variation in the population growth rate only to the extent that changing a_{ij} actually changes the population growth rate; this explains why the eigenvalue sensitivities appear in Equation 7.13. Thus Tuljapurkar's approximation says that environmentally driven variability in matrix element a_{ij} will result in a large increase in τ^2, and thus a large decrease in the stochastic log growth rate, only if the variation in a_{ij} [i.e., $\text{Cov}(a_{ij},a_{ij}) = \text{Var}(a_{ij})$] is relatively large *and* changes in $\bar{a}_{ij}$ have a large effect on $\bar{\lambda}_1$ (i.e., $\bar{S}_{ij}$ is large).

Now consider two different matrix elements that covary positively. Recall from Equation 7.10 that, because sensitivities of eigenvalues to changes in matrix elements are positive, increasing either element would increase the population growth rate (although not necessarily to the same degree). Hence positive covariance means the two elements will either both be high, contributing to a high population growth rate, or both low, leading to a low population growth rate. This scenario will tend to cause high year-to-year variation in population growth, and thus a lower stochastic log growth rate. On the other hand, if the two elements covary negatively, then when one element is high and would thus tend to cause high population growth, the other will be low and "detract" from population growth; that is, the effects of variation in the two elements tend to cancel each other out, leading to less variable population growth and a higher stochastic log growth rate. Thus negative covariances between matrix elements actually *decrease* τ^2 and thus *increase* $\log \lambda_s$ (while the opposite is true of positive covariances between matrix elements). More generally, it makes good biological sense that, to estimate the long-term rate of population growth in a stochastic environment, we must know both the patterns of covariation between matrix elements and how sensitive the population growth rate is to each of those elements.

As we noted above, Tuljapurkar's approximation was derived by assuming that the projection matrix doesn't vary greatly over time (i.e., that the variances and covariances of matrix elements are small). In practice, Tuljapurkar's formula does a reasonably good job of approximating the stochastic log growth rate even when the variances and covariances are fairly large. In addition to calculating $\log \lambda_s$ by simulation, the MATLAB code in Box 7.4 also calculates $\log \lambda_s$ using Tuljapurkar's approximation. In the case of mountain golden heather, computer simulation and Tuljapurkar's approximation yield very similar estimates of $\log \lambda_s$: based on one run of 50,000 simulated time intervals, $\log \lambda_s = -0.0365$ (95% confidence interval: [–0.0371, –0.0360]); based on Tuljapurkar's approximation, $\log \lambda_s = -0.0370$. These estimates translate into stochastic growth rates of $\lambda_s = 0.9641$ and $\lambda_s = 0.9636$, respectively, by means of the formula $\lambda_s = \exp(\log \lambda_s)$.[16] As these estimates are less than 1, they indicate a population that will eventually decline to extinction. By comparison, the dominant eigenvalue of the mean matrix $\bar{\lambda}$ equals 0.9660. Thus variation in the matrix elements causes the stochastic growth rate to be lower than the mean matrix would predict (as we would expect from Equation 7.12), but not dramatically so. If there is any doubt as to whether the variation in matrix elements is sufficiently small to justify the use of Tuljapurkar's approximation, $\log \lambda_s$ should be estimated by both methods (as in Box 7.4), and the simulation estimate used if the two estimates differ substantially. Even though the simulation method is more accurate when there is a large degree of variation in the matrix elements, if the variation is small to moderate, Tuljapurkar's approximation is useful because it can be computed rapidly (using the last few lines of code in Box 7.4), whereas many thousands of runs are needed to compute $\log \lambda_s$ by simulation. In addition, we have already seen the heuristic value of Tuljapurkar's approximation for sharpening understanding of how variation and covariation in matrix elements affects population growth in a stochastic environment.

[16]However, be warned that these estimates use the observed variances of the matrix elements as estimates of environmental stochasticity, without accounting for sampling variation. As we'll see in Chapter 8, a vital-rate approach allows us to make better estimates of true environmental stochasticity.

Calculating the Probability of Hitting a Quasi-Extinction Threshold by Time t

Estimating $\log \lambda_s$ is useful because it allows us to assess whether a population is likely to grow or decline to extinction over the long term in a stochastically varying environment. However, as we saw for unstructured populations, it is also valuable to estimate the probability that quasi-extinction will have occurred prior to a specified future time horizon. Fortunately, there are two methods to calculate the cumulative distribution function, or CDF, for the extinction time of a structured population. The first method uses computer simulation, and is appropriate for any degree of temporal environmental variation in matrix elements. The second method assumes that variation in the matrix elements is small to moderate, and so can make use of the estimates of $\log \lambda_s$ and τ^2 obtained from Equations 7.12 and 7.13. In particular, we can use the same formula for the extinction time CDF that we applied to an unstructured population in Chapter 3 to estimate the cumulative probability of quasi-extinction for a structured population. Although its underlying assumptions render this second method less widely applicable, the results we obtain by applying it to demographic data have important implications for identifying conditions under which the method can be safely applied to count-based data. But before we can discuss those implications, we must first describe the two methods.

METHOD 1: SIMULATING EXTINCTION PROBABILITIES Computing the extinction time CDF by simulation requires only a simple modification of the code in Box 7.3 to keep track of whether the total population density (or the density summed across a subset of the classes with which we are particularly concerned, such as the reproductive classes) has fallen below the quasi-extinction threshold each year. The fraction of realizations that first hit the threshold during or before year *t* gives the cumulative probability of extinction. The MATLAB code in Box 7.5 performs such a calculation using the mountain golden heather matrices from Table 6.7. T9he results of 10 separate runs of this program, each with 5,000 separate realizations of population growth and a quasi-extinction threshold of 500, are shown in Figure 7.5. Note that there is not much variation among the separate runs, indicating that in this case, 5,000 realizations provides a reasonably good estimate of the extinction time CDF.[17] From a starting population density of 4,343, the probability of hitting a threshold density of 500 individual plants (most of which would be seeds in the seed bank) reaches a value of 0.1 after only about 37 years, and exceeds 0.4 by year 50. This analysis supports our con-

[17]In general, we advocate increasing the number of realizations per run until separate runs yield very similar estimates for the CDF.

BOX 7.5 *MATLAB code to simulate the extinction time cumulative distribution function.*

```
% PROGRAM simext
%   Estimates by simulation the quasi-extinction time cumulative
%   distribution function for a structured population in an iid
%   stochastic environment

%************* USER-SPECIFIED PARAMETERS **************
% Enter name of m file containing matrices; e.g. the file
% hudmats.m contains 4 matrices (A85, A86, A87, A88),
% one for each year of Frost's demographic study
hudmats;
% Change names of matrices below: to correspond to the names in
%   the m file from the preceding command:
matrices=[A85(:) A86(:) A87(:) A88(:)];
% Enter probabilities of choosing each matrix below;
% NOTE: sum of penv must equal 1
penv=[.25 .25 .25 .25];
% Enter farthest future time horizon
tmax=50;
% Enter number of times to simulate cdf
maxruns=10;
% Enter number of realization of population growth
% to simulate in each run
numreps=5000;
% Enter initial population vector
n0=[4264; 3; 30; 16; 25; 5];
% Enter quasi-extinction threshold, expressed as a density
Nx=500;
% Enter 0 to omit a class and 1 to include it when computing the
% summed density to compare to the quasi-extinction threshold
sumweight=[1 1 1 1 1 1];
%*****************************************************

rand('state',sum(100*clock));     % seed random number generator
s=sqrt(size(matrices,1));         % s=number of classes
cumdist=cumsum(penv);             % CDF for environmental states

Results=[];           % initialize array to store extinction CDF's
for i=1:maxruns       % calculate CDF "maxruns" times,
    PrExt=zeros(tmax,1);
    for j=1:numreps % with "numreps" realizations per run,
        n=n0;       % starting at the initial population vector,
```

BOX 7.5 *(continued)*

```
        for t=1:tmax                    % for each future time,
            x=sum(rand>=cumdist)+1;  % draw a matrix at random,
            A=reshape(matrices(:,x),s,s); % extract the chosen
                                          % matrix, from "matrices", &
            n=A*n;            % project the population 1 year ahead
            N=sumweight*n;    % compute weighted sum of current
                              % densities
            if N<Nx           % if quasi-extinct,
                PrExt(t)=PrExt(t)+1;    % update counter
                break;             % and start new realization
            end
        end        % for t
    end            % for j
    PrExt=cumsum(PrExt/numreps); % sum extinctions at each
                                 %     time to get cdf
    Results=[Results PrExt];     % store the result
end                % for i

plot(Results)                    % plot all "maxruns" cdf's
xlabel ('Years into the future')
ylabel ('Cumulative probability of quasi-extinction')
meancdf=mean(Results')
figure
plot(meancdf)                     % plot mean cdf
xlabel ('Years into the future')
ylabel ('Cumulative probability of quasi-extinction')
```

clusion based on log λ_s that, without management, this population would face a substantial risk of extinction over the short term.

In Box 7.5, we compute the probability that the total density over all classes falls below 500. However, the code is written so that users can define the quasi-extinction threshold to apply to any subset of the classes. For example, since seeds in the seed bank and seedlings of mountain golden heather are difficult to census, it might make sense to set the threshold in reference to the total density of larger plants, ignoring the seed bank and seedlings. To do so, we would make two small changes in the program. First, we would change the value of `Nx` to represent the quasi-extinction threshold density of plants larger than seedlings. Second, we would change the row vector `sumweights` to [0 0 1 1 1 1]. The zeros in this vector instruct the program to ignore both seeds in the seed bank and seedlings when summing the densities across

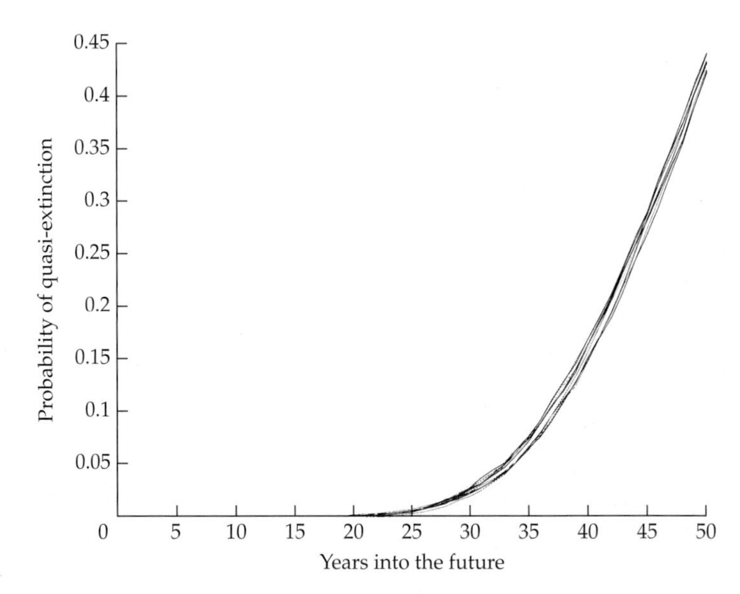

Figure 7.5 Ten simulated cumulative distribution functions (CDFs) for the time to reach a quasi-extinction threshold of 500 individuals for a population of mountain golden heather starting at 4,343 individuals (see Box 7.5).

classes to compare to the quasi-extinction threshold. If we wanted to set the threshold in terms of only the largest plants, we would set all but the last entry in `sumweights` to zero.

At this point, we need to say a word about setting the quasi-extinction threshold for a demographic PVA. As we noted above, the entries in a population vector represent densities, but quasi-extinction thresholds are typically expressed in terms of numbers of individuals. Thus when we ask whether the summed densities across all or a subset of the classes has fallen below the quasi-extinction threshold, we first need to be sure that a numerical threshold has been converted to a density. For example, let's assume that the densities in the population vector represent mean numbers of individuals per hectare, that the population occupies 1,000 hectares, and that we have decided (based on considerations discussed in Chapter 2) to set the quasi-extinction threshold at a total of 100 individuals in the reproductive classes. One hundred reproductive individuals in the entire population represents an average density of 0.1 individuals per hectare summed across the reproductive classes, so we would consider the population to have hit the threshold if this sum hits 0.1, not 100. Alternatively, we could multiply the starting densities by 1,000 to arrive at an estimate of total numbers and set the quasi-extinction threshold to 100.

METHOD 2: CALCULATING EXTINCTION PROBABILITIES FOR STRUCTURED POPULATIONS USING DIFFUSION APPROXIMATIONS Lande and Orzack (1988) showed that under the appropriate conditions, the same diffusion approximation used to derive the extinction time CDF for a count-based model (Equation 3.5) can do a reasonably good job of predicting extinction probabilities for a structured population. Recall from Chapter 3 that the diffusion approximation assumes that the amount of environmental variation is small to moderate. The same assumption underlies Tuljapurkar's approximation for the stochastic log growth rate of a structured population. Not surprisingly, there is a close association between the cumulative probability of quasi-extinction calculated by the diffusion approximation and Tuljapurkar's approximation for the rate of population growth, as we will now see.

To estimate the extinction time CDF with a set of population counts and Equation 3.5, we had to determine the values of three variables: μ, σ^2, and d. Thus to use Equation 3.5 for a structured population, we must decide what to use for each of these variables. Recall that we calculated μ for an unstructured population as the arithmetic mean of the observed log population growth rates (i.e., the mean of $\log \lambda_i = \log(N_{i+1}/N_i)$ for census $i = 1$ to q). This suggests that for a structured population, we should compute the mean of the log ratio of total population size in adjacent years, where total population size is simply the sum of the elements in the population vector. But if you look at Box 7.4, you will see that this is exactly what we did when we calculated $\log \lambda_s$, the stochastic log growth rate, by simulation. Moreover, if we assume environmental variation is small to moderate, we can use Tuljapurkar's approximation for $\log \lambda_s$, instead of the simulated value. Thus to use Equation 3.5 for a structured population, we simply substitute $\log \lambda_s$ calculated using Tuljapurkar's approximation (Equations 7.12 and 7.13) in place of μ. When we discussed Tuljapurkar's approximation, we also noted that the ratio $\tau^2/\bar{\lambda}_1^2$ is approximately the variance in the log population growth rate when the environmentally driven variation in the matrix elements is small to moderate. Because σ^2 in Equation 3.5 measures the variance of the log population growth rate for an unstructured population, we will use $\tau^2/\bar{\lambda}_1^2$ in its place for a structured population.[18]

What to use in place of d requires more thought. For an unstructured population, d represents the difference between the log of the current population size and the log of the quasi-extinction threshold (i.e., $d = \log[N_c] - \log[N_x]$). So one possibility would be to use the sum of the elements in the

[18]Lande and Orzack (1988) suggested that the following alternative method for estimating σ^2 might yield even better estimates of extinction probabilities for a structured population: (1) estimate $\log \lambda_s$ by simulation, as in Box 7.4; (2) after substituting σ^2 in place of $\tau^2/\bar{\lambda}_1^2$ in Equation 7.12, rearrange the equation to obtain an expression for σ^2 as a function of $\log \lambda_s$ and $\log \bar{\lambda}_1$, the log of the dominant eigenvalue of the mean matrix: $\sigma^2 = 2(\log \bar{\lambda}_1 - \log \lambda_s)$. However, as this method requires a long computer simulation to accurately estimate $\log \lambda_s$, yet still relies on an approximation for calculating σ^2, it is not clear that it has any practical advantages over simply computing the extinction time CDF entirely by simulation, as in Box 7.5.

current population vector in place of N_c. However, there is a problem with this approach. We know that individuals in different classes make different contributions to future population growth, yet simply summing up the numbers of individuals in all classes ignores these differences. For example, consider two populations of an organism that takes many years to reach reproductive maturity. Both populations have the same total number of individuals, but one is comprised entirely of newborns and the other entirely of individuals in the prime reproductive classes. Clearly, the first population will have a much higher chance of extinction than the second, because a sequence of low-survival years could drive it to the quasi-extinction threshold before individuals could even begin to reproduce. If we use the sum of the population vector as N_c, we would estimate that these two very different populations have the same probability of quasi-extinction. Thus we need a measure of the current population density that accounts for its structure, particularly as it reflects among-individual differences in potential contributions to population growth. Lande and Orzack (1988) suggested that to calculate N_c, individuals in different classes should be weighted by their respective reproductive values. Recall that reproductive value measures an individual's expected contribution to future population growth. Hence if we multiply the number of individuals in a class by their reproductive value before adding the classes together, individuals in classes with high reproductive value will make a disproportionately large contribution to the weighted population density, whereas classes with low reproductive value will be discounted. Lande and Orzack suggested that the reproductive value vector computed from the mean projection matrix (i.e., the dominant left eigenvector of $\bar{\mathbf{A}}$) would serve well for the purpose of weighting the current population vector.[19]

How well does the analytic CDF (Equation 3.5) derived under the assumption that environmental variation is small to moderate match the more exact simulation-based CDF for mountain golden heather in Figure 7.5? In Box 7.6, we provide code that computes $\log \lambda_s$ and σ^2 using Tuljapurkar's approximation, calculates the current population density weighted by the reproductive values, and plots the extinction time CDF predicted by Equation 3.5. In Figure 7.6, we compare this analytical result to the mean of the ten simulated CDFs from Figure 7.5. The match is fairly close, although the diffusion approx-

[19]More specifically, they suggested computing $\mathbf{v}$ and $\mathbf{w}$ as the dominant left and right eigenvectors of $\bar{\mathbf{A}}$, rescaling $\mathbf{w}$ so that the sum of its elements equals 1 (as in Equation 7.8), and then rescaling $\mathbf{v}$ by dividing it by $\mathbf{v}'*\mathbf{w}$, the scalar product of $\mathbf{v}$ and $\mathbf{w}$ (see Footnote 6), so that the rescaled versions of $\mathbf{v}$ and $\mathbf{w}$ have a scalar product of 1. The scalar product of this rescaled $\mathbf{v}$ times the current population vector is used in place of N_c. The MATLAB code in Box 7.6 performs these operations. Note that this scaling of $\mathbf{v}$ is different than in Equation 7.9.

BOX 7.6 *Fragment of MATLAB code to calculate the extinction time CDF for mountain golden heather using Tuljapurkar's approximation (Equations 7.12 and 7.13) and Equation 3.5.*[1]

```
hudmats;                 % read matrices from file hudmats.m, and
matrices=[A85(:) A86(:) A87(:) A88(:)]; % put them in "matrices"
s=sqrt(size(matrices,1));              % s=number of classes
Abar=reshape(mean(matrices'),s,s);     % calculate mean matrix Abar
C=cov(matrices');                      % calculate covariances of
                                       %    matrix elements
[lams,lam1,W,w,V,v]=eigenall(Abar);    % compute lam1, v, and w for
                                       % Abar using "eigenall" from
                                       % Box 7.1
S=v*w'/(v'*w);    % compute eigenvalue sensitivities for Abar
S=S(:);           %    and put them in vector form
sigma2=S'*C*S/lam1^2; % compute sigma^2 from Eqs. 7.12 & 7.13
loglams=log(lam1)-0.5*sigma2; % compute stochastic log lambda
                              %    from Equation 7.12
n0=[4264; 3; 30; 16; 25; 5];  % current population vector
v=v/(v'*w);    % rescale v so that sum(w)=1 and v'*w=1
Nc=v'*n0;      % starting pop.=initial total reproductive value
Nx=500;        % quasi-extinction threshold
tmax=50;       % farthest time horizon
d=log(Nc/Nx);% log distance to quasi-extinction threshold
cdf=extcdf(loglams,sigma2,d,tmax); % calculate CDF using function
                                   % "extcdf" defined in Box 3.3
plot(cdf)      % plot the CDF
xlabel('Years into the future')
ylabel('Cumulative probability of quasi-extinction')
```

[1]This code assumes that all of the matrices in "matrices" are equally likely. If they are not, you can easily modify it by importing the appropriate code to calculate "Abar" and "C" from Box 7.4.

imation slightly overestimates the probability of extinction, especially as more time elapses. Nevertheless, the overall impression we get from Figure 7.6 is that the analytic formula, which is based only on total population density, provides a close approximation (at least for short times) to the simulated CDF, which in contrast actually accounts for the dynamics of the population's size structure. This result has important implications for count-based PVAs performed on populations that are in reality structured by age, size, or stage, as we now discuss.

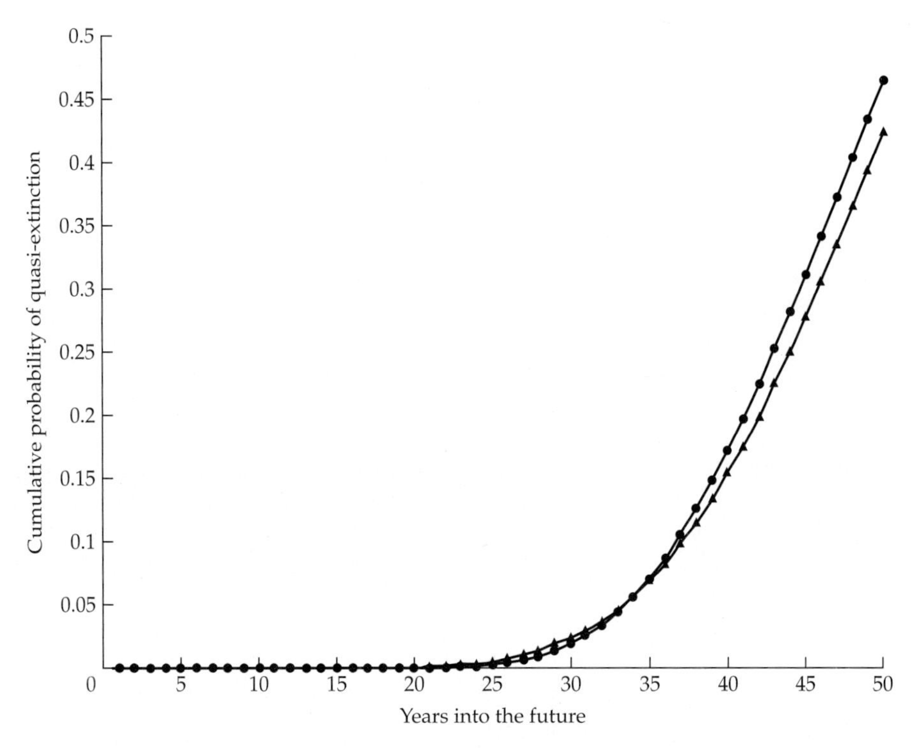

Figure 7.6 A comparison of the extinction time CDF for mountain golden heather computed by simulation (triangles = mean of 10 runs shown in Figure 7.5) and using the analytic expression (Equation 3.5) derived from a diffusion approximation (circles).

IMPLICATIONS OF THE ANALYTIC APPROXIMATION FOR COUNT-BASED PVAS In one way, the close correspondence shown in Figure 7.6 between analytic and simulated CDFs is actually somewhat surprising. We say this because of another assumption of the diffusion approximation that readers will recall from Chapter 3. In addition to assuming that the environmental variation is small to moderate, the diffusion approximation also assumes that the population growth rate is uncorrelated from one time interval to the next. However, in a structured population, successive growth rates should be somewhat correlated *even in the absence of temporal environmental autocorrelation.* The reason is that the population structure itself will generate a correlation between successive rates of population growth. As we noted in Chapter 2, a population that is depauperate in the most highly reproductive classes will show low rates of growth for several years in a row; conversely, a "baby boom" will be followed by a series of "boomlet" years when the "baby boomers" reach the reproductive classes. Thus adjacent values in a series of growth rates of a structured population tend to be more similar to one another than two randomly chosen growth rates would be, the hallmark of autocorrelation.[20] The diffusion approximation assumes this autocorrelation is zero, and yet in the case of mountain golden heather, it produces extinction probabilities similar to those produced by stochastic matrix simulations, which do not make this assumption.

The answer to this apparent paradox lies in the fact that the self-generated autocorrelation in the growth rate of a structured population can sometimes be small and of short duration, so that the growth rate quickly becomes effectively independent of the growth rates in years not very far in the past.[21] In such cases, assuming the autocorrelation is zero is not too egregious, and we can safely apply the analytic approximation to estimate the extinction probability for a structured population. For mountain golden heather, simulating 1,000 replicates of 50 years of population growth gives the seemingly high average correlation between successive log growth rates of 0.43. However, the correlation between two values of the log growth rate declines to the modest levels of 0.06, –0.08, –0.11, and –0.11 as the time elapsed between them increases to 2, 3, 4, and 5 years, respectively. Thus although population structure *does* cause autocorrelation in the growth rate of mountain golden heather, it does not persist over long time periods, so that the assumption of independent growth rates is not too bad, and the diffusion approximation based on that assumption does an adequate job of predicting quasi-extinction probabilities (Figure 7.6).

A more important conclusion to be drawn from Figure 7.6 is that, when the autocorrelation is small, we can apply the diffusion approximation to a series of counts of total population size, even though those counts come from a population that has an underlying age, size, or stage structure. This provides justification for using the count-based methods of Chapter 3 on census data obtained from a population that we know to be comprised of individuals that make different contributions to population growth (e.g., cubs and adult females in the Yellowstone grizzly bear population), provided that we check to see that the autocorrelation is small. We saw in Chapter 3 how to test for significant autocorrelation in the log growth rates estimated

[20]It is for this reason that, whereas it is proper to estimate $\log \lambda_s$ as the arithmetic mean of a large number of values of $\log[N(t+1)/N(t)]$, where $N(t)$ is the sum of the elements in the population vector in year t (see Box 7.4), it is *not* proper in general to estimate σ^2 for a structured population by computing the sample variance of the $\log[N(t+1)/N(t)]$ values (as we did with census counts). Because the variance of the $\log[N(t+1)/N(t)]$ values does not account for the fact that adjacent log growth rates tend to be similar to one another, it would *underestimate* σ^2, and lead to artificially low estimates of extinction probabilities. See Equation 6 in Lande and Orzack (1988).

[21]Whether the auto-correlation is small or large depends in a complex way on the structure of the projection matrix, which reflects the life history of the species it represents.

from census data. We also saw in Chapter 4 how to use alternative, simulation-based methods when we do detect significantly autocorrelated growth rates, which may be driven by *either* environmental autocorrelation *or* population structure.

Thus in a sense, the value of the analytic CDF lies not so much in the fact that it is much faster to compute for a structured population than is a simulated CDF. After all, compared to the investment of time and resources required to conduct a long-term demographic study and to parameterize multiple projection matrices, the additional effort needed to construct a CDF by simulation, which assumes neither that environmental fluctuations are small to moderate nor that the growth rate is uncorrelated, is trivial. Rather, the value of the analytic CDF lies in the fact that we can use it to estimate extinction probabilities for a structured population for which we have census counts but *no detailed demographic data*, provided that the assumptions of small variation and small autocorrelation are met.

ACCOUNTING FOR EXTREME ENVIRONMENTAL CONDITIONS WHEN SIMULATING STOCHASTIC LOG GROWTH RATES AND EXTINCTION PROBABILITIES We end this chapter by pointing out that the programs we have written to simulate the stochastic log growth rate and the extinction time CDF (Boxes 7.4 and 7.5) also provide a way to examine the impact of extreme environmental conditions on the viability of a structured population. Suppose we know that in one or more years of the demographic study, the population experienced an unusual environmental condition (e.g., atypically high or low rainfall, extreme temperatures, very high or very low snow levels, and so on). Furthermore, suppose we are able to estimate the frequency with which such conditions occur (e.g., by using meteorological data). If so, then we can instruct the programs to choose the matrix (or matrices) estimated in the extreme year (or years) less often than others, with the actual frequency estimated from the climatic record. This simply requires changing the entries in the vector `penv`. These extreme matrices summarize all of the demographic effects of catastropes or bonanzas. For an example of this approach in a non-PVA setting, see the work of Åberg (1992a,b), who used meteorological information on the frequency of ice-free, normal-ice, and heavy-ice years to arrive at an estimate of the growth rates of two populations of a nonendangered seaweed species that accounted for the effects of extreme years. Note that if extreme matrices represent large deviations from the range of variation observed among matrices estimated in more typical years, it would not be appropriate to base an estimate of log λ_s or the extinction time CDF on Tuljapurkar's approximation, which assumes such large deviations do not occur. Instead, log λ_s and the extinction time CDF should be calculated by simulation. In the next chapter, we will examine how to account for extreme environmental effects in a different way by simulating variation in the vital rates themselves.

8

Demographic PVAs Based on Vital Rates: Removing Sampling Variation and Incorporating Large Variance, Correlated Environments, Demographic Stochasticity, and Density Dependence into Matrix Models

In Chapter 7 we presented two different approaches to estimating growth rates and extinction probabilities in stochastic environments, based either on the assumption that environmental variation is small to moderate (i.e., Tuljpurkar's approximation for the stochastic log growth rate and the diffusion approximation for extinction probabilities) or on computer simulation. In either case, the smallest units these methods treat are the matrix elements, rather than the vital rates—the demographic "elementary particles"—from which we constructed the matrix elements in Chapter 6. For demographic data sets that lack detailed vital rate estimates, the use of matrix elements as the basic units of analysis is reasonable and justified. It is also much easier to derive and use analytical approximations such as Equation 7.13 in terms of matrix elements than in terms of survival, fertility, and state transition rates. However, as we've emphasized before, vital rates, and not matrix elements, are really the fundamental biological units of demographic analysis, and there are several other factors that commonly make it more natural to recast a demographic

PVA in terms of underlying vital rates. These include phenomena such as demographic stochasticity, catastrophes and bonanzas, and density dependence, all of which operate at the level of vital rates rather than amalgamated matrix elements. For example, in catastrophic years fertility rates may be reduced to zero, but adult survival rates may be relatively unaffected. Similarly, for many species, density dependence may influence growth rates dramatically, but not survivorship or fertility. Shared environmental effects may generate temporal correlations between vital rates, which in turn result in the correlations between matrix elements. Since management actions often influence survival, growth, and fertility rates in very different ways, the analyses we will describe in Chapter 9 are often more useful when performed on vital rates. Finally, it is often possible to estimate and remove the contribution of sampling variation to the observed variance in vital rates, and thus to arrive at better estimates of environmental stochasticity and its effect on viability. For all these reasons, it can sometimes be more informative, and often more accurate, to perform a PVA that is explicitly based on vital rates, rather than subsuming them into matrix elements.

In spite of the arguments for expressing matrix models in the language of vital rates, there are numerous complications and considerations to keep in mind when using this approach (which is in part why the matrix-element approach we covered in Chapter 7 is still useful). We begin by reviewing the basic construction of a stochastic matrix model in terms of vital rates, but we expand on our earlier coverage by describing how to better estimate the means and variances of vital rates by removing the influence of sampling variation. We then move on to the methods needed to simulate stochastic dynamics of a population using vital-rate estimates, including how to incorporate correlations among vital rates within a year, correlations in rates across years, and extreme variability in vital rates. Throughout these sections we build the pieces needed to perform complete vital-rate-based viability analyses for two examples, the desert tortoise and mountain golden heather. We end the chapter with discussions of demographic stochasticity and density dependence and how to include them in stochastic matrix models. Given the diversity of quantitative methods that are covered in the chapter, you may feel that you are getting bogged down in technical details of each section; if so, we suggest that you skip ahead in order to build an idea of the array of considerations and methods for this type of PVA model before returning to specifics of each topic covered.

Throughout this chapter, we will emphasize simulation approaches. Although it is sometimes possible to fit a vital rate-based approach to stochastic matrix models into the analytical framework that we presented in Chapter 7, doing so is algebraically messy at best, and it is frequently impossible without additional approximations that can further compromise the accuracy of the results.

Estimation and Construction of Stochastic Models Based on Vital Rates

In Chapter 6 we discussed the three basic types of vital rates from which matrix elements are constructed: fertility rates (f_i), survival rates (s_i), and state transition, or growth, rates (g_{ij}). We often want to break these into even finer categories (e.g., winter and summer survival, or probability of nesting and clutch size), but for the most part we'll stick to these three basic kinds of rates to simplify our presentation. In this section, we discuss the estimation of these rates and their use in constructing a stochastic matrix model. Even though this occasionally means backtracking to refine some of the analyses described in Chapter 6, for the most part we use the methods described there as the starting point for further analysis. In particular, we assume that you understand the general idea of putting together different vital rates to construct the elements of a matrix model. Following this section, we turn to the problems of simulating the dynamics of a population based on these rates.

The Estimation and Use of Means, Variances, and Correlations of Vital Rates

The key issue in constructing a matrix around vital rates (and the primary reason to do so in the first place) is the accurate estimation of variance and correlation in the demographic rates characterizing a population. To fully parameterize a matrix model, we need to know three things: the mean value for each vital rate, the variability in each rate, and the covariance or correlation[1] between each pair of rates (i.e., the same three descriptors we needed for the matrix element approach in Chapter 7). One way to simulate stochastic population growth is to pick an entire matrix each year from the set of matrices estimated from the data, as we did in Chapter 7. This process simulated the *exact* values observed for each matrix element and their *exact* co-occurrences over time; if one matrix happened to have the highest values for two particular matrix elements among all the estimated matrices, those high values would always co-occur when drawing entire matrices at random. Unfortunately, the assumption that the *precise* combinations of values

---

[1]In Chapter 7, we emphasized covariances between matrix elements, in part because Tuljapurkar's approximation makes direct use of them. In this chapter, we emphasize *correlations* between vital rates. These two measures of the relationship between different variables are closely related, as $r(x,y) = Cov(x,y)/\left(\sqrt{Var(x)}\sqrt{Var(y)}\right)$ where r is the correlation between x and y. Because correlations are scaled by the standard deviations of the two variables, range from 1 to –1, and are thus easier to compare across different pairs of variables. That is, two pairs of variables, each of which vary in perfect synchrony, will both have correlations of 1, even if one pair varies over a wider range of values than the other.

that we observed over the limited duration of a demographic study will always occur is unlikely to be correct. Instead, it is more realistic to estimate the means, variances, and correlations between vital rates, and then simulate a broader (in fact, infinite) range of possible values, while preserving the observed means, variances, and average correlation structure. This is the approach we take throughout this chapter.

One important reason why it is preferable to perform this kind of simulation on vital rates instead of matrix elements has to do with inherent negative correlations between different matrix elements that are caused by shared dependence on the same vital rate (van Tienderen 1995). For example, let's say that an individual of a hypothetical species of snake is currently in size class 3 and has a mean probability $s_3 = 0.95$ of surviving for one year. Furthermore, if it survives it will either stay the same size, or grow to be in size class 4 with mean probability $g_{4,3} = 0.10$. Although the mean matrix elements describing these class transitions will then be $a_{3,3} = s_3(1 - g_{4,3}) = (0.95)(1 - 0.10)$ and $a_{4,3} = s_3\, g_{4,3} = (.95)(0.10)$, the actual survival and growth rates each year will in fact vary due to demographic and environmental stochasticity. Even if growth and survival are not themselves correlated, we would still expect to see a significant negative correlation between these two matrix elements, since one is proportional to the growth rate ($g_{4,3}$) and the other is proportional to the complement of the growth rate ($1 - g_{4,3}$). However, think of using the means of the two matrix elements, along with their estimated correlation, to simulate many pairs of values for the two elements. Even though we will usually come up with reasonable pairs, some of them will be manifestly unreasonable—pairs whose sum is greater than 1, meaning greater than 100% survival (Figure 8.1). These unreasonable results can occur because, even though we estimated that on average $a_{3,3}$ and $a_{4,3}$ must be negatively correlated, in simulating matrix elements and not vital rates we've relaxed the restriction that this pair of matrix elements must *always* reflect a hard-wired negative correlation that guarantees they will sum to 1 or less. The same problem can also arise when performing simulations using random draws of observed individual matrix elements. In contrast, by generating random vital rates instead of matrix elements, we explicitly include biological restrictions on the sums of different matrix elements and can therefore make better predictions of population growth and extinction risk.

With this background, we move on to the actual estimation of vital rates. We assume that the basic steps of deciding on a state variable, breaking it into classes, and estimating vital rates for each class for each annual transition have already been done (see Chapter 6). Throughout this section we will use part of a more extensive data set for the desert tortoise (*Gopherus agassizi*) in the Western Mojave desert of California as our example (Doak et al. 1994). The tortoise is federally listed as a threatened species and has undergone rapid declines over much of its range due to development pressure, crushing

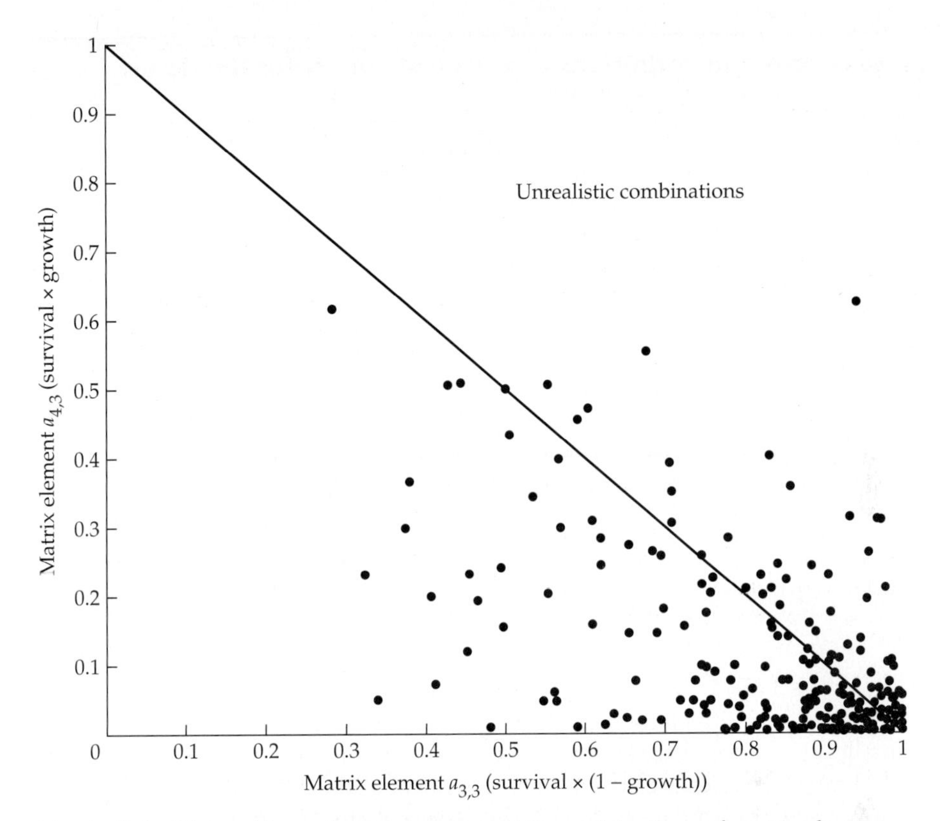

Figure 8.1 The problem of obtaining unrealistic combinations of matrix elements by simulating elements rather than underlying vital rates. The line shows the limit of biologically possible combinations of matrix element values; values of the two elements farther from the origin imply total survival rates greater than one. Pairs of values for the matrix elements $a_{3,3}$ and $a_{4,3}$ were simulated assuming a correlation coefficient of –0.765 and means and standard deviations of {0.8668, 0.1498} and {0.0823, 0.1263}, respectively. These parameters for the matrix elements were estimated by drawing 500 pairs of values for the underlying vital rates (survival, s_3, and growth, $g_{4,3}$), assuming that they are uncorrelated and have means and standard deviations of {0.95, 0.1} and {0.1, 0.15} respectively. Simulating the vital rates directly instead of the matrix elements would prevent generation of the unrealistic combinations of element values that we see here.

by off-road vehicles (and army tanks), destruction of desert vegetation by grazing and vehicle traffic, and introduced disease. In contrast to the complexity of its problems, the desert tortoise life cycle can be portrayed with a fairly simple set of vital rates. In particular, like the hypothetical snake mentioned earlier in this section, individual tortoises in each size class can only stay the same size or grow one class larger in a year, and only the three

TABLE 8.1 Size classes and definitions of matrix elements for the desert tortoise assuming a pre-breeding census[a]

Class name	*Size class in year t+1*	*Size class in year t* 0	1	2	3	4	5	6	7
Yearling	*0*						f_5	f_6	f_7
Juvenile 1	*1*	s_2	$s_2(1-g_2)$						
Juvenile 2	*2*		s_2g_2	$s_2(1-g_2)$					
Immature 1	*3*			s_2g_2	$s_3(1-g_3)$				
Immature 2	*4*				s_3g_3	$s_4(1-g_4)$			
Subadult	*5*					s_4g_4	$s_5(1-g_5)$		
Adult 1	*6*						s_5g_5	$s_6(1-g_6)$	
Adult 2	*7*							s_6g_6	s_7

[a]The vital rates used to define matrix elements are: fertilities (f_i), the number of yearling females at the next census produced by a female at the current census (this value therefore includes survival over the first year of life); survival rates (s_i); and growth, or transition rates, from one class to the next highest (g_i). Note that because there are no vital-rate estimates for growth or survival of classes 0 or 1, we assume that these rates are equal to those for class 2, and that they vary in unison.

largest of the eight classes can reproduce (Table 8.1). Because the data on the smallest two classes are extremely poor, we assume that survival and growth for these classes (classes 0 and 1) are identical to those of class 2.[2]

We use data from a single site, the Desert Tortoise Natural Area (Doak et al. 1994), to estimate the vital rates. There are three time periods over which these data were collected during the 1970s and 1980s. Although the population was censused more than three times, capture-recapture methods require three census dates to obtain each annual survival estimate and resighting rates in some years were too low to use. Therefore, we use just three estimates for each survival and growth rate: late 1970s, early 1980s, and late 1980s (Table 8.2A). The simplest way to estimate the mean and variance of each rate is to assume that the three estimates are equally likely. Similarly, correlations can be estimated for all pairs of rates across the three time periods (Table 8.2B). Note that no estimates of fertilities exist from this site (or any nearby site), so we use mean estimates of fertility from another area (the "medium-low" estimates in Doak et al. 1994), and we don't allow them to vary in the model (hence the correlations among these fertilities and with all other rates are zero).

[2]For the desert tortoise, we use class identification numbers starting with zero to be consistent with the original publication for this analysis. Numbering of stages starting with zero is common in matrix model notation. Note, however, that class 0 animals are one-year-olds for this pre-breeding census matrix.

TABLE 8.2 Estimated vital rates for the desert tortoise in the Desert Tortoise Natural Area[a]

(A) Means and variances in growth and survival[b]

	Growth					*Survival*				
Class	*1970s*	*Early 1980s*	*Late 1980s*	*Mean*	*Variance*	*1970s*	*Early 1980s*	*Late 1980s*	*Mean*	*Variance*
2	0.50	0	0.5	0.33	0.083	0.63	1	0.65	0.76	0.044
3	0.50	0.18	0.177	0.28	0.036	0.91	1	0.98	0.96	0.002
4	0.47	0.067	0	0.18	0.065	0.98	0.59	0.81	0.79	0.039
5	0.23	0.26	0	0.16	0.020	0.98	0.92	1	0.96	0.0018
6	0.063	0.032	0	0.032	0.001	0.99	1	0.68	0.89	0.034
7						0.78	1	0.80	0.86	0.015

(B) Pearson correlation coefficients between growth and survival rates

	g_2	g_3	g_4	g_5	g_6	s_2	s_3	s_4	s_5	s_6	s_7
g_2	1										
g_3	0.469	1									
g_4	0.382	0.995	1								
g_5	−0.597	0.429	0.514	1							
g_6	−0.014	0.877	0.919	0.811	1						
s_2	−1	−0.496	−0.41	0.571	−0.017	1					
s_3	−0.704	−0.957	−0.925	−0.149	−0.7	0.726	1				
s_4	0.898	0.81	0.75	−0.182	0.428	−0.911	−0.945	1			
s_5	0.956	0.189	0.094	−0.806	−0.307	−0.946	−0.465	0.729	1		
s_6	−0.514	0.516	0.597	0.995	0.865	0.487	−0.247	−0.083	−0.743	1	
s_7	−0.994	−0.563	−0.481	0.505	−0.096	0.997	0.778	−0.941	−0.918	0.417	1

[a]These estimates are calculated from data gathered by the BLM under the supervision of Dr. Kristin Berry.
[b]Fertility rates were not estimated at this site, so we use the following constant estimates from the Eastern Mojave for the three reproductive classes: $f_5 = 0.42$, $f_6 = 0.69$, $f_7 = 0.69$. These are the medium-low estimates from Doak et al. (1994).

Discounting Sampling Variation

We now have estimates for each vital rate of the three basic ingredients needed for a stochastic model: means, variances, and correlations. However, as we discussed before (Chapters 2 and 5), observation error due to sampling variation will cause the observed variation in a vital rate to overestimate the variability due to environmental stochasticity. Ideally, we should subtract from the raw estimates of variation the component that was caused by this sampling variation. We will present three of the more useful ways to accomplish this discounting for different vital rates and assumptions about the nature of sampling variation.

Some vital rates, such as fertilities for species with high numbers of offspring, vary more or less continuously, so that offspring number per female can be approximated by a continuous distribution. For these continuously distributed rates, the methods developed by Engen et al. (1998), described in Chapter 4 as a way to remove the contribution of demographic stochasticity from the raw estimate of the parameter σ^2, can be adapted to remove the influence of sampling variation from raw variance estimates. (σ^2 is the environmental variance in the log population growth rate.) The same method works here because demographic stochasticity and sampling variation are two manifestations of the same phenomenon. Demographic stochasticity in the fertility rate of class i is driven by chance variation in offspring production among *all* members of class i in the population. In comparison, sampling variation is the contribution to the raw variance in the fertility rate of class i caused by variation within that *sample* of class i individuals chosen from the population as a whole for inclusion in the demographic study.[3] In turn, this within-year sampling variation will inflate our estimates of the variation in mean fertility rates across years. Knowing $f_{i,j}(t)$, the number of offspring produced by individual j from class i in year t of the demographic study, we can compute a corrected estimate, $V_c(f_i)$ of the environmentally driven variance in the fertility rate of class i. In particular, $V_c(f_i)$ should be the difference between the raw estimate of variation in mean fertilities across years and the average variability caused by sampling variation within each year:

$$V_c(f_i)=\left\{\frac{1}{n-1}\sum_{t=1}^{n}\left[\bar{f}_i(t)-\bar{\bar{f}}_i\right]^2\right\}-\frac{1}{n}\sum_{t=1}^{n}\left\{\frac{1}{m_i(t)\left[m_i(t)-1\right]}\sum_{j=1}^{m_i(t)}\left[f_{i,j}(t)-\bar{f}_i(t)\right]^2\right\} \quad (8.1)$$

Focusing first on the estimate of total between-year variance (the first-term in large curly brackets), $\bar{f}_i(t)$ is the estimated mean fertility in year t, equal to $\frac{1}{m_i(t)}\sum_{j=1}^{m_i(t)} f_{i,j}(t)$, and $\bar{\bar{f}}_i=\frac{1}{n}\sum_{t=1}^{n}\bar{f}_i(t)$ is the grand mean of the annual mean fertility estimates of a demographic study n years long, where $m_i(t)$ is the number of individuals in class i whose offspring production was measured in year t. The second term is the estimated sampling variance. Note that in this expression, the quantity in the curly brackets is the square of the standard deviation in individual fertilities in a single year (i.e., the sum of squares divided by $m_i(t) - 1$) divided by $m_i(t)$ to generate the squared standard error of the mean. Taking the average of this quantity across all years gives an estimate of the mean sampling variance. Subtracting this sampling

[3]Thus in essence, demographic stochasticity is the same as sampling variation at the whole-population scale.

variance from the raw variance in the fertility rate yields the corrected estimate of the environmentally driven variance. One difference between our use of Engen et al's method here and its use to compute the variance due to demographic stochasticity in Chapter 4 (see Equation 4.13) is that here, because we are interested in the sampling variance in fertility only, we do not include the direct contribution of each parent to next year's population (i.e., 1 if it lives, 0 if it dies; we will account for this when we remove sampling variation from survival estimates below).

The only limitation of Engen et al.'s approach to variance discounting for continuously distributed vital rates is its assumption that sampling variance is equal in all years. This assumption is implicit in the calculation of sampling variance as a mean of the estimated variance across years (Equation 8.1). If there is reason to think that variability among individuals is not equal across years, as is likely for many rates and species, we should instead use a variance discounting method recently suggested by White (2000). White's method relies on the same basic reasoning as Engen et al.'s, but incorporates the correct theoretical weighting of different variances in different years to estimate the overall effect of sampling variance. (The White method's weighting depends on each year's observed between-individual variance and the estimated environmental variance.) White arrives at an equation that must be solved iteratively to obtain an estimate of $V_c(f_i)$:

$$1 = \frac{1}{n-1} \sum_{t=1}^{n} \left[\frac{\left(\bar{f}_i(t) - \bar{\bar{f}}\right)^2}{V_c(f_i) + V(f_i(t))} \right] \tag{8.2}$$

where $V(f_i(t))$ is the observed variance in individual fertilities within year t, equal to $\{1/[m_i(t)(m_i(t)-1)]\} \sum_{j=1}^{m_i(t)} [f_{i,j}(t) - \bar{f}_i(t)]^2$, the same expression we see in equation 8.1 for estimated sampling variation for a single year. To find the best estimate of $V_c(f_i)$ we could use a one-dimensional root-finding function (such as the function `fzero` in MATLAB, see Box 5.1), or simply try a large number of values and choose the best one. The program `white.m` in Box 8.1 illustrates the later approach and also provides 95% confidence limits on $V_c(f_i)$. White (2000) shows that lower and upper confidence limits for $V_c(f_i)$ can be found by setting the right side of Equation 8.2 to $\chi^2_{n-1,\alpha_L}/(n-1)$ and $\chi^2_{n-1,\alpha_U}/(n-1)$, respectively, and then solving for the value of $V_c(f_i)$. Here, χ^2_{n-1,α_L} and χ^2_{n-1,α_U} are values from the chi-square distribution with $n-1$ degrees of freedom and cumulative distribution function values of α_L and α_U, respectively; for a 95% confidence interval, we would choose $\alpha_L = 0.025$ and $\alpha_L = 0.975$.

BOX 8.1 *MATLAB code to use White's method to correct for sampling variation.*

```
% Program white.m
% This program finds the best estimate of environmental variance
% for annually collected vital rates using White's (2000) method.
% To use the method, you must have an estimate of the mean and
% variance in each rate for each year (the program will do the
% analysis for each of many different rates or classes).
% This program does a brute force search for the best variance
% estimates from a very large number of values.

%****************Simulation Parameters*********************
% Copy in the basic data to use: mean and sampling variance
% estimate for each year for several different classes.
% The data below are for mule deer fawn survival, used in
% White (2000). The data columns are:
% Class identifier, Year identifier,
% Mean survival rate, and Within-year-sampling-variance:
rates=...
[1      81      0.326 0.0047773
1       82      0.333 0.0019493
1       83      0.0424 0.0003439
1       84      0.179 0.0013879
1       85      0.381 0.001521
1       86      0.379 0.0014617
1       87      0.129 0.0009706];

times = 7;
 % How many time periods (must be identical for all rates)
classes = 1; % how many classes or separate rates are there?
minvar = 0.0; % minimum estimate of true environmental variation
maxvar = 0.25; % Maximum possible variance estimate (0.25 for
       % survival rates, but it will often be much smaller)
trys = 10000; % how many variance estimates to try at one time
%**********************************************************

for class = 1:classes %looping through the different classes
  minrow = (class-1)*times +1;
     % find the min and max rows of the data matrices to use
  maxrow =  class*times;
  data = rates(minrow:maxrow,:); % fetch the data to use
  meanvar = linspace(minvar,maxvar,trys);
       % generate a set of guesses for true env. variance
```

BOX 8.1 *(continued)*

```
  meanrate = mean(data(:,3));
       % calculate the grand mean rate over years

  for ii = 1:length(meanvar) % a loop to calculate the sum
       % that should equal 1 (see White 2000 for details)
     sumup(ii) = ...
       sum( ((data(:,3)-meanrate).^2)./( (data(:,4) ...
       + meanvar(ii)).*(times-1)) );
  end;  % ii loop

  [bestdiff,jj] = min(abs(sumup - 1)); %find the best sum:
       % the one with the minimum difference from one
  estvar = meanvar(jj);
       % then find the corresponding best env. variance estimate

  % below, find the Chi^2 values for 95% confidence limits
  % If you don't have the Statistics Toolbox
  % look these up in a table of chi-square values
  lowChi=chi2inv(0.975,(times-1));
  hiChi = chi2inv(0.025,(times-1));
  % below, find the differences from the correct chi^2 value
  CLdiffslow=abs(sumup*(times-1)-lowChi);
  CLdiffshi=abs(sumup*(times-1)-hiChi);
  %find differences of sums from the correct chi-square values
  [bestdifflow,jjlow] = min(CLdiffslow);
  bestCLlow = meanvar(jjlow); % best lower CL estimate
  [bestdiffhi,jjhi] = min(CLdiffshi);
  bestCLhi = meanvar(jjhi);% best upper CL estimate

% lines below display the best variance estimate and CLs for it
disp('For class or rate =');
disp(class);
disp('Best variance estimate and lower and upper CLs =');
disp([estvar, bestCLlow,bestCLhi]);
disp('Differences of rt. side of equ. 8.2 and one, ');
disp('for best, lower, and upper conf. limit estimates =');
disp([bestdiff, bestdifflow,bestdiffhi]);

% Using the estimates above, you should narrow the bounds
% on possible variances (minvar, maxvar) and rerun
% to see if the estimates or the differences from one
% change substantially. If not, then the estimated
% variance values are good.
end; % class loop
```

The preceding two methods describe how to discount for sampling variation using the known offspring numbers for reproductive individuals in a single class, and the method must be repeated separately for each class. However, as we described in Chapter 6, when the numbers of individuals per class are small the optimal way to estimate size- or age-specific fertility rates is to fit a continuous fertility function using the data for all individuals simultaneously and then use it to create separate fertility estimates in each year for each class, the $\bar{f}_i(t)$ values. But if we use this approach, we don't have a clear way to say what sample size was used to estimate $\bar{f}_i(t)$ and thus what value should be used for the first $m_i(t)$ in the denominator of the second term in Equation 8.1 and in the expression for $V(\bar{f}_i(t))$ in White's formula [in both cases the second $m_i(t)$ is part of the calculation of the standard deviation and does not need correction]. It is certainly greater than the number of individuals in that class in year t (because information on individuals of other nearby sizes or ages contributed to the estimate), but it is less than the total sample of individuals of all ages or sizes (because those of very different sizes or ages will have little effect on fertility estimates for class i). This is currently an unsolved problem. If you want to do a variance correction in this circumstance using either method, we tentatively suggest that you use the numbers of individuals in a class plus the two adjoining classes (or the single adjoining class for the first and last class) as the sample size to discount the estimate of individual variability in order to estimate sampling variances. Thus, the quantity $1/(m_i(t)[m_i(t)-1])$ in Equation 8.1 (and in the expression for $V(f_i(t))$ below Equation 8.2) becomes $1/(k_i(t)[m_i(t)-1])$, where $k_i(t)$ is the sum of individuals in class i plus its adjoining classes. This is basically an assumption that each class's fertility estimate is close to a running average across sets of three adjacent classes. Alternatively, you could use the sample size for class i alone, knowing that this will inflate the role of sampling variation and therefore lead to underestimates of true environmental variability (and hence optimistic viability assessments).

Although fertility rates can be more or less continuous for some species, many vital rates are binomially distributed, where each individual has one of two fates. Survival always falls into this category, as do other vital rates used in some matrices, such as attempted nesting (yes or no) or even total fertility for organisms with low fecundity (0 or 1 offspring per year). In addition, for some size-structured matrices, including the one for the desert tortoise, growth is a binomial process, since there are only two possibilities for each animal: staying the same size or growing one size class. White's method can still be used for such rates, by substituting each year's calculated sampling variance into Equation 8.2.[4] However, for such two-outcome

[4]For a binomial rate, such as survivorship, the sampling variance in one year is $\hat{p}(1 - \hat{p})/n$, where $\hat{p}$ is the estimated survival probability and n is the sample size (the number of individuals that lived plus the number that died).

vital rates, both Gould and Nichols (1998) and Kendall (1998) have developed more specific approaches to the estimation and removal of sampling variance when estimating environmental variation. We will now illustrate how to use Kendall's method, which not only accounts for sampling variation in estimating the environmental variance, but also provides corrections for estimates of mean survival (or other binary rates), which can also be influenced by unequal sample sizes and sampling variation.

If the probability that an individual in class i survives from census t to census $t+1$ is $s_i(t)$ and there are $m_i(t)$ marked individuals in class i in the demographic study at census t, then the probability that $k_i(t)$ of the $m_i(t)$ individuals survive, $P[k_i(t)|m_i(t), s_i(t)]$, is given by the binomial distribution:

$$P[k_i(t)|m_i(t), s_i(t)] = \left\{\frac{m_i(t)!}{k_i(t)![m_i(t)-k_i(t)]!}\right\}[s_i(t)]^{k_i(t)}[1-s_i(t)]^{m_i(t)-k_i(t)} \quad (8.3)$$

The values of $P[k_i(t)|m_i(t), s_i(t)]$ will be highest for values of $k_i(t)$ near $s_i(t) \times m_i(t)$, the expected number of survivors, but other values of $k_i(t)$ may have quite high probabilities, especially if the number of marked individuals is small. (The same phenomenon underlies the results in Table 2.2.) In essence, the binomial probabilities of different $k_i(t)$ values give a prediction of the extent of sampling variation in year t. If the number of marked individuals $m_i(t)$ is small in year t, then we expect that the observed survival rate $[k_i(t)/m_i(t)]$ may be quite far from the true rate, $s_i(t)$. Variability in the observed survival rates across years exceeding that predicted by this binomial variation is due to real differences among the $s_i(t)$'s, which is truly caused by environmental stochasticity.

Variation in the $s_i(t)$'s among years is modeled with the beta distribution. The beta distribution is extremely handy for this purpose because beta-distributed random variables are confined to the interval from 0 to 1 (for more on this distribution, see Key Distributions for Vital Rates later in the chapter). This makes the beta distribution ideal for modeling variability in survival rates or other probabilities, which can never be negative and can never exceed 1. Unlike the zero-to-one uniform distribution (i.e., the basis for the `rand` function in MATLAB), the beta distribution can also take on a variety of forms that range from a bell-shaped curve to a U-shaped distribution of extreme events, with most years either very good or very bad (Figure 8.2). In addition, the beta distribution can be skewed either to the left (mostly good years for survival with infrequent catastrophes; Figure 8.2B) or to the right (mostly bad years with infrequent bonanzas). The probability density[5]

[5]Since the beta is a continuous distribution it has a probability density function. We introduced the idea of probability density functions in Chapters 2 and 3 when discussing extinction times.

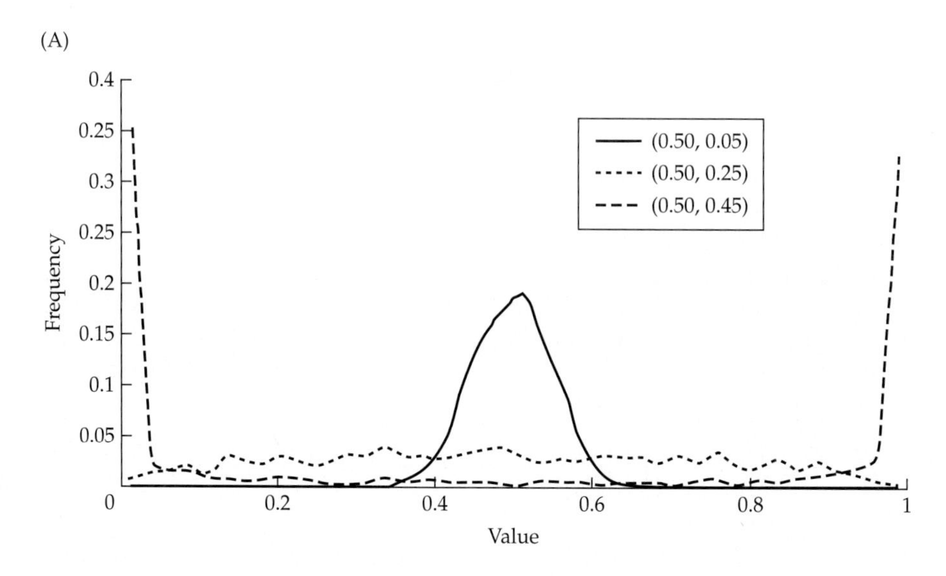

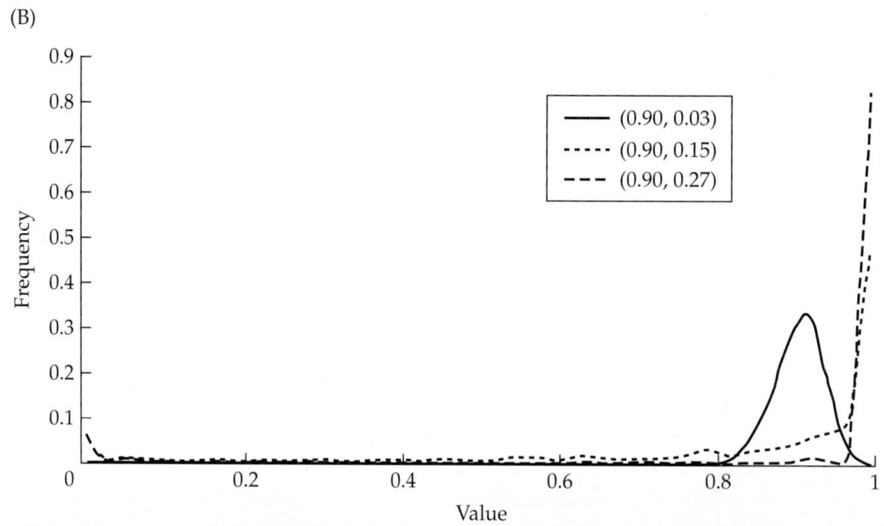

Figure 8.2 Examples of beta distributions. (A) Three distributions, all with a mean of 0.5, but with increasing variation (mean and standard deviation shown in legend). (B) Differing variance with a mean of 0.9. With high variance, beta values allow modeling of extreme environmental variation. Each line shows the result of drawing 1,000 random values.

$f(x)$, which is proportional to the probability that the random variable $s_i(t)$ lies in a small interval near x, depends on the across-year mean, $\bar{s}_i$, and variance, var(s_i), of the s_i values, and is equal to:

$$f(x) = x^{a-1}(1-x)^{b-1}/Beta(a,b) \tag{8.4}$$

where *Beta(a,b)* is the beta function (which is a built-in function in some statistical and mathematical packages,[6] including MATLAB) and a and b are transformations of the mean and variance of $s_i(t)$, $\bar{s}_i$ and $\text{var}(s_i)$, that are commonly used in calculations involving the beta distribution:

$$a = \bar{s}_i \left[\frac{\bar{s}_i(1-\bar{s}_i)}{\text{var}(s_i)} - 1 \right] \text{ and } b = (1-\bar{s}_i) \left[\frac{\bar{s}_i(1-\bar{s}_i)}{\text{var}(s_i)} - 1 \right] \tag{8.5}$$

Together, the beta [which predicts the chances of seeing different values for $s_i(t)$] and the binomial [which predicts the chances of seeing $k_i(t)$ survivors given $s_i(t)$ and $m_i(t)$] allow the calculation of the likelihood of different values for the parameters of the beta distribution, $\bar{s}_i$ and $\text{var}(s_i)$, given the observed numbers of marked individuals and survivors in a set of years. As in our previous use of likelihoods for parameter estimation (Chapters 4 and 6), we actually want to calculate log-likelihoods, which will allow us to conveniently estimate confidence limits around the maximum likelihood estimates for $\bar{s}_i$ and $\text{var}(s_i)$. For year t, the log likelihood, $\log L_t$, is:

$$\log L_t(\bar{s}_i, \text{var}(s_i)) = \log\left\{ \left[\frac{m_i(t)!}{k_i(t)!(m_i(t)-k_i(t))!} \right] \frac{Beta\big((k_i(t)+a),(m_i(t)-k_i(t)+b)\big)}{Beta(a,b)} \right\} \tag{8.6}$$

where $\bar{s}_i$ and $\text{var}(s_i)$ are embedded in a and b, according to Equation 8.5; the notation $\log L_t(\bar{s}_{i,} \text{var}(s_i))$ emphasizes that the log likelihood is a functin of $\bar{s}_i$ and $\text{var}(s_i)$. Over all years, the total log likelihood, $\log L$, is :

$$\log L(\bar{s}_i, \text{var}(s_i)) = \sum_{t=1}^{n} \log L_t[\bar{s}_i, \text{var}(s_i)] \tag{8.7}$$

where n is the total number of census intervals over which you have survival data. The $\bar{s}_i$ and $\text{var}(s_i)$ estimates with the highest log-likelihood are the best parameter estimates, and the combinations of values for $\bar{s}_i$ and $\text{var}(s_i)$ with log likelihood values equal to or greater than the maximum log likelihood minus 3.0 fall within the 95% confidence limits of the parameters

[6]The beta function can also be represented by a combination of gamma functions, that are more widely available (e.g., Excel has a log gamma function):

$$Beta(a,b) = \frac{\Gamma(a)\Gamma(b)}{\Gamma(a+b)}$$ where $\Gamma(a)$ is the gamma function with parameter a (defined in footnote 6 in Chapter 5).

(Hilborn and Mangel 1997).[7] The parameter var(s_i) provides an estimate of the environmental variance in the survival rate of class i that is uncontaminated by sampling variation. Moreover, if the confidence interval for var(s_i) overlaps zero, then we cannot rule out the possibility that *all* of the observed variance in the $s_i(t)$'s is actually due to sampling variation, and that the true survival rate doesn't vary due to environmental stochasticity at all. (Obviously, all rates in fact vary over time, but not necessarily to a statistically significant degree.) If this is the case, we still suggest using the best (maximum likelihood) estimates of temporal variation in your models, although you should check the importance of this variation by also constructing models with variation in some or all vital rates set to their minimum values, including zero.

Using Kendall's method to determine the best estimates of the true mean and variance for a stochastic survival rate is fairly straightforward. The program `Kendall` in Box 8.2 uses a brute-force approach to find the maximum likelihood estimates, searching the range of possible values of $\bar{s}_i$ and var(s_i) with an accuracy users can define.[8] A important point to note here is that the variance as well as the mean of a beta random variable are constrained to a limited range of values. Because no value of a beta random variable can be outside (0,1), for a given mean, $\bar{s}_i$, the variance is also limited: var(s_i) $\leq \bar{s}_i(1 - \bar{s}_i)$. One final aspect of Kendall's method is that the maximum likelihood estimates of both var(s_i) and its confidence limits are biased (Kendall 1998), and should be corrected by multiplying them by $n/(n - 1)$.[9] The program `kendall` makes this correction. Also note that Kendall's method not only improves our estimates of var(s_i), but also gives the correct weighting to each $s_i(t)$ when estimating $\bar{s}_i$, which can be important if sample sizes vary substantially from year to year.

In the case of the desert tortoise, we have exact numbers for the sample sizes used to estimate growth rates in each year, but the capture-recapture method used to estimate survival doesn't yield a directly observed sample size. Instead, we used a rounded estimate of the total number of individuals that

---

[7]This critical value, 3.0, comes from the chi-square distribution, which is used to find significance values using log likelihoods. In particular, twice the difference between the log likelihood values of two nested models (or the same model with two different sets of parameters) has a chi-square distribution with degrees of freedom equal to the number of parameters that differ between the models. In this case, we are varying two parameters, and we want to find the chi-square value with a cumulative probability of 0.95 (since we want to encompass the 95% confidence limits). This value is about 5.99, and rather than doubling all the log-likelihoods, we halve this value to get 3.0.

[8]The method used in the program `Kendall` can't directly test a model with variance set to zero, but if the lowest variance tried is very small and the lower confidence limit includes this value, it indicates weak to nonexistent temporal variation.

[9]The same is true in estimating σ^2, as we discussed in Chapter 3.

BOX 8.2 ***MATLAB code to use Kendall's method to correct for sampling variation.***

```
% Program Kendall.m
% This program finds the best estimates of mean and environmental
% variance for beta-binomial vital rates (see Kendall 1998),
% using a brute force search for the best adjusted
% estimates from a very large number of combinations of different
% possible mean and variance values.
% Note that it will deliver messages of 'divide by zero,'
% indicating use of very low values for variance but not
% a malfunction.
clear all;
% clears all pre-existing variables; this speeds up
% the program and can prevent errors.

%****************User-defined Parameters*********************
% First, copy in the basic data to use. Here we show data for
% only the last three desert tortoise growth rates (g4
% g5 g6), coded as rates (1 2 3).
% Enter the data in five columns: Rate identifier,
% Year, Total number of starting individuals, Number
% growing (or surviving):
grow=...
[4 1        17    8    ;
4 2         15    1    ;
4 3         7     0    ;
5 1         22    5    ;
5 2         19    5    ;
5 3         4     0    ;
6 1         32    2    ;
6 2         31    1    ;
6 3         10    0    ];

rates = grow; % rates is the data set used below (you
% could input several different data sets above here and
% this allows a choice of which to use).
times = 3; % How many time periods for each rate (must be
% the same for all rates for the program as written
classes = 3; % how many different rates or classes for the
% rates are there?
grades = 1000; % the number of different levels of means and
% variances to try; more levels is better: 1000 is quite good
```

BOX 8.2 *(continued)*

```
% and doesn't take too long
maxvar = 0.2; % the maximum variance to search over. The maximum
% ever possible is 0.25, but a reasonable maximum will often be
% much smaller. Searching a narrower range will improve the
% accuracy of the answer.
minvar = 0.00001; % The minimum variance to search. You can't set
% this minimum to zero, as this is un-computable.
maxmean = 1; % maximum limit on the mean values to search
minmean = 0.010; % minimum limit on the mean values to search
%*********************************************************

results = []; results2 = [];
means = repmat(linspace(minmean,maxmean,grades),grades,1);
% makes all the sets of mean values to search over
vars =  repmat(linspace(minvar,maxvar,grades)',1,grades);
% makes all the sets of variance values to search over

aa = means.*( ((means.*(1-means))./vars) - 1);
bb = (1 - means).*( ((means.*(1-means))./vars) - 1);
% using the means and variances to compute the a and b
% (aa and bb) parameters of a beta distribution

aa(find(aa<=0)) = NaN;
bb(find(bb<=0)) = NaN;
% eliminate impossible combinations of mean and variance values

for class = 1:classes % loop through each rate or class
      % find the min and max rows of the data matrices to use
    minrow = (class-1)*times +1;
    maxrow =  (class)*times;
    data = rates(minrow:maxrow,:); % fetch the data to use
    estmn = mean(data(:,4)./data(:,3)); % raw estimate of mean
    estvar = var(data(:,4)./data(:,3)); % raw est. of variance
    loglikli = zeros(grades,grades); % initialize -loglik.'s
% As is often done, we use the negative of log likelihoods,
% or -log likelihoods, in the following computations.
    for tt = 1:times % calculate -loglikelihood for each year
       newlogL = -log(beta( aa + data(tt,4),data(tt,3) ...
          - data(tt,4)+ bb)./beta(aa,bb));
         % this is uses the -log-likelihood formula from Kendall
       loglikli = newlogL + loglikli;
```

BOX 8.2 *(continued)*

```
        % add up the -log-likelihoods for each year
    end; % tt loop

disp('Class is');, disp(class); % track where the program is

% the lines below summarize the results
minLL = min(min(loglikli)); % what is the best log-liklihood?
[minvars,ii] = min(loglikli);  % finding best var
[minii, jj] = min(minvars);
MLEvar = vars(ii(jj),1);
[minvars,ii] = min(loglikli');  % finding best mean
[minii, jj] = min(minvars);
MLEmean = means(1,ii(jj));

clLL = loglikli - minLL;  % differences from best -LL
hivar = max(max(vars(find(clLL<3.0))));% find conf. limits for var
lowvar = min(min(vars(find(clLL<3.0))));
himean = max(max(means(find(clLL<3.0)))); % conf limits for mean
lowmean = min(min(means(find(clLL<3.0))));

% perform correction for max. likelihood estimate of variance
% (see Kendall 1998 for explanation)
corrMLEvar = MLEvar*times/(times-1);
corrlowvar = lowvar*times/(times-1);
corrhivar = hivar*times/(times-1);

% storing the results
results = ...
 [results; data(1,1), estmn,MLEmean,estvar,MLEvar, corrMLEvar];
results2 = ...
 [results2; data(1,1),lowmean, himean, corrlowvar, corrhivar];

end; % class

% now, display the results:
disp('Basic results are...')
disp('class, estimated mean, MLE mean, estimated var,');
disp('MLE variance, corrected MLE variance');
disp(results);
disp('The likelihood 95% conf. limits for the mean');
disp('and corrected variances are:');
```

BOX 8.2 *(continued)*

```
disp('class, low and hi mean limits, low and hi variance limits')
disp(results2);

% finally, make a 3d mesh diagram for relative -LL for
% the last vital rate class, using values only within the
% 95% conf. limits for variance and mean
clLL = -(loglikli - minLL) + 3.0;
%standardize and reverse the orientation of the -2logL's
clLL(find(clLL<0)) = 0;
vars = vars.*(times/(times-1));
mesh(means,vars,clLL);
xlabel('Mean');
ylabel('Variance (corrected)');
zlabel('2Log-Likelihood (only values within 95% C.L.s');
axis([0,1,0,maxvar]);    %AXIS([XMIN XMAX YMIN YMAX])
view(50,20);    %VIEW([AZ,EL]) AZ is the azimuth and EL is
 % the vertical elevation (both in degrees)
```

would have produced the observed number of live tortoises seen at the end of each time period, given the estimated survival rate. Using `kendall` on the data for desert tortoises provides corrections for our raw estimates of environmental variance and mean survivorship (Table 8.3). For the most part, the raw means for growth and survival rates agree well with the maximum likelihood estimates from Kendall's method, although the broad confidence limits reinforce how uncertain vital-rate estimates can be with limited sample sizes. However, the best estimates of the environmental variances of these rates are often substantially different from the raw estimates. Sometimes the unbiased estimates of the variances are higher than the raw estimates and sometimes they are lower. When the true variance is low, raw estimates will tend to be biased upwards, whereas when the true variance is high, raw estimates will usually be biased downwards (Kendall 1998). In all cases, the confidence limits around the best estimates are extremely large, reflecting the relatively low sample sizes in the data set.

As for Engen et al.'s and White's methods to account for sampling variation for continuous vital rates, Kendall's approach assumes that each vital rate is calculated separately and thus is difficult to use if you have estimated survival using the logistic regression method we advocate in Chapter 6. We again make a tentative suggestion on how to deal with this problem. First, you could use the number, n_i, of individuals that were initially alive in a

TABLE 8.3 Comparison of raw estimates of vital rates for the desert tortoise with those from Kendall's method[a]

	Mean rate		Temporal variance		
Vital rate	*Raw estimate*	*Kendall's estimate (95% conf. limits)*	*Raw estimate*	*Kendall's estimate (biased)*	*Unbiased Kendall's estimate (95% conf. limits)*
s_2	0.759	0.77 (0.35, 0.97)	0.0436	0.001	0.0015 (<0.0001, 0.25)
s_3	0.9633	0.9534 (0.48, 1.0)	0.002	0.001	0.0015 (< 0.0001, 0.25)
s_4	0.7917	0.79 (0.30, 0.98)	0.039	0.0359	0.054 (0.0004, 0.25)
s_5	0.964	0.93 (0.68, 0.99)	0.0018	0.001	0.0015 (< 0.0001, 0.15)
s_6	0.8907	0.91 (0.32, 1.00)	0.0339	0.0323	0.048 (0.0004, 0.25)
s_7	0.859	0.87 (0.38, 0.98)	0.0151	0.0135	0.020 (0.0004, 0.25)
g_2	0.3333	0.30 (0.04, 0.78)	0.0833	0.0064	0.0096 (0.0004, 0.25)
g_3	0.281	0.31 (0.09, 0.66)	0.036	0.0116	0.017 (0.0004, 0.23)
g_4	0.1791	0.19 (0.02, 0.69)	0.0648	0.0321	0.048 (0.0006, 0.25)
g_5	0.1635	0.22 (0.07, 0.46)	0.0204	0.001	0.0015 (0.0004, 0.25)
g_6	0.0316	0.052 (0.01, 0.23)	0.001	0.001	0.0015 (< 0.0001, 0.11)

[a]Note that 0.25 is the maximum possible variance for any beta variable, and can only occur when the mean is 0.5.

class i plus the adjoining classes as the starting number for class i, and then use the already estimated mean survival rate, s_i, for class i to determine an effective number of survivors for the class, $n_i s_i$. After doing this for each class, you can use Kendall's method to estimate true temporal variation in mean survival rates. Alternatively, you can use the actual numbers for each class separately, probably leading to underestimates of true variation most of the time.

Although the methods we've just described are useful for demographic data collected in many different ways, you may be able to more directly estimate true environmental stochasticity in vital rates if your data are amenable to capture-recapture analysis. Several methods are now available to estimate temporal variation, discounted for sampling variation, as part of a capture-recapture analysis. In particular, the variance discounting method of Gould and Nichols (1998) is presented in the context of capture-recapture methods.

A final concern with sampling variation is how it will influence estimates of correlations between demographic rates. In the tortoise data set, we can clearly see some spuriously large correlations—both negative and positive—between pairs of vital rates (Table 8.2B). This is a result both of small samples of individuals in each year and also of having few years of data. However, there appears to be no established way to correct these estimates for low sample size. The best solution to this problem is probably to run each PVA both with the estimated correlations and again with all correlations set to zero to determine the effects of the observed correlation patterns on the PVA results. Note that the corrected variances we get from any of the variance discounting methods will not influence our estimates of the *correlations* among vital rates (which are standardized so that they are independent of variance estimates), but will influence the estimates of *covariances*.

Extreme Variation

Because we will ultimately use simulations to estimate stochastic population growth and extinction risk, high variation in vital rates is not in and of itself a problem for constructing and using a matrix model with stochastic vital rates. As we've seen, the beta distribution, like the lognormal, can model extreme events (Figure 8.2). The only problem comes when truly unusual events fall well outside any reasonable distribution of vital rate values for more typical years (Chapter 2). When this occurs, the most reasonable approach is to leave out these years when estimating the mean and variance of each vital rate, and include them as atypical years occurring with a specified frequency. This is basically the same approach we took in Chapter 7 to deal with unusual events by choosing extreme matrices with some low probability.

More Complicated Growth Rates

The most difficult problem in estimating vital rates and then using them for stochastic growth models comes from growth or transition rates for which there are more than two possible outcomes. Mountain golden heather provides a clear example of this problem; individuals in classes 4 or 5 have four different classes in which they can appear in the following year (Table 6.5). The most natural probability distribution to describe these multiple possible outcomes is the multinomial distribution, which generalizes the binomial by allowing many different fates for each individual. However, there are two problems with including multinomial vital rates in a stochastic simulation. First, although there is no difficulty in generating random variables from a multinomial distribution (see Caswell 2001, p. 459), we know of no way to make those random variables exhibit a specified correlation with other vital rates, which can be a critical component of a stochastic simulation (as we describe later in the chapter). Second, because the multinomial distribution is not governed by a single parameter describing the probabilities of all outcomes, as is the binomial, it is not amenable to the use of Kendall's method to correct for low sample sizes.

The best solution to this problem is to convert multinomial growth rates into binomial ones. At first glance, this may seem unduly complicated, but the procedure actually doesn't result in any increase in the total number of parameters that need to be estimated (the true measure of model complexity), it allows us to better calculate and simulate correlations in vital rates, and it means we can use Kendall's corrections to improve our estimates of environmental variation. To illustrate this method, we'll use the mountain golden heather growth rates. Table 8.4A shows the growth rates as we estimated them in Chapter 6 (see Table 6.5), with $g_{i,j}$ representing the probability that a surviving individual that was in class j in one year will be in class i the following year. In Table 8.4B we convert these into $g_{\geq i,j}$ rates, the probability that a surviving individual that was in class j will be in class i *or any larger class* the following year. Thus, each of these new growth rates is a binary "choice" with only two possible results. For any starting size class, the probability of being in size class i the next year can be represented by subtracting the probability of a transition to size class $i + 1$ or higher from the probability of a transition to size i or higher. Note that, because these are

TABLE 8.4 The correspondence between multinomial growth probabilities and binomial growth choices

(A) Multinomial growth probabilities

Size class in year t + 1	***Size class in year t***			
	3	***4***	***5***	***6***
3	$1 - g_{4,3}$	$1 - g_{4,4} - g_{5,4} - g_{6,4}$	$1 - g_{4,5} - g_{5,5} - g_{6,5}$	0
4	$g_{4,3}$	$g_{4,4}$	$g_{4,5}$	$1 - g_{5,6} - g_{6,6}$
5	0	$g_{5,4}$	$g_{5,5}$	$v_{5,6}$
6	0	$g_{6,4}$	$g_{6,5}$	$g_{6,6}$

(B) Growth probabilities as binomial choices[a]

Size class in year t + 1	***Size class in year t***			
	3	***4***	***5***	***6***
3	$(1 - g_{\geq 4,3})$	$(1 - g_{\geq 4,4})$	$(1 - g_{\geq 4,5})$	0
4	$(g_{\geq 4,3})$	$(g_{\geq 4,4} - g_{\geq 5,4})$	$(g_{\geq 4,5} - g_{\geq 5,5})$	$(1 - g_{\geq 5,6})$
5	0	$(g_{\geq 5,4} - g_{\geq 6,4})$	$(g_{\geq 5,5} - g_{\geq 6,5})$	$(g_{\geq 5,6} - g_{\geq 6,6})$
6	0	$(g_{\geq 6,4})$	$(g_{\geq 6,5})$	$(g_{\geq 6,6})$

[a]As an example, $g_{\geq 4,3}$ is the probability that an individual in size class 3 grows to be in size 4 *or larger* over one year.

probabilities of growth *given* survival, the probability of a transition to class i_{min} (the smallest class to which an individual in the focal class can move in a year) or higher is always 1. This allows us to represent any of the original multinomial outcomes by combinations of these binomial rates. There are other ways to break apart the multinomial growth probabilities that result in the same number of binomial vital rates[10], but the method we just described is likely to be among the simplest and most generally applicable.

After calculating the sample sizes for each of these binomial growth rates for each year of the mountain golden heather study, we can input the data into the program `Kendall` (Box 8.2) to estimate the correct mean and variance for each rate (Table 8.5; if you're unsure on how we got these numbers, refer back to the sample sizes given in Table 6.5). We will return to these estimates when we simulate the mountain golden heather matrix later on.

Simulations to Estimate Population Growth Rate and Extinction Risk

There are two major issues in simulating a stochastic matrix using the means and variances of vital rates and their correlations. The first is which probability distribution to use to simulate each rate. That is, we need to know not only the mean and variance for a vital rate, but also the shape of the probability distribution from which it is to be drawn in order to generate reasonable values in each year of the simulation. The second issue involves the simulation of correlations in vital rates. Picking sets of vital rates, each of which is random but also properly correlated to the others, requires considerably more complicated methods than does generation of uncorrelated rates. However, the ability to simulate biologically reasonable correlations is one of the major reasons to base stochastic models on vital rates rather than on a small set of directly estimated matrices, as we did in the last chapter. In this section, we first give brief introductions to the most useful statistical distributions for vital rates and how to simulate them. (For more detailed discussion of statistical distributions and their manipulation, see Fishman 1973, Hastings and Peacock 1974 and Press et al. 1996.) We then discuss how within-year correlations can be included in stochastic models, as well as correlations in vital rates between years. Finally, we put these pieces together to construct simulation models for both the desert tortoise and mountain golden heather.

---

[10]For example, we can describe the four outcomes for class 4 individuals of mountain golden heather with the following vital rates: the probability of staying in the same size class or changing classes; the probability an individual grows rather than shrinks, given that it changes; and the probability an individual grows more than one size class given that it grows.

TABLE 8.5 Data and estimates of binomial growth rates for *Hudsonia montana*[a]

	Transition period[b]					
Rate	*1985–1986*	*1986–1987*	*1987–1988*	*1988–1989*	*Kendall's mean (95% conf. limits)*	*Kendall's unbiased variance (95% conf. limits)*
$g_{\geq 4,3}$	4/25 (0.16)	7/19 (0.37)	2/10 (0.20)	3/13 (0.23)	0.24 (0.12, 0.44)	0.0013 (0.0003, 0.09)
$g_{\geq 4,4}$	13/15 (0.87)	13/17 (0.77)	10/15 (0.67)	12/14 (0.86)	0.79 (0.60, 0.90)	0.0013 (0.0003, 0.08)
$g_{\geq 5,4}$	6/15 (0.40)	5/17 (0.29)	1/15 (0.067)	6/14 (0.43)	0.30 (0.12, 0.58)	0.010 (0.0003, 0.16)
$g_{\geq 6,4}$	1/15 (0.067)	0/17 (0)	0/15 (0)	0/14 (0)	0.024 (0.01, 0.39)	0.0013 (<0.0001, 0.27)
$g_{\geq 4,5}$	22/24 (0.92)	19/20 (0.95)	12/14 (0.86)	13/13 (1)	0.0013 (0.76, 0.98)	0.92 (<0.0001, 0.08)
$g_{\geq 5,5}$	17/24 (0.71)	16/20 (0.80)	10/14 (0.71)	13/13 (1)	0.79 (0.51, 0.94)	0.0017 (0.0003, 0.18)
$g_{\geq 6,5}$	2/24 (0.083)	7/20 (0.35)	3/14 (0.21)	5/13 (0.39)	0.25 (0.10, 0.53)	0.0095 (0.0003, 0.14)
$g_{\geq 5,6}$	4/5 (0.80)	6/6 (1)	12/13 (0.92)	10/10 (1)	0.93 (0.62, 0.99)	0.0013 (<0.0001, 0.20)
$g_{\geq 6,6}$	3/5 (0.60)	6/6 (1)	7/13 (0.54)	9/10 (0.90)	0.75 (0.38, 0.94)	0.027 (0.0003, 0.23)

[a]Numbers under each transition period are (number growing)/(total surviving), and the direct estimate of the growth rate (in parentheses). Mean and variance are from Kendall's method.
[b]Data from Frost 1990.

Key Distributions for Vital Rates

THE BETA DISTRIBUTION We have already discussed the shape of the beta distribution and some of the basic equations governing its use (Figure 8.2, Equations 8.4 and 8.5). This is the appropriate distribution to use when simulating variation in vital rates that are probabilities of binary events such as survival, or growth, when growth has only two possible outcomes. As we discussed above, in almost all circumstances more complicated possibilities for growth can also be described by a set of binary possibilities, which are also best modeled as beta-distributed variables.

To simulate a beta-distributed vital rate, we need to convert the uniform random numbers produced by most statistical and mathematical software into beta random variables with a specified mean and variance. The

MATLAB Statistics Toolbox includes the function `betarnd` that generates beta-distributed random numbers with specified values of the parameters a and b (which are related to the mean and variance according to Equation 8.5). MATLAB users who do not have access to the Statistics Toolbox can use the function `betarv` defined in Box 3.3 to generate vital rate values that follow a beta distribution with a specified mean and variance.

Although these two methods allow easy simulation of random beta values, they don't work to simulate beta values that are correlated with other vital rates. Because of this limitation, we generally use a more cumbersome approach to simulate beta values, but one that allows simulation of correlated values. We could wait to describe this method until the section *Including correlations in vital rates*, but we present it here since it is the most common method that we ourselves use.

This third, rather indirect route to picking random beta values relies on the formula for the cumulative distribution function of the beta, which is relatively simple to calculate. (Although not a simple closed form equation, it is included in many spreadsheets and statistical programs.) This distribution function is often called the incomplete beta function. For a beta distribution with parameters a and b (which are transformations of the mean and variance; Equation 8.5) and some value p between 0 and 1, this function gives the probability that a randomly chosen value from the distribution will be less than or equal to p, a probability we'll call $F(p \mid a,b)$. Here is how to pick a random value from a beta distribution with parameters a and b:

1. Choose a uniform random number between 0 and 1. This is not a random value of p, but of $F(p \mid a,b)$. The problem now is to figure out what value p from the beta distribution actually has this randomly chosen value for $F(p \mid a,b)$.
2. Now, pick another random number between 0 and 1, a number we'll call x_1. Use the incomplete beta function to calculate $F(x_1 \mid a,b)$.
3. If $F(x_1 \mid a,b)$ is greater than $F(p \mid a,b)$, then pick a random number x_2 between 0 and x_1. If $F(x_1 \mid a,b)$ is less than $F(p \mid a,b)$, then pick a random x_2 between x_1 and 1.
4. Now, set $x_1 = x_2$, and recalculate $F(x_1 \mid a,b)$.
5. Repeat steps 3 and 4 until $F(x_1 \mid a,b) - F(p \mid a,b)$ is less than some acceptable amount of error (0.001 works well for all but very high variances). The final x_1 is a good approximation of p, and is the random beta value we desire.

Even though it is rather slow, this method works well to generate random beta values. Box 8.3 shows the function `betaval` that takes as input a random $F(p \mid a,b)$ value, along with the mean and standard deviation of the beta distribution to simulate, and finds the corresponding random beta value p. At the end of the Box is a small program that shows how this function is called to generate random values.

BOX 8.3 ***A second MATLAB function to make beta-distributed random numbers (see Box 3.3 for a different method). "betaval" returns a beta-distributed value with the specified CDF value. The program BetaDemo is also included, showing the use of betaval (BetaDemo was used to make Figure 8.2).***

```
function bb = betaval(mn,sd,fx)
% betaval (mean, sd, Fx)
% This function calculates a random number
% from a beta distribution with mean mn, st, deviction
% sd, and cum. distr. function value fx.
% This function uses the MATLAB function betainc(x, vv,ww),
% where x is the value of beta. v,w are beta parameters
% that are called a and b in the text
if sd == 0; bb = mn;
else
 toler = 0.0001; % this is tolerance of answer: how close
 % the CDF value of the answer must be to the input value (Fx)
 var = sd^2;
 if var >=(1-mn)*mn disp('sd too high for beta'), pause, end;
 % this checks that the input mean and st. deviation
 % are possible for a beta.

 vv = mn*((mn.*(1-mn)/(var))-1); % calculate the beta parameters
 ww = (1-mn).*((mn.*(1-mn)/(var))-1);

 upval = 1; lowval = 0; x = 0.5+ 0.02*rand;
 % start with a beginning guess x; the use of rand
 % adds wiggle to the search start to avoid pathologies
 i = betainc(x,vv,ww); % find the CDF value for x

 % the following while loop searches for ever better
 % values of x, until the value has a CDF within the
 % toler of Fx (unless the value
 % is very close to 0 or 1, which will also terminate
 % the search)
  while ( (toler < abs(i-fx))&(x >1e-6)&((1-x)>1e-6) )
    if fx > i
       lowval = x; x = (upval+lowval)/2;
    else
       upval = x; x = (upval+lowval)/2;
    end; % if
       i = betainc(x,vv,ww);
```

BOX 8.3 *(continued)*

```
  end; % while

 % This makes values of x somewhat random to eliminate
 % pathologies when variance is very small or large.
 % It also truncates values of x, with the
 % smallest values equal to toler and the biggest
 % equal to 1 - toler.
 bbb = x + toler*0.1*(0.5-rand);
 if bbb < toler; bbb = toler; end;
 if bbb > 1; bbb = 1- toler; end;

 bb=bbb;
end; % else

-----------------------------------------------------------------

%BetaDemo
% This program calls betaval.m, a function to return a beta value
% from a Beta distribution with Mean = mn and standard
% deviation = sd. The other input is fx, a value for the
% cumulative distribution function. To pick a random value, make
% fx = rand.
% This program calculates random values for each of several mean
% and st. deviation values

clear all;
%****************Simulation Parameters*********************
means = [0.5,0.9] ; %  means to use;
psds = [0.1,0.5,0.90]; %percentage of max possible s.d. to use
reps = 1000; % how many values to make
%*********************************************************
results = [];
rand('state',sum(100*clock)); % seeds random number generator
meannum = length(means); % how many means
psdnum = length(psds);   % how many s.d. values

% now, 2 loops to get all the combinations of means and s.d.s
count = 1
for meani =1:meannum
 for psdi = 1:psdnum
    mn = means(meani);
```

BOX 8.3 *(continued)*

```
        sd = ((mn*(1-mn))^0.5)*psds(psdi); % calculate sd
        for fxi = 1:reps
            temp(fxi) = betaval(mn,sd,rand); % find random values
        end; % fxi
        results(:,count) = [mn; sd; temp']; % store the values
        count = count+1
    end; % psdi
end; % meani

% make a set of frequencies and bin means for the results
[freqs,bins] = hist(results(3:(reps+2),:),40);
disp('In the following histogram results')
disp('the first two rows of each column are')
disp('the mean and standard deviation, and')
disp('the first column has the bin means')
finalresults = [zeros(2,1),results(1:2,:); bins, freqs]
hist(results(3:(reps+2),:),40) % make a histogram of all results
```

The disadvantage of this way of simulating beta values is the time needed to perform an iterative search to generate each beta-distributed vital rate in each year of each simulation. Therefore, we frequently use a short-cut that can dramatically speed up simulations. Rather than searching for a new value for each beta-distributed rate in each year, we start off by storing a set of 99 (or more) beta values for each vital rate at the beginning of a simulation. These values are uniformly spaced by their cumulative distribution function values [$F(p \mid a,b) = 0.01, 0.02, \ldots, 0.99$]. Then we draw random numbers for the $F(p \mid a,b)$ value of each rate in each year and choose the stored rate with the closest F-value. With a large number of stored values, this simplification doesn't seem to influence the outcome of a stochastic simulation in any appreciable way, and it greatly speeds it up. This method also provides the right set-up to simulate correlations using beta-distributed vital rates.

It is not difficult to generate beta-distributed vital rates by using any of the three functions we've described, yet many PVAs have used the normal distribution to simulate stochastic variation in survival and other rates that are probabilities. We strongly discourage you from doing this. The problem with this procedure is that a normal distribution can take on values anywhere between positive and negative infinity. Since probabilities are bounded by 0 and 1, to use the normal distribution, you must truncate all values outside of this range. Even though this may be acceptable for vital rates with very small variances and means far from zero or one, truncation will change both the mean and the variance of the simulated numbers. The normal distribution is also incapable of simulating the same non-bell-shaped curves as the beta distribution (see Figure 8.2).

THE LOGNORMAL DISTRIBUTION The most common distribution to use in the simulation of fertilities is the lognormal.[11] Because the lognormal is bounded by zero and positive infinity, it is a reasonable distribution to use for fertilities, which must be positive but can take on values greater than one (see Figure 3.2). The easiest way to generate random numbers from a lognormal distribution is to first generate normal random numbers and then exponentiate them. If a normally distributed variable X has mean M_n and variance V_n, then $Y = e^x$ is lognormally distributed with mean M_{ln}, and variance V_{ln} given by:

$$M_{ln} = \exp\left(M_n + \frac{1}{2}V_n\right), \text{ and } V_{ln} = \exp\left[2\left(M_n + \frac{1}{2}V_n\right)\right]\left[\exp(V_n) - 1\right] \quad (8.8)$$

Usually, we have estimated the mean and variance of a lognormally distributed fertility and we want to know what mean and variance to use for X, so that the exponentiated values will have the proper mean and variance (that is, we know M_{ln} and V_{ln} and we want to calculate M_n and V_n). It is straightforward to solve Equations 8.8 for M_n and V_n:

$$M_n = \log(M_{ln}) - \frac{1}{2}\log\left(\frac{V_{ln}}{M_{ln}^2} + 1\right), \text{ and } V_n = \log\left(\frac{V_{ln}}{M_{ln}^2} + 1\right) \quad (8.9)$$

Using Equation 8.9 and the ubiquitous functions in computer packages to generate random values from a standard normal distribution,[12] it is easy to create lognormally distributed random variates. Box 8.4 gives a MATLAB function to create lognormal numbers in this way

THE STRETCHED BETA DISTRIBUTION Although the lognormal distribution is most often used to simulate random variation in fertility rates, it is not ideal

---

[11]Strictly speaking, the fertility of individuals is always discrete, not continuous (because there is no such thing as fractional offspring). However, since the fertility rate represents the *average* offspring number per individual in a given class, little error is introduced into a population projection by drawing fertilities from a continuous (e.g., lognormal or stretched beta) rather than discrete (e.g., Poisson or negative binomial) distribution, so long as the number of individuals in the class is high. When the number is low, the discrete number of offspring produced by each parent can be simulated to incorporate demographic stochasticity.

[12]The only other trick needed is a way to convert standard normal numbers to normally distributed values with a different mean and variance. If Z is a number from a standard normal distribution, then $X = M_n + \sqrt{V_n}Z$ will represent a random number drawn from a normal distribution with mean M_n and variance V_n.

BOX 8.4 ***A MATLAB m-file for the function lnorms, which returns random lognormal values.***

```
function lns = lnorms(means,vars,rawelems)
% lnorms (mean,vars,rawelems)
% Converts standard normal random values (rawelems) to
% lognormals with the defined means and variances (means
% and vars). For 1 random value you would call:
% lnorms(mean, var, randn)
% Inputting vectors makes multiple lognormals simultaneously
% Note that vars are Variances, not st. deviations
nmeans = log(means)-0.5.*log(vars./means.^2 + 1);
nvars = log(vars./means.^2 + 1);
normals = rawelems.*sqrt(nvars) + nmeans;
lns = exp(normals);
```

for this purpose. In particular, there is no upper limit to the values of a lognormal random variable, while biologically there is almost always an upper limit to fertility. For example, if the maximum clutch size ever observed for a passerine species is 6, then the fertility rate can be no greater than this number (assuming only a single clutch is produced each year). If the mean and variance of a lognormal distribution are fairly small, few random values you generate will be absurdly big, but some will be, especially if you simulate tens of thousands of years of population growth to estimate quasi-extinction probabilities. Another limitation of the lognormal distribution is that it cannot be skewed downwards to reflect a vital rate with mostly high values but occasional low ones.

A solution to this problem is to use a "stretched" beta distribution to model fertilities. This is simply a rescaled beta distribution that has a minimum and maximum value that you define, expanding the range of the usual beta (0 to 1) to fit the interval that makes biological sense for the vital rate being modeled. Because it is bounded by these limits, this version of a beta can allow more realistic simulations of fertility rates.

Creating stretched betas relies on some simple rules of probability distributions, along with the methods described above to find beta values. If you've estimated that the mean and variance of a fertility rate are M_f, and V_f, and that the maximum and minimum values possible are f_{max}, and f_{min}, then the mean and variance of a beta-distributed variable with the same shape, but that is bounded by 0 and 1, are M_{beta} and V_{beta}:

$$M_{beta} = \left(\frac{M_f - f_{min}}{f_{max} - f_{min}}\right) \text{ and } V_{beta} = V_f\left(\frac{1}{f_{max} - f_{min}}\right)^2 \tag{8.10}$$

These values are then used to generate a beta-distributed random value B_i, which is then converted into the value S_i from a stretched beta:

$$S_i = B_i\,(f_{max} - f_{min}) + f_{min} \tag{8.11}$$

Box 8.5 contains a function to make stretched betas in this way.

In principle, one could use maximum likelihood methods to determine whether the lognormal or the stretched beta distribution better describes a set of estimated annual fertilities. In practice, when the number of such estimates is small, the fact that it disallows infinitely large values makes the stretched beta a better distribution to use on *a priori* grounds.

Including Correlations in Vital Rates

Beyond having a method to generate random values from a reasonable distribution for each vital rate, it is also necessary to make these random values correlated. The method of picking an entire matrix in each year of a simulation from the set of separately estimated matrices (Chapter 7) provides a kind of correlation, but one that is quite rigid—the exact combinations of values observed in the demographic study always occur together. A more realistic kind of correlation is one in which there is statistical correlation but not absolute rigidity. The problem lies in generating this kind of correlation while also letting each vital rate vary according to its own probability distribution. To do this, we rely on some of the special features of normally distributed variables, along with the use of cumulative distribution functions, which we described and used above under *The beta distribution*. Doak et al. (1994) used a simple version of the method we now describe (with all vital rates correlated to a single environmental driver), while John R. Lockwood and Kevin Gross developed and used the more complete procedure described below to model mountain golden heather (Gross et al. 1998). You don't have to understand in detail how this method works in order to use it, but we provide a nontechnical explanation below. (Burgman et al. 1993 and Fieberg and Ellner 2001 also give partial descriptions of the same basic method.) Box 8.6 is a MATLAB file that follows the steps of this procedure to generate sets of correlated random variates for three vital rates with different probability distributions (a beta, a stretched beta, and a lognormal distribution). Following this description of how to simulate correlated rates, we will discuss the important fact that some sets of estimated correlations are impossible to simulate, and we discuss how to deal with this problem.

STEP 1: MAKING CORRELATED NORMAL VALUES Generating sets of standard normal values that have a specified matrix of pairwise correlations is not difficult. We start with the correlation matrix estimated from a set of vital rate data, which we'll call **C**. Most statistical packages provide only the upper or

BOX 8.5 ***A MATLAB m-file defining the function stretchbetaval, which returns stretched beta-distributed values. Note that this program uses betaval, defined in Box 8.3.***

```
function bb = stretchbetaval(mn,sd,minb,maxb,fx)
% STBetaval(mean, sd, minb, maxb, fx)
% This routine generates a stretched beta number with
% mean mn, standard deviation sd, minimum and maximum
% values (minb, maxb) and CDF value (fx).
% This function calls the function betaval. m.

if sd == 0; bb = mn; % with no variation, then the value = mean
else
    % convert the stretched beta parameters to corresponding
    % ones for a {0,1} beta
       mnbeta = (mn - minb)/(maxb-minb);
       sdbeta = sd/(maxb-minb);
     % next, check for undoable parameter combos
       if sdbeta < (mnbeta*(1-mnbeta))^0.5
       bvalue = betaval(mnbeta,sdbeta,fx); % find beta value
       bb=bvalue*(maxb-minb) +minb; % convert to stretched value
    else
       disp('the sd is too high for the mean');
       disp('for a vital rate with the following')
       disp('mean, sd, and min and max values')
       disp([mn,sd,minb,maxb])
       disp('the maximum sd possible is:')
       maxsd = ((mnbeta*(1-mnbeta))^0.5)*(maxb-minb);
       disp(maxsd);
       disp('you should abort the program (control C)')
       disp('and reset the limits or the sd of this rate');
       pause;
       bb=NaN;
    end; % else
end; % else
```

lower triangular part of the correlation matrix (i.e., the diagonal plus elements above or below it), but the correlation matrix must be entirely filled for the procedure we use below. (However, since the correlation coefficient between vital rates a and b is the same as the correlation between b and a, we

BOX 8.6 ***A simple example program to generate correlated random vital rates using an estimated correlation matrix between vital rates.***

```
%CorrRates: a program to generate sets of correlated vital rates
clear all;

%****************Simulation Parameters**********************
% parameters for 3 vital rates:
% a beta, a lognormal, and a stretched beta.
vrmeans= [0.77,2,5]; % means
vrvars= [0.02,0.5,2]; % variances (not standard deviations)

% minimum and maximum values for each vital rate:
% zeros are place holders for rates that are not stretched betas
vrmins= [0 0 0];
vrmaxs= [0 0 24];
%then a full correlation matrix.
corrmx = ...
[1      0.3    -0.2;
0.3     1      -0.6;
-0.2    -0.6   1];

tmax = 200;    % number of sets of vital rates to simulate
%***********************************************************

np = length(vrmeans); % finds the number of vital rates
results = [];
rand('state',sum(100*clock)); randn('state',sum(100*clock));

% find the eigenvalues (D) and eigenvectors (W)
% of the correlation matrix
[W,D] = eig(corrmx);

%Calculate C12, the matrix to use to make correlated
% standard normal variates from uncorrelated ones; see
% the text in Chapter 8 for an explanation of this process
C12 = W*(sqrt(abs(D)))*W';

for tt=1:tmax % loop to do each year's vital rates
 normvals = randn(np,1);
 % make a set of random standard normal values.
```

BOX 8.6 *(continued)*

```
 corrnorms = C12*normvals; % make correlated normals
 % get the beta vital rate with the same Fx as the first normal:
 vrate1 = ...
   betaval(vrmeans(1),(vrvars(1))^0.5,stnormfx(corrnorms(1)));
 % convert the second normal into a corresponding log normal:
 vrate2 = lnorms(vrmeans(2),vrvars(2),corrnorms(2));
 %then a stretched beta corresponding to the 3rd normal:
 vrate3 = stretchbetaval(vrmeans(3),(vrvars(3))^0.5,...
             vrmins(3), vrmaxs(3),stnormfx(corrnorms(3)));
 results = [results;vrate1,vrate2,vrate3]; % store the values
end; % tt

meanrates = mean(results) % the means of simulated vital rates
variances = var(results)  % variances of simulated rates
correlations = corrcoef(results) % and the correlation matrix
```

can easily fill the entire correlation matrix if we know either its upper or lower portion.)

Any symmetrical matrix[13] (such as a correlation matrix) can be "decomposed" into matrices made up of its eigenvalues and right eigenvectors. As we discussed in Chapter 7, eigenvalues and their corresponding right eigenvectors can serve to predict the ultimate growth rate and structure of a population in a constant environment. Although in Chapter 7 we emphasized the dominant eigenvalue, we also noted that there are as many eigenvalues and right eigenvectors as there are rows (or columns) of a matrix. For our current purpose, it is important to know all the eigenvalues and their associated right eigenvectors. With the eigenvalues and right eigenvectors of **C**, we can construct two other matrices to decompose **C**: **W** is a matrix whose columns are the right eigenvectors of **C**, and **D** is a matrix with diagonal elements that are the eigenvalues of **C** (and all other elements equal to 0), where the eigenvalue in the *i*th position along the diagonal of **D** corresponds to the right eigenvector in column *i* of **W**. (The MATLAB statement `[W,D]=eig(C)` generates **W** and **D** for matrix **C**.)

[13]A symmetrical matrix yields the same matrix when transposed (i.e., when rows become columns and vice versa).

W and **D** can be combined to rebuild **C**: $\mathbf{C} = \mathbf{W} * \mathbf{D} * \mathbf{W}'$. Here, $\mathbf{W}'$ is the transpose of matrix **W**, and $*$ indicates matrix multiplication.[14] However, we want to generate a matrix we'll call the "square-root of **C**" or $\mathbf{C}^{1/2}$:

$$\mathbf{C}^{1/2} = \mathbf{W} * \mathbf{D}^{1/2} * \mathbf{W}' \qquad (8.12)$$

where $\mathbf{D}^{1/2}$ is a matrix with the square-roots of each eigenvalue on the diagonal (and the other elements zero).[15]

With $\mathbf{C}^{1/2}$ in hand, it is possible to take a set of uncorrelated standard normal values and convert them into a set of correlated values. Specifically, if **m** is a column vector of k *uncorrelated* random values from a standard normal distribution (where k is the number of vital rates used to compute **C** and $\mathbf{C}^{1/2}$), then

$$\mathbf{y} = \mathbf{C}^{1/2} * \mathbf{m} \qquad (8.13)$$

is a column vector that also contains k standard normal values, but now ones that are correlated according to the correlation matrix **C**. Even though this method may seem complex, it is routinely used in statistics. Indeed, this is basically the opposite process to that used in Principal Components Analysis, which takes a set of correlated variables, assumes that they are normally distributed, and finds the set of uncorrelated variables (the principal components) that can be recombined to describe the original data.

STEP 2: USING CORRELATED NORMAL RANDOM VARIABLES TO GENERATE CORRELATED VITAL RATES WITH OTHER DISTRIBUTIONS Since we used the original matrix of vital rate correlations to generate the vector **y** of correlated standard normals, this vector has the same number of elements as there are vital rates that we want to simulate. Therefore, by generating multiple **m** vectors with different random values and then applying Equation 8.13 to each, we can simulate a new vector of correlated values, $\mathbf{y}_t$, for each year of a simulated stochastic population trajectory. However, this gives us sets of *standard normal* values, and what we need are correlated values that are *not* normally distributed, but instead come from beta, lognormal, or other distributions. The trick is to convert these correctly correlated standard normal values into

[14]For the transpose of a matrix, see Footnote 13. To multiply two matrices, we treat the matrix on the right as a set of column vectors, use the rules in Chapter 7 to multiply the matrix on the left by each vector, and then use the resulting vectors as the columns of the matrix that is the result of the multiplication. For worked examples of matrix multiplication, see the MATLAB manual, an introductory linear algebra book, or the appendix of Caswell 2001.

[15]Note that you could generate $\mathbf{C}^{1/2}$ using the MATLAB command `C^(0.5)`, which produces the same matrix that this decomposition does. However, the decomposition is more useful because, as we'll see in a few pages, it allows us to adjust correlation matrices that, as estimated directly, are unusable.

values that have the same correlations, but are each drawn from the appropriate distribution for one of the vital rates. For lognormally distributed vital rates (e.g., fertilities), this is easily done using the procedure described earlier under *The lognormal distribution*, that coverts a standard normal variable into a lognormal with a specified mean and variance.

Vital rates with distributions other than the lognormal are not so easily converted from normal random variables. Doing so involves using the standard normal values in each $\mathbf{y}_t$ vector to make a vector of corresponding values from the cumulative distribution function for a standard normal,[16] F_x. There are several good, fast approximation formulae for F_x for standard normals; we showed one in Box 3.3 and give a second, `stnormalfx`, in Box 8.7. In calculating these F_x values, we are essentially computing the relative position of each random value along the range of possible values that a normal random variable could exhibit. Next, we use these F_x values to find corresponding values for each of the vital rates we want to simulate. For example, we might find a value of 1.4 for the standard normal that "represents" a beta-distributed vital rate. F_x for this standard normal value is 0.9192. If the vital rate is beta-distributed with a mean of 0.77 and a variance 0.2, the value of the vital rate with F_x equal to 0.9192 is approximately 0.8244. To find vital rate values, we can either search for each value from each distribution in each year using `betaval` (Box 8.3), or use a pre-determined list of values and their corresponding cumulative probabilities F_x for each vital rate, as we described earlier under *The beta distribution*. If you have access to MATLAB's Statistical Toolbox, you could also use the function `betainv`.

In sum, the basic method is to generate a set of standard normal values each year, multiply these by $\mathbf{C}^{1/2}$ to make a set of correlated normal random numbers, and then either convert these directly into numbers from the correct distribution (for lognormal vital rates) or find their cumulative distribution function values and then identify the number in the distribution of a vital rate with the same cumulative distribution function value (for beta and stretched beta vital rates). This method does not reproduce exactly the Pearson correlation matrix of the original vital rate estimates, because it is essentially recreating rank correlations. Thus, it may be better to estimate the initial correlation matrix using Spearman rank correlations. However, for most purposes the estimated and simulated correlation matrices will be very similar, and in our experience the small errors introduced by simulating correlated vital rates in this way will not influence the results in any significant way.

THE PROBLEM OF SPARSE AND INCOMPLETE SAMPLING Aside from mastering computational tricks, the greatest problem to overcome when simulating correlated vital rates is that the estimated correlation matrix is often invalid. If the data used to estimate a set of correlations are incomplete, with some

[16]In Chapter 3, we used the symbol $\Phi(x)$ for this same quantity; both symbols are in common use.

BOX 8.7 ***A MATLAB file for the function stnormalfx, which provides an approximation formula for cumulative distribution values, F_x, for standard normals (Abramowitz and Stegun 1964).***

```
function ff = stnormfx(xx)
%stnormfx takes a value from a str. normal distr., xx,
% and returns its cumulative distribution function value
% (Abramowitz and Stegun 1964).
   ci = 0.196854; % these are approximation constants
   cii = 0.115194;
   ciii = 0.000344;
   civ = 0.019527;

   if xx >= 0; z= xx; else z = -xx; end;
   a = 1 + (ci*z) + (cii*z*z);
   b = (ciii*z*z*z) + (civ*z*z*z*z);
   w = a+b;
   if xx >= 0
        ff= 1 - 1/(2*w*w*w*w);
   else
         ff = 1 - (1 - 1/(2*w*w*w*w));
   end;
```

rates not measured in some years (and those years omitted from the calculation of correlations involving the unmeasured rates), or if one data set is used to estimate some correlations and other data sets used to estimate others (as might happen if the correlation matrix is compiled from different literature sources), the entire set of correlations can be invalid. An extreme example of an invalid set of correlations would be if three vital rates (X, Y, and Z) all have perfect negative correlations. Clearly, this combination of correlations must be wrong, since X and Y cannot be strongly negatively correlated if they are both perfectly negatively correlated with a third rate. When this problem arises in real estimates of correlations, it is usually not so extreme, but the basic problem is the same: negative correlations can not be both strong and common among a set of correlated rates.

Formally, the problem is that a valid correlation matrix must be positive semidefinite, which means (among other things) that all the eigenvalues of the matrix should be positive or zero. If data are incomplete or if the estimated correlations come from multiple sources, it is possible to have a correlation matrix for which some of the eigenvalues are negative; this is clearly a problem, since our decomposition in Step 1 above (making correlated normal values) involves taking the square roots of all the eigenvalues. Importantly, even if the correlations are based on a single data set in which every vital rate was measured in every year, the correlation matrix may still not be positive semidefinite. This can happen if the number of years of data used to calculate the vital rate correlations is less than 1 plus the number of vital rates (such limited data are the norm in conservation, and in ecology in general). In this case, some of the eigenvalues should be zero, but the numerical algorithm used to compute them may yield very small positive or (more problematically) negative values instead. Specifically, if there are, say, data for four transitions, then only three eigenvalues should be nonzero, no matter how many vital rates were measured.

There are various ways to test for these problems, but the easiest is to see if the diagonal of **D** contains only positive numbers, or, equivalently, if the elements of $\mathbf{C}^{1/2}$ contain only real numbers. If some of the elements of $\mathbf{C}^{1/2}$ are complex,[17] then the original correlation matrix was invalid. We have three recommendations for how to avoid this situation, and how to deal with it should it occur:

1. When possible, avoid using estimated correlations compiled from multiple sources.
2. If all rates were measured for all n transitions, and n is less than the number of vital rates being correlated plus 1, then modify the matrix **D** by setting all but the largest $n - 1$ eigenvalues to zero. The eigenvalues should show a clear break in magnitude between the $n - 1$ largest values and the remaining, effectively zero, values.
3. Set all negative eigenvalues in **D** to zero. You should also set all positive eigenvalues that are smaller than the largest absolute value of the negative eigenvalues (which will be small) to zero. This is essentially doing what we did in alternative (2), but without a clear knowledge of exactly how many eigenvalues to eliminate. For example, if different correlations were estimated with data from different numbers of years, it would not be clear what to use for n in the preceding suggestion.

If you modify **D** by using recommendations (2) or (3), you will also have to do two more steps. After you modify **D** to $\mathbf{D}_m$ by zeroing negative and small positive eigenvalues, recombine it with **W** to create the now-modified correlation matrix: $\mathbf{C}_m = \mathbf{W} * \mathbf{D}_m * \mathbf{W}'$. This should look only slightly different from the original **C** matrix of correlations. However, this is now a covariance matrix, with variance estimates on the diagonal that are not all equal to 1. To use it as we've described to generate correlated standard normal variables, it is necessary to convert it back into a correlation matrix, $\mathbf{C}_{final}$, with each new

[17]See Footnote 4 in Chapter 7.

element, $C_{final}(i,j) = C_m(i,j)/\sqrt{C_m(i,i)C_m(j,j)}$[18]. Finally, you decompose this new correlation matrix to determine the $\mathbf{C}^{1/2}$ matrix to use in your stochastic simulations.

Analysis of the correlations in the vital rates for the desert tortoise show how these corrections work (Box 8.8). With only three transitions of data to estimate the correlations for tortoise growth and survival, only two of the eigenvalues should be nonzero if there were no rounding errors in their calculation. In fact, the two largest eigenvalues are 4.5942 and 6.4061, while the others are less than 0.002, and five of these are negative. In this case, it seems clear that we should set all but the two largest eigenvalues to zero. Doing this, the program in Box 8.8 goes through the steps in the previous paragraph to arrive at a new correlation matrix. One way of checking how much the correlations differ from those estimated originally is to calculate the proportional change in each correlation. As the largest of these proportions is less than 0.01, none of the changes in the correlation matrix were at all substantial. With more vital rates and a very small number of years of data, much larger changes from estimated rates can occur.

AUTOCORRELATION AND CROSS-CORRELATION A final issue regarding correlation of vital rates is between-year correlation. Autocorrelation is correlation in the sequential values of a single vital rate caused by environmental factors that are correlated between years (we dealt with autocorrelation in the population growth rate in Chapter 4). Similarly, cross-correlation is correlation of different rates *across* time steps, such as a correlation in adult survival in one year with juvenile survival the following year. (We dealt with cross-correlation in matrix elements in Chapter 7.) Autocorrelation and cross-correlation have rarely been added to a PVA in any but the simplest ways (such as forcing drought years to come in series). However, if there are enough data to measure correlations among the values of different vital rates within a year, there are often enough data to estimate between-year correlations as well. Because the sample size for estimating between-year correlations with a one-year time lag is one less than the number of transitions, four or more estimates of the vital rates are needed to obtain any value other than +1, 0, or –1. If you have the data from five or more censuses, it is wise to at least estimate between-year correlations. The simplest procedure to do so is to create a spreadsheet with a block of values for each vital rate (columns) by year (rows) combination. Next, compute all pairwise correlations among the columns; these estimate the within-year correlations. Then, copy all but the first row of this block to the adjacent columns in the spreadsheet, moved one row up; delete the last row of the first block; and again compute all of the

[18]See Footnote 1 for the relationship between covariance and correlation.

BOX 8.8 *A MATLAB program going through steps needed to calculate a correlation matrix and look for the problems caused by sparse sampling or small numbers of observations, using correlations for the desert tortoise.*

```
%AnalyzeCorrs
% This program goes through the steps of checking whether a
% correlation matrix has impossible combinations of values
% due to missing data or small numbers of observations,
% and then correcting these problems.

clear all;
%****************Simulation Parameters**********************
%Define the correlation matrix: you can use a 1/2 matrix as below
% (with zeros in upper triangle) and fill in the other 1/2
% (see below). This is the matrix for the desert tortoise
corrmx = [...
1    0   0   0   0   0   0   0   0   0   0;
0.469   1  0   0   0   0   0   0   0   0   0;
0.382   0.995 1  0   0   0   0   0   0   0   0;
-0.597  0.429  0.514  1  0   0   0   0   0   0   0;
-0.014  0.877  0.919  0.811  1  0   0   0   0   0   0;
-1  -0.496 -0.41  0.571  -0.017 1  0   0   0   0   0;
-0.704 -0.957 -0.925 -0.149 -0.7   0.726  1  0   0   0   0;
0.898  0.81   0.75   -0.182 0.428  -0.911 -0.945 1  0   0   0;
0.956  0.189  0.094  -0.806 -0.307 -0.946 -0.465 0.729  1  0   0;
-0.514 0.516  0.597  0.995  0.865  0.487  -0.247 -0.083 -0.743 1  0;
-0.994 -0.563 -0.481 0.505  -0.096 0.997  0.778  -0.941 -0.918 0.417  1];
%**********************************************************
np = max(size(corrmx)); % how many vital rates?
corrmx = corrmx + corrmx' - eye(np); % make a full matrix

[W,D] = eig(corrmx);  % find the eigenvalues and eigenvectors

disp('this is the eigenvalue matrix: if any elements are')
disp('negative, then some adjustment is needed')
disp(D)
disp('the corresponding eigenvectors are:')
disp(W)
disp('  ');
disp('the C12 matrix is:')

C12 = W*(sqrt(abs(D)))*W'
```

BOX 8.8 *(continued)*

```
disp('for this correlation matrix, based on only three')
disp('time periods, all but the last 2 eigenvalues are')
disp('small, so we should redo the calculation of the')
disp('correlation matrix, eliminating all the others:')
newD = D;
newD(:,1:(np-2)) = zeros(np,(np-2));
newcorrmx = W*newD*W';

disp('but this newcorrmx must now be corrected to be')
disp('a proper correlation matrix, not a covariance matrix')
for ii=1:np
 for jj=1:np
  newcorrmx(ii,jj) = ...
     newcorrmx(ii,jj)/((newcorrmx(ii,ii)*newcorrmx(jj,jj))^0.5);
 end;
end;

[W,D] = eig(newcorrmx); % corrected eigenvectors and values
C12 = W*(sqrt(abs(D)))*W';
disp('The new, corrected, correlation matrix is: ')
disp(newcorrmx)
disp('And the new C12 is: ')
disp(C12)
disp('Finally, the proportional differences between the old')
disp('correlations and the new ones are:')
disp([(corrmx - newcorrmx)./corrmx])
```

correlations among the columns. These will include both with-year and between-year (auto- and cross-) correlations, but discard the former and use the ones you've already calculated above (which were based on all the data). In looking at the cross-correlations, note that the correlation between vital rate a in year t and vital rate b in year $t + 1$ need not be the same as the correlation of rate b in year t and rate a in year $t + 1$. Between-year correlations over more than a one year time lag can be estimated as well, but since more and more data are needed to test for correlations with increasingly long time lags, we'll restrict our consideration to one time step correlations (although these will also drive correlations at longer time lags, as we'll see).

Although looking for between-year correlations with a data set spanning several years is easy, adding these correlations to a simulation in a way that accurately reflects the measured correlation structure is not. The basic problem is that the procedure we used in the last section to simultaneously generate a set of correlated normal variables can not be modified in a simple way to *sequentially* generate correlated numbers. One possibility is to treat each vital rate in each year as a separate random variable and then to generate all the vital rates for all years of an entire simulation in one iteration of the procedure we used above to calculate the vital rates for a single year. Then the problem is simply one of defining the correlations between all pairs of vital rates across all years. However, this tactic may be slow or impossible for long time series and multiple vital rates. Thus, we are usually in the position of wanting to first generate a set of vital rates for one year, and then generate a new set for the next year with a certain pattern of correlations with the first set (not to mention having correlations between the rates within each year), as we were able to do for the population growth rate in the count-based case (see Equation 4.16).

A successful, if less than obvious, way to simulate between-year correlations uses an elaboration of the same general approach we've taken for generating within-year correlations. To begin with, we need to extract and rearrange two parts of the correlation matrix we generated earlier. If we estimated the correlations for two vital rates, a and b, and we had a column for $a(t)$, a column for $b(t)$ and then two columns for the lagged values, $a(t + 1)$ and $b(t + 1)$, the matrix of correlations and autocorrelations will be:

$$\mathbf{E} = \begin{bmatrix} 1 & r(b_t, a_t) & r(a_{t+1}, a_t) & r(b_{t+1}, a_t) \\ r(a_t, b_t) & 1 & r(a_{t+1}, b_t) & r(b_{t+1}, b_t) \\ r(a_t, a_{t+1}) & r(b_t, a_{t+1}) & 1 & r(b_{t+1}, a_{t+1}) \\ r(a_t, b_{t+1}) & r(b_t, b_{t+1}) & r(a_{t+1}, b_{t+1}) & 1 \end{bmatrix} \tag{8.14}$$

where, for example, $r(a_t, b_{t+1})$ is the correlation coefficient between parameter a in one year, a_t, and b in the next year, b_{t+1}. Now, notice that the upper-left and lower-right quarters of **E** contain only within-year correlations. However, the upper-left estimates use the full data set, while the lower-right estimates use the lagged data, and hence are based on one fewer years of observations. Therefore, we replace these values with the ones from the upper-left elements of the matrix. After doing this, you should use the methods in the last section to make an $\mathbf{E}_{final}$ matrix that is positive semidefinite (no negative eigenvalues). This step will nearly always be necessary, since the within-year correlations will always be based on one less transition than are the between-year correlations.

Next, extract two pieces of $\mathbf{E}_{final}$, the upper-right and the lower-right quarters:

$$\mathbf{C} = \begin{bmatrix} 1 & r(b_t, a_t) \\ r(a_t, b_t) & 1 \end{bmatrix} \text{ and } \mathbf{B} = \begin{bmatrix} r(a_t, a_{t+1}) & r(b_t, a_{t+1}) \\ r(a_t, b_{t+1}) & r(b_t, b_{t+1}) \end{bmatrix} \tag{8.15}$$

Using these two pieces of the full (and corrected) correlation matrix, we can make a matrix of correlations that span many years, assuming that all the correlations between variables that are more than one year apart are caused by the correlations across a single year. (This assumption is not necessarily true, but as we've explained earlier, we will rarely have enough data to make any estimate of these longer-term effects.) If this is the case, then a matrix for four years, **M**, can be written in terms of **C** and **B**:

$$\begin{array}{l} \text{Year 1} \\ \text{Year 2} \\ \text{Year 3} \\ \text{Year 4} \end{array} \quad \mathbf{M} = \begin{array}{c} \begin{array}{cccc} \text{Year 1} & \text{Year 2} & \text{Year 3} & \text{Year 4} \end{array} \\ \begin{bmatrix} \mathbf{C} & \mathbf{B}' & (\mathbf{B}')^2 & (\mathbf{B}')^3 \\ \mathbf{B} & \mathbf{C} & \mathbf{B}' & (\mathbf{B}')^2 \\ \mathbf{B}^2 & \mathbf{B} & \mathbf{C} & \mathbf{B}' \\ \mathbf{B}^3 & \mathbf{B}^2 & \mathbf{B} & \mathbf{C} \end{bmatrix} \end{array} \quad (8.16)$$

Here, each submatrix, **C** or **B**, is a n by n matrix, where n is the number of vital rates. **B**′ is the transpose of **B**, which is the same matrix as the lower-left quarter of $\mathbf{E}_{final}$. For our example of two vital rates and four years, **M** is an 8 × 8 matrix. The higher powers of **B** or **B**′ for pairs of rates in years that are increasingly distant from one another reflect the decay of correlation over time. Again, we are assuming that correlations between multiple years are generated solely by one-year correlations, rather than effects operating over longer time periods that might directly cause correlations between rates more than a year apart.[19]

To simulate autocorrelations and cross-correlations with high accuracy, we need to construct **M** for dozens or more years, which the program in Box 8.9 automates, given estimates of the **C** and **B** matrices. Let's say we choose to make an **M** matrix for 50 years and that there are three vital rates in each year (this is the situation in the Box 8.9 program), making **M** a 150 × 150 matrix. Once we've made this large, multi-year **M** matrix, we must again use the methods in the last section to make sure it is positive semidefinite. We then find its square root, $\mathbf{M}^{1/2}$. If we wanted to simultaneously simulate all the vital rates for all the years represented in **M** and $\mathbf{M}^{1/2}$, we would generate a column vector of 150 uncorrelated standard normal values **m** and then multiply $\mathbf{M}^{1/2}$ by **m** to generate 150 correlated normal values, one for each rate in each year. This will work if the number of vital rates and the length of the simulation are reasonably small. However, if this direct approach would result in a truly massive **M** matrix, it may be more feasible to instead use a trick to calculate one year of vital rates at a time. Here is how it works:

1. Choose a set of n (the number of vital rates) rows that are in the very middle of $\mathbf{M}^{1/2}$, say, the ones for year 25 for our $\mathbf{M}^{1/2}$ of 50 years. For our three

[19]If you have the data to directly estimate correlations over longer lags, they can also be used in building **M**.

BOX 8.9 *A program to demonstrate the simulation of within-year correlations, autocorrelations, and cross-correlations in vital rates.*

```
% BetweenYrCorr: this program generates sets of within-year
% auto- and cross-correlated (with one time step)
% vital rates.
clear all;

%****************Simulation Parameters*********************
% parameters for 3 vital rates:
% a beta, a lognormal, and then a stretched beta.
vrmeans= [0.77,2,5]; % means
vrvars= [0.02,0.5,2]; % variances (not standard deviations)
% minimum and maximum values for each vital rate:
% zeros are place holders for rates that are not stretched betas
vrmins= [0 0 0];
vrmaxs= [0 0 24];
% pos. semi-definite matrix for within year correlations
% (corrected if original was not good):
corrin=...
[1      0.3    -0.2;
0.3     1      0.6;
-0.2    0.6    1];

% then the auto- and cross-correlations for one step.
% The form should be columns of v1(t), v2(t), etc...
% and rows of v1(t+1), v2(t+1), etc.., where v1(t) is
% vital rate 1 in year t
corrout=[...
0.5 0.3 -0.1;
0.3 0.2 0.3;
-0.1 0.3 0.7];

yrspan = 50;
 % this is the number of years of correlation info to use to
 % simulate the correlation pattern —
 % more years are better. 50 is quite accurate.
tmax = 100;   % number of years of vital rates to simulate;
%*********************************************************

np = length(vrmeans);
results = [];
```

BOX 8.9 *(continued)*

```
rand('state',sum(100*clock)); randn('state',sum(100*clock));
%------creating and using the big correlation matrix M-----
% this set of loops makes the big correlation matrix (M)
% with multi-year correlations: the if statements are used to
% estimate the correct correlations with increasing time lags,
% always assuming that all long-time-lag correlations are only
% caused by within-year and one-time step correlations
for ii = 1:yrspan;
 for jj=1:yrspan;
       if ii==jj, litmx = corrin;
          else litmx =corrout; end;
       expo=1;
       if ii > jj, expo = ii-jj; litmx = (litmx)^expo; end;
       if ii < jj expo = jj-ii;  litmx = (litmx')^expo; end;
       for ip=1:np;
       for jp = 1:np;
       M(ip+np*(ii-1),jp+np*(jj-1)) = litmx(ip,jp);
       end; % jp
       end; % ip
 end; % jj
end; % ii

% get the eigenvalues for calculating the M12 matrix
[W,D] = eig(M); % getting the eigenvalues and vectors

% now, a check for negative eigenvalues – if you have them,
% it sets negatives and small positive values = 0
check = min(min(D)); % are the smallest eigenvalues negative?
if check < 0
    disp('The min eigenvalue is < 0. Eigenvalues are:')
    disp(diag(D))
    disp('hit enter to continue with approximation')
    pause
    maxneg = max(max(abs(D(find(D<=0)))));
    % maxneg is the largest negative eigenvalue
    D(find(abs(D<=maxneg))) = 0; % set the negatives = 0
    disp('Corrected eigenvalues are:')
    disp(diag(D))
    newfullmx = W*newD*W'; % make a corrected matrix
    for ii=1:np % change matrix from covariances to correlations
     for jj=1:np
```

BOX 8.9 *(continued)*

```
        if newfullmx(ii,ii) ==0 | newfullmx(jj,jj)==0;
               newfullmx(ii,jj) =0;
        else
               newfullmx(ii,jj) = ...
newfullmx(ii,jj)/((newfullmx(ii,ii)*newfullmx(jj,jj))^0.5);
        end;
      end;
    end;
    [W,D] = eig(newfullmx);
end; %check < 0

M12 = W*abs(D.^0.5)*W';
sz = length(zfull); % the total number of lines in M12

% get the lines from the middle of M12
% to use to generate correlations
startcase = (round(yrspan/2)*np +1);
zvalold=real(M12(startcase:(startcase+np-1),:));
zvalnew=real(M12((startcase+np):(startcase+2*np-1),:));
clear M12 W D; % clearing memory
% Zvalold and Zvalnew are each one year of rows of M12
% to start the whole thing off, calculate a first set of
% normals, then multiply with Zvalold to get correlated normals:
newns = randn(sz,1);
oldxy = zvalold*newns;
%-----end of: creating and using the big correlation matrix------

normresults = [];  vitalresults = [];

for tt=1:tmax % a loop to make multiple sets of rates
    disp('time is'); disp(tt);
    %update the uncorrelated random normals
    newns = [newns((np+1):sz); randn(np,1)];
    % make the new set of correlated normals
    newxy= zvalnew*newns;
    normresults = [normresults; oldxy', newxy']; % save results
    oldxy = newxy;  % make the new correlated rates old

% now convert correlated normals to the correct distributions:
    vrate1 = ...
     betaval(vrmeans(1),(vrvars(1))^0.5,stnormfx(newxy(1)));
```

BOX 8.9 *(continued)*

```
    vrate2 = lnorms(vrmeans(2),vrvars(2),newxy(2));
    vrate3 = stretchbetaval(vrmeans(3),(vrvars(3))^0.5,...
             vrmins(3), vrmaxs(3),stnormfx(newxy(3)));
    vitalresults = [vitalresults;vrate1,vrate2,vrate3];
end; % tt

disp('The input correlations (all rates and one step')
disp('correlations are:');
disp([corrin corrout';corrout,corrin])
disp('The correlations of the normals are:')
disp(corrcoef(normresults));
disp('The within-year correlations of the vital rates are:');
disp(corrcoef(vitalresults));
disp('The input means and variances were');
vrmeans, vrvars
disp('Means and variances of the simulated rates are:')
meanrates = mean(vitalresults)
variances = var(vitalresults)
```

vital rates, this is a 3 x 150 matrix that we can call $\mathbf{M}^{1/2}_{1\,\text{year}}$. Once we have this piece of $\mathbf{M}^{1/2}$, we have no further need of any of the rest of the matrix and can clear it from the computer's memory to increase simulation speed.

2. For year 1, make a column vector $\mathbf{m}_1$ of standard normal random numbers with the same number of values as $\mathbf{M}^{1/2}$ has rows (the number of vital rates times the number of years), and then multiply $\mathbf{M}^{1/2}_{1\,\text{year}}$ by $\mathbf{m}_1$. This gives a set of vital rate estimates for year 1. Notice that although we are only using a small part of $\mathbf{M}^{1/2}$, these three lines are the only part of the matrix that would multiply with a vector to produce the year 25 vital rates if we were doing the full multiplication.
3. Now, make a new vector $\mathbf{m}_2$ that is largely, but not completely, the same as $\mathbf{m}_1$. Take all the values of $\mathbf{m}_1$ except the first n (3, for our example), and put them in the first $n(y - 1)$ rows of $\mathbf{m}_2$, where y is the number of years (50 in this example). Now, draw n new standard normal random numbers and make these the last values in $\mathbf{m}_2$. In our example, this means that we throw out the first 3 random numbers in $\mathbf{m}_1$, move the remaining numbers into rows 1 to 147 of $\mathbf{m}_2$, and then place three new standard normal random numbers into rows 148–150.
4. Multiply $\mathbf{M}^{1/2}_{1\,\text{year}}$ by $\mathbf{m}_2$ to get the vital rates for Year 2.
5. Repeat steps 3 and 4, each time converting $\mathbf{m}_i$ to $\mathbf{m}_{i+1}$.

The trick works for two reasons. First, **M** (and hence $\mathbf{M}^{1/2}$) is a completely symmetrical matrix (Equation 8.16), so the numbers in the block of rows corresponding to one year's vital rates are the same as those one block of rows below, but offset by a number of columns that corresponds to the number of vital rates. To see this, just look at the position of **C** within the different rows of **M**. Thus, if we multiplied a vector of random numbers times **M** to get a series of vital rates for multiple years, for each year the entries in the random number vector would mostly be multiplied by the same numbers as in the previous year, but offset by one block. This pattern only breaks down at the right and left edges of the matrix. Other than at the edges, multiplying the vector and the matrix to get one year's vital rates could just be done by shifting over the matrix entries or shifting up or down the numbers in the vector. This shifting is exactly what we do in step (3) above. The second key feature of making the trick work is making a matrix for enough years that the matrix entries at the right and left edges all represent very small correlations, so that this business of shifting over and using the same row to calculate vital rates in different years introduces only very small errors.[20] Because of this, we can use one small part of the matrix over and over again, and simply shift over the random normal numbers a small amount each year to create auto- and cross-correlated variables. All that said, remember that **M** is *not* the matrix we use to do any of this—we use its square-root, $\mathbf{M}^{1/2}$. Unlike **M**, the rows and columns of $\mathbf{M}^{1/2}$ do not have a clear interpretation as relating one particular year to another, but the size and pattern of its matrix entries generally follow those of **M**. In particular, the magnitudes of its elements at the right and left edges get very small as the number of years in **M** becomes large.

This solution to the simulation is particularly appealing, because it allows the simultaneous simulation of within-year and between-year correlations, and other than the initial process of constructing **M** and generating $\mathbf{M}^{1/2}$, it adds very little time to a simulation.

Model Results for Two Examples

To put the various tools we've gone over into action, we will use vital rate simulations to estimate stochastic growth rates and the cumulative distributions of quasi-extinction times for *H. montana* (mountain golden heather; see Chapter 6) and the desert tortoise. Table 8.5 gives the re-estimated growth rates for *H. montana*, and we similarly used Kendall's method to reanalyze survival rates. However, in constructing a vital-rate-based matrix for *H. montana*, we used the raw numbers for the first two rows of the matrix (which represent seed and seedling demography), rather than breaking these elements into separate vital rates. We did this because there was little informa-

[20]How many years are "enough" will depend on the strength of the estimated between-year correlations. While we have not examined this question, it seems likely that constructing **M** to include only 5 to 10 years of lagged correlations will be enough to yield accurate simulations for most sets of vital rates.

tion about the temporal variation of these vital rates. (See Chapter 6 for a description of the vital rates contributing to these matrix elements.) For the same reason, we did not attempt to estimate correlation between these rates and rates for the aboveground plants in the population. However, for all the growth and survival rates of classes 3 through 6, we did estimate both within-year correlations and also auto- and cross-correlations. The four years of data available for this population are the minimum that can be used to estimate between-year correlations. As such, little faith should be placed in the exact estimates of auto- and cross-correlations. Nonetheless, it is worth including these correlations to see how they influence the model results. After making these estimates, we corrected the matrix of correlation values, and then broke it into within and between year correlations as in Equation 8.15, which are then used by the program `VitalSim.m` in Box 8.10 to construct a $\mathbf{M}^{1/2}$ matrix in order to simulate all the estimated correlations. The program then simultaneously calculates stochastic growth rate and extinction probabilities.

The changes in estimated vital rates led to a somewhat lower deterministic growth rate for the mean matrix than for the mean matrix used in Chapter 7. For the re-calculated mean rates, $\bar{\lambda} = 0.9586$, while it equaled 0.9660 for the uncorrected mean values. The stochastic growth rate we estimate is considerably lower, $\lambda_s = 0.9449$, than the $\lambda_s = 0.9641$ that we obtained using whole matrices. Not surprisingly, this lower stochastic growth rate translates into substantially greater extinction risk at 50 years (Figure 8.3; compare with

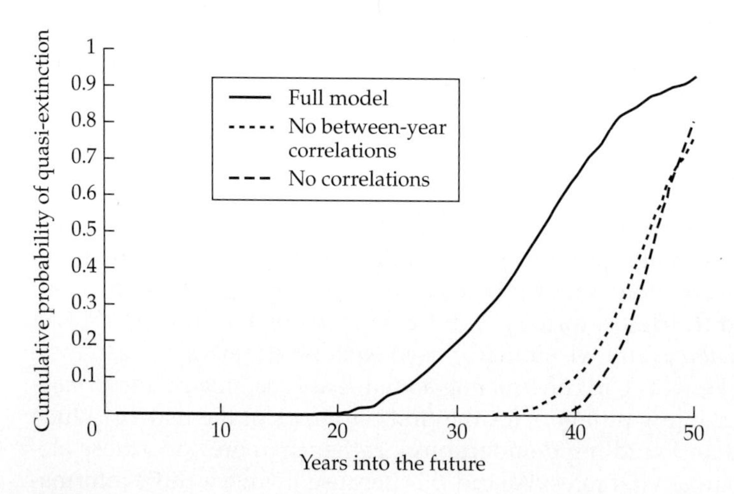

Figure 8.3 Cumulative distribution of quasi-extinction times from stochastic vital rate simulations of a model for *H. montana*. The three lines show results for simulations that include both within-year and between-year correlations (labeled "Full model"), within-year correlations only ("No between-year correlations"); or neither type of correlation ("No correlations"). Results are based on 500 trajectories.

BOX 8.10 ***A program to calculate the extinction time CDF and the stochastic growth rate for Hudsonia, using simulations that include correlation, autocorrelation, and cross-correlation.***

```
%VitalSim:
% This is a program to do a general stochastic PVA simulation
% including within-year, auto-, and cross-correlations
% and a choice of different distributions for the vital rates.
clear all;

%****************Simulation Parameters**********************
% the data shown here are for H. montana (best Kendall estimates):
% the parameters are defined in a set of arrays
% with elements corresponding to these vital rates or matrix elements:
%1,     2,     3,     4,     5,     6     7     8     9     10   11  12  13
%g>4,3 g>4,4 g>5,4 g>6,4 g>4,5 g>5,5 g>6,5 g>5,6 g>6,6 S3   S4  S5  S6
% and these:
%14  15  16  17  18  19  20  21  22  23  24
%a11 a13 a14 a15 a16 a21 a23 a24 a25 a26 a32

% vrtypes identifies the distribution for each rate:
% 1 = beta, 2 = stretched beta, 3 = lognormal
vrtypes= [ones(1,13),3,3,3,3,3,1,1,1,1,1,1];
% now means and variances of these rates or elements
vrmeans= [0.24    0.79    0.30    0.024    0.92    0.81    0.25    0.93 ...
    0.75   0.78798799    0.96452906     0.99519038     0.98076152...
    0.4995 4.5782 12.1425   22.3167     50.1895   0.0004  0.0033 ...
    0.0088 0.0162 0.0364 0.4773];
vrvars=[0.00018836    0.00016164    0.00026851      0.00002805 ...
    0.00005477    0.00184484    0.00937915    0.00002805 ...
    0.01333333    0.01790190    0.00000963    0.00000028 ...
 0.00000322 0 0.012 0.0843 0.2848 1.4403 0  0  0  0 0.0003];
% minimum and maximum values for each vital rate:
% put in zeros for rates that are not stretched betas
vrmins=zeros(1,24); vrmaxs=zeros(1,24);
np = length(vrmeans); % how many parameters are there?
% you must load a file with a pair of correlation matrices:
% corrin is the within yr correlation part of the matrix,
% corrout is the between yr correlations. See Box 8.9 for a
% description of the right format
load hudcorrs
% NOTE: In this example, we use correlations between the first 13
```

BOX 8.10 *(continued)*

```
% parameters, but not the other 11 reproductive parameters, which
% vary little.
yrspan = 20; % the number of years of correlations to
% build into the M12 matrix

% you must also have a separate m-file that uses a vector of vital
% rates (vrs) of the same form as vrmenas, above, to define the
% elements of the population matrix (mx) (see Table 8.1 for example)
makemx='hudmxdef'; % hudmxdef.m defines the elements of mx

n0=[4264; 3; 30; 16; 25; 5];% initial population vector
Nx=500;                     % quasi-extinction threshold
tmax = 50;     % number of years to simulate;
np   = 13;     % number vital rates in correlation matrix;
np2 = 11;      % number of uncorrelated rates.
dims = 6;      % dimensions of the population matrix;
runs = 100;    % how many trajectories to do
%*****************************************************

rand('state',sum(100*clock)); randn('state',sum(100*clock));
Nstart = sum(n0);              % starting population number
vrs=vrmeans;  % set vital rates to their mean values
eval(makemx)  % use matrix definition file to make mean matrix
lam0=max(eig(mx)); %find the deterministic population growth rate

%-----------------------------------------------------
% this section makes sets of beta or str. beta values to choose
% from during the simulations; it makes 99 values for 1%
% increments of Fx for each parameter - if you already have made
% a set of beta values for this life history, you can recall
%  these, and save the time of recalculating them.
yesno = ...
 input('type 0 to calculate betas, or 1 to get a stored set');
if yesno == 1
  betafile = ...
input('type filename with stored betas; put in single quotes');
  load betafile parabetas
else    % make a set of values for each beta or stretched beta
  parabetas=zeros(99,np+np2);
  for ii = 1:(np+np2)
    if vrtypes(ii) ~= 3
     for fx99 = 1:99
```

BOX 8.10 *(continued)*

```
      if vrtypes(ii) ==1;
         parabetas(fx99,ii) = ...
          betaval(vrmeans(ii),sqrt(vrvars(ii)),fx99/100); end;
      if vrtypes(ii) ==2;
         parabetas(fx99,ii) = ...
          stretchbetaval(vrmeans(ii),sqrt(vrvars(ii)),...
          vrmins(ii), vrmaxs(ii), fx99/100); end;
     end; % fx99
    end; % if vrtypes(ii)
  end; % ii loop
  yesno = input('type 1 to store the betas, or 0 if not');
  if yesno ==1
    betafile = ...
            input('type filename to store betas, put in single quotes');
    save betafile, parabetas;
  end; %if yesno
 end; %else

% the next section is identical to that in the program
% BetweenYrCorr in Box 8.9 - simply substitute in the lines
% from that program that are delineated by the same
% comment lines as the two that follow here:
%--------creating and using the big correlation matrix, M--------
%-----end of: creating and using the big correlation matrix------

% finally, do sets of runs to get growth rate and extinction risk
results = []; normresults = [];
PrExt = zeros(tmax,1);     % the extinction time tracker
logLam = zeros(runs,1);    % the tracker of log-lambda values
stochLam = zeros(runs,1); % tracker of stochastic lambda values
for xx = 1:runs;
 if round(xx/10) == xx/10 disp(xx); end; % displays progress
 nt = n0; % start at initial population vector
 extinct = 0;
 for tt = 1:tmax
    % make random normals
    newns = [newns((np+1):sz,1); randn(np,1)];
    newxy= zvalnew*newns;  % make new set of corr'ed normals
    normresults = [normresults; oldxy', newxy']; % these lines
    oldxy = newxy;          % save normals to check correlations
    yrxy=[newxy;randn(np2,1)];
              %adds in randoms for uncorrelated vital rates.
```

BOX 8.10 *(continued)*

```
    for yy = 1:(np+np2)   % loops to find vital rate values
       if vrtypes(yy) ~= 3 % if not a lognormal rate
          index = round(100*stnormfx(yrxy(yy)));
          if index == 0 index = 1; end; % round at extremes
          if index ==100 index = 99; end;
          vrs(yy) = parabetas(index,yy); % find stored value
       % else, calculate a lognormal value:
       else vrs(yy) = lnorms(vrmeans(yy),vrvars(yy),yrxy(yy));
       end;% if vrtypes(yy) ~= 3
    end; % yy loop
    eval(makemx);      % make a matrix with the new vrs values.
    nt = mx*nt;        % multiply by the population vector

    if extinct == 0 % check for extinction
       Ntot = sum(nt);
       if Ntot <= Nx
             PrExt(tt) = PrExt(tt) +1;
             extinct = 1;
       end; % if Ntot
    end; % if extinct

 end % time (tt) loop
logLam(xx) = (1/tmax)*log(sum(nt)/Nstart); % calculate loglambda
stochLam(xx) = (sum(nt)/Nstart)^(1/tmax);  % and stoch. lambda
end; % runs (xx) loop

CDFExt = cumsum(PrExt./runs); % make the extinction CDF function
disp('This is the deterministic lambda value'); disp(lam0);
disp('And this is the mean stochastic lambda'); disp(exp(mean(logLam)));
disp('Below are mean and standard deviation of log lambda');
disp(mean(logLam)); disp(std(logLam));
disp('Next is a histogram of logLams'); Hist(logLam);
disp('And then, the extinction time CDF'); figure; plot(CDFExt);
```

Figure 7.6), using the same starting population size and quasi-extinction threshold as before. These differences are due in part to the simulation of both within- and between-year correlations in vital rates rather than whole matrices, but, as the difference in deterministic growth rates shows, some of the changes in results also come from the different mean vital rate values used.

To investigate the extent to which the correlation structure influenced model results, we also ran simulations that included only within-year correlations and simulations without any correlations between vital rates. Removing auto- and cross-correlations substantially increased stochastic growth (λ_s = 0.9546) and decreased extinction risk (Figure 8.3). However, simulations without any correlations had very similar results to those with only within-year correlations (λ_s = 0.9549) and in fact showed somewhat higher extinction risk at 50 years, suggesting that for *H. montana* between-year correlations are more important for viability than are within-year correlations.[21] The different predictions we obtained with these different methods of analysis are important cautions against placing too much faith in exact predictions of viability results, but are also striking in their similarity. In particular, all these vital rate models, as well as the models from Chapter 7, predict substantial rates of decline and agree that in the absence of management, immediate extinction risk is small, but that it would rapidly increase from about 25 years in the future onwards.

We also constructed a vital-rate simulation model for the desert tortoise. Although there are not enough data to estimate between-year correlations for this population, we did include estimates of within-year correlations (Table 8.2) as well as Kendall-corrected mean and variance estimates for growth and survivorship. Again, we have only point estimates of reproductive rates, and use one set of unvarying values for all years (matrix elements $a_{0,5}$ = 0.42, $a_{0,6}$ = 0.69, and $a_{0,7}$ = 0.69). This stochastic model (the program is not shown here as it is largely the same, though simpler, than the one for *H. montana* in Box 8.11) yields λ_s = 0.9634. In the absence of any data on current size distributions in the field, we started simulations with 500 animals arranged in the stable stage distribution for the mean matrix, and set the quasi-extinction threshold at 50 individuals in all classes. As Figure 8.4 shows, there is a substantial risk of extinction given these starting conditions, with a predicted median time to quasi-extinction of 61 years. In contrast to the predictions for *H. montana*, extinction risk is predicted to rise more gradually, although there is again a predicted lag of several decades before a substantial risk of extinction occurs.

Simulating Demographic Stochasticity

Another issue in constructing a simulation-based PVA by using vital rates is how best to include the effects of demographic stochasticity. As discussed in Chapters 2 and 3, in many PVAs the quasi-extinction threshold is set high enough to obviate the need to include demographic stochasticity. However,

---

[21]This result is not too surprising, given that *H. montana* shows a mixture of strong positive and negative within-year correlations between vital rates.

BOX 8.11 ***A function to decide the fates of a set of individuals simultaneously, given a set multinomial probabilities of different outcomes (after Caswell 2001, pg. 459).***

```
function outcome = multiresult(N,p);
% This function takes as arguments
% N, the number of individuals in a class,
% and p, a vector of the probabilities of
% different fates, including death
% It returns a vector of the number of individuals
% with each of the possible fates.
ind = ones(1,N);
pp = cumsum(p/sum(p)); % find cumulative probabilities
rnd = rand(1,N)      % make a uniform random for each individual
for ii = 1:length(p); % find each individual's random fate
    ind = ind + (rnd > pp(ii));  end;
for ii = 1:length(p); % add up the individuals in each fate
    outcome(ii) = sum(ind==ii); end;
```

you can also directly include its effects on population growth. In general, demographic stochasticity is simulated by performing so-called Monte Carlo simulations, in which the fate of each individual in a certain class and a certain year is decided by a set of independent random choices, all of which are based on the same set of mean vital rates. This approach directly simulates the randomness caused by small population sizes. However, doing Monte Carlo simulations for all years of all simulated population trajectories can

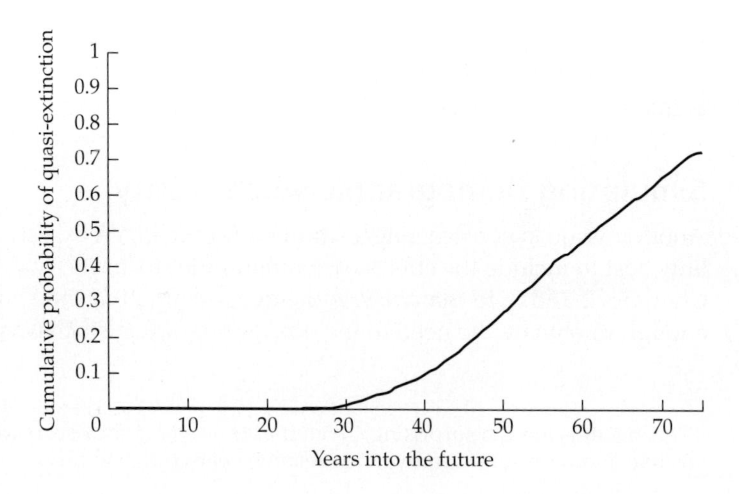

Figure 8.4 Cumulative distribution of quasi-extinction times from stochastic vital rate simulations of a model for the desert tortoise. The simulation uses Kendall's corrected estimates of environmental variation, and includes correlations between rates within each year (Tables 8.2 and 8.3). These results are based on only 200 trajectories, but have been smoothed to emphasize the general pattern.

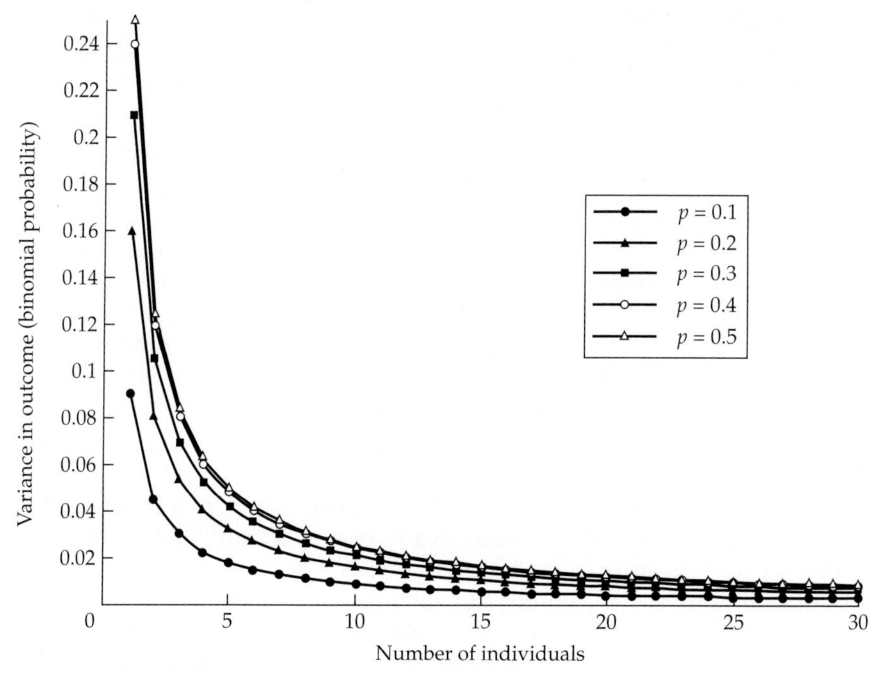

Figure 8.5 Variability caused by demographic stochasticity in binomial vital rates. As population size increases, variance in the outcomes of a vital-rate process such as growth or survival declines rapidly. Each line shows results for a different mean probability.

greatly slow a program. As a consequence, most PVAs use some numerical cutoff in total population size or the size of one or more classes, above which the model will simply multiply the number of individuals in a class by that year's mean vital rates.

Even though using a cutoff to determine when to simulate demographic stochasticity is reasonable, the diminution of demographic stochasticity as population size increases is gradual, and exactly where this boundary should be set therefore isn't entirely clear. A common practice is to use a cutoff of somewhere between 20 and 100 individuals in either the whole population or in a particular class.[22] The relationship between the variance in survival outcomes generated by demographic stochasticity and the number of individuals shows why a cutoff as low as 20 is reasonable (Figure 8.5). While for a given number of individuals, the exact variance created by demograph-

[22]Here it is important to assure that the entries in the population vector represent numbers of individuals, not densities. See *Simulating extinction probabilities* in Chapter 7.

ic stochasticity will depend on the mean vital rate, with 20 or more individuals, demographic variability in survival (or anther binomial vital rate) is always very small. However, if there are more than two alternative fates, considerably higher population sizes may still allow substantial demographic stochasticity in at least some of the rarer possible fates. A fairly safe rule is to perform Monte Carlo simulations for a particular class or the entire population when its numbers fall below 50.

Techniques differ as to how to perform Monte Carlo simulations for vital rates that are probabilities, such as survival and growth, and ones that are not, such as fertilities. For survival and growth rates, Monte Carlo simulations can be performed by picking a uniform random number and comparing its value to the probabilities of different fates an individual might experience. This can also be done after multiplying all survival and growth rates together to get the matrix elements determining the probabilities of all fates (including death, which is one minus the summed probabilities of any other fate), which represent a set of multinomial probabilities. For example, in any year of a simulation, that year's mean values of survival and growth rates would be picked. Next, these would be multiplied together to calculate the matrix elements. Then, class by class, the fate of each individual would be determined using these matrix elements. For example, for an individual mountain golden heather plant in size class 4, there are five fates: shrinking, remaining the same size, growing to size class 5, growing to size class 6, and dying, the probabilities of which are given by $a_{3,4}$, $a_{4,4}$, $a_{5,4}$, $a_{6,4}$ and $1 - (a_{3,4} + a_{4,4} + a_{5,4} + a_{6,4})$, respectively. To choose the fate of one plant, a single random number is drawn, and then compared with these five possible outcomes, using the probabilities of each to divide a number line from 0 to 1 into the five different outcomes (Figure 8.6). This time-consuming procedure can be made somewhat faster in MATLAB by picking random numbers in sets (for all the individuals in a class in one year) and comparing them *en mass* to the appropriate probabilities. A function from Caswell (2001) to perform this mass comparison in shown in Box 8.11.[23]

Adding demographic stochasticity to reproduction is not as clear-cut. First of all, the "reproduction" elements on the top row of a population matrix include both fertility and survival rates. If these are estimated and simulated separately, then we can apply the procedure just described to the survival part of these elements. For example, if a matrix is constructed assuming a post-breeding census (see Figure 6.3), then a reproductive

[23]Note, however, that Caswell is mistaken when he claims that this way of simulating demographic stochasticity results in drastic savings in the number of random variables that must be picked, and that these are independent of the number of individuals (Caswell 2001, pp 459–459). A quick look at Box 8.9 will show that a number must still be picked for each individual simulated.

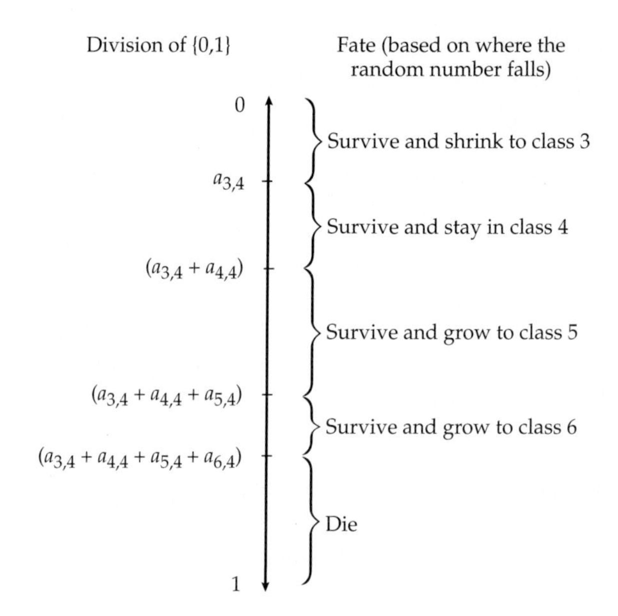

Figure 8.6 How to use a uniform random number to decide between fates in a Monte Carlo simulation of demographic stachasticity. The probabilities of several alternative fates are used to divide the number line between zero and one into different zones corresponding to different fates. The value of a single uniform random number is then used to decide which fate an individual experiences. The fates used here are for a class 4 *H. montana* individual.

element is the probability that a female survives for one year times her fertility if she is alive at the next census. In this case, we can first use the results of the Monte Carlo simulations described in the last paragraph to determine if each individual lives. Then we need to simulate the realized fertility of each survivor. With data on individual fertility in one or more years, we can estimate the individual variance around each year's mean. This can be used to simulate individual fertilities, which will be best modeled in different ways depending on how fertility is defined. If data are available for the number of newborns produced, then individual fertility is a discrete variable, and it can be modeled either as a multinomial variable (with the classes defined by the numbers of offspring produced), or by the Poisson or several other discrete distributions.[24] If fertilities include some element of survival (such as production of fledglings, for example), then using a lognormal or stretched beta distribution may be more appropriate.

In some cases you can use the properties of sums of individual random fates to determine the distribution of the summed fertilities of all members of a whole class, allowing you to pick just one random variable for fertility,

[24]The Poisson distribution can be very useful in some circumstances. It describes a random variable that can only take on discrete values and has only one parameter (the mean equaling the variance). It is more appropriate than a continuous distribution such as the lognormal when reproductive rates are low. However, its properties arise from the independence assumption (in this case, that births are independent of one another) and it is probably less generally useful than the multinomial.

rather than one for each individual, but this short-cut only works for some distributions. (See Ratner et al. 1997, Caswell 2001:458–459 for some of these relationships.) Also, before using any of these methods, you should look carefully at the data to see if individual variation in fertilities is dependent upon the annual mean fertility. For example, it may well be that very good or very bad years for reproduction have low variance between individuals, whereas intermediate years show high variation. If this is the case, the variance of the distribution of individual fertilities should be modelled as a function of the mean fertility, thus adjusting each year's variance in conjunction with changing mean values.

In the following section on density dependence we present a simulation that also includes demographic stochasticity.

Including Density Dependence in Matrix Models

The absence of density dependence is the most glaring omission in what we have discussed about demographic PVAs up to this point. There is no question that density dependence can strongly impact the viability of a structured population, just as including it can radically change the predictions of count-based PVAs. In principle, we can include density dependence in a demographic PVA by making the vital rates respond to population density (or size), just as we can make the population growth rate change with density in count-based PVAs (Chapter 4). However, two factors make it more difficult to account for density dependence in a demographic PVA than in a count-based PVA. The first is that we will rarely have as many years of vital-rate estimates from a demographic study as we will have counts from a typical long-term population census. In an ideal world, we would base demographic PVAs on as many years of data as we would a typical count-based PVA, so that we could do as good a job of estimating both the magnitude of environmental stochasticity and the shape of density dependence. But the reality is that the cost of marking and following multiple individuals to measure their vital rates makes long-term demographic studies much rarer than long-term population censuses. With fewer years of data spanning a smaller range of densities, it will be more difficult to detect density dependence and, if it is detected, to accurately estimate the shape of any density-dependent relationships.

The second factor that complicates the inclusion of density dependence in a demographic PVA is the fact that there are many more variables that are potentially density-dependent than there are in a count-based PVA. In the later, all we have to do to incorporate density dependence is to make the population growth rate a function of density. In contrast, any or all of the vital rates of a structured population could be density-dependent. In fact, when we contemplate including density dependence in a demographic PVA, there are three questions (at a minimum) that we must consider:

1. Which vital rates are density-dependent?
2. How do those rates change with density?
3. Which classes contribute to the density that each vital rate "feels"?

Thus in addition to considering which rates increase or decrease with density and in what way, we must also decide whether, for example, only the density of juveniles affects juvenile survival (as might make sense if juveniles use different resources than adults and so compete only among themselves), or if the combined density of both juveniles and adults influences juvenile survival. The number of possible combinations of the answers to these three questions generate complications for density-dependent demographic PVAs well beyond those created by the typically short-term nature of the data.

Given the manifold ways that density dependence could act on the growth of a structured population and the limited information that will typically be available to estimate its effects, we can rarely (if ever) adopt a fully scaled-up version of the careful model-fitting approach we took in building density-dependent count-based PVAs (Chapter 4). That is, we will not usually be in a position to compare multiple density-dependent functions fit to each of the vital rates, using many different combinations of the classes to measure density. However, there are two more limited approaches to building density-dependent projection matrix models that are often more practical. The first is to assume that there is a maximum number or density of individuals in one or more classes, or of the population as a whole, that the available resources can support, and to construct a simulation model that prevents the population vector from exceeding this limit. This is the structured-population equivalent of the ceiling model we discussed in Chapter 4 (see Equation 4.1). The second approach is to choose one or at most a few vital rates that, on the basis of direct data, limited experiments, natural history knowledge, or comparison with similar taxa, are suspected to be strongly density-dependent. We then model these particular rates with density-dependent functions, while treating all other vital rates as density-independent. We now illustrate both of these approaches to constructing a density-dependent projection matrix model.

Placing a Limit on the Size of One or More Classes

Estimating a reasonable cap on total population size or density can sometimes be done using historical records of the largest number of individuals ever observed in a given area. However, this solution may result in unrealistically high population caps if the habitat has undergone recent deterioration. Caps or ceilings on population density are most often used to introduce density dependence when the focal species is territorial. In this case, information on the average size of territories for the focal species or related taxa is frequently available and can be used to set a cap on the number of territory-holding individuals, which are often the breeding adults in a population. Dividing the total area of habitat available to the population by the average territory size yields an estimate of the maximum number of breeding adults the area can support. Placing this cap on the breeding classes effectively makes the rate of transition from the pre-reproductive to the reproductive classes density-dependent. Even if no other vital rates or classes experience density dependence, capping the number of breeding adults at the number of available territories will restrict the maximum size the population as a whole can reach.

An example of a territorial species for which this approach has been used is the Iberian lynx, *Lynx pardinus*. Comprising fewer than 1,000 individuals in total and restricted to nine isolated populations in Spain and Portugal, this species has been labeled the most vulnerable cat species in the world (Nowell and Jackson 1996). Gaona et al. (1998) built a density-dependent projection matrix model for the lynx population in Doñana National Park in Spain by placing a cap on the number of breeding adults. Here, we present a simplified version of their model to illustrate how the maximum number of territories affects the probability of quasi-extinction. The Doñana population comprises 50 to 60 individuals dispersed across a set of semi-isolated patches of suitable habitat, the largest of which can support five territories. The model Gaona et al. constructed was based on a projection matrix that included 13 age classes (including post-reproductive classes); it tracked both males and females with and without territories (so that reproduction depended on the ratio of females to males); and it explicitly represented dispersal of individuals among habitat patches. Because the total number of individuals is small, Gaona and colleagues included demographic stochasticity in the model, and they assumed (on the basis of little or no data) that periodic droughts would occasionally depress the survival of the youngest age class only.

We will simplify Gaona et al.'s model to represent a post-breeding census of a birth-pulse population in which we will track four stages of females only: (1) cubs (born in the pulse just before the census); (2) juveniles (at the end of their first year of life); (3) floaters (females old enough to breed but not in possession of a territory); and (4) breeders (territory-holding adult females). However, we follow Gaona et al. by including demographic stochasticity in survival of all classes and environmental stochasticity in class 1 survival. More importantly for our present purposes, we retain their assumption that the number of breeding adults is limited by the maximum number of territories in a habitat patch. Therefore, a floater will only advance to the breeder class if a previously breeding female vacates a territory.[25] Our projection matrix has the following structure

[25]Gaona et al. (1998) assumed that females older than 9 years of age were permanently evicted from their territories by younger females. Because they contribute neither to population growth nor to the strength of density dependence experienced by the other classes, we omit these post-reproductive females from our model, but we decrease slightly the estimated "survival" of breeding females to account for this loss to an unrepresented post-reproductive stage.

$$\mathbf{A} = \begin{bmatrix} 0 & 0 & 0 & b \times c \times p \times s_4 \\ s_1 & 0 & 0 & 0 \\ 0 & s_2(1-g) & s_3(1-g) & 0 \\ 0 & s_2 g & s_3 g & s_4 \end{bmatrix} \quad (8.17)$$

where s_i is the survival rate of individuals in class i, b is the probability that a territory-holding female will breed in a given year, c is the number of cubs produced by females that do breed, and p is the proportion of cubs that are female (assumed to be 0.5).[26] Density dependence acts on g, which represents the probability that a surviving juvenile or floater will acquire a territory next year. Specifically, let K represent the maximum number of territories. Just before the birth pulse that precedes census $t + 1$, there will be $s_4 n_4(t)$ breeding females still in possession of a territory and $K - s_4 n_4(t)$ vacant territories available for floaters or juveniles to occupy. The number of surviving floaters and juveniles[27] just before the birth pulse is $s_3 n_3(t) + s_2 n_2(t)$, so the probability that a particular one of these survivors acquires a territory to become a member of class 4 at census $t + 1$ is $g = [K - s_4 n_4(t)]/[s_3 n_3(t) + s_2 n_2(t)]$. A fraction $1 - g$ of the surviving floaters from census t remain floaters at census $t + 1$ and the same function of surviving juveniles become floaters. Notice that g declines with increasing numbers of individuals in classes 2, 3 and 4, i.e., it is negatively density-dependent. All we need to do to set the strength of this density dependence is to specify the maximum number of territories, K. A MATLAB program to perform stochastic iterations of the model using the matrix in Equation 8.17 is shown in Box 8.12. Although the preceding discussion treated survival as though it were deterministic, in the actual model we introduce demographic stochasticity into survival by Monte Carlo simulation, and we add environmental variability to s_1 which equals 0.5 in most years, but is set to 0.2 in drought years, which occur with a 10% probability (following Gaona et al.).

Not surprisingly, the maximum number of territories strongly affects the probability that an initial population of 4 cubs, 1 juvenile, no floaters, and 5 breeding females (the numbers Gaona et al. observed in the largest habitat patch) will drop to only 2 females[28] in 50 years (Figure 8.7). At 5 territo-

---

[26]We assume that a floater does not have a chance to breed in the same year that she gains a territory; hence the element $a_{1,3} = 0$.

[27]In our model, juveniles can acquire territories without spending a year as floaters. Although juveniles have lower dominance status than floaters, and dominance determines the order of territory acquisition (Gaona et al. 1998), this ability to quickly move into breeding status is important for the tiny populations of this species.

[28]We use such a low quasi-extinction threshold to emphasize the effects of demographic stochasticity. However, for a model such as this one, which ignores genetics and mate-finding, a higher threshold is much preferable in practice.

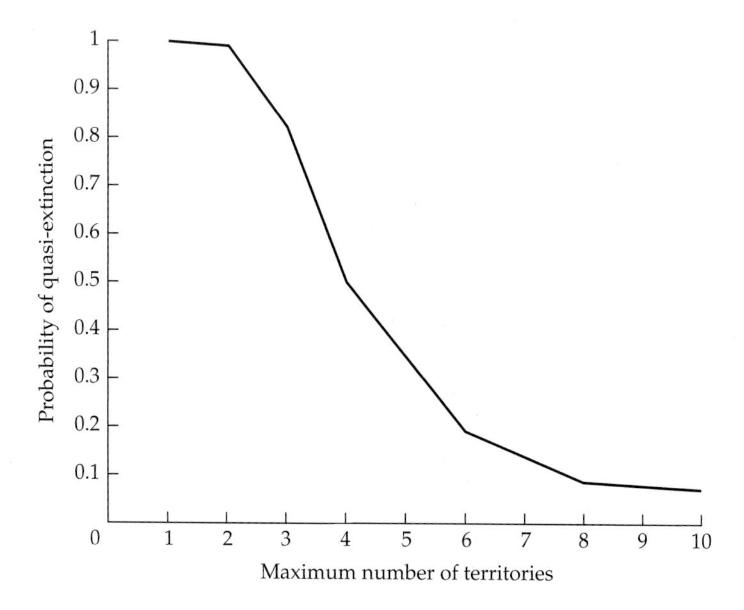

Figure 8.7 Quasi-extinction probabilities at 50 years for the Iberian lynx as a function of the maximum number of territories the habitat can support. See Box 8.12 for simulation code.

ries, the actual number in the largest habitat patch, the probability of quasi-extinction exceeds 30% by year 50; a patch with only 1 or 2 territories would face virtually certain quasi-extinction. In contrast, patches even slightly larger than the biggest existing habitat area are predicted to have a fairly low probabilities of hitting the threshold in 50 years.[29]

A low number of territories results on high extinction risk because the total number of individuals in the population (and in the breeding class especially) can never escape the zone in which demographic stochasticity is strong; this effect can lead to a large proportion of the individuals dying by chance in a single year. This effect of density dependence is further exacerbated by environmental stochasticity, which repeatedly pushes numbers even lower than the density cap itself. This example illustrates how simple information ancillary to the demographic study, in this case the average territory size, can allow us to include a simple form of density dependence in a demographic PVA.

---

[29]Note that in this discussion we are assuming that patches are completely isolated—unlike in the original model of Gaona et al. If we lumped all patches into a single population, applied the survival and fertility rates from the largest patch to all of them, and ignored dispersal mortality, the entire Doñana population with 16 territories would have a quasi-extinction probability of less than 10 percent over 50 years. In Chapter 11, we will take up the issue of how to include dispersal among habitat patches into models such as this one.

BOX 8.12 *MATLAB code to perform a density-dependent demographic PVA for the Iberian lynx.*

```
% PROGRAM lynx
%   Performs a density-dependent demographic PVA for the Iberian
%   lynx, with demographic stochasticity (survival of all
%   classes), environmental stochasticity (class 1 survival
%   only), and a maximum number of breeders set by the number of
%   territories, K.
%   Classes are:
%   (1) cubs;
%   (2) subadults;
%   (3) floaters (adults without territories);
%   (4) breeders (territory-holding adults).
%   Assumes a post-breeding census, and tracks females only.
%   Parameters modified from Gaona et al., 1998

% ************  USER-DEFINED PARAMETERS  ************************
s1min=0.2;       % class 1 survival in bad years
ds1=0.3;         % class 1 surv. in good years=s1min+ds1
freqbad=0.1;     % frequency of bad years
s2=0.7;          % class 2 survival
s3=0.7;          % class 3 survival
s4=0.86;         % class 4 survival
f4=0.5*0.6*2.9;  % reproduction of breeders=sex ratio *
                 %  breeding propensity * mean litter size
Ks=[1 2 3 4 6 8 10];  % max. number of territories
imax=length(Ks);
n0=[4 2 0 5];    % starting population vector
Nx=2;            % quasi-extinction threshold
tmax=50;         % time horizon
numreps=1000;    % no. of replicates for prob.(ext.) calculation
% ************************************************************

rand('state',sum(100*clock)); % seed random number generator

for i=1:imax        % For each possible no. of territories,
    K=Ks(i);
    Ns=[];
    for rep=1:numreps       % for each replicate population,
        n=n0;               % starting at the initial vector,
        for t=1:tmax        % for each year
```

BOX 8.12 *(continued)*

```
            s1=s1min+ds1*(rand>freqbad); % draw class 1 survival
                                         % at random given
                                         % environmental
                                         % stochasticity, then
            n(2)=sum(rand(1,n(1))<s1);   % determine no. of
            n(3)=sum(rand(1,n(2))<s2)... % survivors in all
                + sum(rand(1,n(3))<s3);  % classes by Monte Carlo
            n(4)=sum(rand(1,n(4))<s4);   % simulation.
            n(1)=round(f4*n(4));         % Reproduction
            newbreeders=min([n(3) K-n(4)]); % New breeders take
                                     % empty territories, if any,
            n(3)=n(3)-newbreeders;   % leaving class 3
            n(4)=n(4)+newbreeders;   % and joining class 4.
            N=sum(n);                % Sum total population size &
            if N<=Nx break; end;     % stop if threshold is hit.
        end; % for t
        Ns=[Ns N];                   % Store final population size.
    end; % for rep
    probext(i)=sum(Ns<=Nx)/numreps; % Calculate quasi-ext. prob.
end; % for i

plot(Ks,probext);  % Plot quasi-ext. prob. vs. no. of territories
axis([0 max(Ks) 0 max(probext)])
xlabel('Maximum number of territories')
ylabel('Probability of quasi-extinction')
```

Estimating and Incorporating Density-Dependent Functions for Particular Vital Rates into a Demographic PVA

The second practical approach to incorporating density dependence begins by hypothesizing that one or more vital rates critically depend on the number of individuals in one or more particular classes. If we have estimates of population densities at the times when the vital rates were measured, we can use regression to fit a function describing how each vital rate changes with density. Statistical analysis of the regression slope also provides a (weak) test for density dependence. Substituting these density-dependent functions into the projection matrix then allows predictions of future population sizes that include at least some effects of density.

To illustrate this approach, we'll use the survival of salmon from the egg stage to the first year of life. There are two reasons to suspect that first-year salmon survival may depend on the initial number of eggs in a spawning stream (and thus indirectly on the number of spawning adults). First, if the number of suitable sites for nests (called redds) is limited, late-arriving spawners will often destroy the redds of earlier spawners when creating their own (Ratner et al. 1997). High densities of eggs due to overlapping redds can also result in oxygen depletion and hence increased egg mortality. Thus we would expect that the more eggs are laid in a stream bed, the smaller the proportion of those eggs that will be in undisturbed redds, and thus able to hatch successfully. Second, the more eggs that do hatch, the greater the competition for food among the hatchlings is likely to be, and food limitation may depress first-year survival. If our demographic study provides estimates of both survival from egg to age 1 and the initial number of eggs laid in a stream (which could be estimated by multiplying the number of female spawners entering the stream by the average number of eggs laid per female), we can use regression to see if survival is related to the initial number of eggs.

Two density-dependent functions have commonly been used as the basis for such regressions.[30] One is the Ricker function we encountered in Chapter 4:

$$s_0[E(t)] = s_0(0)\exp[-\beta E(t)] \tag{8.18}$$

where $s_0[E(t)]$ represents the survival of eggs to become one-year old fish as a function of the initial density of eggs at time t, $E(t)$, $s_0(0)$ is the survival rate when the number of eggs is close to zero, and β describes the decline in survival as egg number increases. The easiest way to fit Equation 8.18 to data on egg survival versus number of eggs is to perform a linear regression of log survival against egg number.[31] The slope of the regression is $-\beta$ and the intercept is $\log[s_0(0)]$. A regression slope that differs significantly from zero is evidence that survival is density-dependent. The second function commonly used to represent density dependence comes from the Beverton-Holt model:

$$s_0[E(t)] = \frac{s_0(0)}{1+\beta E(t)} \tag{8.19}$$

Equation 8.19 can also be fit by linear regression, but in this case using the inverse of survival as the dependent variable and $E(t)$ as the independent variable. The inverse of the intercept is an estimate of $s_0(0)$, and the slope divided by the intercept is an estimate of β.

[30]Given that our current example focuses on salmon, it is interesting to note that both models were initially developed by fisheries modelers but have subsequently been applied to many other species.

[31]Equations 8.18 and 8.19 can be fit using linear regression only if all survival rate estimates are greater than zero; if not, nonlinear regression must be used.

How do we decide whether to use the Ricker function, the Beverton-Holt function, or some other function (perhaps with more than two parameters, such as the theta logistic function in Equation 4.2) to represent a density-dependent vital rate? As we did in Chapter 4, we could use the corrected Akaike Information Criterion, AIC_c, to identify the best function. But in practice, if the duration of the demographic study is short and therefore few vital rate estimates are available, even AIC_c will be unable to usefully distinguish among potential density-dependent functions (or even between density dependence and density independence). In this situation, there is little point in trying to fit functions with more than two parameters (the minimum needed to represent density dependence). One way to choose between two-parameter functions such as the Ricker and Beverton-Holt is to consider how the predicted number of survivors differs between the two functions when the initial number of eggs is high. The number of survivors is simply the survival rate (given by Equations 8.18 and 8.19) multiplied by the initial number of eggs. The Beverton-Holt function predicts that the number of survivors will reach an asymptote as the initial number increases (Figure 8.8); this is so-called compensatory density dependence in which each increase in egg number is exactly compensated by an increase in the number of eggs dying once the asymptote has been reached. In contrast, the Ricker function predicts that the number of survivors will increase and then decrease as initial egg number increases without ever reaching a stable number (Figure 8.8), a pattern known as overcompensation.

In our salmon example, density dependence would be compensatory if there is a maximum number of suitable sites for redds in a stream, if a redd that displaces an earlier one produces just as many hatchlings on average as a redd constructed in a virgin site, and if displacement of earlier redds is

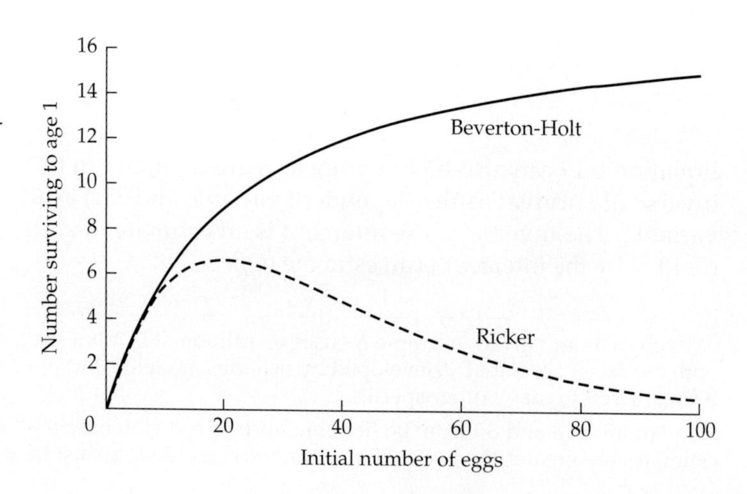

Figure 8.8 Density-dependent survival as represented by the Ricker (Equation 8.18) and Beverton-Holt (Equation 8.19) functions. Here, $s_0(0) = 0.9$ and $\beta = 0.05$ in both functions.

the only source of density dependence in egg survival. Under this scenario, there is a maximum number of redds that can be packed into a stream bed, and once this number is reached, the number of surviving eggs will be constant. For this reason, the Beverton-Holt function is often used to represent density-dependent reproduction in territorial species, as a limit to the number of territories has the same effect as a limit to the number of redds in our salmon example. In contrast, if density dependence is caused by oxygen depletion or competition for food among hatchlings, high egg numbers mean that most eggs die of anoxia or that all hatchlings obtain a suboptimal amount of food. In this case, the number of survivors might actually decline as egg number increases; that is, density dependence may be overcompensatory.[32] Thus we can use knowledge of the likely sources of density dependence to help us to decide which density-dependent function to use. Of course, if we have estimates of the survival rate when the initial egg number is high, we can look directly at whether survival declines at high numbers, and choosing between these two functions will be easy. Unfortunately, if all our data come from years with low densities (as is typical of endangered species), it may be quite difficult to distinguish visually between the two functions (Figure 8.8), leaving us no choice but to choose between them on biological grounds alone.

The final step in this process is to substitute the density-dependent function into the correct places in the projection matrix. Let us finish our salmon example by constructing such a density-dependent projection matrix. We will build a model that is a variant of one analyzed by Levin and Goodyear (1980). Our matrix also closely resembles one used in a PVA performed by Ratner et al. (1997) on a declining population of chinook salmon (*Oncorhynchus tshawytscha*) in the South Umpqua River in Oregon, although they relied on long-term spawner counts to estimate density-dependence parameters rather than regressing vital rate estimates against numbers, as we discussed above. Our matrix will assume a pre-spawning census, classes representing females of ages 1 through 5, and that a female can choose to spawn at age 3, 4, or 5 (following Ratner et al.). As salmon are semelparous, all individuals die after spawning. Thus all females that reach age 5 must spawn and then die. Let b_i represent the probability that a female of age i chooses to spawn, f_i the average number of eggs produced by spawners of age i, and s_i the survival rate of females of age i. Finally, let

$$E(t) = \sum_{i=3}^{5} b_i f_i n_i(t) \tag{8.20}$$

[32]If at high density a fixed number of hatchlings obtain adequate food regardless of the initial number of hatchlings, the Beverton-Holt function would again be more appropriate than the Ricker.

represent the total number of eggs produced by all spawning females during the birth pulse that occurs immediately after census t. A projection matrix that describes this population is

$$\mathbf{A} = \begin{bmatrix} 0 & 0 & b_3 f_3 s_0[E(t)] & b_4 f_4 s_0[E(t)] & f_5 s_0[E(t)] \\ s_1 & 0 & 0 & 0 & 0 \\ 0 & s_2 & 0 & 0 & 0 \\ 0 & 0 & s_3(1-b_3) & 0 & 0 \\ 0 & 0 & 0 & s_4(1-b_4) & 0 \end{bmatrix} \quad (8.21)$$

There are two important points to note about this matrix. First, the transition probabilities of age classes 3 and 4 to the next-older age classes are reduced by the probability of breeding, as all females that choose to spawn must die. Second, the density-dependent function relating the egg-to-age-1survival rate to the initial egg number, $s_0[E(t)]$, appears in all of the reproduction elements in the matrix. Because this matrix models a pre-spawning census, the number of eggs produced by females in each age class must be multiplied by the density-dependent egg survival rate to accurately predict the resulting number of age 1 fish at the next census (see Figure 6.3A). Since either version of $s_0[E(t)]$ (i.e., Equation 8.18 or 8.19) predicts that egg survival declines as egg number increases, and as egg number is positively related to spawner numbers according to Equation 8.20, per-capita production of age 1 fish in the projection matrix in Equation 8.21 is a negatively density-dependent function of the number of fish in the spawning age classes.

Despite the fact that they may predict quite similar numbers of survivors when egg density is low, the Ricker and Beverton-Holt functions can produce dramatically different population dynamics when substituted into the projection matrix. We used the parameter estimates from Ratner et al. (1997) to simulate a salmon population described by the matrix in Equation 8.21, including density dependence in egg survival following either the Ricker or the Beverton-Holt functions, but including no demographic or environmental stochasticity (Box 8.13). We then increased the estimated fertilities of the potentially spawning age classes by successive factors of two. With egg survival described by the Beverton-Holt function, the spawner classes reach an equilibrium regardless of the magnitudes of the fertilities (Figure 8.9). As very large numbers of eggs are produced, the number of surviving eggs approaches a constant value, and hence the numbers of returning spawners those eggs produce over the subsequent years is stable. In contrast, with the Ricker function governing egg survival, spawner numbers are stable when fertilities are low, but they become cyclic at higher fertilities and even begin to undergo chaotic fluctuations when fertilities are very high. Note that there are no stochastic forces whatsoever influencing the dynamics shown

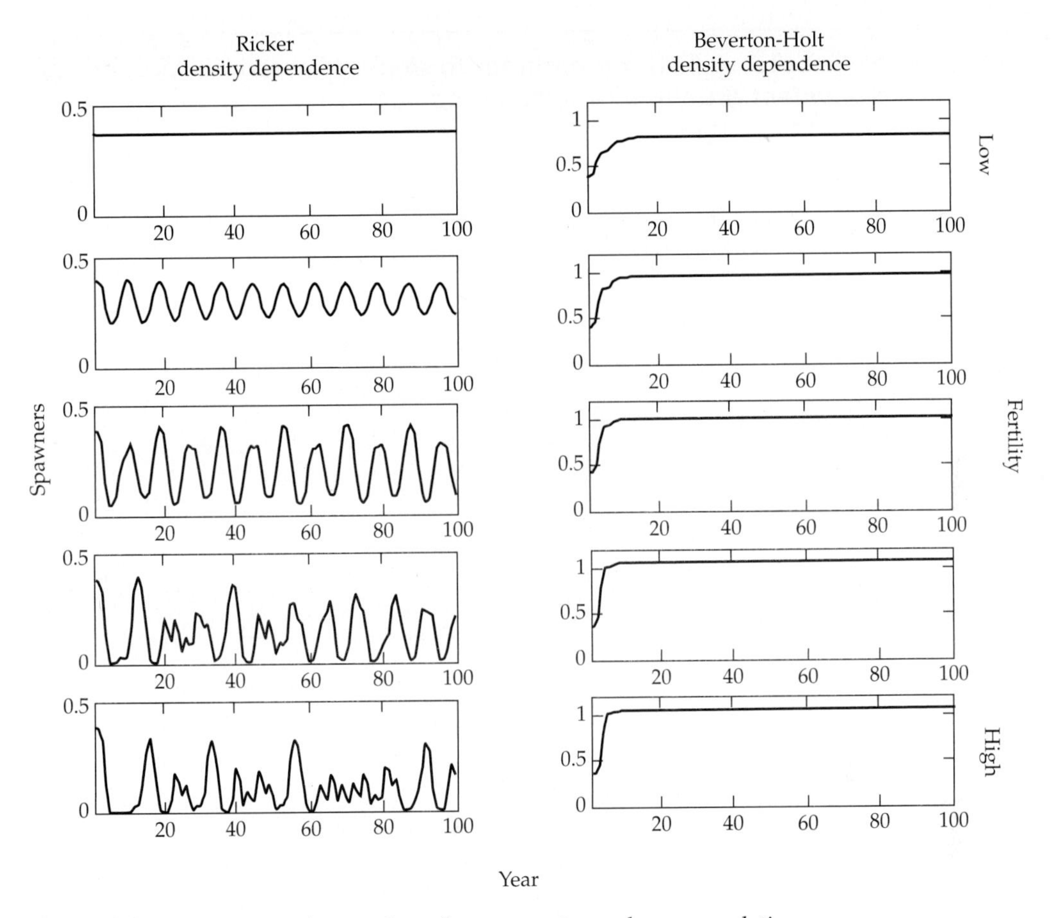

Figure 8.9 Dynamics in the number of spawners in a salmon population governed by a density-dependent projection matrix (Equation 8.21) in which survival from egg to age 1 is determined by the Ricker function (left panels) or the Beverton-Holt function (right). Successive rows of graphs represent a sequential doubling of the fertilities estimated by Ratner et al. (1997). There is no demographic or environmental stochasticity in this example.

in Figure 8.9. But when density dependence is overcompensatory, very complex population dynamics can result as high density leads to very low survival, followed by low density, which allows survival to improve, and so on in a pattern that repeats itself, but not quite exactly, indefinitely.

Figures 8.8 and 8.9 illustrate three important points. First, for either of the two density-dependent functions, population size is bounded, never exceeding a certain maximum level determined by the vital rates. Thus, unlike density-independent models of structured populations, the growth rate (λ_1 or

BOX 8.13 ***A MATLAB program to simulate deterministic, density-dependent growth of a salmon population.***

```
% PROGRAM salmon_dd
%   Uses parameters from Ratner et al. (Conservation Biology
%   11:879-889 (1997)) to perform a purely determinstic, but
%   density-dependent, simulation of salmon population growth

n0=[0.7667; 0.6163; 0.4945; 0.3524; 0.1309]; % initial population
                                              %    vector
s0=0.002267;    % maximum egg survival
beta=0.001;     % density-dependent egg survival parameter
s=[0.8 0.8 0.8 0.8];     % survival probabilities
b=[0 0 0.112 0.532 1];  % probability of breeding at each age
f=[0 0 3185 3940 4336]; % eggs per breeding female
tmax=100;        % time horizon
eggsurv=1;     % Set eggsurv=1 for Ricker egg survival,
               %    eggsurv=2 for Beverton-Holt egg survival

n=n0;            % Starting at initial pop. vector,
for t=1:tmax     % for each year
    eggs=(b.*f)*n;  % compute total eggs produced from Eq. 8.20,
    if eggsurv==1                      % compute surviving eggs
        n(1)=eggs*s0*exp(-beta*eggs); %    using Ricker or
    elseif eggsurv==2
        n(1)=(eggs*s0)/(1+beta*eggs); %    Beverton-Holt function
    end;
    n(2:5)=( s.*( 1-b(1:4) ) )'.*n(1:4); % and update older age
                                          %   classes.
    spawners(t)=b*n;  % Observed spawners=breeders of all ages
end;

plot(spawners)  % plot no. of spawners vs. year
xlabel('Year');
ylabel('Spawners');
```

log λ_s) cannot exceed 1 indefinitely. Moreover, there is no simple metric of stochastic (or nonstochastic) growth rate, since the matrix changes as a function of both overall density and the relative numbers in different classes. As with count-based models, a population restricted to finite numbers has an ultimate extinction probability of 1 when we add stochastic forces to the results shown in Figure 8.9 (although the time to reach extinction may be large). Second, overcompensatory density dependence can contribute a complex but deterministic component to population dynamics that is very difficult to distinguish from the effects of inherently unpredictable forces such as environmental stochasticity.[33] By causing the population to visit low numbers, such complex dynamics can also greatly increase a population's risk of extinction due to stochastic forces (Ripa and Lundberg 1996, Ripa and Heino 1999). Third, it can be very difficult to distinguish whether density-dependence is occuring, or which of several density-dependent functions best fits a limited set of data, yet those functions can lead to very different viability predictions. Given this, we will repeat a by-now familiar mantra: when the data are ambiguous about the form or existance of density-dependance, try all likely possibilities and see how much the results are affected.

Summary

In this chapter we've discussed various complications that can be added to a matrix model PVA. At the end of this litany of considerations and methods, it is important to emphasize that you don't have to include all (or any!) of these difficulties in a PVA. More often than not, you will only have enough data to cobble together a single mean matrix of results, without any data on environmental stochasticity, density dependence, individual variation in fertility, and so on. A simple model such as this can still be very useful in understanding population viability and especially the likely effects of management alternatives. Indeed, we spend much of the next chapter dealing with the analysis of a simple deterministic matrix model. However, if you do have more information, you should carefully consider adding complexities such as environmental variability that can strongly influence population growth and extinction risk. To leave these processes out of a model will always result in a poorer understanding of viability problems and what to do about them.

---

[33]We have not discussed adding environmental stochasticity directly to a vital rate that is also density-dependent. However, with data to do so, this is not difficult. For example, Equation 8.18 can be modified to $s_0[E(t)] = s_0(0)\exp[-\beta E(t) + \varepsilon(t)]$, where $\varepsilon(t)$ is a normally distributed with mean zero and a variance that reflects environmental stochasticity.

9

Using Demographic Models in Management: Sensitivity Analysis

Although projection matrix models, like count-based PVAs, provide estimates of population growth and extinction risk, they differ in allowing analyses that can suggest different management possibilities and evaluate the effectiveness of these alternatives. In essence, any management for improved population viability is an effort to change the mean and variance of population growth (and hence extinction risk). Specific interventions, such as control of competitive invasive species or creation of nesting habitat, almost always target a subset of the life stages of a species and even particular vital rates for that subset (e.g., survival of adults as opposed to their reproduction), hoping to influence overall population performance by changing these particular parts of the life cycle. A frequent complication is that management interventions can have both positive and negative effects on population viability through effects on different vital rates (e.g., fires will kill adult trees but also promote recruitment of seedlings). Since a projection matrix integrates specific vital rates to yield an estimate of population growth, it provides a natural way to ask how effective different management efforts are likely to be in changing population growth, *before* going to the time and expense of initiating them. This kind of analysis can also be used to assess the importance of uncertainty in the estimates of different vital rates for predictions about population viability. If population growth is highly influenced by the exact value of, say, adult survival, then having an accurate estimate of this rate is key for proper viability assessment.

The use of matrix models to assess management options depends on the use of sensitivity analysis, a suite of methods to ask, "How sensitive is population growth (or extinction risk or other measures of population behavior) to particular demographic changes?" In this chapter, we'll first present the basic idea of sensitivity analysis. This section builds on the definition of eigenvalue sensitivities from Chapter 7, but we review the important concepts here. Next, we discuss the different ways to perform sensitivity analysis for deterministic matrix models. We then use these methods as the basis upon which to build the more complicated sensitivity analyses needed for stochastic models, including estimation of sensitivities for both stochastic growth rates and extinction probabilities. The chapter ends with a discussion of sensitivity analysis for density-dependent models.

The Basic Idea of Sensitivity Analysis

What sensitivity analysis is and how to use it does not depend on the exact calculations used, each of which have their own advantages and limitations. A huge array of mathematical methods can be used to address the importance for population viability of changes in a particular matrix element or vital rate. All these methods can be termed *sensitivity analyses* because they seek to answer how sensitive, or changeable, one variable is to changes in another. In most of what we present below, we will ask how sensitive population growth is to changes in vital rates. But, as we'll see, the general idea of sensitivity analysis can also be applied to extinction risk and other measures of viability and population performance.

The idea of sensitivity measurement is easiest to understand graphically, and also easiest to learn about with a fairly simple matrix model. We will use a population matrix for the emperor goose population that breeds in the Yukon-Kuskokwim delta of Alaska (Schmutz et al. 1997). This population declined precipitously in the 1960s and has been more or less stable since that time. Although sport hunting of the emperor goose was halted in 1986, subsistence hunting by Alaskan natives still occurs, and one of the primary management questions is how important this harvest has been, or may in the future be, in preventing the population from returning to its former numbers. Schmutz et al. (1997) provide several estimates of the mean vital rates for the population, but there are no estimates of temporal variability, so only a deterministic matrix can be constructed. This matrix represents a post-breeding census and projects the population from fall to fall, with the youngest age class (class 0) the newly fledged young of the year. The matrix has four age-based classes (0-, 1-, and 2-year-olds, and birds three years of age or older; compare the very similar matrix for the spotted sandpiper in Equation 7.5) and depends on only four vital rates: the annual survival of juvenile birds in their first year of independence (s_0), the survival rate of all older birds ($s_{\geq 1}$), reproduction of two year olds (number of female fledglings, f_2) and reproduction of all older birds ($f_{\geq 3}$) (Table 9.1):

TABLE 9.1 Estimated current values, sensitivities, and elasticities for vital rates of the emperor goose[a]

Vital rate	*Estimated value*	*Sensitivity*	*Elasticity*
s_0, survival of zero-year-olds[b]	0.1357	0.58	0.080
$s_{\geq 1}$, survival of one-year-old and older birds	0.8926	1.02	0.92
f_2, reproduction of two-year-olds[c]	0.6388	0.0088	0.0057
$f_{\geq 3}$, reproduction of three-year-old and older birds	0.8943	0.082	0.074

[a]The vital rate estimates used here come from Schmutz et al. (1997) Table 1, Survey Set 2. However, Schmutz et al. added several estimation and modeling complexities to their models that we do not include here, limiting specific comparisons with their results.

[b]Schmutz et al. break both reproductive and survival rates into several distinct rates (e.g., winter, spring, and fall survival rates) that we amalgamate here. In addition, for some parameter sets, they provide separate survival estimates for one-year-olds versus older geese.

[c]Although Schmutz et al. use a simulation model that accounts for both males and females, we follow only females, and hence these reproduction estimates use half the clutch size reported in their paper, assuming an equal sex ratio.

$$\mathbf{A} = \begin{pmatrix} 0 & 0 & f_2 s_{\geq 1} & f_{\geq 3} s_{\geq 1} \\ s_0 & 0 & 0 & 0 \\ 0 & s_{\geq 1} & 0 & 0 \\ 0 & 0 & s_{\geq 1} & s_{\geq 1} \end{pmatrix} = \begin{pmatrix} 0 & 0 & 0.570 & 0.798 \\ 0.136 & 0 & 0 & 0 \\ 0 & 0.893 & 0 & 0 \\ 0 & 0 & 0.893 & 0.893 \end{pmatrix} \tag{9.1}$$

For this matrix, $\lambda_1 = 0.9887$, suggesting slow decline in the population (not too far off from the recent stability or slight increase in emperor geese numbers; Schmutz et al. 1997). To see how "sensitive" population growth is to changes in a particular vital rate, say, survival of older birds, we could systematically vary $s_{\geq 1}$ from values substantially lower than its current estimate up to values much higher than this estimate. For each value, we make a new matrix (leaving all the other vital rates constant) and recalculate λ_1 (for example, using the MATLAB function `eigenall` in Box 7.1). Plotting the pairs of ($s_{\geq 1}$, λ_1) values then provides a way to see how rapidly population growth changes with changing values of adult survival. Going through this process for each vital rate in turn allows a comparison of the effects of each rate on λ_1 (Figure 9.1). The sensitivity of λ_1 to a vital rate is shown by the slope of the curve relating λ_1 to that rate. As the high average slope for λ_1 versus adult survival shows, a change in this vital rate will have a much larger effect on λ_1

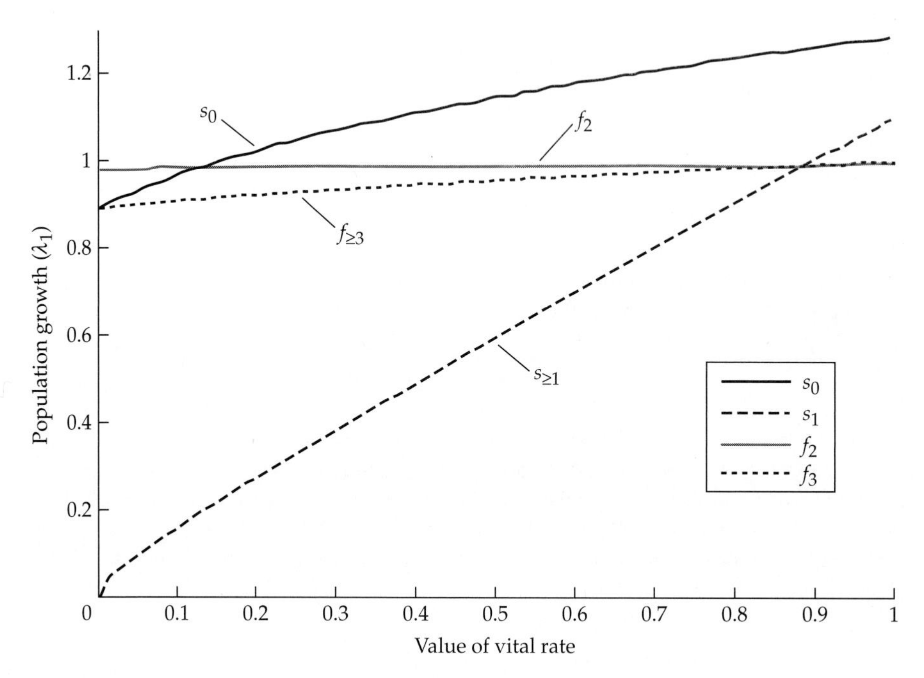

Figure 9.1 The effects of changing vital rates on population growth rate (λ_1) for the emperor goose. Labels indicate the vital rate corresponding to each line.

than will a change of the same magnitude in any of the other three vital rates. The survival rate of juveniles (s_0) has the second highest sensitivity, while, at least over the range of values we plotted, the sensitivity of population growth to either reproductive rate is very low, with almost no change in population growth even with large changes in reproduction.

These results allow two basic kinds of conclusions for management. First, if the survival values, especially adult survival, can be increased, this will result in much greater enhancement of λ_1 than would similar increases in reproduction. Second, our predictions of current population growth hinge critically on the estimate of adult survival. If this estimate is even somewhat inaccurate, growth could be much higher or lower than we think. For example, if adult survival is only about 10% higher than Schmutz et al. estimated it to be, then λ_1 would be greater than 1, and the population would be predicted to grow rather than decline.

A graph such as Figure 9.1, with curves showing the relationship of each vital rate to λ_1, gives the best picture of sensitivity patterns. However, it is much more convenient to use a single number to summarize the sensitivity of λ_1 to each rate. To do this, the most common basic measure of sensitivity is the slope of the tangent to the curve describing the population growth rate as a function of the vital rate, evaluated at the current value of the vital rate (Table 9.1; Figure 7.2 illustrates what this means for sensitivity to a matrix element, and the same relationship applies for sensitivity to a vital rate). Said differently, the sensitivity of λ_1 to changes in a particular vital rate r_i is S_{r_i}, which equals the partial derivative of λ_1 with respect to r_i, or $\partial\lambda_1/\partial r_i$, evaluated at the estimated current value of r_i.

The basic idea of sensitivity analysis is simple, but there are a variety of complications that make doing and interpreting sensitivity analyses a little less straightforward. Two are worth mentioning before we begin to delve into the different ways to conduct sensitivity analyses. First, nonlinearity in the relationship between λ_1 and a vital rate makes it less accurate to summarize sensitivity with a single number, as is the usual practice. This is particularly a concern when doing a sensitivity analysis in order to infer the effects of management, which hopefully will result in big changes in vital rates and hence in λ_1. For example, if we took the slope of the tangent lines at the current values in Figure 9.1 as the sensitivities of λ_1 to adult and juvenile survival, we get 1.02 and 0.58, respectively. Now, adult survival is already reasonably high, so we might want to manage for increased survival of juveniles, which is currently very low (0.136). With intensive management, we might estimate that we could raise s_0 to 0.70. From just a point estimate of the sensitivity of λ_1 taken at the current value of s_0, we would estimate that this change would result in a new λ_1 of about 1.32: the current value (0.989) plus the change in s_0 (0.70 – 0.136) times the sensitivity (0.58). However, directly calculating the growth rate of the new matrix with $s_0 = 0.70$ gives $\lambda_1 = 1.21$. Even though it still is a very rapid growth rate, it is more than 10% less growth per year than that predicted by the sensitivity value. The discrepancy comes from using the sensitivity value in a way that assumes a linear relationship between vital rates and λ_1. In this case we can see from Figure 9.1 that this assumption is incorrect, but most methods of calculating sensitivities use shortcuts that give *only* the slope of the tangent line, making it impossible to see how badly this assumption might lead us astray if we are trying to predict the effects of large management changes. As you'll see in the following sections, there is often a tradeoff between the elegance of a sensitivity analysis method and its robustness to this kind of problem.

A second major issue is in the scaling of sensitivity values. Consider the sensitivities for adult reproduction and juvenile survival for the emperor goose, which are 0.082 and 0.58, respectively. These sensitivities indicate that a change in juvenile survival has a much more important effect on λ_1 than does adult reproduction. But remember what a sensitivity value is: the slope of the line between λ_1 and a vital rate, or (change in λ_1)/(change in vital rate). It seems biologically unrealistic to compare these two slopes directly. For example, a change of 0.15 in adult reproduction is a small increase relative to the current value (0.8943), while a change of 0.15 more than doubles the current value of juvenile survival (0.1357). In other words, a *biologically* small change in reproduction is *mathematically* equivalent to a huge change in survival. This kind of "unfairness" in the comparison of sensitivity values is even more striking for highly fecund species. For example, in a PVA for Chinook salmon, the mean fertilities of three adult classes ranged from 3,257 to 5,149 eggs per female (Kareiva et al. 2000; also see Ratner et al. 1997). A change of, say, 1.0 in the value of one of these fertilities is very different than a hypothetical change of 1.0 in a survival rate, which must be bounded by 1 and thus can never actually change by even this amount. It would be biologically meaningless to say that a change of this magnitude in fertility had a smaller effect on λ_1 than an identical change in a survival rate—that is, that they had the same sensitivity values, given the very different amounts of change that each rate could be expected to undergo.

To make meaningful comparisons between different sensitivity values it is thus necessary to somehow rescale the S_{r_i} values to reflect these differences. The standard way to do this is to calculate elasticities, which measure the *proportional* change in λ_1 resulting from a *proportional* change in a vital rate (Caswell et al. 1984, de Kroon et al. 1986). Mathematically, the elasticity of λ_1 to changes in a vital rate r_i, E_{r_i}, is:

$$E_{r_i} = \frac{r_i}{\lambda_1}\frac{\partial \lambda_1}{\partial r_i} \tag{9.2}$$

where $\partial\lambda_1/\partial r_i$ is just S_{r_i}, the sensitivity of λ_1 to r_i. The way to understand the logic of this adjustment is to rearrange it this way:

$$E_{r_i} = \frac{\partial \lambda_1/\lambda_1}{\partial r_i/r_i} \tag{9.3}$$

Here the numerator is change in population growth divided by the current mean growth rate—in other words, the proportional change in λ_1. The denominator is similarly the proportional change in r_i. We can recalculate the sensitivity results for the emperor goose as elasticities, and see that the relative importance of changes in different rates for λ_1 now appears somewhat different (Table 9.1). On the proportional scale of elasticities, the difference between adult and juvenile survival is even more pronounced, while adult reproduction and juvenile survival are almost equal. As with sensitivities, elasticities are usually shown as point values for each vital rate. However, elasticities can also be thought of as the slopes of tangent lines, but this time tangents to curves relating proportional changes (rather than actual values) of λ_1 and each vital rate. We can also plot full sets of proportional changes of different magnitudes to get a fuller picture of the effects of proportional change on population growth (Figure 9.2). Equal elasticities would mean that the same proportional change in two different vital rates would result in the same effect (the same proportional change) in λ_1, as is nearly true for $f_{\geq 3}$ and s_0 for the emperor goose (Figure 9.2).

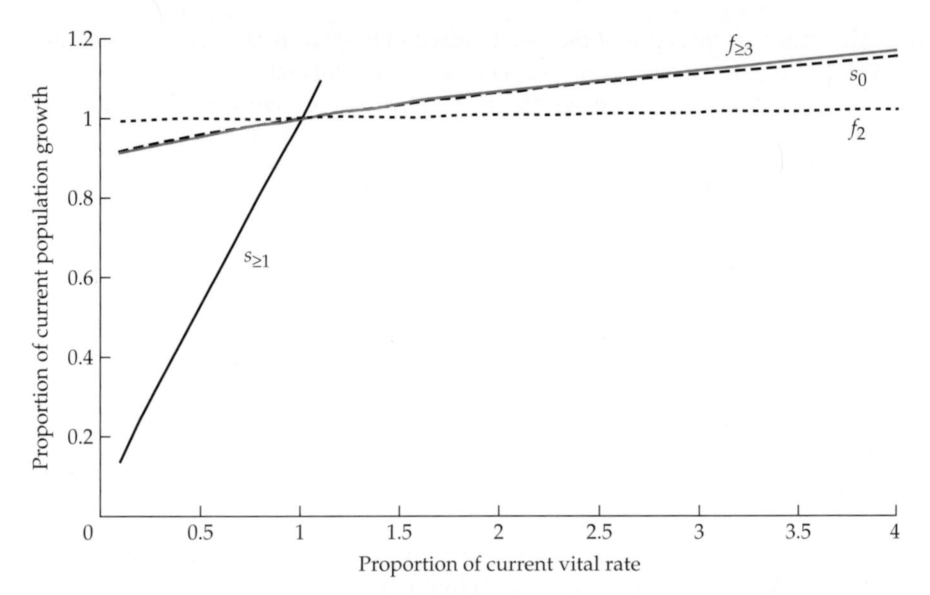

Figure 9.2 Influence of proportional changes in vital rates on population growth of the emperor goose. Labels indicate the vital rate corresponding to each line. Note that current estimates of all vital rates and of current population growth correspond to a value of 1. The line for $s_{\geq 1}$ stops at a small proportional change, as this change leads to a survival value of 1.0, the maximum possible.

Elasticities are the standard way to recalibrate sensitivity analysis to account for differences in the scale of measurement of different vital rates. However, they are still subject to a variety of interpretation problems (Mills et al. 1999). We'll discuss more about these complications and how to account for or avoid them as we move through the rest of the chapter.

Sensitivity Analysis for Deterministic Matrices

Although we've repeatedly emphasized the need for estimates of temporal variation in PVAs, the simplest sensitivity analysis methods—and not coincidentally, the ones most often used—are based upon a simple matrix model of mean vital rates. As you'll see, we recommend considerable caution in using these deterministic sensitivity results to make management judgments, but they are still the starting point for most sensitivity analyses and are worth understanding thoroughly.

Calculation of Simple Sensitivities and Elasticities: The Easy, Brute Force Method

There are two ways to calculate sensitivity and elasticity values, one easy and intuitive, and one more exact but difficult to understand. Not surprisingly, we'll present the easy method first. As we noted above, the sensitivity of λ_1 to a vital rate r_i is the slope of the line tangent to the curve of λ_1 versus r_i, evaluated at the current value of r_i (Figure 9.1). The easiest way to estimate the slope of this tangent is to estimate the slope between two sets of values of λ_1 and r_i, both very close to the current values. The simplest way to do this is to use the current values of r_i and λ_1, and a pair that are very nearby. To get this second pair, we start with a new value for the vital rate, obtained by adding a small number, Δr_i to r_i, put this new value into the matrix, and calculate λ_1 for the matrix using this new vital-rate value (but unchanged values of all other vital rates). The difference between this new growth rate and the original one is $\Delta\lambda_1$, and an estimate of the slope of the tangent line is:

$$S_{r_i} = \frac{\lambda_{1,new} - \lambda_{1,original}}{r_{i,new} - r_{i,original}} = \frac{\Delta\lambda_1}{\Delta r_i} \tag{9.4}$$

The key to obtaining an accurate assessment of the sensitivity value using this method is to make Δr_i small and to calculate the λ_1 values with high precision. For example, Heppell et al. (1996) used Δr_i values equal to 1% of the current value for each vital rate.[1] In general, if you use perturbation (Δr_i) values equal to 5% of the original r_i's or less, you will get quite accurate estimates of sensitivity values by this method (Mills et al. 1999).

Calculation of Simple Sensitivities and Elasticities: The Elegant Method

The second method to estimate sensitivity is more mathematically elegant, but has no appreciable gain in practical utility. Its main advantage is that it can be simpler to do, providing you have access to a program that will calculate the dominant right and left eigenvectors of a matrix. However, there is also an added step in this method that can make it considerably more complicated (and will help you exercise your calculus skills) for most PVA applications. The heart of this approach was introduced to ecologists by Caswell (1978, 2001), who showed how, using Equation 7.10, the sensitivity of λ_1 to changes in matrix element a_{ij}, S_{ij}, can be calculated from the eigenvectors of the projection matrix. For convenience, we repeat Equation 7.10 here:

$$S_{ij} = \frac{\partial\lambda_1}{\partial a_{ij}} = \frac{v_i w_j}{\sum_{i=1}^{s} v_i w_i} \tag{9.5}$$

[1]In addition, these authors actually calculate two new matrices, one with r_i 5% less than the estimated value and the other with r_i 5% greater. By using the growth rates of these two new matrices, they "center" the calculation used in Equation 9.4 on the current value of r_i, rather than perturbing r_i in only one direction as we do. Although this modification would matter for large values of Δr_i, for small values it is not necessary.

Here, s is the number of classes in the matrix, w_j is the jth element of the dominant right eigenvector (the stable distribution value for the jth class), and v_i is ith element of the dominant left eigenvector of the projection matrix (the reproductive value for the ith class). If you aren't familiar with right and left eigenvectors, you may want to review the explanations of them in Chapter 7. As noted there, estimating the stable distribution is done simultaneously with estimating λ_1, and it is also easy to estimate the dominant left eigenvector as the dominant right eigenvector of the transpose of the original matrix. Finally, the denominator in Equation 9.5 is the scalar product of the two eigenvectors. The function `eigenall` in Box 7.1 will calculate all the correctly scaled eigenvalues and eigenvectors, and the code in Box 7.2 uses these to calculate the sensitivity values of all matrix elements. Table 9.2 shows the sensitivities to the matrix elements for the emperor goose.

This method is simpler than the brute force estimation technique because all the numbers needed to calculate all sensitivity values are at hand after you've calculated the eigenvectors. However, there is a problem: All the sensitivity values we can calculate using Equation 9.5 are for *matrix elements*, not for *vital rates*, which usually contribute to more than one element and/or combine with different vital rates to make a single element. To estimate sensitivity values for vital rates, we need to use the chain rule for differentiation[2] (Caswell 2001). We know the derivatives of λ_1 with respect to each of the matrix elements (i.e., the sensitivity values for the elements), and we want the derivatives with respect to vital rates, so we need only calculate the derivatives of the matrix elements with respect to the vital rates. The only complication is that changes in a vital rate may influence multiple matrix elements, which requires us to sum up the effects on λ_1 of changes in a vital rate via all these elements:

$$S_{r_k} = \sum_{i=1}^{s}\sum_{j=1}^{s} \frac{\partial \lambda_1}{\partial a_{ij}} \frac{\partial a_{ij}}{\partial r_k} = \sum_{i=1}^{s}\sum_{j=1}^{s} S_{ij} \frac{\partial a_{ij}}{\partial r_k} \tag{9.6}$$

Here, we are summing up the derivatives of λ_1 with respect to a_{ij} times the derivatives of a_{ij} with respect to r_k for all matrix elements (all combinations of i and j). Actually, $\partial a_{ij}/\partial r_k = 0$ for all matrix elements to which r_k does not contribute, but it is simpler to express the result in this general way. Using this equation, we can now estimate the sensitivity of λ_1 to changes in vital rates. For example, the sensitivity of λ_1 to adult goose survival is:

$$S_{s_{\geq 1}} = (S_{32} \times 1) + (S_{43} \times 1) + (S_{44} \times 1) + (S_{13} \times f_2) + (S_{14} \times f_{\geq 3}) \tag{9.7}$$

---

[2]The chain rule states that if both Y and Z are functions of X, then= $\frac{\partial Z}{\partial X} = \frac{\partial Z}{\partial Y}\frac{\partial Y}{\partial X}$

TABLE 9.2 Sensitivity and elasticity values for matrix elements of the emperor goose[a]

Sensitivity values			
0.0797	0.0109	0.0099	0.0917
0.5807	0.0797	0.0719	0.6678
0.6432	0.0883	0.0797	0.7398
0.6616	0.0908	0.0820	0.7609
Elasticity values			
0	0	0.0057	0.0740
0.0797	0	0	0
0	0.0797	0	0
0	0	0.0740	0.6869

[a]Values correspond to the matrix elements in Equation 9.1.

where the values 1, 1, f_2 and $f_{\geq 3}$ are the partial derivatives of each of the matrix elements that contain $S_{\geq 1}$ with respect to $S_{\geq 1}$ (see Equation 9.1 for the definitions of each element in terms of vital rates). Most matrices, like the one for the emperor goose, are relatively simple, with each matrix element a simple linear combination of different vital rates, making Equation 9.6 easy to use. However, sometimes vital rates combine in much more complicated ways. For example, in constructing a matrix model for the cheetah, Crooks et al. (1998) used a function for fertility that accounted for the rapid return to estrus of females whose litters are predated. This results in a matrix element for adult reproduction that involves six different vital rates (plus several others that combine to make the ones shown)[3]:

$$a_{1i} = \frac{(365/2)s_{ad}l}{\sum_{t=1}^{120}\left[(1-\rho_1)^{t-1}(\rho_1)(t+R)\right] + (1-\rho_1)^{120}\sum_{t=1}^{420}\left[(1-\rho_2)^{t-1}(\rho_2)(120+t+R)\right] + (1-\rho_1)^{120}(1-\rho_2)^{420}(540+R_i)} \qquad (9.8)$$

Especially in these cases, it may be much simpler to use the first method we described, estimating sensitivities to vital rates by simulating matrices with small changes in each rate in turn. However, as long as you can write out a formula for each matrix element, you can use symbolic logic programs to estimate the derivatives of each matrix element with respect to each vital rate, and hence estimate sensitivity values. Box 9.1 provides a general program to do this for any matrix, using the symbolic logic functions of MATLAB.

[3]See Crooks et al. (1998) for definitions of the symbols and for the reasoning behind the structure of the equation.

BOX 9.1 ***A MATLAB program to find the sensitivities and elasticities of λ_1 to vital rates.***

```
% Vitalsens.m Takes vital rate means and a matrix
% definition to calculate deterministic sensitivities and
% elasticities of lambda to vital rates, using symbolic
% functions to take derivatives. This example uses emperor
% goose demography: fall to fall, with the youngest class as
% fledglings, using Schmutz et al. 1997 survey set 1
% The program calls the function eigenall.m (Box 7.1)
clear all;
%*************** SIMULATION PARAMETERS ********************
vr = [0.1357    0.8926    0.6388    0.8943];  % vital rates
syms  Ss0 Ss1 Sf2 Sf3  % vital rates as symbolic variables
Svr = [Ss0 Ss1 Sf2 Sf3]; % vector of symbolic vital rates

% Next, a symbolic definition of the matrix
mx =   [  0,     0,      Sf2*Ss1,  Sf3*Ss1;
          Ss0,   0,      0,        0;
          0,     Ss1,    0,        0;
          0,     0,      Ss1,      Ss1];
%**********************************************************

% make a matrix of the mean numerical values using subs
realmx = subs(mx,Svr,vr);
% use eigenall.m to get eigenvalues
[lambdas,lambda1,W,w,V,v]= eigenall(realmx);
sensmx = v*w'/(v'*w);   % sensitivities of matrix elements
elastmx = (sensmx.*realmx)/lambda1; % element elasticities
numvrs = length(vr); % how many vital rates?
vrsens = zeros(1,numvrs); % initialize vital rate sens.
% a loop to calculate sensitivity for each vital rate
for xx=1:numvrs
   % derivatives of elements with respect to vital rates
   diffofvr = subs(diff(mx,Svr(xx)),Svr,vr);
   % sum up to get vital rate sensitivities
   vrsens(xx) = sum(sum(sensmx.*diffofvr));
end; % xx

vrelast = ((vrsens.*vr)/lambda1); % calculate elasticities
disp('Matrix element sensitivities and elasticities:')
disp(sensmx);
disp(elastmx);
disp('Below are the vital rate results');
```

BOX 9.1 *(continued)*

```
disp('vital rates:');
disp(Svr)
disp('sensitivities');
disp(vrsens)
disp('elasticities');
disp(vrelast)
```

Converting sensitivity values to elasticities is simple, and does not depend on how the sensitivity values were estimated. Just use Equation 9.2 along with estimates of each vital rate and current population growth to do the transformation. The code in Box 9.1 automatically does this after calculating the vital rate sensitivities.

Sensitivity to the Effects of Management

Until now, we've only considered sensitivity and elasticity values for either the three basic vital rates (survival, growth, and fertility) or for entire matrix elements. However, management actions (as well as negative factors such as invasive species or habitat degradation) will usually influence multiple vital rates simultaneously, although not always to the same degree. Furthermore, many types of management lead to a mixture of negative and positive effects. Since a major goal of most sensitivity analyses is to gauge the importance of management actions or of negative impacts on a population, we would like to summarize these multiple effects to arrive at the total influence of management on population growth.

One way to arrive at an estimate of the total effects of a management action is to simply add up the sensitivity or elasticity values of the rates that are affected. That is, if multiple rates are changing together, their joint effects on λ_1 can be approximated by adding up their individual effects (either sensitivities or elasticities). For example, if a partial ban on subsistence hunting of emperor geese increases both juvenile and adult survival, to a first approximation the elasticity of λ_1 to a hunting ban is $0.08 + 0.92 = 1.00$. Alternatively, a change in the allowable times for hunting might have positive effects on juvenile survival but might decrease adult survival. In this case, we'd sum the effects, but change the sign of the effect on adults: $0.08 + (-0.92) = -0.86$. The way to interpret negative elasticity values is to remember that they are the slopes of lines relating proportional changes in vital rates (or combinations of rates) and λ_1. Thus, –0.86 means that a greater shift in hunting times leads to a larger decrease in λ_1.

The above method can work reasonably well when few vital rates are affected by management, but a more comprehensive solution is to make management itself a "rate" in the population matrix, the sensitivity of which we can directly estimate. In particular, simply adding up the sensitivity values for different vital rates assumes that they will all be influenced to the same absolute (sensitivities) or proportional (elasticities) amounts by management, while in reality these effects are likely to differ dramatically in magnitude. For example, let's say that the duration of the hunting season has much bigger effects on the survival of young, inexperienced geese than it does on the wily adults. We might estimate that the same decrease in the number of hunting days has a threefold greater effect on juvenile survival than on adult survival, but that both are influenced by the length of the hunting season. With these estimates of the magnitudes of hunting effects, we can rewrite the matrix for emperor geese, adding in a variable h for the length of the hunting season as a proportion of the period currently allowed:

$$\mathbf{A} = \begin{pmatrix} 0 & 0 & f_2 s_{\geq 1}(2-h) & f_{\geq 3} s_{\geq 1}(2-h) \\ s_0(4-3h) & 0 & 0 & 0 \\ 0 & s_{\geq 1}(2-h) & 0 & 0 \\ 0 & 0 & s_{\geq 1}(2-h) & s_{\geq 1}(2-h) \end{pmatrix} \quad (9.9)$$

The terms $(2 - h)$ and $(4 - 3h)$ show the effects on hunting, with h a variable that represents hunting effects as a proportion of current rates. With the current value of $h = 1$, the matrix collapses to that shown in Equation 9.1. As h decreases due to reduced season length, the effective survival of both adults and juveniles goes up, but at different rates. Conversely, the two survival rates decline, again at different rates, as h increases. This way of including hunting effects is a reasonable way to gauge the influence of changing season length on goose demography. [4] Modifying the program `vitalsens` (Box 9.1) to use this new matrix gives sensitivity and elasticity values for h of –1.15 and –1.16, respectively, showing the extremely strong, positive effects that a decrease in hunting would have on population growth.

There can be many alternative ways to put management into a matrix model. How best to do so will depend on your knowledge of the species and the different effects of management. For example, if we knew the exact

---

[4]However, any value of h greater than 1.33 (a dramatic increase in current hunting pressure) will lead to negative survival values. To avoid this problem, hunting pressure could be put into matrix elements as a multiplicative effect, such as multiplying current survival rates by $\exp[a(1 - h)]$, so that increasing hunting leads to decreasing survival but never to values less than 0.

amount of juvenile and adult goose mortality that is currently due to hunting versus natural causes, we could add hunting effects in this way:

$$\mathbf{A} = \begin{pmatrix} 0 & 0 & f_2 s_{\geq 1,n}(1 - hm_{\geq 1,h}) & f_{3+} s_{\geq 1,n}(1 - hm_{\geq 1,h}) \\ s_{0,n}(1 - hm_{0,h}) & 0 & 0 & 0 \\ 0 & s_{\geq 1,n}(1 - hm_{\geq 1,h}) & 0 & 0 \\ 0 & 0 & s_{\geq 1,n}(1 - hm_{\geq 1,h}) & s_{\geq 1,n}(1 - hm_{\geq 1,h}) \end{pmatrix} \quad (9.10)$$

Here, we've divided current survival and mortality rates into two pieces, natural survival ($s_{0,n}$ and $s_{\geq 1,n}$) and survival of hunting, which we include as the mortality rate due to hunting ($m_{0,n}$ and $m_{\geq 1,n}$). Again, h is the proportion of current hunting effort.

Finally, putting management directly into a matrix allows you to introduce conflicting management effects that are based on life history tradeoffs or interactions between individuals in a population. For instance, conditions that allow improved growth also create reduced survivorship for some plants. Similarly, efforts to enhance breeding habitat might increase successful fledging for some species, but these juveniles might then compete for new nest sites, reducing first year reproduction or post-fledging survival. Putting these effects into a matrix model allows a better representation of the true effects of management actions. Importantly, with conflicting effects, some intermediate level of management may be optimal so that a simple elasticity or sensitivity value is meaningless. In this case, the calculation of λ_1 over a range of management values is key to properly interpreting sensitivity patterns.

In general, we strongly encourage you to formulate and analyze matrix models that explicitly include management effects on different vital rates. Even though some of these effects will often need to be based on guesswork, directly incorporating them in a demographic model sharpens your sense of which effects will be most important and which need better estimation to really be understood.

Caveats for Using Deterministic Sensitivities and Elasticities

The general goals of a sensitivity analysis are to: (1) determine the effects on population growth of different amounts and combinations of changes in vital rates; (2) indicate the *relative* importance or potential of changes in different demographic rates for improving population growth and reducing extinction risk; and (3) pinpoint vital rates for which it is especially desirable to have highly accurate estimates. It should be obvious by now that you could try to answer any of these questions in one of two general ways: by calculating λ_1 for many different matrices to make graphs such as Figure 9.1, or by using formal sensitivity and elasticity values (Equations 9.2 and 9.6) to less directly infer the consequences of different changes in demography. In practice, PVA practitioners have usually used a combination of these approaches, with some direct exploration of major changes in specific vital rates and also calculation of sensitivity and elasticity values to summarize major patterns important for management concerns. Elasticity values, in particular, have become by far the most commonly used outputs of a PVA sensitivity analysis.[5] This is largely because, as we discussed above, elasticities are seen as a fair way to compare the importance for population growth of changes in vital rates with very different current values. The key question to ask in deciding whether to use direct manipulation of vital rates (and recalculation of λ_1 for different versions of the projection matrix) versus the more convenient summaries that elasticities provide is: When will elasticities be inaccurate or misleading, negating their obvious advantage of conciseness? Correctly answering this question for any particular PVA requires understanding the limits and assumptions of formal elasticity values, which we discuss in the remainder of this section. The following two sections then provide additional sensitivity methods that expand on simple deterministic elasticities and sensitivities and straightforward matrix manipulations, in order to solve some of the limitations we bring up below.

LINEARITY AND SMALL CHANGES When using elasticities to compare and contrast the importance of manipulating or accurately knowing different vital rates, the first limitation to keep in mind is that elasticities predict the response of λ_1 to very small changes in vital rates. This is another way of stating the linearity assumption mentioned above: Because elasticities assume a linear relationship between proportional changes in λ_1 and in vital rates, they are only guaranteed to be accurate for small changes. But how small is small? While a comprehensive survey of matrix models for real species hasn't been conducted, Mills et al. (1999) found that in most but not all cases, elasticities made good to adequate predictions even for fairly large changes in vital rates (20% or more). Still, if you are interested in evaluating specific management plans to improve population growth, it is wise to take the approach used in the preceding section in including management effects in the matrix, and then to construct a range of matrices

[5]It is a confusing convention that *sensitivity analysis* refers to any analysis of how changes in different vital rates or matrix elements will influence population growth, extinction times, and so on, whereas *sensitivity values* are, for the most part, strictly used to mean the values defined by Equations 9.5 and 9.6. In many PVAs, the sensitivity analysis will report only *elasticity* values.

with different strengths of management impacts to directly gauge the effects on λ_1.[6]

CHANGES IN MULTIPLE RATES A second potential problem with the accuracy of elasticity values also involves the assumption that changes in rates are very small. If changes are small, then the change in λ_1 that results from changes in multiple matrix elements can be approximated by summing up the effects of changes in each element multiplied by their sensitivity values. This assumption forms the basis of the double summation in Equation 9.6 and it can also be used to estimate the total effect on population growth of changes in multiple matrix elements. In general:

$$\Delta\lambda_1 \cong \sum_{i=1}^{s}\sum_{j=1}^{s} S_{ij}\Delta a_{ij} \tag{9.11}$$

where $\Delta\lambda_1$ is the change in λ_1 resulting from changes in each of the matrix elements, Δa_{ij}. The same summation of effects can be used on the proportional scale of elasticity values, and it can also be used for the effects of changes in vital rates, rather than matrix elements. In particular, the proportional change in λ_1 can be approximated by summing up the proportional changes in each vital rate multiplied by their elasticity values:

$$\frac{\Delta\lambda_1}{\lambda_1} \cong \sum_{i=1}^{n} E_{r_i}\frac{\Delta r_i}{r_i} \tag{9.12}$$

where n is the number of vital rates. However, these approximations all ignore any multiplicative effects of changing more than one matrix element at a time. In other words, they assume that there are no interactions between the effects of different vital rates on λ_1, which is clearly not true. Still, there have been no systematic surveys of matrix models to see how egregious such effects are.

---

[6]An alternative way to check for nonlinearities is to calculate the second derivatives, or second-order sensitivities, of λ_1 with respect to a vital rate or matrix element (Caswell 1996, 2001). We discuss second-order sensitivities in the stochastic analysis section of the chapter, and Box 9.3 gives MATLAB code to calculate these second derivatives if you wish to use this method to check for nonlinearities. The second derivatives measure the changes in sensitivities that will result from changing vital rates or matrix elements; values near zero indicate that sensitivities will change little in response to changes in a matrix element or vital rate and hence will give robust predictions. However, second derivatives are still point estimates that may not give the full picture of the responses of sensitivities to large changes in rates.

THE APPROPRIATENESS OF PROPORTIONAL CHANGE COMPARISONS Converting sensitivity to elasticity values is designed to get rid of the problem of comparing relationships between λ_1 and vital rates that are measured on very different scales. Although it is clear that this is an important problem to overcome, it is less clear that elasticities always are the right solution. The implicit assumption in comparing elasticity values is that the same magnitude of proportional change is equally likely (or possible, or costly to accomplish) for rates with different current values. For example, one could use the elasticity values for a salmon matrix, in which elasticities of adult survivorship will almost always be higher than those for fertilities, to argue that population growth will be more strongly influenced by changes in survivorship than reproduction. However, to use elasticities in this way (which is extremely common), one has to assume that a change from current adult survivorship of, say, 0.6 to 0.66 is "equivalent to" a fertility change from 4,000 to 4,400 eggs. Although this is not an unreasonable starting point in a sensitivity analysis, this assumption should be carefully evaluated before it becomes the basis of management interventions.

One particularly worrisome aspect of elasticity comparisons is that they will always have the tendency to identify vital rates with higher means as being the most important, since a vital rate's elasticity is its sensitivity value times its current value (and divided by the current λ_1; Equation 9.2). This effect, and the confusion it can cause, is most starkly seen in comparing the importance of different survival rates. For the emperor goose, the sensitivity value for adult survival is about double that of juvenile survival, while its elasticity is ten times greater, with this drastic difference due to the low current value of juvenile survival. Now, we could just as easily have constructed the matrix, and the sensitivity analysis, using *mortality* rates (i.e., 1 minus the survival rate), instead of *survival* rates. The sensitivity values for a survival rate and its corresponding mortality rate are of identical magnitude but opposite signs, so the sensitivities of adult and juvenile mortality are –1.02 and –0.58, respectively. The elasticity values are then the current mortality rates (adults: 1 – 0.893, juveniles: 1 – 0.136) times the elasticities, divided by the current λ_1, giving elasticities of –0.11 for adult mortality and –0.51 for juvenile mortality. The absolute value of the elasticity is so much higher for juvenile mortality than for adult mortality due to its much higher current value.

Thus, these mortality elasticities indicate that a given proportional reduction in juvenile mortality will be about five times more beneficial for population growth than will a similar proportional reduction in adult mortality. Neither this or the opposite conclusion about the importance of changes in adult and juvenile *survival* are wrong. They differ because proportional changes in a survival rate and its corresponding mortality rate are only equal if survival is 0.5. These results highlight the problem of assuming—as is usually done—that proportional changes in survival rates are comparable, and—as a result—implicitly assuming that proportional changes in mortality rates are *not* comparable. As we show with the emperor goose example, the assumption that is used can drastically change the conclusions of a sensitivity analysis.

The best lesson to draw from these problems is that any comparison of the elasticity values of different vital rates should be accompanied by some justification that proportional changes in these rates are in some sense comparable to one another. Indeed, this need for clarity also applies to direct matrix manipulations to gauge effects of management actions. For example, Kareiva et al. (2000) manipulated a matrix model for Snake River Chinook salmon to discover the magnitude of changes in first year or early ocean/estuarine mortality rates needed to stabilize the population. The decrease in *mortality* rates needed for $\lambda_1 = 1$ that they report are only 6% and 5%, respectively. However, had the same results been reported as changes in *survival* rates, they would have shown the need for increases above current survival rates of 267% and 289 % of the estimated values. In this case, with extremely low current survivorships, reporting changes in terms of mortality rates makes the alterations needed to stabilize the population seem trivial, while the exact same change, expressed as a shift in survivorship, appears massive. As with a comparison of elasticity values, which perception is correct depends upon the biology of the species and the ways in which management alters vital rates, which together will determine whether the mortality or survival "scale" is most reasonable.

THE ABILITY OF RATES TO CHANGE As we've repeatedly mentioned, the calculation and hence the interpretation of elasticity and sensitivity values are based on the notion of small changes in rates. However, almost any change in vital rates that we are interested in for conservation purposes will be at least moderately large—either large drops in performance due to increasing impacts or large improvements in vital rates that we hope to effect through management shifts. Large changes may influence the accuracy of elasticity predictions for any of the reasons already listed. They also make it hard to guess at the effectiveness of different management changes because vital rates we want to change (e.g., those with high elasticities) may quickly hit biological limits that prevent us from altering them by more than moderate amounts. In other words, our ability to influence λ_1 by changing a vital rate depends not only on the sensitivity or elasticity value of that rate, given its current value, but also on how much room there is for improvement, or how feasible it is to make this improvement. An example of this issue can be seen in Figures 9.1 and 9.2. Although λ_1 is most sensitive to changes in adult survival, the current estimate for this rate is already quite high, and since a survival rate can't be greater than 1, there is relatively little potential to improve population growth by increasing its value. In contrast, there is a much greater scope to improve juvenile survival, and this results in a greater ability to influence population growth in spite of juvenile survivorship's relatively low sensitivity and elasticity values.

Thus, to accurately assess how much you can influence λ_1 by influencing a particular vital rate, you need to know *both* the sensitivity of λ_1 to changes in that rate *and* how much it is feasible to change that vital rate in the first place. Estimating the limits to changes in different vital rates can be done in various ways. Most obvious are the limits of zero and one on any survival rate. However, for most species narrower upper limits can be estimated, based on temporal or spatial variation in demography across populations or for related taxa. For example, Wisdom and Mills (1997) used all the information they could gather on temporal and spatial variation in the vital rates of the greater prairie chicken, including rates for both growing and shrinking populations. They then used the extreme estimates for each vital rate to establish likely upper and lower bounds on the rates for this species in natural conditions. For instance, the bounds on survival of fertile eggs were 0.20 and 0.80. Without the wealth of studies that these authors had available, one could, for example, use the maximum clutch size ever observed for any individual as an estimate of the maximum limit on a population mean clutch size, or use the observed temporal variation in vital rates within one population to estimate the 95% confidence limits for each rate as their practical limits (Wisdom et al. 2000). If a species is in captivity or has intense management (e.g., eliminating all natural predators of a population), better survival, growth, or fertility rates might be possible to achieve. This general approach seems a reasonable way to estimate natural limits on vital rates. A similar approach is to use information on related, nonthreatened species to guess at upper bounds on the possible vital rate values of a species of concern.

With both elasticity values and estimates of the limits of each vital rate, it is possible to make better estimates of which parts of the life cycle might be best to target for management. In particular, the maximum proportional change in λ_1 that we can hope for by altering a vital rate r_i is:

$$\frac{\Delta\lambda}{\lambda_1} \cong E_{r_i} \frac{\left(r_{i,max} - r_i\right)}{r_i} \tag{9.13}$$

where r_i is the current value of the vital rate and $r_{i,max}$ is the maximum possible value of this rate. Equation 9.12 can also be used with maximum possible changes to estimate best-case changes in λ_1 from alterations of multiple rates. However, these estimates are still prey to all of the other potential problems we list above, as they still use elasticity values with all their attendant assumptions (such as additive and linear effects from changes in different vital rates). One way around these problems is to perform targeted alterations of the matrix, using specific combinations of maximum vital rates to calculate the resulting population growth rates directly. In the next section, we briefly sketch another approach to the inclusion of biological limits to vital rates that uses the basic ideas of sensitivity analysis while avoiding the assumptions inherent to formal sensitivity and elasticity values.

Sensitivity Analysis Including Biological Limits and Avoiding Small Change Assumptions

The approach we describe here allows a better estimate of how realistic changes in vital rates, either separately or together, will influence population growth (Wisdom and Mills 1997, Mills et al. 1999, Wisdom et al. 2000). The basic idea is to generate a large set of matrices for a species or population, each using different values for each vital rate, picked from within the reasonable limits for that rate. We calculate λ_1 for each matrix and then perform simple and multiple regressions on the results, with λ_1 as the dependent variable and the vital rates as the independent variables. From the simple regressions, which separately use each vital rate as a single independent variable (even though all vital rates are varried), the percentage of the explained variation in λ_1 is an estimate of the relative importance of a vital rate in generating variation in population growth, *contrasted* with changes generated by simultaneous variation caused by other rates. Similarly, a multiple regression λ_1 on two or more vital rates and their interactions yields an estimate of how important changes in these rates are in influencing growth, again in contrast to possible changes in other rates. Below, we illustrate this method using the emperor goose matrix.

First, we must estimate the reasonable limits on the four vital rates used in this matrix (Equation 9.1). Depending on the question we are asking, we can use different limits for the following analyses. In past applications of this method, the question has generally been: Variations in which rates are the most important in influencing population growth? This question has led to setting limits at the maximum and minimum values that seem likely to occur in nature (e.g., Wisdom and Mills 1997, Crooks et al. 1998). However, if we are mostly interested in how management to *improve* a particular rate will influence λ_1, it may be more reasonable to use current values as minimum values and only set maximum values at their biological limits. We take this approach with the emperor goose example, setting the upper bounds well above the best estimates for the current population (Table 9.3).

To simulate different matrices, we next need to decide what distribution to use to pick different values of each rate. For our purposes here, it is most reasonable to use a uniform distribution bounded by the upper and lower values of each rate. Note, however, that using different distributions, or randomly picking from a set of separate estimates for each vital rate, could change your results. In any case, once the decision is made, a simple program can be written to generate a suite of matrices constructed with different values of each rate (Box 9.2).

TABLE 9.3 Results of simulations to determine the sensitivity of population growth to possible increases in vital rates for the emperor goose

Vital rate	*Maximum vital rate value*[a]	*Maximum potential value of λ_1*	*Percentage variation in λ_1 explained*
s_0, survival of zero-year-olds	0.70	1.21	76
$s_{\geq 1}$, survival of one-year-old and older birds	0.95	1.05	6
f_2, reproduction of two-year-olds	2	1.00	2
$f_{\geq 3}$, reproduction of three-year-old and older birds	2	1.06	7

[a]Maximum vital rate values were roughly estimated by assuming that the current minimum estimates of mortality risks could be halved for all age classes and that the propensity of two-year-olds to breed could equal that of older birds. Minimum vital rate values used were the current estimated rates from Schmutz et al (1997) Survey Set 2.

Outputs of this simulation-based sensitivity analysis can be summarized in several ways. First, we can simply show the maximum value of λ_1 that could be generated by increases in each vital rate (with all other rates unchanged), and contrast these values to the simple elasticities for each rate (Table 9.3 versus Table 9.1). By this metric, changing juvenile survival has a larger potential for enhancing population growth than does changing adult survival. As we've mentioned before, this contrast shows that the highest elasticity values do not necessarily correspond to the largest possible effects on population growth, once limits on vital rate values are taken into account. Calculating the percentage of the variation in λ_1 explained by each vital rate (which is given by the squared correlation coefficient between λ_1 and each rate) provides another way to gauge each rate's importance, and this again shows that juvenile survival has a much greater potential to change λ_1 than does alteration of adult survival (Table 9.3). Indeed, increases in adult reproduction also have more potential to influence population growth, in spite of its low sensitivity and elasticity values. It is key to realize, however, that these conclusions are highly dependent on the upper bounds placed on each vital rate. If these are inaccurate, the results could shift substantially. It is also important to generate a sufficient number of matrices to capture the full range of possible values of each vital rate, and hence of λ_1. This analysis could be extended to examine each two-way interaction term, looking for significant multiplicative effects on population growth of simultaneous changes to different vital rates. However, for the emperor goose, the one-way effects together account for 91% of the variation in λ_1, suggesting little need to look further.

BOX 9.2 *MATLAB code to simulate random matrices between user-defined limits.*

```
% Limitsens.m  Calculates sensitivities over a range of values
% for each vital rate, accounting either for variation or
% uncertainty in rate values.
% This example assumes a uniform distribution of values for
% each rate between their high and low bounds
clear all;

%****************Simulation Parameters*********************
% Emperor goose data from Schumtz et al. 1997
meanvr = [0.1357    0.8926    0.6388   0.8943]; % best estimates
vrhi = [0.38 0.90  0.64   1.16]; % highest estimates
vrlo = [0.018 0.50 0.57 0.80];   % lowest estimates
syms  Ss0 Ss1 Sf2 Sf3 % symbolic variable definitions
Svr = [Ss0 Ss1 Sf2 Sf3]; % vector of symbolic variables
% symbolic definition of matrix:
mx =   [  0,      0,      Sf2*Ss1,  Sf3*Ss1;
          Ss0,    0,      0,        0;
          0,      Ss1,    0,        0;
          0,      0,      Ss1,      Ss1];
reps = 500; %number of replicate matrices to make and analyze
%**********************************************************

rand('state',sum(100*clock)); % random seed
numvrs = length(meanvr);      % how many vital rates?

% make uniform randoms and convert to random vital rates
allvrs  = rand(reps,numvrs);
allvrs =allvrs.*repmat(vrhi-vrlo,reps,1) + repmat(vrlo,reps,1);

for rr = 1:reps  % do calculations for each random matrix
 disp(rr);       % display progress
 realmx = subs(mx,Svr,allvrs(rr,:)); % Make a random matrix
 [lambdas,lambda1,W,w,V,v]= eigenall(realmx);
 sensmx = v*w'/(v'*w);    % sensitivities of elements
 elastmx = (sensmx.*realmx)/lambda1;  % element elasticities
 alllams(rr,1) = lambda1;             % save lambda value
 for xx=1:numvrs % loop to make sensitivities for vital rates
      diffofvr = subs(diff(mx,Svr(xx)),Svr,allvrs(rr,:));
      vrsens(xx) = sum(sum(sensmx.*diffofvr));
 end; % xx
```

BOX 9.2 *(continued)*

```
 allelasts(rr,:) = ((vrsens.*allvrs(rr,:))/lambda1);
end; % rr

% now get sensitivities and elasticities for the best estimates
realmx = subs(mx,Svr,meanvr);
[lambdas,lambda1,W,w,V,v]= eigenall(realmx);
meanlam1 = lambda1;    sensmx = v*w'/(v'*w);
elastmx = (sensmx.*realmx)/lambda1;
meansens = zeros(1,numvrs);
for xx = 1:numvrs
   diffofvr = subs(diff(mx,Svr(xx)),Svr,meanvr);
   meansens(xx) = sum(sum(sensmx.*diffofvr));
end;
meanelast = ((meansens.*meanvr)/lambda1);

% ask in turn how much the maximum value for each rate would
% change lambda, if all other rates are at best estimate
maxlams=zeros(1,numvrs);
for rate = 1:numvrs
   vrates = meanvr;  vrates(rate) = vrhi(rate);
   realmx = subs(mx,Svr,vrates);
   [lambdas,lambda1,W,w,V,v]= eigenall(realmx);
   maxlams(rate) = lambda1; % save resulting lambdas
end ;

% Now display the results
disp('Below are the maximum lambdas and max proportional')
disp('change in lambdas from changing each vital rate')
disp(Svr);   disp(maxlams)
disp((maxlams-ones(1,numvrs)*meanlam1)/meanlam1)


disp('Below are the r^2 values for lambda and each')
disp('vital rate, a measure of influence on population')
disp('growth for the random simulated matrices')
correls = corrcoef([allvrs,alllams]);
disp(Svr); disp((correls(numvrs+1,1:numvrs)).^2);

disp('Below are the elasticities for the mean vital rates,')
disp('and then the min, max, mean, and st. deviation of the')
disp('elasticity values from the random matrices')
```

BOX 9.2 *(continued)*

```
disp(Svr); disp(meanelast);
disp(min(allelasts)); disp(max(allelasts))
disp(mean(allelasts)); disp(std(allelasts))

disp('Next, upper and lower 95% confidence limits for the')
disp('elasticities calculated from the random matrices')
disp(Svr); CLup=zeros(1,numvrs); CLlo=zeros(1,numvrs);
for vr=1:numvrs
   x=sort(allelasts(:,vr));
   CLup(vr)=x(1+round((reps-1)*0.975));
   CLlo(vr)=x(1+round((reps-1)*0.025));
end;
disp(CLup); disp(CLlo);

disp('Finally, squared correlations between the elasticity')
disp('of each rate and the values of other vital rates,')
disp('a measure of the influence of each vital rate on the')
disp('elasticity results')
for rate = 1:numvrs
   disp('For elasticity of rate:');
   disp(Svr(rate));
   correls = corrcoef([allvrs,allelasts(:,rate)]);
   disp(Svr);
   disp((correls(numvrs+1,1:numvrs)).^2);
end;
```

Incorporating Uncertainty into Sensitivity Results

A final topic in sensitivity analysis for deterministic matrix models is the reliability of the results obtained. Elasticity and sensitivity results are dependent upon all the vital rate estimates that are used to make the matrix model in the first place. Therefore, inaccuracy in these estimates may have substantial influence on the sensitivity results. This influence is particularly of concern for rare species, given that some vital rates may be difficult or impossible to estimate with high accuracy. The simplest way to explore the importance of this uncertainty is to modify the method described in the preceding section to address uncertainty in vital rate estimates rather than potentially real changes in mean rates.[7] To do this, set the minimum and maximum values in the analysis to, say, the 95% confidence limits around the mean vital-rate esti-

TABLE 9.4 Percentage of variation in elasticity values explained by variation in each vital rate for simulation tests of the emperor goose matrix model[a]

Variation in value of:	Minimum and maximum values	Elasticity of rate			
		s_0	$s_{\geq 1}$	f_2	$f_{\geq 3}$
s_0, survival of zero-year-olds	0.018, 0.38	86%	86%	88%	83%
$s_{\geq 1}$, survival of one-year-old and older birds	0.50, 0.90	3	3	5	2
f_2, reproduction of two-year-olds	0.57, 0.64	0	0	1	0
$f_{\geq 3}$, reproduction of three-year-old and older birds	0.80, 1.16	1	1	0	1

[a]Each of the four columns of numbers on the right show the percentage of the variation in the elasticity values for the corresponding rate that was explained by variation in the values of the rates listed in the leftmost column.

mates calculated from your data. Then, you can create a set of random matrices that incorporate the uncertainty in the vital-rate estimates based upon current data and then ask how this uncertainty influences elasticity results. This is essentially a parametric bootstrap method to put confidence limits on the elasticities, akin to the bootstrap we used in Chapter 3 to place confidence limits on the extinction time CDF for a count-based PVA by sampling from the distributions of key parameters (see Box 3.4). The code in Box 9.2 includes calculation of elasticity values for each random matrix and then quantifies the uncertainty in them, asking how much variation there is in each elasticity value across matrices, and also how much of the variation in each elasticity value is due to uncertainty in each vital rate. These results can then be used to determine which vital-rate estimates have such broad confidence limits that they lead to important uncertainty in elasticity values, which in turn could call into question any management recommendations based on these values.

To illustrate this method, we again use the emperor goose model, this time using the maximum and minimum estimates of each vital rate obtained by Schmutz et al. using different parameter estimation methods (Table 9.4) to span the range of likely values for each vital rate. The elasticity values for each vital rate are quite consistent across the 500 randomly generated matrices, indicating that the rankings of elasticities are robust to parameter uncer-

[7]Another way to approach the same question is by using the second-order derivative of population growth with respect to two matrix elements or vital rates, which shows how sensitive the sensitivity to one rate is to the value of another (Caswell 1996, 2001). Still, we prefer to use the more direct simulation approach we describe in the text, which has the advantage of including simultaneous changes in multiple vital rates.

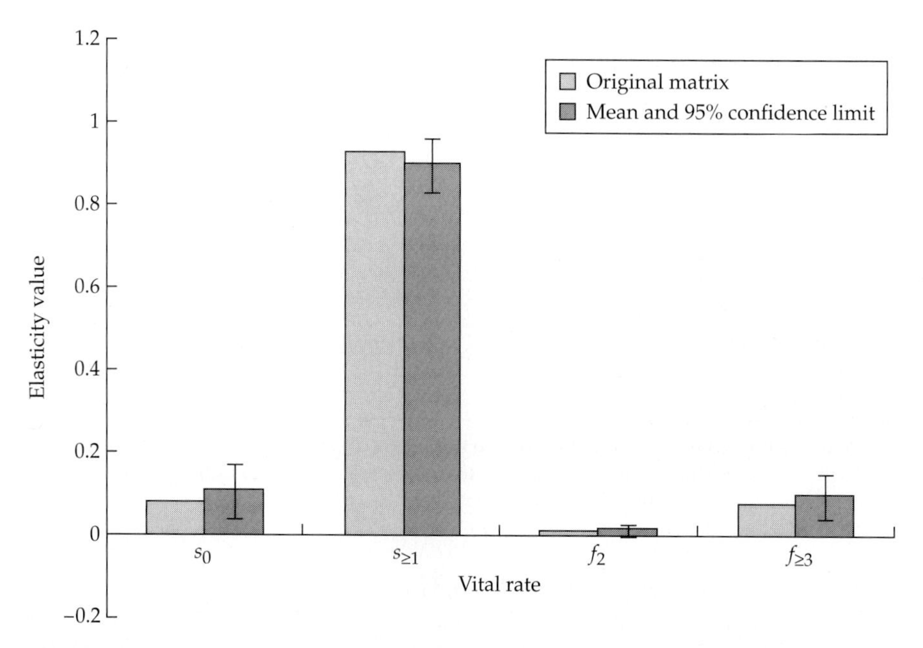

Figure 9.3 Variation in elasticity values due to parameter uncertainty for the emperor goose. The elasticity values from the original matrix are shown next to the mean and 95% confidence limits for 500 randomly generated matrices using minimum and maximum values of vital rates given in Table 9.4.

tainties for this model (Figure 9.3). However, the variation that does occur in all four elasticity values is largely generated by variation in juvenile survival values (Table 9.4), showing that uncertainty in the estimation of this rate is by far the most important in influencing the results of this sensitivity analysis.

Summary of Deterministic Sensitivity Analysis

A careful approach to sensitivity analysis is essential if the results will be used to justify management decisions—decisions that will often be costly or unpopular. Sensitivity analysis is a powerful tool to help evaluate management methods, but there are limitations, assumptions, and complications to sensitivity results that must be recognized and evaluated before reaching firm conclusions about the best way to get a demographic bang for a management buck. With this in mind, we advocate going through the following steps in any sensitivity analysis that is conducted as part of a PVA:

1. Calculate and present both sensitivity and elasticity values.
2. For vital rates with high elasticity and/or sensitivity values, make plots of changes in population growth versus each rate to look for nonlinearities and to show limits to their potential effects resulting from biological bounds on their values.

3. Determine if switching from survival to mortality rates drastically changes the ranking of elasticities. If it does, report both and justify which scaling is most biologically appropriate to use.
4. Use the simulation approach in Box 9.2 to ask both how large and simultaneous changes in vital rates will influence population growth, and also whether uncertainty in your vital rate estimates mean that some results of the sensitivity analysis must be considered preliminary or highly suspect.
5. Finally, reformulate the matrix to include parameters (e.g., harvest rate) that can portray the effects of one or more clearly defined management actions, and explore how the combined effects of management at different intensities will influence population growth.

Although these steps describe a considerably more complicated sensitivity analysis than those included in most PVAs, they also allow a more informative and justifiable set of conclusions to be reached.

Sensitivity Analysis for Stochastic Matrix Models

There is little difference in concept between sensitivity analysis for stochastic matrix models and that for deterministic models. The basic goal in both cases is to determine which vital rates have the most influence on population-level behavior. Indeed, Caswell (2001) has argued that in general the deterministic sensitivity values for a mean matrix are very good approximations of the sensitivity values for stochastic models. This generalization is probably true for relatively long-lived species that are well-buffered against environmental variability (which include most species for which matrix models have been made), but is less likely to apply to species with other, more variable life histories. In either case, if you have the information to build a stochastic model in the first place, you will obtain more accurate, defendable sensitivity results by directly addressing the sensitivity of stochastic population behavior, which is not hard to do.

However, sensitivity analyses for stochastic models are more complicated than those for deterministic matrices, for three reasons: (1) two different population-level phenomena, stochastic growth and extinction risk, are important to consider in a sensitivity analysis; (2) not only the mean vital rates but also their variances and covariances can influence population growth and extinction, and hence should be considered in a sensitivity analysis; and (3) the methods of calculating sensitivity and elasticity values must include temporal variation and hence are more involved. We first consider how to perform a sensitivity analysis for stochastic growth rate, and then move on to extinction risk sensitivity.

Sensitivity Analysis for Stochastic Growth Rate

ANALYTICAL APPROXIMATIONS We start by showing analytical formulas for stochastic sensitivities. As with stochastic population growth formulas

(Chapter 7) these equations are as or more useful for sharpening insight as they are for yielding practical estimates. And as in deterministic sensitivity analysis, it is easier to find stochastic sensitivity values for matrix elements than for vital rates. One approach to estimating stochastic sensitivity and elasticity values is to find an analytical approximation by taking the derivative of Tuljapurkar's formula for stochastic log growth rate with respect to different mean vital rates (Equations 7.12 and 7.13). For sensitivity of log λ_s to the mean value of a matrix element, a_{mn}:

$$\frac{\partial \log \lambda_s}{\partial a_{mn}} \approx \frac{\bar{S}_{mn}}{\bar{\lambda}_1} + \frac{\bar{S}_{mn}}{\bar{\lambda}_1^3}\tau^2 - \frac{1}{2\bar{\lambda}_1^2}\sum_{i=1}^{s}\sum_{j=1}^{s}\sum_{k=1}^{s}\sum_{l=1}^{s} Cov(a_{ij}, a_{kl})\left(\bar{S}_{kl}\frac{\partial \bar{S}_{ij}}{\partial a_{mn}} + \bar{S}_{ij}\frac{\partial \bar{S}_{kl}}{\partial a_{mn}}\right) \quad (9.14)$$

where τ^2 is a function of deterministic sensitivities and covariances defined in Equation 7.13,[8] and $\bar{S}_{mn}$ is the sensitivity of the deterministic growth rate of the mean matrix, $\bar{\lambda}_1$, to a_{mn}. $Cov(a_{ij}, a_{kl})$ is the covariance between a_{ij} and a_{kl}; remember that $Cov(a_{ij}, a_{ij})$ is the variance of a_{ij}. $\partial \bar{S}_{ij} / \partial a_{mn}$ is a second-order partial derivative of the deterministic population growth rate, or equivalently, a second-order sensitivity (Caswell 1996): $\partial \bar{S}_{ij} / \partial a_{mn} = \partial^2 \bar{\lambda}_1 / \partial a_{ij} \partial a_{mn}$.

Equation 9.14 shows why deterministic sensitivity values may be good approximations to stochastic ones for species with low temporal variation in vital rates (and hence, in matrix elements). If covariances are small, the second two terms on the right-hand side of Equation 9.14 will also be small, leaving only the first term. And if this is the only substantial term, the stochastic sensitivity of log $\bar{\lambda}_1$ is almost equal to the determistic sensitivity of log $\bar{\lambda}_1$ because $\bar{S}_{mn}/\bar{\lambda}_1 = (1/\bar{\lambda}_1)(\partial \bar{\lambda}_1 / \partial a_{mn}) = \partial \log \bar{\lambda}_1 / \partial a_{mn} \approx \partial \log \lambda_s / \partial a_{mn}$ since, with very small environmental variation, $\lambda_s \approx \bar{\lambda}_1$.[9]

You can also take the derivative of Tuljapurkar's approximation with respect to the covariance between two matrix elements or the variance of a single element to obtain a sensitivity value (Caswell 2001):

$$\frac{\partial \log \lambda_s}{\partial Cov(a_{ij}, a_{kl})} \approx -\frac{1}{\bar{\lambda}_1^2}\bar{S}_{a_{ij}}\bar{S}_{a_{kl}} \text{ and } \frac{\partial \log \lambda_s}{\partial Var(a_{ij})} \approx -\frac{1}{2\bar{\lambda}_1^2}\left(\bar{S}_{ij}\right)^2 \quad (9.15)$$

Note that in this approximation, the sensitivity of stochastic growth to a variance or covariance term is *only* a function of the deterministic growth rate of the mean matrix and the deterministic sensitivity values of the mean

---

[8]For reference, $\tau^2 = \sum_{i=1}^{s}\sum_{j=1}^{s}\sum_{k=1}^{s}\sum_{l=1}^{s} Cov(a_{ij}, a_{kl})\bar{S}_{ij}\bar{S}_{k}.$

[9]Here we use the definition of the derivative of the log of a variable: $\partial \log Y / \partial X = (1/Y)(\partial Y / \partial X)$.The same definition implies that if $\partial \log \bar{\lambda}_1 / \partial a_{mn} \approx \partial \log \lambda_s / \partial a_{mn}$, then the sensitivities of $\bar{\lambda}_1$ and $\bar{\lambda}_3$ (not on a log scale) must also be approximately equal.

matrix elements whose covariance is being considered. In other words, the importance of variance or covariance of matrix elements depends upon how sensitive deterministic population growth is to the values of those elements. Also note that, just as sensitivities to the means of matrix elements are always positive, sensitivities of $\log\lambda_s$ to variances and covariances are always negative—which reinforces the lesson from previous chapters that increasing levels of environmental stochasticity (in this case affecting matrix elements for a structured population) decrease population growth rate over the long term.

Although Equations 9.14 and 9.15 are the sensitivities of $\log\lambda_s$, to calculate elasticities for the stochastic growth rate, it is more sensible to use the sensitivities of λ_s, which can only take positive values—an important feature when comparing proportional changes, as we do with elasticities. The sensitivity of λ_s to a mean matrix element or a covariance is simply the sensitivity of $\log\lambda_s$ to that element or covariance multiplied by λ_s (see Footnote 9). We will use the symbol S^s_{mn} for the stochastic sensitivities of λ_s to the mean matrix element a_{mn}. Therefore, the elasticity of λ_s to changes in the mean of a matrix element, E^s_{mn}, is:

$$E^s_{mn} = \frac{a_{mn}}{\lambda_s} S^s_{mn} = \frac{a_{mn}}{\lambda_s} \frac{\partial \lambda_s}{\partial a_{mn}} = a_{mn} \frac{\partial \log \lambda_s}{\partial a_{mn}} \tag{9.16}$$

with the same relationship applying to elasticities and sensitivities of λ_s to variances and covariances of matrix elements.

To calculate these stochastic sensitivity and elasticity values we only need a mean matrix and the variance and covariance values for the matrix elements. Box 9.3 provides MATLAB code to calculate second-order sensitivities, and Box 9.4 uses this function in the estimation of stochastic sensitivities. We used this program to estimate the stochastic sensitivity and elasticity values for the means and variances of matrix elements for the desert tortoise. The sensitivity values for mean elements (at least those that are nonzero; see Table 8.1) show a strong pattern of higher sensitivites for survival and growth elements of older individuals than for reproduction or survival of the very youngest tortoises (Table 9.5). In addition, until reaching reproductive sizes, elements for advancement from one class to the next (i.e., those on the subdiagonal of the matrix) generally have higher sensitivity than do elements for survival but no growth (i.e., those on the diagonal). Somewhat in contrast to these results, elasticities suggest that changes in elements governing the survival of reproductive adults have by far the greatest proportional influence on stochastic growth, and that elements for stasis (which are always much larger in magnitude) have higher values than those for advancement. Both sensitivity and elasticity values for the variances of the matrix elements show somewhat greater importance (i.e., more negative

BOX 9.3 ***A MATLAB function to calculate the second derivatives of deterministic growth rate (or first derivatives of sensitivities) with respect to matrix elements (modified from Caswell 2001). This function is used to calculate stochastic sensitivities (see Box 9.4).***

```
function d2 = secder(A,k,l)
% this function finds the second derivatives of lambda
% with respect to one element (at row k, column l) of the
% matrix A in combination with each of the other elements.
% The function returns a matrix of second derivatives:
% d^2(lambda)/d(aij)d(akl)
n = size(A,1);
% find and sort the eigenvalues and vectors of a
[w,lambda]= eig(A);
v=inv(conj(w));
lambda = diag(lambda);
[lambda,index] = sort(lambda);
lambda = flipud(lambda);
index = flipud(index);
w=w(:,index);
v=v(index,:);
v=v.';
% calculate sensitivities
for i=1:n;
   senmat=conj(v(:,i))*(w(:,i).');
   svec(:,i) = senmat(:);
end
s1=svec(:,1);
for m = 2:n
   scalesens(:,m-1)=svec(:,m)/(lambda(1)-lambda(m));
end
% the line below is modified from Caswell 2001
if n>2 vecsum=sum(scalesens.'); else vecsum = scalesens.'; end;
%compute second derivatives
for i=1:n
   for j=1:n
      x1=(l-1)*n+1 + (i-1);
      x2=(j-1)*n+1 + (k-1);
      d2(i,j)= s1(x1)*vecsum(x2)+s1(x2)*vecsum(x1);
   end
end
d2=real(d2);
```

BOX 9.4 *A MATLAB program to calculate the sensitivities and elasticities of stochastic growth rate (λ_s) to means, variances, and covariances of matrix elements.*

```
% Stochsens
% This is a set of MATLAB commands to calculate the
% sensitivities and elasticities of stochastic growth rate,
% stochastic lambda.
% You need to have computed and have available the mean matrix
% and also the covariance matrix for the matrix elements:
% the covariance matrix should be arranged in order of,
% for example, a11, a21, a12, a22. It will be an n by n matrix,
% where n is the total number of elements in the population
% matrix.
% This program uses the function secder.m that calculates
% second derivatives of the deterministic matrix lambda.

load tortmxs % load the stored population and covariance matrix
mx = meanmx; % set mx equal to the name of stored pop'n matrix
% if the stored cov matrix is not called covmx, rename it here
n = length(mx);
nn=length(covmx);
[lambdas,lam,W,w,V,v]= eigenall(mx); % use eigenall.m (BOX 7.1)
sensmx = v*w'/(v'*w); % to get the sensitivities of mx elements
elastmx = (sensmx.*mx)/lam; % find mx element elasticities
vecSS = reshape(sensmx,nn,1); % make the sensitivities a vector

% use Tuljapurkar's approximation for stochastic-lambda
stochlam=exp(log(lam)-((vecSS')*covmx*vecSS)/(2*(lam^2)));

% following loops calculate the sensitivities of
% log(stochlambda) to the mean matrix elements:
  for kk=1:n
     for ll =1:n
        DD=secder(mx,kk,ll);
        vecDD = reshape(DD,nn,1);
        p1 = ...
     (sensmx(kk,ll)/lam)*(1+ ((vecSS')*covmx*vecSS)/(lam^2));
        p2 = ...
(((vecDD')*covmx*vecSS) + ((vecSS')*covmx*vecDD))/(2*(lam^2));
        mnsens(kk,ll) = stochlam*(p1 - p2);
      end; % ll
  end; % kk
```

BOX 9.4 *(continued)*

```
% converting to sens. and elast. of stochastic lambda
mnsens = mnsens*stochlam;
mnelast = mnsens.*mx/stochlam;

disp('sensitivities of mean matrix elements are:')
disp(mnsens)
disp('elasticities of mean matrix elements are:')
disp(mnelast)

% results for variances and covariances of matrix elements
covsens = -repmat(vecSS,1,nn).*repmat(vecSS',nn,1)/(stochlam);
covsens = covsens./(eye(nn)+ones(nn)); % adjust variances
covelast = covsens.*covmx/stochlam;
% the following displays can be used, but the covariance
% matrices are 64 by 64 for tortoises
%disp('Sensitivities of element covariances are:')
%disp(covsens)
%disp('Elasticities of element covariances are:')
%disp(covelast)

% the following shows sens. and elast. for the variances:
varsens = reshape( diag(covsens),8,8);
varelast = reshape(diag(covelast),8,8);
disp('Sensitivities to the matrix element variances are:')
disp(varsens)
disp('Elasticities for matrix element variances are:')
disp(varelast)
```

values) for elements governing advancement until reaching class 6 (Table 9.6). However, the sensitivity values for the variances of nonzero matrix elements are all close to zero and the elasticities even more so due to the very low estimated values for temporal variation (Table 8.3). We don't show the sensitivity values for covariances, but they too are all very small.

If you are basing an analysis on matrix elements, as we did in Chapter 7, the matrix element sensitivities and elasticities we've just calculated are the end of the sensitivity analysis. However, if we are trying to estimate sensitivity to vital rates, we need a further set of steps, again parallel to those taken in a deterministic analysis. In particular, to calculate the sensitivity of λ_s to changes in r_k, the mean of the kth vital rate, we need to sum up its effects on the growth rate through all matrix elements it influences:

TABLE 9.5 Sensitivity and elasticity of stochastic growth rate to changes in mean matrix elements for the desert tortoise[a]

SENSITIVITY VALUES		*Class in year t*							
Class name	*Class in year t+1*	*0*	*1*	*2*	*3*	*4*	*5*	*6*	*7*
Yearling	*0*	0.038	0.069	0.038	0.030	0.026	**0.016**	**0.028**	**0.015**
Juvenile 1	*1*	**0.048**	**0.086**	0.048	0.037	0.032	0.020	0.035	0.019
Juvenile 2	*2*	0.086	**0.155**	**0.086**	0.067	0.058	0.036	0.063	0.034
Immature 1	*3*	0.155	0.280	**0.155**	**0.122**	0.105	0.066	0.114	0.061
Immature 2	*4*	0.160	0.290	0.161	**0.127**	**0.110**	0.068	0.118	0.064
Subadult	*5*	0.368	0.666	0.369	0.289	**0.250**	**0.156**	0.271	0.145
Adult 1	*6*	0.344	0.622	0.345	0.271	0.235	**0.146**	**0.252**	0.136
Adult 2	*7*	0.299	0.542	0.299	0.233	0.202	0.126	**0.222**	**0.118**
ELASTICITY VALUES		*Class in year t*							
Class name	*Class in year t+1*	*0*	*1*	*2*	*3*	*4*	*5*	*6*	*7*
Yearling	*0*	0	0	0	0	0	0.007	0.020	0.011
Juvenile 1	*1*	0.038	0.048	0	0	0	0	0	0
Juvenile 2	*2*	0	0.037	0.048	0	0	0	0	0
Immature 1	*3*	0	0	0.037	0.083	0	0	0	0
Immature 2	*4*	0	0	0	0.039	0.073	0	0	0
Subadult	*5*	0	0	0	0	0.039	0.118	0	0
Adult 1	*6*	0	0	0	0	0	0.031	0.226	0
Adult 2	*7*	0	0	0	0	0	0	0.011	0.106

[a]Sensitivity values that apply to nonzero matrix elements are shown in bold.

$$S_{r_k}^s \cong \sum_{i=1}^{s}\sum_{j=1}^{s} S_{ij}^s \frac{\partial a_{ij}}{\partial r_k} \tag{9.17}$$

Similarly, the influence of a vital rate variance or covariance between two vital rates is

$$S_{Cov(r_m,r_n)}^s \cong \sum_{i=1}^{s}\sum_{j=1}^{s}\sum_{k=1}^{s}\sum_{l=1}^{s} S_{Cov(a_{ij},a_{kl})}^s \frac{\partial\left[Cov(a_{ij},a_{kl})\right]}{\partial\left[Cov(r_m,r_n)\right]} \tag{9.18a}$$

where

$$S_{Cov(a_{ij},a_{kl})}^s = \frac{\partial\lambda_s}{\partial Cov(a_{ij},a_{kl})} = \lambda_s\left(\frac{1}{\lambda_s}\right)\frac{\partial\lambda_s}{\partial Cov(a_{ij},a_{kl})} = \lambda_s\frac{\partial\log\lambda_s}{\partial Cov(a_{ij},a_{kl})} \tag{9.18b}$$

can be calculated using Equation 9.15.

TABLE 9.6 Sensitivity and elasticity of stochastic growth rate to changes in the variance of matrix elements for the desert tortoise[a]

SENSITIVITY VALUES		*Class in year t*							
Class name	*Class in year t+1*	*0*	*1*	*2*	*3*	*4*	*5*	*6*	*7*
Yearling	*0*	–0.001	–0.003	–0.001	<–0.001	<–0.001	**<–0.001**	**–0.001**	**<–0.001**
Juvenile 1	*1*	**–0.001**	**–0.004**	–0.001	–0.001	–0.001	<–0.001	–0.001	<–0.001
Juvenile 2	*2*	–0.004	**–0.014**	**–0.004**	–0.002	–0.002	–0.001	–0.003	–0.001
Immature 1	*3*	–0.015	–0.047	**–0.014**	**–0.008**	–0.006	–0.002	–0.009	–0.002
Immature 2	*4*	–0.016	–0.053	–0.015	**–0.008**	**–0.007**	–0.003	–0.010	–0.002
Subadult	*5*	–0.078	–0.250	–0.073	–0.040	**–0.032**	**–0.012**	–0.047	–0.011
Adult 1	*6*	–0.074	–0.239	–0.070	–0.038	–0.031	**–0.012**	**–0.045**	–0.011
Adult 2	*7*	–0.039	–0.125	–0.036	–0.020	–0.016	–0.006	**–0.024**	**–0.006**

ELASTICITY VALUES		*Class in year t*							
Class name	*Class in year t+1*	*0*	*1*	*2*	*3*	*4*	*5*	*6*	*7*
Yearling	*0*	0	0	0	0	0	<–0.0001	<–0.0001	<–0.0001
Juvenile 1	*1*	<–0.0001	<–0.0001	0	0	0	0	0	0
Juvenile 2	*2*	0	–0.0001	<–0.0001	0	0	0	0	0
Immature 1	*3*	0	0	–0.0001	–0.0001	0	0	0	0
Immature 2	*4*	0	0	0	–0.0001	–0.0005	0	0	0
Subadult	*5*	0	0	0	0	–0.0012	<–0.0001	0	0
Adult 1	*6*	0	0	0	0	0	<–0.0001	–0.0023	0
Adult 2	*7*	0	0	0	0	0	0	<–0.0001	–0.0001

[a]Sensitivity values that apply to nonzero matrix elements are shown in bold.

These formulas are easy to write, but using them is, in practice, difficult. If multiple vital rates combine to form each matrix element, calculating the variances and covariances of the elements can be difficult to do accurately-without simulating random matrices[10]. Similarly, determining the derivatives of the covariances between matrix elements as functions of the covariance between vital rates is not easy. Therefore, a matrix analysis based on vital rates essentially must resort to simulations in order to get the inputs for these analytical approximations for stochastic sensitivities. (Indeed, this is how we generated the covariance matrix used in Box 9.4 to generate the ana-

[10]Two alternatives are to use the delta method to estimate variances (see Footnote 7 in Chapter 5) or, if we have full demographic information from each year of our study, to use the matrix element approach of Chapter 7 to directly estimate element variances and covariances.

lytical results for the desert tortoise.) Even though in some cases it may be worth to taking this route, more often it is easier to simply use simulation methods directly to estimate vital-rate sensitivities. This is especially true since calculation of extinction risks will also require simulation methods, as we'll see below.

A SIMULATION APPROACH Using stochastic simulations to estimate sensitivity and elasticity values follows directly from the tactics we used to directly estimate sensitivities for deterministic matrix models. Essentially, we perturb each mean, variance, or covariance (or correlation) term one at a time, simulate the model long enough to arrive at a good estimate of λ_s and then estimate the stochastic sensitivity as

$$S_x^s = \frac{\lambda_{s,new} - \lambda_{s,original}}{x_{new} - x_{original}} = \frac{\Delta\lambda_s}{\Delta x} \quad (9.19)$$

where x is the entity (mean, variance, or correlation of the matrix element(s) or vital rate(s)) being perturbed[11]. After calculating the S_x^s values, stochastic elasticities are found by the usual conversions. The only caution in using this approach to calculate stochastic sensitivities is that many years of simulated population growth rates are needed to obtain an accurate estimate of λ_s for each perturbation of the vital rates (see Chapter 7). Box 9.5 shows how to use this approach to calculate the vital rate sensitivities and elasticities for the desert tortoise. It is best to use a small and consistent perturbation; we use a 5% change in each rate here. The results show several important patterns. First, as for deterministic sensitivities, stochastic population growth has, overall, higher sensitivity to the mean values of survival rates than to the means of growth rates, and fertilities have the lowest sensitivities of all (Figure 9.4), as is true for many long-lived species. A key advantage of conducting sensitivity analysis on the vital rates instead of the matrix elements is the ability to distinguish the very different effects of growth versus survival on the rate of population growth (contrast the patterns in "growth" versus "stasis" matrix elements in Table 9.5 with the growth and survival vital rates in Figure 9.4). Also, in contrast to the sensitivities of the matrix elements, survival of the youngest animals has a large effect on stochastic growth. This is largely because, as we noted about this example in the last chapter, we have grouped the survival of the three youngest classes together, so that a change in this rate simultaneously influences all three classes. These results also confirm the very low effects on the stochastic growth rate of the variance in vital

---

[11]In estimating sensitivities by this simulation approach, it can be more straight forward to use correlations than covariances, since correlations are distinct from and uneffected by the variances of the vital rates.

BOX 9.5 *A MATLAB program that performs stochastic simulations to estimate the sensitivities and elasticities of stochastic growth rate (λ_s) and extinction probabilities to mean, variance, and covariance of matrix elements.*

```
%StochSensSim: this program performs simulations to estimate
% stochastic growth rate and extinction risk sensitivity
% analyses for vital rate means, variances, and correlations.
% Notes in the Box 8.9 program explain the stochastic
% simulation methods used and the way to create some of the
% files that must be pre-made.
clear all;

%****************Simulation Parameters**********************
%the data shown here are for the desert tortoise:
% order of rates: survival2-7, growth2-6, fertility5-7
vrtypes= [ones(1,11),3,3,3]; % types of distributions to use
vrmeans= [0.77 0.95    0.79      0.93      0.91      0.87 ...
0.3   0.31      0.19      0.22      0.052 0.42   0.69  0.69];
vrvars=[0.0006       0.0006    0.053     0.0006     0.049 ...
0.02      0.0096 0.017     0.048     0.0006     0.0006 0  0  0];
vrmins=zeros(1,14);
vrmaxs=zeros(1,14);
load tortcorrmx; % a .mat file that contains a premade,
% corrected correlation matrix (corrmx) for the vital rates.
% You must also have a file with a MATLAB function to create
% the matrix elements from the vital rates.
% The program assumes that the function uses a vector 'vrs' to
% build the matrix. For tortoises, this file is maktortmx.m
makemx = 'maketortmx';
% Also, you must have a set of premade beta values to load:
load tortbetas  % (Box 8.10 has a procedure to make these)
Nstart = 500;   % total initial population size;
Nx = 50;         % quasi-extinction threshold.
tmax = 75;      % number of years to simulate;
np = 14;        % number of vital rates;
dims = 8;       % dimensions of the population matrix;
runs = 1000;    % how many trajectories to do
change = 0.05;  % the proportional magnitude of the
                % perturbations used to estimate sensitivities
%**********************************************************
rand('state',sum(100*clock)); randn('state',sum(150*clock));
[W,D] = eig(corrmx); % making the M12 matrix
```

BOX 9.5 *(continued)*

```
M12 = W*(sqrt(abs(D)))*W';
vrs = vrmeans;
eval(makemx)            % make a matrix of mean values
[uu,lam1] = eig(mx);    % find the eigenvalues and vectors
lam1s = max(lam1);
[lam0,iilam] = max(lam1s);    % find dominant eigenvalue
uvec = uu(:,iilam)/sum(uu(:,iilam));% dominant right e-vector
n0 = Nstart*sum(uvec)*uvec; % make the initial pop'n vector
ExtResults = [];
Lresults = [];
crow = 1; ccol = 2; % used to pick correlation mx elements
pertnum=(2*np + 0.5*(np^2 -np));
% initialize results variables
Extsens = zeros(tmax,pertnum); ExtElast = zeros(tmax,pertnum);
Lsens = zeros(pertnum,1); Lelast = zeros(pertnum,1);

% loop one-by-one through perturbations of each mean, var,
% correlation. For each type of rate, a small change is made,
% the appropriate set of parameters are redone, and then the
% new model is stochastically simulated.
for pert = 0:(2*np + 0.5*(np^2 -np));
  % set temporary parameters to estimated values
  Tvrmeans =vrmeans; Tvrvars=vrvars;
  Tparabetas=parabetas; TM12 = M12;
  %now, modify values, depending on what is changed:
  if pert > 0 & pert < np+1 % change a mean rate
   Tvrmeans(pert) = (1+change)*vrmeans(pert);
   % to guard against too high a mean:
   if Tvrmeans(pert)*(1-Tvrmeans(pert))<= vrvars(pert)
      Tvrmeans(pert) = (1-change)*vrmeans(pert); end;
   vrdiff = Tvrmeans(pert) - vrmeans(pert);
   Tparabetas(:,pert) =betaset(vrtypes(pert),Tvrmeans(pert),...
   Tvrvars(pert),vrmins(pert),vrmaxs(pert));
  elseif pert > np & pert < 2*np+1 % change a variance
   Tvrvars(pert-np) = (1+change)*vrvars(pert-np);
   vrdiff = Tvrvars(pert-np) -vrvars(pert-np);
   Tparabetas(:,pert-np) =betaset(vrtypes(pert-np), ...
   Tvrmeans(pert-np), Tvrvars(pert-np),vrmins(pert-np), ...
   vrmaxs(pert-np));
  elseif pert > 2*np % change a correlation
   Tcorrmx = corrmx;
```

BOX 9.5 *(continued)*

```
   Tcorrmx(crow,ccol) = corrmx(crow,ccol)*(1+change);
   Tcorr(ccol,crow) = Tcorrmx(crow,ccol);
   vrdiff = Tcorrmx(crow,ccol) - corrmx(crow,ccol);
   [W,D] = eig(Tcorrmx);
   TM12 = W*(sqrt(abs(D)))*W';
   % advancing counters thru indices for the upper 1/2
   % of the correlation matrix
   if ccol == crow+1;          % advancing counters thru
      ccol = ccol+1; crow = 1;   % indices for upper 1/2
      else crow = crow+1;       % of the correlation matrix
   end
end; % if pert

% now do a set of random runs for the perturbation
PrExt = zeros(tmax,1); % extinction time tracker
logLam = [0];          % log-stochastic lambda tracker
for xx = 1:runs;
 if round(xx/10) == xx/10 disp([pert,xx]); end;
 nt = n0; extinct = 0;
 for tt = 1:tmax
   m = randn(1,np); rawelems = (TM12*(m'))';
   for yy = 1:np % find vital rates for this year
       if vrtypes(yy) ~= 3
          index = round(100*stnormfx(rawelems(yy)));
          if index == 0 index = 1; end;
          if index ==100 index = 99; end;
       vrs(yy) = Tparabetas(index,yy);
       else
        vrs(yy) = lnorms(Tvrmeans(yy),Tvrvars(yy),rawelems(yy));
       end;% if vrtypes(yy)
   end; %yy
   eval(makemx); %make a matrix with the new vrs values.
   nt = mx*nt;   %multiply by the population vector
   if extinct == 0
       Ntot = sum(nt);
       if Ntot <= N
              PrExt(tt) = PrExt(tt) +1;
              extinct = 1;
       end; % if Ntot
   end; %if extinct
 end %time (xx)
 logLam = logLam +(1/tmax)*log(sum(nt)/Nstart);
 end; % xx
```

BOX 9.5 *(continued)*

```
if pert == 0 % establish the base-line results
   baseext= cumsum(PrExt./runs);
   baselam = exp(logLam/runs);
   origbaselam = baselam;
   % smooth the extinction CDF function:
   baseext=([baseext]+[baseext(1);baseext(1:tmax-1)] ...
       + [baseext(2:tmax);baseext(tmax)])/3;
else % summarize results for each perturbed model
   cumext=cumsum(PrExt./runs);
   origcumext = cumext;
   cumext=([cumext]+[cumext(1);cumext(1:tmax-1)] ...
       + [cumext(2:tmax);cumext(tmax)])/3;
   ExtResults(1:tmax,pert)= cumext;
   Lresults(pert,1) = exp(logLam/runs);

   if vrdiff ~=0
    % calculate extinction time sensitivities and elasticities:
    ExtSens(1:tmax,pert)=(cumext - baseext)/vrdiff;
    % next line will generate divide by zero errors: its OK
    ExtElast(1:tmax,pert)=((cumext - baseext)./baseext)/ ...
       (change*vrdiff/abs(vrdiff));
    % calculate stoch-lambda sensitivities and elasticities:
    Lsens(pert,1) =(Lresults(pert,1)-baselam)/vrdiff;
    Lelast(pert,1)=((Lresults(pert,1)-baselam)/ ...
       (baselam))/(change*vrdiff/abs(vrdiff));
   end; %vrdiff
   % smooth extinction elasticities:
   sExtElast(1:tmax,pert)=([ExtElast(1:tmax,pert)]+ ...
       [ExtElast(1,pert);ExtElast(1:tmax-1,pert)]...
       + [ExtElast(2:tmax,pert);ExtElast(tmax,pert)])/3;
end; %if, else
end; % pert loop
% ExtSens and sExtElast hold extinction sensitivities and
% elasticities for each year (rows) for each rate (columns,
% going through the vital rate means, variances, and then
% correlations in order).
% Lsens and Lelast hold the stochastic growth rate
% sensitivities and elasticities in the same order.
% Given their size, it is best to pull out pieces of these
% results for display. For example:
disp('Below are the stochastic growth elasticities for')
disp('survival rates from classes 2 to 7')
disp([Lelast(1:7)])
```

BOX 9.5 *(continued)*

```
disp('The figure shows extinction elasticities for')
disp('survival rates from years 25 to tmax');
figure
plot([25:tmax], sExtElast(25:tmax,1:7));
xlabel('Time');
ylabel('Extinction Elasticities');
```

rates for this long-lived, low-variance species (Figure 9.5). Indeed, the scattered pattern in these elasticity values is largely due to error from inadequate replication; a number of trajectories that is sufficient to estimate the stronger effects of mean vital rates may be too low to accurately gauge the much weaker effects of variances and correlations for species with this type of life history. However, if elasticity values are this low, they may not be worth the time to pinpoint more accurately. Finally, note that while we have just conducted this analysis on the basic vital rates, it is no more difficult to estimate sensitivity and elasticity for parameters describing management effects, just as we did in deterministic sensitivity analysis.

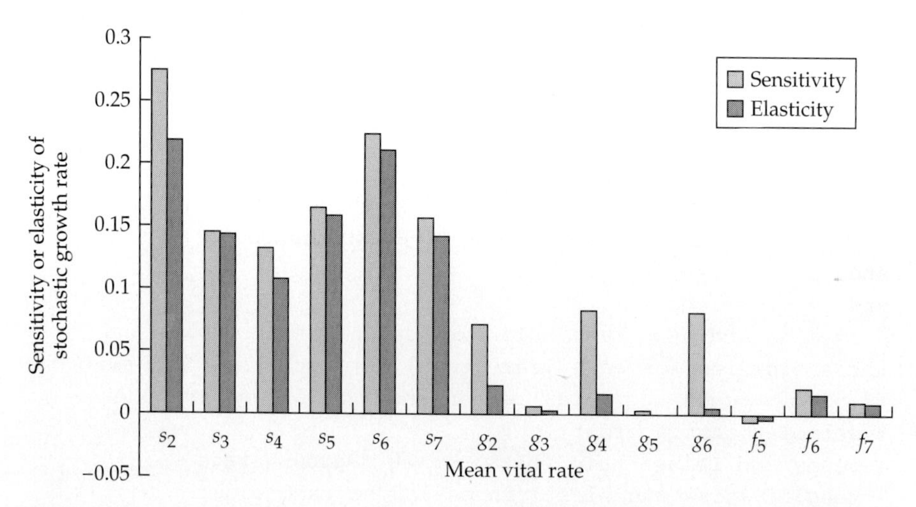

Figure 9.4 Sensitivity and elasticity of stochastic population growth rate to mean vital rates for the desert tortoise. These estimates were generated by the code in Box 9.5. (The negative values for f_5 are simply errors resulting from simulation-based estimation of very small true values.)

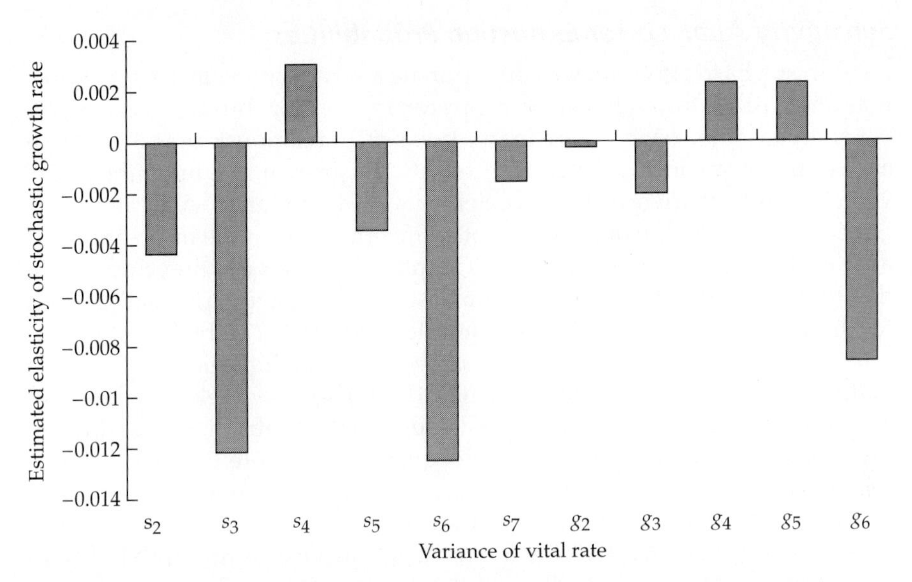

Figure 9.5 Estimated elasticity values of stochastic population growth rate in response to variances of vital rates for the desert tortoise.

Besides estimates of the simple stochastic sensitivity and elasticity values, a sensitivity analysis should also include more detailed consideration of the pattern in which the stochastic population growth rate changes over a range of values for important vital rates, and it should include some exploration of the potential for change in stochastic growth rates given biological limits to each vital rate. The only thing that makes these analyses different from the ones described for deterministic models is the need to run many years of simulations to generate reliable results. In general, these more in-depth explorations of the effects of different rates will focus on changes in mean vital rates, rather than variance and covariance of rates. It is rare that variation and correlation of vital rates can be altered through management of a population in the wild,[12] and typically (but not always) these variance terms will have lower sensitivity values than will the mean rates.

[12]Possible methods to reduce variation include supplemental feeding during food shortages and provision of shelter to mitigate extremes in environmental conditions. However, note that most strategies for reducing the variance in vital rates will probably also change the means. Correlation patterns are also unlikely to be subject to management, as they will often reflect immutable life history tradeoffs or common responses of individuals in different life history stages to strong environmental drivers.

Sensitivity Analysis for Extinction Probabilities

As with stochastic growth, we could pursue a sensitivity analysis for extinction times either through analytical approximations or through simulations. An analytical approach would marry two sets of relationships: the dependence of the mean and variance of the stochastic growth rate upon the mean, variance, and covariance of matrix elements (and ultimately, vital rates) (e.g., Equation 7.13) and the dependence of extinction risks on the mean and variance of the population growth rate (Chapter 3). It then requires differentiation of the formulas for the different measures of extinction risk with respect to different vital rates. Although a simple expression can be found for the sensitivity of the ultimate extinction probability (see Equation 3.7) to mean matrix elements (Caswell 2001, Equation 14.147), for more relevant measures of extinction risk, such as the time-dependent probabilities of extinction described by the extinction time CDF (Equation 3.5), there is no such simple relationship. Therefore, we will pursue a simulation approach to estimating the sensitivity values for extinction risks instead.

As just implied, we will focus on the sensitivity of the probability of extinction by a certain time, $G(t)$, as the most useful measure of extinction risk. Even here, there are two ways to proceed. One is to set a single time horizon over which to evaluate sensitivities. In this case, the task is no more difficult than it was for estimation of sensitivities for stochastic population growth rates. However, it is likely that the sensitivity of extinction risk to changes in matrix elements may be quite different depending upon the time horizon chosen, and plotting the time-dependence of sensitivity results could be a useful way to ferret out this effect. In other words, unlike sensitivities for growth rates, sensitivities for extinction times can be time-dependent, since extinction risks themselves are time-dependent functions. In practice, a combination of setting *a priori* time horizons of interest and plotting the full sensitivity functions against time makes the most sense, since some quick summary of the vital rates that have the highest sensitivities can be used to pick out those rates are worth examining in more detail for time-dependent patterns.

The code in Box 9.5 is designed to estimate sensitivities to extinction risk as well as population growth, running replicate populations for short time periods to calculate the extinction time CDF for each perturbation of a mean or covariance of vital rates. (Box 7.5 shows a similar procedure for obtaining the CDF using random draws of entire matrices.) The result of this simulation is a matrix with a separate column of $G(t)$ values for each perturbation of a vital rate. We can then estimate the sensitivity of the extinction probability as:

$$\frac{G(t)_{new} - G(t)_{original}}{x_{new} - x_{original}} = \frac{\Delta G(t)}{\Delta x} \tag{9.20}$$

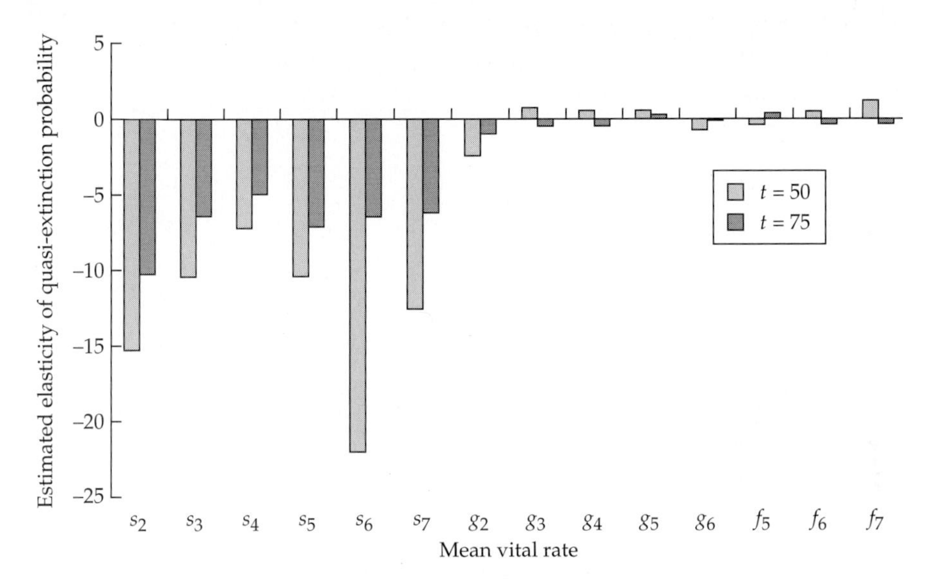

Figure 9.6 Estimated elasticity values for cumulative quasi-extinction probabilities at 50 and 75 years, in response to mean vital rates for the desert tortoise. These estimates were generated by the code in Box 9.5. Populations were started with 500 individuals and the quasi-extinction threshold was set at 50.

where x stands for the mean, variance or covariance being perturbed. Before applying Equation 9.20 to the CDF estimates, it is a good idea to smooth these values. If a very large number of replicate simulations have been done, smoothing should not be needed, but in general taking a three-year running average of the extinction probability values will yield better results. (If this smoothing function still results in a very jagged CDF, you need to run more replicate simulations to estimate extinction risks.) These sensitivity values can, in the usual way, be converted to proportional changes to yield elasticities.

We ran this procedure for the desert tortoise, generating the estimates of elasticity values for $G(50)$ and $G(75)$ shown in Figure 9.6. The pattern of sensitivity and elasticity values for these extinction probabilities are very similar to those for the stochastic growth rate, with the lion's share of the effects on extinction risk coming from changes in mean survival rates. Indeed, we can plot elasticity values for the stochastic growth rate versus those for $G(75)$ and see a very strong correlation (Figure 9.7). Another similarity is that none of the variance or covariance terms have high elasticity (or sensitivity) values for extinction probability (results not shown). However, we caution yet again that species with more variable demography may show strikingly different patterns.

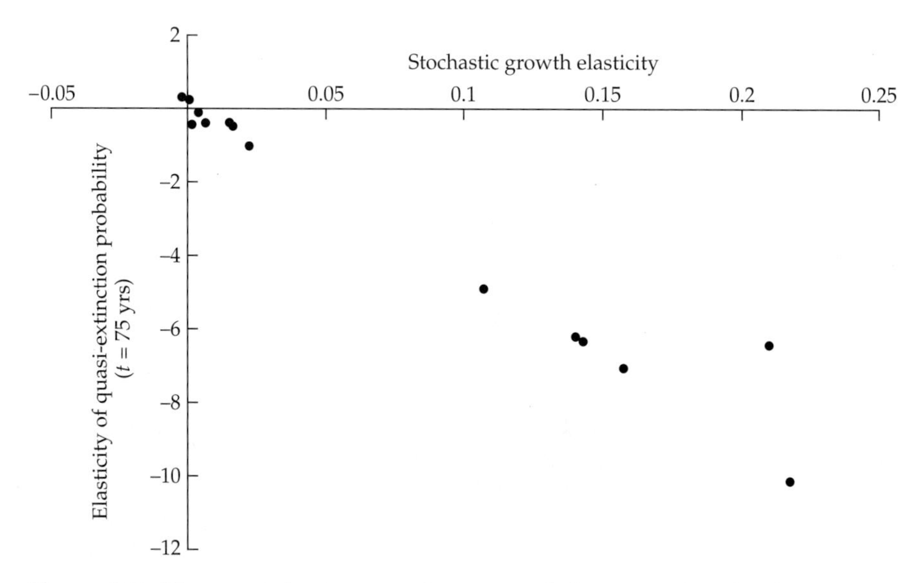

Figure 9.7 The correlation between the estimated elasticity values for the stochastic population growth rate and those for the cumulative quasi-extinction probability. Each point is a pair of elasticity values for a mean vital rate for the desert tortoise.

On the basis of these values, it seems worth plotting the full time-dependent elasticities for the mean survival rates (Figure 9.8). Note that all these elasticity values are negative, since increasing survival rates will decrease extinction probabilities. These time-dependent plots show that the sensitivity of extinction risk can be dramatically different, depending on the time horizon we look at. For this model, survival rates have elasticities that are stronger (more negative) for short-term extinction risk than for longer-term risk. This pattern is due to the overall negative growth rate of the population: Small increases in survival rates can delay extinction, but cannot stave it off forever. The pattern for s_6, survival of the major reproductive class, is most interesting. Although this class comprises a relatively small fraction of the entire population, it produces the majority of offspring. Because of their small numbers, change in their survival rate doesn't strongly change very near-term extinction risk, but because their survival leads to more reproduction over the medium-term, there is a strong effect in the range of 35–45 years in the future. Overall, plots of extinction sensitivities as a function of time can give considerable insight into how and why changes in vital rates may influence viability.

Finally, it can also be useful to ask the remaining questions we addressed in sensitivity analyses of the deterministic population growth rate: How dependent on particular vital rate estimates are the sensitivity results we found? And, how much can extinction risks be changed by biologically real-

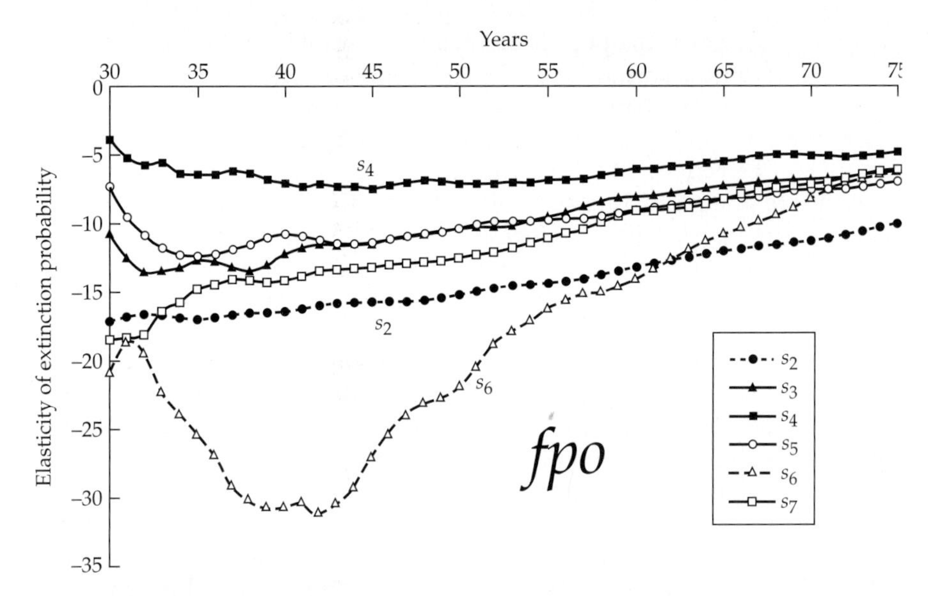

Figure 9.8 Changes in the estimated elasticity values for cumulative extinction probabilities in response to mean survival rates for the desert tortoise. Elasticities for <30 years are not plotted; so few extinctions occur before this time that extinction probabilities, and hence elasticity values, are highly variable and unreliable.

istic alterations in means, variances, or covariances in vital rates, including management that changes multiple rates? Answering these questions for extinction probability is identical to doing so for population growth, simply requiring simulation of multiple trajectories for each set of parameter values to ascertain average behavior in the face of environmental (and, if included in the model, demographic) stochasticity.

Sensitivity Analysis for Density-Dependent Models

As we saw in Chapter 8, the dynamics of a structured population when one or more of the vital rates are density-dependent differs dramatically from the density-independent case. Most importantly, we can no longer summarize the behavior of the population by a single metric describing its growth rate (i.e., λ_1 in the deterministic case and $\log\lambda_s$ in the stochastic case). Instead, the population growth rate (either its perfectly predictable value in a deterministic model or its average value in a stochastic one) changes in response to population density. In this setting, it is meaningless to ask about the sensitivity of the population growth rate to changes in vital rates or matrix elements because there is no single-most-important population growth rate about which to ask this question. In fact in the deterministic case, if the population reaches a stable equilibrium size and structure under the influence of density dependence, its growth rate will (by definition) be 1; changing the vital rates may shift the position of this equilibrium, but once the new equilibrium is reached, the growth rate will again be 1. Effectively, the deterministic sensitivities of the population growth rate at equilibrium (or the average population growth rate for a cyclic or chaotic population once it has settled into a steady pattern of oscillation) to changes in vital rates are always zero, because the equilibrium or long-term average growth rate is always 1 regardless of the values of the vital rates.

However, in the deterministic, density-dependent case, we *could* ask about the sensitivity of the equilibrium population size itself to changes in the vital rates. In a sense, this is the obverse of the density independence coin: On the density-independent side, the long-term population growth rate has a meaningful interpretation, but there is no equilibrium population size (that is, barring the unlikely case of λ_1 being exactly equal to 1, the population either grows indefinitely or declines to extinction), whereas on the density-dependent side there is a (possibly unstable) equilibrium but the long-term growth rate is of no real interest to us. An analytical formula for the sensitivity of the equilibrium population size to a change in a matrix element is available (see Equation 16.111 in Caswell 2001; also see Grant and Benton 2000), assuming that the density-dependent elements in the matrix are functions of a weighted sum of the numbers of individuals in each class (for example, the salmon matrix in Chapter 8 fits this description; see Equations 8.20 and 8.21). In this case, an equilibrium sensitivity is proportional to the corresponding sensitivity of the population growth rate of the density-dependent projection matrix evaluated at the equilibrium (i.e., the familiar sensitivity S_{ij}, but calculated with the equilibrium population size substituted into the density-dependent functions in the matrix). Using that formula, we could always calculate the sensitivities of the equilibrium population size to the underlying vital rates by using the chain rule (see Footnote 2).

One might be tempted to believe that the deterministic sensitivities of equilibrium population size in a density-dependent model would be useful in the context of PVA, on the basis of the following argument. The larger the size of a population, the farther it will be from a realistically low quasi-extinction threshold. Hence it might be advantageous to know which vital rates would, if altered, cause the largest increase in the equilibrium size toward which the population will tend, which is just what the deterministic equilibrium sensitivities tell us. There is, however, one problem with this argument. Large population size confers safety from extinction because it reduces the chances that the population will be carried to the quasi-extinction threshold *by stochastic forces*, such as environmental and demographic stochasticity. In a purely deterministic world, the equilibrium population size is not terribly valuable as a surrogate for population viability. *Any* stable equilibrium above the threshold is immune from extinction. And if the equilibrium is unstable, leading to population fluctuations, we are less interested in the size of the equilibrium than we are in knowing whether changing the vital rates will cause the fluctuations to be severe enough that they will carry the population across the threshold. So what we would really like to have for the purpose of assessing how changes in vital rates would change population viability are not the *deterministic* sensitivities of the equilibrium population size, but the *stochastic* sensitivities of the probability of quasi-extinction by a specified future time.

Unfortunately, there are no analytical formulas for the stochastic sensitivities of the extinction probability for density-dependent projection matrices. Just as in the density-independent case, we must rely on simulation to approximate these sensitivities. However, once we have written computer code to calculate the density-independent sensitivities of the extinction probability (see Box 9.5 and Equation 9.20), the small modification of replacing some of the vital rates with density-dependent functions that update the matrix on the basis of the current population size at each time step renders the code usable for calculating the density-dependent sensitivities. Thus if we have the data to fit density-dependent functions describing the vital rates and to estimate the magnitude of the environmentally driven variation around the vital rate values those functions predict, we can perform simulations to ask how the quasi-extinction probability would change with changes in the vital rates. For those vital rates that are density-independent, it makes the most sense to ask about how changing the mean values of those rates would alter the extinction probability. For the density-dependent vital rates, the mean value changes with density. Therefore, it makes more sense to ask how changing the parameters in the density-dependent functions describing those rates (e.g., the parameter β in the salmon matrix in Chapter 8) would change the extinction probability. For example, let's say that the first-year survival of salmon eggs declines as the total number of eggs increases because redds (nests) produced by early-spawning adults are displaced by those of later spawners, and that we are using a Beverton-Holt function (see Equation 8.19) to describe egg survival. Increasing the amount of spawning habitat would increase the proportion of eggs that survive at any given egg number; that is, it would *decrease* the value of β. Thus by simulating the probability of extinction at the estimated value of β and at smaller values, and then using Equation 9.20 (with β in place of x) to quantify the sensitivities, we could make management recommendations regarding how effective increasing the amount of spawning habitat would be in reducing extinction risk.

Summary

Discussion of sensitivity analysis for matrix models has burgeoned in the last ten years, with the introduction of increasing numbers of methods and concerns, including many techniques we have not covered here (e.g., sensitivity of the short-term population structure). Instead of covering all this work, we've emphasized the methods that are most useful in a conservation setting, and in particular those that are of clear relevance in deciding among different management alternatives. Furthermore, with the techniques and ideas we've presented you should be well-equipped to understand the new sensitivity techniques that will undoubtedly emerge over the coming years.

We have now finished with the major methods used in single population PVAs, both count-based and demographic. In the final set of chapters, we address the problems of viability analysis for multiple populations, an area that will use many of the methods we've already covered, but usually with some modifications and added difficulties.

10

Population Dynamics across Multiple Sites: Interactions of Dispersal and Environmental Correlation

In Chapters 3 through 9 we have considered the methods available for estimating the viability of single populations. As the term *population* viability analysis implies, this focus reflects the general emphasis in the PVA literature on single population models and predictions. In spite of this, and even though the great majority of data sets available to conservation biologists are for only one site or population, many of the most pressing questions that PVA can help to address involve multiple populations and spatially complex situations. How many populations are needed to ensure a high probability of survival for a species? To what extent should multiple populations be clumped together in space versus spread apart? Can small populations or those occupying sites with low habitat quality substantially add to regional viability of a species? Even the most endangered of species often occur in more than one—or even more than five—sites (Figure 10.1, Press et al. 1996, Dobson et al. 1997). Thus, the question of how best to protect species from extinction frequently involves the analysis of populations arrayed across multiple sites with real, idiosyncratic spatial configurations. In this and the following chapter, we address the analysis of these potentially complicated situations.

Multi-site situations are usually quite complex, with each site having different demographic rates, movement probabilities, and correlations with the dynamics of other sites. Reflecting this complexity, spatial population dynamics is an area of ecology rife with complicated models. However, the data available for most species of concern do not include enough or the right kinds of information to support such complicated modeling approaches. Therefore, even more than in previous chapters, we will take an approach here and in

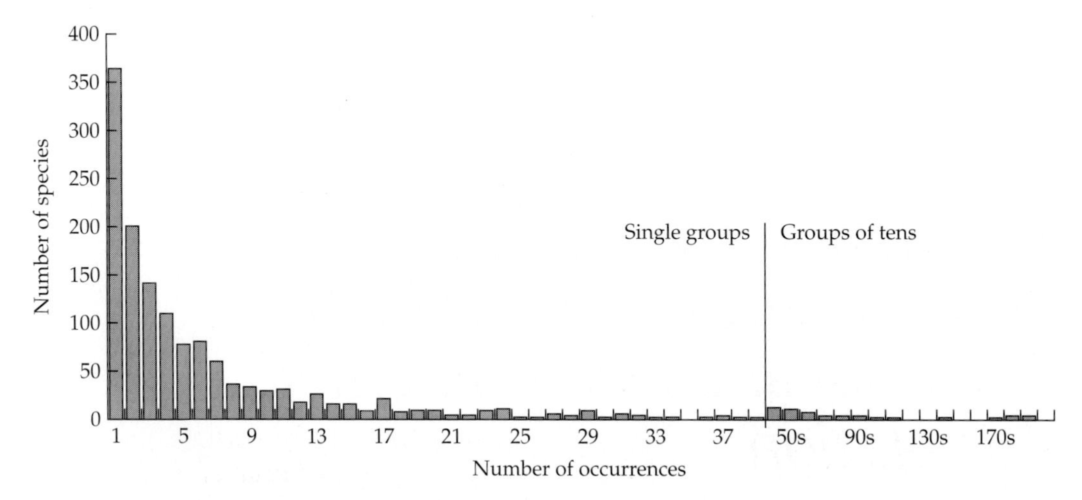

Figure 10.1 The numbers of individual occurrences (sites or populations) of high priority rare species in the United States, compiled for The Nature Conservancy by Mark Shaffer (after Morris et al. 1999).

Chapter 11 that distinguishes useful methods for viability assessment in the face of realistic data limitations from more ideal analyses that simply require too much information.

Just as for a single population PVAs, there are different approaches to the analysis of multi-site viability. Two of these approaches are comparable to the count-based and demographic approaches to single population PVA that we have already discussed in detail. For a multi-site analysis, a *count-based approach* is a direct extension of the single population count-based approach, but with the added complication that population growth rates (and any other parameters, such as those defining density dependence) are now site-specific, and that probabilities of movement between sites and environmental correlations between population growth rates at different sites must be determined. A *demographic approach* to multi-site PVA, like that for a single population, tracks the numbers of individuals in each of several classes defined by age, size, or stage, but also classifies individuals by the site in which they are currently located. For this approach, we must also quantify class-specific probabilities of moving between sites, and the environmental correlations in vital rates among sites. For multi-site PVAs, there is also a third, more broad-brush class of models that are included in the so-called *patch-based approach* (Kareiva 1990), in which we keep track only of whether each site has an extant population or not, without explicit regard to the details of population dynamics, size, or structure within each site. For these models, the main parameters governing population behavior are an annual extinction probability and the probability that an unoccupied site is colonized by dispersers from some other population.

We begin this chapter by defining terminology we will need to consistently describe multi-site PVAs. The bulk of the chapter is then devoted to a review of the three main processes governing multi-site viability, with an emphasis on the two "emergent properties" of multiple populations that we briefly introduced in Chapter 2, namely among-population correlations in environmental fluctuations and movement of individuals among populations. Movement between populations is the glue that binds together otherwise separate populations inhabiting different sites. Perhaps less obviously, correlations in the vital rates and population growth rates at different sites can also link the fates of separate populations (even in the absence of movement), with important consequences for multi-site viability. At the end of the chapter, we use the degrees of correlation and movement among populations to classify different multi-site situations that require more or less complicated data collection and modeling approaches. In Chapter 11, we present the details of the most useful modeling methods for multi-site PVAs, giving examples of their use and outputs.

Terminology for Multi-Site PVAs

The terminology used to describe multi-site populations is not standardized and can be quite confusing. We will use the term *site* to refer to a discrete piece of habitat that has some potential (not necessarily very high) of maintaining a population of the species of interest. These sites may be juxtaposed to one another, as in the case of two dissimilar but adjacent habitats, or they may be separated by nonhabitat. The group of individuals living on a site will be called a *population,* while the individuals across all sites will be called the *total* or *multi-site population*. Any collection of populations is often generically called a *metapopulation*, but we'll reserve this term for its original and more specific meaning: sets of discrete, largely (but not entirely) independent populations whose dynamics are driven by local extinction and recolonization via movement from other populations (Levins 1969). We use *movement* to mean any form of dispersal, colonization, etcetera that results in an individual from one population arriving at another one and surviving long enough to breed or be censused there. *Dispersers* are simply the individuals who enact this movement. *Colonizers* are the special group of dispersers that reach an unoccupied site and found a new population.

Multi-Site Processes and Data Needs

Fully understanding and modeling the operation of a set of populations requires all the information to do a good job of a single-population PVA for each site, plus data on movement rates between populations, plus estimates of how temporal fluctuations in population processes (e.g., birth and survival rates, movement probabilities, etc.) are correlated between populations. In the following sections, we discuss why each of these three types of data is needed, and we review some of the basic methods for obtaining them. We focus on methods that can be used for demographic and count-based PVAs. With only patch-based information, estimation of all three types of data is highly convolved, relying on methods we will discuss in Chapter 11.

Data Requirement 1: Site-Specific Population Dynamics

Understanding how a single population will contribute to the viability of a collection of populations requires knowing something about its inherent viability, or, more generally the "quality" of the population. For example, regardless of any fancy spatial population dynamics, if a single population has a probability of extinction of 0.1% over 500 years, it is of high quality and will, if preserved, ensure that total extinction of a collection of populations is very unlikely. However, it is important to emphasize that the converse is *not* necessarily true: A population that in isolation has a high chance of extinction may still play a key role in the persistence of a suite of sites, especially if they function as a true metapopulation (Pokki 1981, Crone et al. 2001). But all in all, the higher the inherent viability of a population, the more likely that it will serve to bolster multi-site viability. *Ideally*, this means that for every population being considered in a multi-site PVA, we would have the data needed to conduct a separate single-site PVA, either using count-based methods (Chapters 3–5) or a demographic analysis (Chapters 6–9). Since we have covered the gathering and use of these types of data already, we don't repeat ourselves here.

But how likely is it that for every population of an endangered species, many years of census data, let alone estimates of all vital rates, will be available? Even for common species, the norm is for data to have been collected at only one or at best a handful of sites, over a small fraction of a species' range, and for only one or a few years (Figure 10.2). Although this is not surprising (good data, especially demographic data, are extremely time consuming and expensive to collect at even one site), it means that the multi-site PVA practitioner usually has very few data with which to quantify, or even rank, the demographic quality of different sites. A common approach to this problem is to assume that population growth rates or vital rates (or their means and variances in a stochastic setting) are identical at all sites, but that carrying capacities differ among sites (e.g., Lahaye et al. 1994, Akçakaya and Atwood 1997). For example, Lahaye et al. (1994) had demographic data for only one of 22 local populations of the California spotted owl in southern California, and thus assumed that demographic rates were identical in each location. However, Lahaye et al. also had estimates of the current carrying capacity of each habitat area, and so used different local caps to territory number. Although we often must resort to such simplifying assumptions, we should be aware of

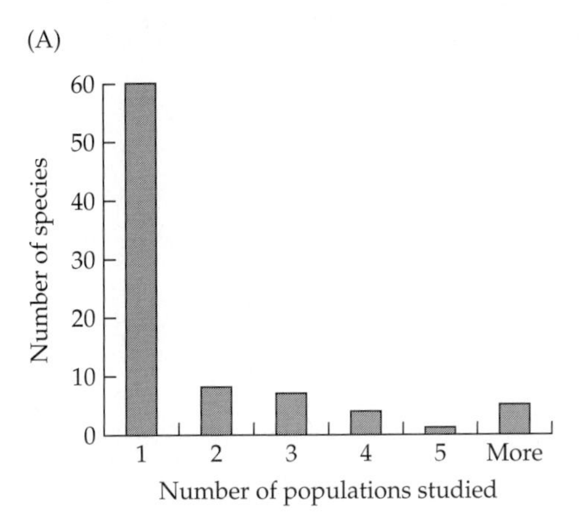

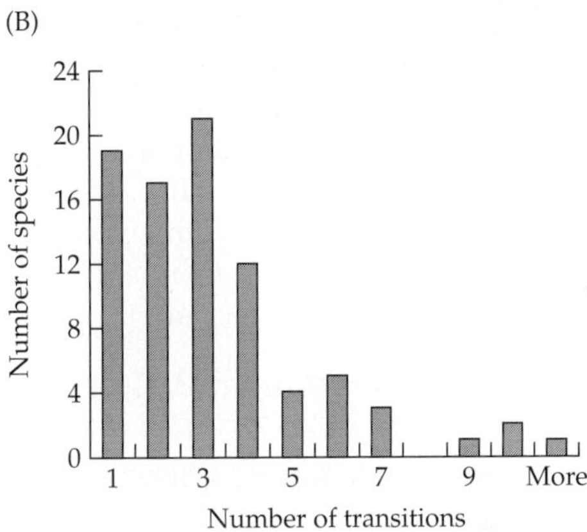

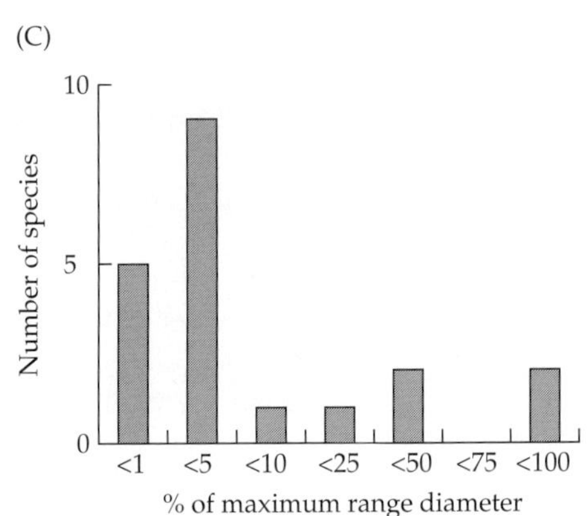

Figure 10.2 Spatial and temporal extent of published demographic studies for 85 species of terrestrial plants. (A) number of populations studied per species; (B) maximum number of annual transitions per species; (C) maximum distance between sites for species studied at more than one site (as a percentage of maximum range diameter, as stated in the source or determined from floras). The two species with the widest geographic coverage are Furbish's lousewort, an endemic found only along a single river in Maine (Menges 1990), and royal catchfly (Menges and Dolan 1998).

their potential consequences. If larger populations are actually better off demographically, this approach will underestimate the contribution of large sites to multi-site viability. Although perhaps less likely, the reverse pattern could lead us to underestimate the importance of smaller populations.

A second approach to limited data is to assume that most demographic rates are identical across sites, but to allow a handful of rates, about which more information is available, to differ. For example, we may know that certain demographic rates show a strong correlation with site and/or population size. For many forest-nesting birds, reproduction is much poorer in small fragments than in large ones. An especially well-studied example is the ovenbird, an eastern North American warbler that breeds in deciduous forest habitats. Porneluzi and Faaborg (1999) found that ovenbirds in fragmented and unfragmented forest landscapes had similar mating success, survival, and territory sizes, but that forest fragmentation strongly decreased reproductive success, with a lower percentage of pairs successfully raising young in small fragments due to increased cowbird parasitism. Furthermore, this general pattern of fecundity effects also holds for a diversity of other forest-nesting neotropical migrants (Donovan et al. 1995, Robinson et al. 1995). With knowledge of this general relationship between fragment size and reproductive success, one could assess the multi-site viability of sets of populations occupying forest fragments of many different sizes, without comprehensive demographic data from each site. (Although one would of course need to know the size of each site).

Data requirement 2: Correlations in Population Growth Rates or Vital Rates across Sites

THE IMPORTANCE OF CORRELATIONS Through "safety in numbers," multiple populations can strikingly decrease the risk of total extinction of a species. However, this benefit critically relies on a lack of correlation in the dynamics, and hence risks of extinction, of the different populations. By analogy, consider the risk that multiple towns in the southeastern United States will all be hit by a hurricane in one year. If you consider towns that are all quite distant from one another, and thus are unlikely to encounter the same hurricane (i.e., that have uncorrelated fates), then adding more towns will lead to a rapid decline in the chances that all will be damaged. In contrast, if you choose towns that are relatively close together, their chances of being hit in any one year are strongly correlated, and increasing the number of sites will only modestly decrease the risk of comprehensive damage. For the same basic reason, positive correlations between the population growth rates (or vital rates) in separate sites increase the chances of total extinction of a species.

As an example of such correlations, consider the threatened California clapper rail. From 1991 to 1996, the U.S. Fish and Wildlife Service conducted winter censuses of the rail in four marshes in San Francisco Bay.[1] Using the resulting population estimates, we calculated λ_t for each marsh in each year (Figure 10.3).[2] Because there is little or no movement of rails between these now isolated marshes (Harding et al. 2001b), it is reasonable to use the methods in Chapter 3 to calculate the probability of extinction of rails in each marsh separately, based upon the mean and variance in population growth

---

[1]Even though four marshes were surveyed, we use only the three with the best data for this example.

[2]There are two complications with this example that we will ignore here. First, we do not correct our estimate of σ^2 to account for sampling variability (see Chapter 5) because estimation of population sizes was not done using a straightforward counting procedure. Second, these data were collected while active management to reduce predation of clapper rails (primarily by introduced red fox) was being initiated. Thus, the mean and variability in population growth rates were in part driven by active management efforts that varied across time and between marshes.

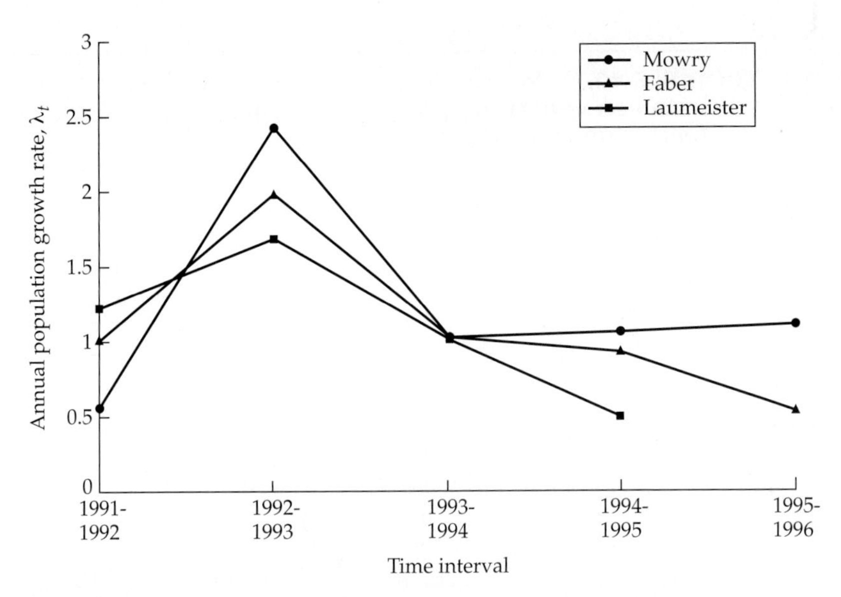

Figure 10.3 Annual population growth rates for populations of California clapper rails inhabiting three different marshes on the perimeter of San Francisco Bay.

and the estimated numbers in 1996 (Table 10.1). Although the Mowry population has by far the highest viability, it still has a significant extinction risk (0.06 over 50 years). How much added safety is gained by protecting the less viable Faber and Laumeister populations as well? The most obvious way to estimate the chance that all three populations will become extinct is to calculate the product of their individual extinction risks: $0.06 \times 0.79 \times 0.72 = 0.034$. In other words, while they individually have low viability, the Faber and

TABLE 10.1 Log population growth rates and extinction risk for three populations of the California clapper rail[a,b]

Population	*Numbers in 1996*	μ	σ^2	*Probability of quasi-extinction by 50 yrs*
Mowry	70	0.043	0.051	0.06
Faber	29	−0.002	0.041	0.79
Laumeister	33	0	0.051	0.72

[a]Data from Harding et al. 2001b.

[b]Extinction risk calculated for a 50-year time horizon, using an extinction threshold of 20 and four estimates of annual growth for Laumeister and five for Mowry and Faber.

TABLE 10.2 Pearson correlation coefficients between values of λ_t for three populations of the California clapper rail[a]

	Mowry	*Faber*
Mowry	1.000	
Faber	0.995	1.000
Laumiester	0.896	0.938

[a]Data from Harding et al. 2001b.

Laumeister populations would appear to nearly halve the total risk of extinction that would be faced by the Mowry population alone. However, this safety in numbers argument rests on an implicit assumption that the dynamics of the populations are uncorrelated, so that their probabilities of extinction are likewise uncorrelated. If that were true, we would be justified in calculating the risk of total extinction as the product of the three risks, just as we can calculate the probability of getting three heads when flipping three fair coins as ½ × ½ × ½. In fact, the dynamics of all the populations are highly correlated (Figure 10.3, Table 10.2), meaning that the same run of poor years that would drive one population extinct over our 50-year time horizon would also be very likely to cause the extinction of the other populations. In cases such as this, when population dynamics are correlated, estimating combined extinction risk is considerably more complicated, as we discuss below and in Chapter 11.

In general, populations of the same species are likely to be affected in the same way by widespread fluctuations in the weather or in the abundances of resources or natural enemies, or by broad-scale anthropogenic effects (e.g., lake acidification or hunting pressures; Novaro et al. 2000). Such factors lead to positive correlations in population dynamics. Indeed, surveys for the strength and frequency of spatial correlations in population counts have found moderately positive correlations at fairly extensive spatial scales for many taxa (Ranta et al. 1995, 1998), although at more than 100–200 km these correlations often disappear (Koenig 1998).

In contrast to these widespread positive correlations, strong negative correlations in population dynamics can also occur in some circumstances, dramatically reducing overall extinction risks. With negative correlations, a particularly poor set of years for one population is likely to be very good for another, so that simultaneous extinction is much less likely. For the threatened bay checkerspot butterfly, winter rainfall interacts with the aspect of hillsides (i.e., the compass direction in which the slope faces) to determine the most successful sites for growth, survival, and timely adult emergence (Weiss et al. 1988, Weiss et al. 1993). Because of this strong interaction, there is a negative correlation in the relative success of larvae on south- and north-facing slopes across years, with a likely reduction in extinction risk due to variable weather conditions.

ESTIMATING CORRELATIONS IN VITAL RATES If you have estimates of vital rates for multiple populations in multiple years, it is a simple matter to calculate between-site correlations for all pairs of rates. The resulting set of correlation coefficients can then be used to simulate correlated changes in demography in different populations through time, using exactly the same approach described in Chapter 8 for correlation of within-population rates (we will see how to do so in Chapter 11). As an example of correlations in vital rates, we can use data collected by Schmalzel et al. (1995) on the demography of *Coryphantha robbinsorum*, a threatened cactus in Arizona. Schmalzel et al followed individual growth and survival for three size classes, using three study plots with different aspects (e.g., north-facing versus south-facing) and collecting data from 1989 to 1993, resulting in four estimates for each vital rate. The resulting 153 correlations in survival and growth probabilities, both within and between plots, includes almost equal numbers of positive and negative values, showing that variation in vital rates among the plots is not highly

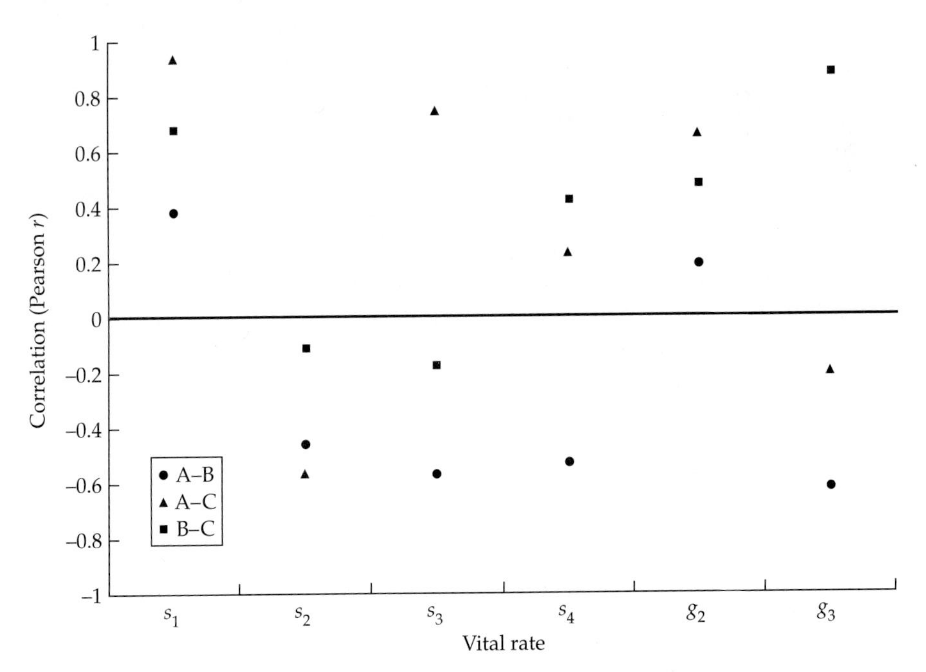

Figure 10.4 Correlations in the vital rates for *Coryphantha robbinsorum*, a threatened cactus, measured in three study plots across four annual transitions. Correlations of the same rates in different plots are shown (e.g., the correlation between size class 4 survival in plots A and B). s_i and g_i are survived and growth rates of class i, respectively. Data from Schmalzel et al. 1995.

synchronized. Considering just the correlations between the same vital rates in different study plots, it appears that the survival rates of the smallest plants (class 1) are more positively correlated across sites than are the survival rates of larger individuals (Figure 10.4). This pattern is probably explained by the greater effects of rainfall on the performance of smaller individuals (Schmalzel et al. 1995). In any case, the many near-zero and even negative correlations indicate that considerable safety from extinction may come from different local populations living at sites with different aspects. We will use these results to parameterize a demographic multi-site PVA in the next chapter.

ESTIMATING CORRELATIONS IN POPULATION GROWTH RATES As we illustrated above with data from the California clapper rail, if we are starting with census data for multiple populations, we can simply calculate the annual population growth rate for each site, and then estimate the correlation in annual growth rates for each pair of populations. However, the interpretation of mean growth rates and especially of correlated dynamics can be seriously complicated if movement of individuals occurs between populations (Ranta et al. 1995, Sutcliffe et al. 1996, Lande et al. 1999, Kendall et al. 2000). Ideally, we would calculate the correlation in the intrinsic population growth rates at different sites, unaffected by movement. If movement rates are low, they are unlikely to have a large effect on estimated growth rates, and hence on correlations in these rates through time. However, higher movement probabilities can substantially alter population growth. This is particularly true if some populations are strongly subsidized by others, making much of their observed population growth a product of immigration from other sites. This in turn can lead to stronger positive correlations in observed population changes than exist between the intrinsic local population growth rates themselves. Dispersal could also generate negative correlations, especially if the probability of an individual dispersing is density-dependent, so that high density in one population leads to large numbers of dispersers moving to neighboring populations.

One way to deal with this problem is to estimate the number of individuals emigrating from and immigrating into a population in each year. We can then calculate population growth rates that account only for within-population reproduction and survival. (For an example of such an analysis for one population, see Middleton and Nisbet 1997.) However, movement information of this quality is difficult to get. Furthermore, even with perfect data on dispersers, this approach will work properly only if population dynamics are density-independent. As Kendall et al. (2000) show, with density dependence, dispersal, and environmental correlation interact nonadditively to create correlated population dynamics. Similarly, Lande et al. (1999) emphasize that the strength of density dependence will strongly influence dispersal's role in generating synchronous population dynamics. Lele et al. (1998) have developed a model-fitting approach to properly account for these complexities by simultaneously estimating density-dependent population growth, dispersal, and spatial correlation from spatially arrayed count data taken through time. However, although it is not yet thoroughly tested, it seems unlikely that this method will be able to untangle dispersal effects and environmental correlation patterns in the scanty data sets usually available for rare species. Thus, for the moment, the method of "undoing" dispersed by reassigning migrants to their original populations before calculating growth rates may be the only practical way to try to reduce bias in correlation estimates in most conservation situations.

Finally, we should note that for both demographic and count-based data sets, we are suggesting the estimation of separate correlations for each pair of populations, rather than using formal spatial correlation methods (Sokal and Oden 1978a,b, Scheiner and Gurevitch 1993) . Spatial correlation analyses and other model-fitting procedures, (e.g. Lele et al. 1998), seek to estimate how correlated different variables are as a function of distance, whereas we are usually interested in the correlations between *particular places*, not in the estimation of general distance effects on correlation.

ESTIMATING CORRELATIONS WITH PATCH-BASED DATA In contrast to demographic and count-based data, with only patch-level data on presence-absence patterns or extinction and colonization events, general patterns of correlation with distance are all we can estimate. Furthermore, with only patch-level data the problems of disentangling effects due to dispersal between populations versus environmental correlation are much worse. To see this, think of a set of habitat sites, some with and some without extant populations. Perhaps we have only static data on the presence or absence of a population at each site at one point in time, and there is strong aggregation of extant populations. This pattern could have arisen from high movement rates, with colonists from a few founding sites creating some lucky "neighborhoods" with a high proportion of the patches occupied, while other unlucky areas have few or no extant populations. However, just as reasonably, there may be strong spatial correlations in extinction risk, leading the neighborhoods in good regions to have many populations, while others have few.

The same problem applies to data on extinction and colonization rates, because both dispersal and correlated environmental factors can generate spatial correlations in population behavior. For example, spatial correlation in extinction risk over time can create some regions with few populations and a high risk of extinction, and other regions with many populations and low observed extinction risk. Unfortunately, this same pattern is usually attributed to colonization and rescue effects (i.e., immigrants from nearby populations reducing extinction risks at neighboring sites) when incidence functions or logistic regressions are used to estimate colonization and extinction rates for metapopulations (which we discuss in detail in Chapter 11).

Because of this ambiguity in the interpretation of patch-level information, we suggest starting any analysis with an exploration of spatial correlations in the presence-absence patterns and the extinction and colonization rates without initially attributing these correlations to any particular process. The method of join-counts is usually used to estimate spatial correlation in binary data sets of this kind (Sokal and Oden 1978a). The basic approach is to look for the nonrandom association of events at pairs of sites that are at different distances from one another. To do this, you must first define M distance classes with boundaries $d_0, d_1, d_2, \ldots, d_M$. That is, distance class k includes all distances greater than d_{k-1} but less than or equal to d_k. It is best to make these classes broad enough that an absolute minimum of ten pairs of sites are in each, but more is better. The tradeoff between finer distance classes and more robust samples with which to test for correlations within each class is reminiscent of the constraint faced when choosing class boundaries in a projection matrix model (see Chapter 6).

For each type of data (presence-absence, extinction-no extinction, or colonization-no colonization), each site i has a binary value x_i (yes or no). If there are multiple time intervals, it is generally best to perform the correlation analysis separately for each, as a site's fates over two time periods are very likely to be correlated.[3] Using just pairs of sites whose distances from one another are in distance class k (that is, that form a "join" with each other in that distance class), we can create a 2 x 2 contingency table, recording each combination of fates. For example, with presence-absence data, the rows of the contingency table would be labeled "yes" and "no" to reflect whether a population is present at one site (say site A) and the columns labeled "yes" and "no" to indicate the presence of a population at a second site (site B). Note that in such a table, the normal procedure is to record each pair of site-specific fates twice; for each pair, each site is first used as site A and then as site B (although joins of a site with itself are not counted; Sokal and Oden 1978a). Thus, we get a total sample size that is twice the actual number of pairings and that can be reduced to the number of (yes-yes), (yes-no + no-yes), and (no-no) pairs observed for sites at a particular range of distances from one another.

Testing the significance of the association seen between pairs of outcomes is not straightforward, because each site is likely to take part in a unique set of pairings with other sites. Although an appropriate significance test has been developed for join-count associations, with the small sample sizes typical of most patch data, a randomization test is recommended instead (Sokal and Oden 1978a). This test is based on randomly reshuffling all the observed fates among the sample of sites that are involved in at least one join in distance category k. To make these random rearrangements it is generally most appropriate to keep the total number of yes and no fates the same as in the original sample (that is, we randomize without replacement; Sokal and Oden 1978a).

[3]While we discuss analyzing join-count data only for binary variables, the same method can be used to look for spatial correlations in multi-outcome processes, including different combinations of presences and absences at multiple censuses (Sokal and Oden 1978a).

For each of the randomized data sets we can again count up the number of pairs of each type (yes-yes, etc.) and use these numbers for each random arrangement to form a distribution of expected numbers of pairs of each type given no real association. We then compare the rank (or percentile) of the observed value for each type of pair with its corresponding random distribution. Quite high or quite low percentiles (e.g., <5% or > 95%) indicate significant correlations for a distance and type of pair. For example, if the percentile for yes-yes pairs at small distances is 97%, there is only a 3% chance that the observed pattern would have arisen randomly. Box 10.1 contains the MATLAB code to do these analyses.

To illustrate this method we have used data on *Delphinium uliginosum* from Harrison et al.'s (2000) study of a suite of rare plants restricted to small, scattered serpentine seeps in Northern California. Harrison et al. had data for a total of 57 seeps on occupancy in two different time periods (the mid-1980s and the mid-1990s), from which they could identify extinction and colonization events that occurred between these two time periods. To look for spatial correlations, we created five distance categories and tested associations in four binary variables: occupancy in the first time period, occupancy in the second time period, extinction of initially extant populations, and colonization of initially unoccupied sites.

The results show a high degree of spatial correlation at the shortest distance class for occupancy at both times and for extinction risk (Figure 10.5A,B,C), although not for colonization rate (Figure 10.5D). These patterns could be due either to spatial correlations unrelated to dispersal or to localized movement of seeds among seeps, creating localized rescue effects. More surprisingly, there are also significant correlations at longer distances. In particular, at the highest distance class (>7 km) all four analyses show signs of strong correlation. This pattern probably results from larger spatial gradients in environmental factors, perhaps due to correlated differences in seep quality or regional climate.

In interpreting join-count statistics, it is important to note that while a high value of one type of pairing must indicate a low value of some other type of pair, both homogeneous pairs (e.g, extinct-extinct and not extinct-not extinct) do not have to show identical responses; one can be high and the other low. Another factor that must be remembered in using this method is that it involves interpreting a large number of correlations, with the potential for spuriously significant results. For example, in the *Delphinium uliginosum* data, extinction correlations flip-flop at intermediate distances, suggesting that there may be no true spatial correlations at these distances. Reanalyzing the data using both finer and broader distance classes is a good way to see if complex patterns in correlations are robust.[4]

[4]For example, Harrison et al. did a join-count analysis of nearest-neighbor sites and found results quite similar to those for our shortest distance classes.

BOX 10.1 ***A MATLAB program to calculate join-count statistics for spatial correlations between binary data.***

```
% Program joincount.m. This program estimates the correlation
% in binary conditions or fates in different distance
% categories, using a randomization procedure to test
% significance. See Sokal and Oden 1978a for background
% The program requires information on a binary variable
% (occupancy at one time, or colonization or extinction
% across a time step) and the location of each site
% in the sample.

clear all;
%****************User-defined Parameters******************
distcats = [ 1 3 5 7 12]; % the upper limit of distance
             % categories with which to partition the data
(km)
% Load stored information on each site's location and fate or
% state. Each row of data should be for one site, and the
% columns should be: X-coordinate of site, Y-coordinate, fate
% or state (0 = no, 1 = yes). We use data on D. uliginosum
load delug90s.txt;     % from Harrison et al (2000)
siteinfo = delug90s; % assign to the siteinfo variable
randruns = 1000; % number of random rearrangements to
% use for significance estimation
%*******************************************************
rand('state',sum(100*clock)); randn('state',sum(100*clock));
XX = siteinfo(:,1); % making variables to use in calculations:
YY = siteinfo(:,2); % XX,YY coordinates
FF = siteinfo(:,3); % FF fates
sitenum = length(XX); % how many sites
x1 = XX(:,[ones(1,sitenum)]); % making matrices of
x2 = x1';                     % x and y values
y1 = YY(:,[ones(1,sitenum)]); % to calculate distances
y2 = y1';
Dists = sqrt((x1-x2).^2 + (y1-y2).^2); % pairwise distances

for dd =1:length(distcats) % loop thru distance classes
 disp('dd ='); disp(dd);
 updist = distcats(dd); % Upper and lower distance bounds
 if dd == 1;   downdist = 0;
     else downdist = distcats(dd-1); end;
 ii=[]; % initialize the list of sites to include
 for jj=1:sitenum   % find coordinates of sites to include
```

BOX 10.1 *(continued)*

```
    ind = ...
     find((Dists(:,jj) <= updist)&(Dists(:,jj) >downdist));
    if ~isempty(ind) ii = [ii;jj];   end;
end;
XXd = XX(ii); % define the reduced sample that have joins
YYd = YY(ii); % of right length.
FFd = FF(ii);
sitenumd = length(XXd); % number used in this distance class

x1 = XXd(:,[ones(1,sitenumd)]); % making matrices of
x2 = x1';                        % x and y values,
y1 = YYd(:,[ones(1,sitenumd)]); % then all the pairwise
y2 = y1';                        % distances between sites
Distsd = sqrt((x1-x2).^2 + (y1-y2).^2);

f1 = FFd(:,[ones(1,sitenumd)]); % matrix of combined fates,
f2 = f1';  % (0,0)=0, (0,1) and (1,0)=1, and (1,1)=2
fatesmx = f1+f2; % making combined fates matrix
fates=fatesmx(find((Distsd <= updist)&(Distsd >downdist)));
 % a column of the combined fates with correct distances
fatehist = hist(fates,3);  % find numbers of (0,0),(0,1),
realhist(dd,:) = fatehist; % and (1,1) combo's of fates.

% Now, generate random rearrangements of fates to make
% distributions of expected joins with no correlations
for pp = 1:randruns % make all the random rearrangements
   FFr = FFd(randperm(sitenumd));   % randomize fates
   f1 = FFr(:,[ones(1,sitenumd)]);
   f2 = f1';
   fatesmx = f1+f2;                % combined fates matrix
   fates=fatesmx(find((Distsd <= updist)&(Distsd >downdist)));
   fatehist = hist(fates,3);
   randhist(pp,:) = fatehist; % store results
end; % pp
 % sort random results in ascending order for 00 pairs
 randnum00 = sort(randhist(:,1));
 if realhist(dd,1) <= min(randnum00)  % find ranking of real
       probs(dd,1) = 0;               % result
   elseif realhist(dd,1) >= max(randnum00)
       probs(dd,1)=1;
   else rank = find(randnum00 < realhist(dd,1));
        probs(dd,1) = max(rank)/randruns;
```

BOX 10.1 *(continued)*

```
  end; % if realhist

  randnum01 = sort(randhist(:,2)); % repeat for 01 pairs
  if realhist(dd,2) <= min(randnum01)
        probs(dd,2) = 0;
    elseif realhist(dd,2) >= max(randnum01)
        probs(dd,2)=1;
    else rank = find(randnum01 < realhist(dd,2));
         probs(dd,2) = max(rank)/randruns;
  end;

  randnum11 = sort(randhist(:,3)); % repeat for 11 pairs
  if realhist(dd,3) <= min(randnum11)
        probs(dd,3) = 0;
    elseif realhist(dd,3) >= max(randnum11)
        probs(dd,3)=1;
    else rank = find(randnum11 < realhist(dd,3));
         probs(dd,3) = max(rank)/randruns;
  end;
  expected(dd,:) = [mean(randhist)]; % expected results
end; % dd loop

disp('Below are the observed numbers of 00 prs, the expected')
disp('number, and the percentile rank for the observed');
disp('(Rows are for each distance class, short to long)')
disp([realhist(:,1),expected(:,1),probs(:,1)]);

disp('Below are the observed numbers of 01 prs, the expected')
disp('number, and the percentile rank for the observed');
disp('(Rows are for each distance class, short to long)')
disp([realhist(:,2),expected(:,2),probs(:,2)]);

disp('Below are the observed numbers of 11 prs, the expected')
disp('number, and the percentile rank for the observed');
disp('(Rows are for each distance class, short to long)')
disp([realhist(:,3),expected(:,3),probs(:,3)]);

disp('Finally, the numbers of pairs of 00,01,11');
disp('and the total in each distance category');
disp([realhist,sum(realhist')']);
```

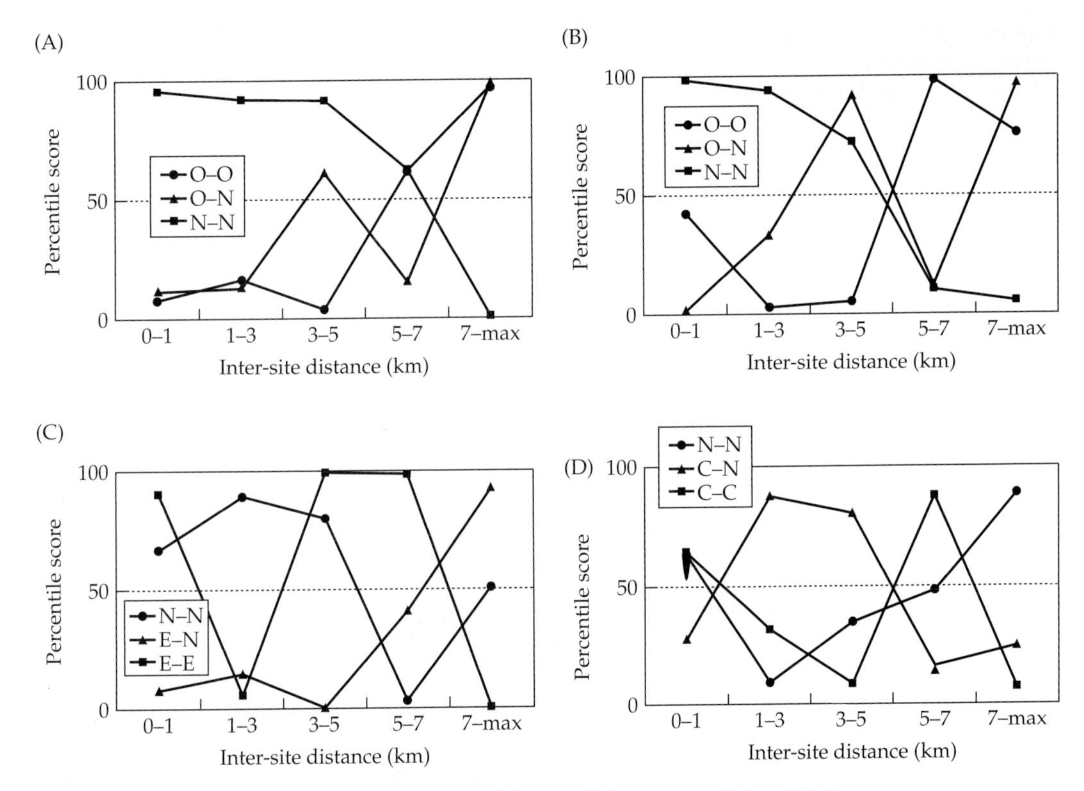

Figure 10.5 Join-rank correlations for *Delphinium uliginosum* patch-level data. Graphs show the percentile scores for four different data types and five different distance classes. High percentiles indicate more joins (e.g., occupied-occupied, or extinction-extinction) than expected, whereas low percentiles indicate fewer than expected joins. Results are shown for: (A) Occupancy in the 1980s (O–O = occupied-occupied, N–N = not occupied-not occupied, O–N = occupied-not occupied). (B) Occupancy in the 1990s. (C) Extinction or nonextinction from the 1980s to the 1990s (E–E = extinct-extinct, N–N = not extinct-not extinct, E–N = extinct-not extinct). (D) Colonization or noncolonization from the 1980s to the 1990s (C–C = colonized-colonized, N–N = not colonized-not colonized, C–N = colonized-not colonized).

More generally, join counts are best thought of as an exploratory tool to look for the scale and type of *pattern* in patch-level data, prior to fitting *process*-based models to assess viability. In this example, we see strong indications of large-scale patterns that should be considered when performing a PVA. That is, although this analysis cannot tell us whether the observed significant correlations are due to correlated environmental factors or to movement, their presence tells us that the PVA model we build should include one or both of these factors. We will return to this example in Chapter 11 to build such a patch-based model.

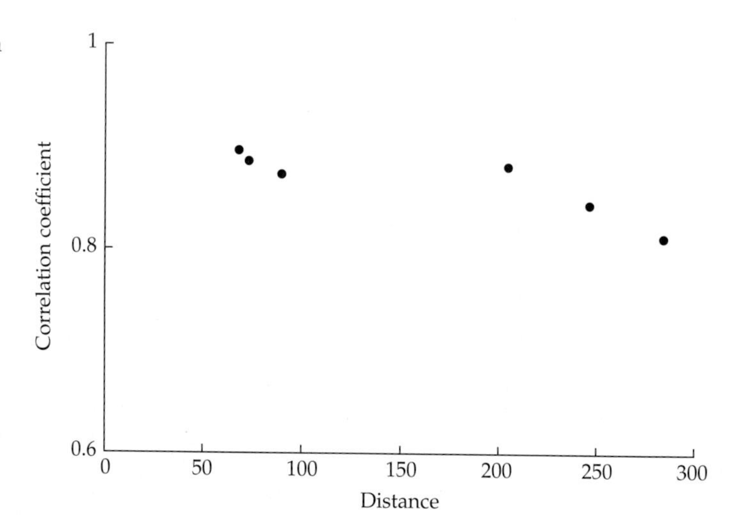

Figure 10.6 The correlation between annual rainfall for four weather stations near spotted owl populations in Southern California, plotted against distance between stations (after Lahaye et al. 1994).

ESTIMATING CORRELATIONS IN POPULATION GROWTH WITH LITTLE OR NO DIRECT DATA Unfortunately, there are usually too few years of data to make even slightly useful estimates of between-site correlations or there are simply no data at all for some populations.[5] There are several ways to deal with these situations. The first is to find data on a surrogate variable that may drive correlations in demography. The most obvious surrogate is the weather, which may vary similarly over wide areas, generating positive correlations in vital rates and hence population growth across sites. Taking the correlation in, say, yearly rainfall at two sites and using this to infer the strength of correlation in population dynamics is not necessarily reliable (Swanson 1998, Benton et al. 2001). However, such information can at least allow a rough guess as to which populations are most correlated and which are less so. For example, to estimate the degree of correlation in growth rates for a set of spotted owl populations, Lahaye et al. (1994) used the pattern of correlation in rainfall with distance (Figure 10.6). With no information on how rainfall influences owls, they assumed that the strongest positive correlation in population performance would be identical to the correlation for this important environmental variable, but they also conducted simulations with much weaker correlations, including no correlation at all. With knowledge of a species' biology and of population differences, you might also be able to use this approach to infer negative correlations, such as when a drought may benefit a population in a wet area and harm another population living in a dry microhabitat.

---

[5]Or, more aggravatingly, there are data for some populations in some years and for other populations only in different years, precluding the estimation of correlations.

A slightly different way of approximating correlations is with data on the sizes of catastrophes or bonanzas. Quite often the most important correlated responses are those driven by unusual events with geographic ranges that can be estimated fairly accurately: fires, freezes, or other extreme weather events such as El Niño flooding. If you assume that correlation in demography is driven solely by these extreme events, then information on their size and frequency can be used to define the added probability of an event hitting one population or part of a population if it has already hit another. Especially for fires, the data sets and analytical methods exist to predict the size and frequency distribution of future impacts based upon past records (e.g., Lindenmayer and Possingham 1996, Mortiz 1997, Richards et al. 1999), as well as to more mechanistically predict the probabilities of different fire-spread patterns over complex landscapes (e.g., Perry et al. 1999, Hargrove et al. 2000).

Finally, in the absence of any other information, it is usually a reasonable guess that the closer together two populations are to one another, the more correlated their demography will be through time. This approach is taken by Akçakaya and Atwood (1997) in a multi-site model for the California gnatcatcher. With no data on correlations, these authors assumed that the correlation in the temporal variations of vital rates for a pair of populations would fall off exponentially with distance, and they ran models using three different guesses as to the rate of this decline. If proximity, rather than habitat type or some other factor (e.g., aspect), is the dominant force generating correlations in dynamics, then we can roughly rank the expected correlations between all pairs of sites by their distances. Inter-site distances *per se* can't be used to estimate correlation coefficients quantitatively, but in this way they can be used to add a range of conceivable correlation values to a multi-site PVA or simply to interpret the results of a PVA that assumes no correlation. For example, say, three sites appear to have equal viability, but we can only afford to purchase two of them. If two are very close together, and the third is quite distant, it may make sense to prioritize purchase of the third, as its population dynamics are likely to be least correlated with the other two (although consideration of movement may lead to the opposite conclusions: see the next section). Note that for some organisms, distance along a corridor through which environmental effects are likely to propagate (e.g., along streams for flooding effects) may be a more appropriate surrogate for likely correlations than is straight-line distance.

A final note on this strategy of guessing at plausible correlation structures is that it is safest to assume no negative correlations and then vary the strength of positive correlations over a fairly wide range. Negative correlations in population growth rates or in vital rates that have positive sensitivity values (see Chapter 9) will always lessen the risk of extinction. So, in the absence of any estimates, not including negative correlations will yield more cautious assessments of population viability.

Data Requirement 3: Rates of Movement between Sites

THE IMPORTANCE OF MOVEMENT Movement is the more obvious way that that fates of local populations can be linked. If movement rates are quite high, then multiple sites do not truly harbor multiple populations, but instead host a single population that, by effective interchange of individuals, utilizes a dispersed set of habitat patches. If movement occurs, but only at low rates, it may nevertheless play an important role in supporting multi-site viability by allowing "rescue" of populations on the verge of extinction (Brown and Kodric-Brown 1977) and recolonization of suitable habitat patches with no extant populations—the extinction/recolonization process that drives true metapopulation dynamics (Levins 1969, Gilpin and Hanski 1991). Even quite low movement rates may still be sufficient to offset genetic differentiation of populations and loss of fitness due to inbreeding (reviewed by Mills and Allendorf 1996). If no movement among populations occurs, this too is important to know. Lack of movement greatly simplifies conducting a multi-site PVA, and it also means that a suite of populations will be much more extinction-prone than we might think if we were to assume that metapopulation dynamics are occurring. In spite of its importance, quantifying movement is difficult, and we now review four of the major approaches to doing so.

HOW TO QUANTIFY MOVEMENT: CAPTURE-RECAPTURE ANALYSIS The simplest and most useful estimate of movement is the proportion of a population at one site that successfully moves to another site in a given year (with some mean and variance). The key word here is *successfully*, and the most direct

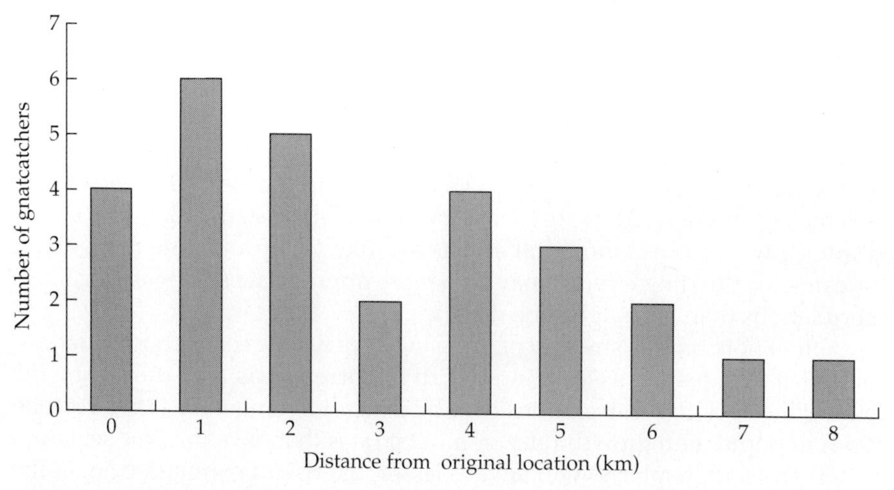

Figure 10.7 The number of resighted California gnatcatchers found at different distances from their original location (redrawn from Bailey and Mock 1998).

way to document successful movement is to mark individuals in different populations and then see how many are recaptured in other populations, clearly showing successful movement. For example, Bailey and Mock (1998) banded 100 juvenile California gnatcatchers from 1988 to 1992 and then surveyed for them over subsequent years. The resulting data document the proportion of resighted birds that had moved different distances from their banding site (Figure 10.7). There is a clear decline in numbers with distance; Bailey and Mock are careful to note that their data may underestimate dispersal to far sites due to the difficulty of thoroughly searching the increasingly larger areas at farther distances.

The first step in estimating movement this way is to mark individuals, using any of the methods we outlined in Chapter 6. After marking, you must try to recapture or resight marked individuals across multiple sites, and again a diversity of obvious methods exists to maximize the chances of finding the rare colonist in a sea of nonmovers, including use of high-visibility tags and intense trapping effort. If apparently suitable but unoccupied sites exist, they should also be searched, as one key aspect of estimating movement is to locate colonists where there haven't been any individuals living previously.

As in Bailey and Mock's gnatcatcher study, once dispersers have been found, the most common way to estimate movement probabilities has been to directly use of the percentage of marked dispersers observed at different locations (Bailey and Mock 1998, Harrison 1989, Olea 2001). This approach is especially feasible when the resighting probability is high and/or when colonization of small, specific areas is of most interest. However in many cases the majority of dispersers will not be recaptured, and we ideally would like to estimate the total number of successful dispersers, rather than relying on raw numbers of observed dispersers. If each individual has a unique mark, the best way to analyze the resulting data is with formal capture-recapture analysis. The industry standard for performing these analyses is the program MARK by Gary White of Colorado State University.[6] In Chapter 6, we briefly discussed the use of capture-recapture analysis to estimate survival rates, but these methods can also yield estimates of other parameters, including rates of movement between populations. Capture-recapture methods use resighting data to estimate the actual numbers of individuals in each class (e.g., successful dispersers from population A to B), *including those not directly seen*. To accomplish this, you must not only count and relocate the marked animals, but also estimate the number of unmarked animals in each site.

With direct observations or capture-recapture analysis, you can either estimate probabilities of individuals moving between each separate pair of sites or estimate a more general dispersal curve as a function of distance. If distance

[6]As we noted in Chapter 6, the program, manuals, and other information can be downloaded at http://www.cnr.colostate.edu/~gwhite/mark/mark.htm.

has a strong effect on dispersal probabilities, estimating such a curve will yield better estimates of movement than will many separate estimates of dispersal between specific locations. Moreover, having a dispersal curve in hand would allow us to explore how changes in the spatial arrangement of sites (e.g., due to ongoing habitat fragmentation) would affect multi-site viability. Although there are many possible shapes for a dispersal curve, data for rare species are rarely extensive enough to distinguish between them. The most common assumption is that dispersal probability declines as a negative exponential function of distance (Turchin 1998, Hanski 1999), which can be fitted using simple regression methods. Specifically, if y represents numbers of successful dispersers and x represents distances from the site of marking, then the slope b and intercept a of a linear regression of the log y against x, provides estimates for the parameters of the negative exponential dispersal curve $y = a \exp(-bx)$. However, this dispersal curve only yields estimates of the distance successful dispersers move once they have decided to do so. To fully account for inter-site movement rates, we must combine the information contained in the dispersal curve with estimates of the fraction of individuals leaving the release site and the fraction of *those* that survive to reach another site; estimating these probabilities requires more detailed demographic—and usually capture-recapture—methods than provided by simple observations of successful dispersers. For example, Hanski et al. 2000 and Petit et al. 2001 develop a capture-recapture method explicitly designed to estimate movement distances and survival rates of dispersers simultaneously.

In estimating and using movement data for PVAs, it is crucial to distinguish between movement of males and females (or hermaphrodites). A single pregnant female (or a single hermaphrodite, if self-fertile) can found a new population, whereas a single male (or pollen grain) cannot. Nevertheless, male movement can be important, for two reasons. First, male movement alone is often sufficient to prevent inbreeding. Second, if males become limiting to local reproduction, facilitation of increased male movement alone can prevent problems with local demographic performance. For example, male mountain lions have much larger territories than do females, leading to a high chance of male limitation to reproduction in the very small population of mountain lions in the Santa Ana Mountains of California (Beier 1993). However, juvenile males are also much more likely to move long distances and to cross urban barriers than are females, with important consequences for management of dispersal corridors (Beier 1993, 1996). Still, as we've noted before, most PVAs focus solely on females; unless most sites can only harbor very small populations, male movement is likely to be of secondary importance in determining multi-site viability.

Several other complications should also be considered in estimating movement. First, for animals that disperse at multiple ages or sizes, it can be important to distinguish between types of dispersers that may have very different reproductive values and movement behaviors (see Chapter 6).

Second, variation in intervening habitats can strongly influence movement rates and should be considered when analyzing movement distances and probabilities (Bailey and Mock 1998, Ricketts 2001). You should also consider the possibility of density-dependence in the probability of leaving the natal population or in the decision of where to stop moving (Stamps 1988, Stamps and Krishnan 1990). Finally, if possible, it is ideal to estimate the variation in movement rates across time, allowing the modeling of temporal variation in dispersal distances and rates. All of these complications can be built into a capture-recapture analysis, provided sufficient data have been collected.

HOW TO QUANTIFY MOVEMENT: EXTRAPOLATION FROM INDIVIDUAL BEHAVIOR Very often, direct estimation of successful dispersal is difficult or impossible. The most common difficulty (even when time and money are not critically limited) is that even when movement is suspected or known to occur, it happens at such low rates that no reasonable study will be able to find even a handful of dispersers. Especially for long-lived species, movement rates on the order of one successful disperser per decade may be quite important for viability, but our chances of documenting it are approximately nil. In this case, data on individual movement behavior can provide an alternative way to estimate movement distances and hence rates of movement between sites. Techniques used to extrapolate rates of inter-site movement from individual behavior are highly technical, and we make no attempt to provide a "user's guide" to them here. Instead, we refer readers to other sources that describe the methods and their application.

There are three basic strategies for collecting movement behavior data, any of which can (with sufficient additional information) be converted into an estimate of movement rates between sites:

1. The simple distance an individual animal is located from its starting point at increasing times can contribute to an estimate of movement rates. With data on many individuals, the change in dispersal distance (i.e., the mean squared distance from the starting point) as a function of time can be directly estimated, and this relationship can be used to estimate a diffusion coefficient. This, in turn, can be used to estimate a dispersal curve (see Turchin 1998 for an advanced review of these methods). A key point (one that is often forgotten when performing movement calculations) is that dispersal is usually dangerous, and the longer a disperser searches, the greater the chance that it dies before reaching a new home. Thus, some estimate of the mortality rate during dispersal is usually needed to turn observed distances from the starting point into realistic estimates of the dispersal curve for successful dispersers. For species in which dispersers clearly change state to become "settled" individuals at the end of dispersal (e.g., seeds on the ground and animals that acquire permanent

territories), only settled, former dispersers need to be censused. In this case it is also more straightforward to estimate the shape of the dispersal curve, although this will not always be a simple function (e.g., Turchin 1998, Bullock and Clarke 2000).

2. A second approach relies on much more detailed information about individual behavior. At the least, individual animals must be carefully followed to quantify their turning angles, move lengths, and the time elapsed between turns. These data can be used—again with an estimate of the mortality rate during dispersal—to generate predictions of distance from the starting point as a function of time (Turchin 1998). These predictions can be generated from random-walk models or by brute-force computer simulation of a set of "virtual" dispersers following the observed movement behaviors (e.g. Schultz and Crone 2001). The advantage of this method is that it can use fairly short-term, small-scale data to predict larger scale movements that are very difficult to see. However, as with the preceding method, a key problem is the uncertainty in predicting the extreme tail of the dispersal curve. That is, most individuals move only small distances, and we mostly care about the few that move very far, which is the hardest thing to estimate or confirm (e.g., Bailey and Mock 1998). Our ability to extrapolate from short-term, small-scale movements to the extreme tails of the dispersal curve is often suspect. Another issue concerns differences in movement behaviors in distinct types of habitats or through habitat corridors. Very often, movement between suitable sites involves crossing different types of habitats or depends on the ways that animals respond to habitat borders. These effects have been shown to strongly influence the movement rates of butterfly species (Haddad 1999, Haddad and Baum 1999, Ricketts 2001, Schultz and Crone 2001) and are likely to be important for other groups as well. Thus we may need to estimate not only movement parameters (i.e., move lengths and turning angles) within each type of habitat, but also the behaviors individuals employ when they encounter the ecotone between two habitat types.
3. Finally, for passively dispersed propagules such as seeds and many planktonic larvae, models of air or water motion can yield predictions of expected dispersal distances (e.g., Okubo and Levin 1989). To make these estimates, it is necessary to measure the settling velocity of propagules in a lab and have some estimates of the speed, direction, and degree of turbulence of the dispersal medium (i.e., air or water). However, the tail-of-the-distribution problem still exists. These models don't necessarily do a good job predicting extreme dispersal distances, which are likely to be determined by large air movements or the mud on ducks' feet—factors that are not usually considered in the equations.

Finally, note that all three of these methods must also be supplemented with some estimate of the number or fraction of individuals initiating dispersal.

Even with all the caveats given above, movement analyses are certainly underused in efforts to understand how strongly linked different populations are. At their best, these methods can indicate situations in which extensive movements between populations are likely, or when any movement is exceedingly improbable. Either of these conclusions have a great deal of practical value. If intermediate amounts of movement are predicted, then a reasonable range of possible rates can be included in PVA simulations to see how much the exact values matter.

HOW TO QUANTIFY MOVEMENT: GENETIC DATA The use of genetic data to infer the rates and directions of individual movement is a highly complex field that we will not discuss in any detail. Although using genetic markers has great potential to discern both short- and long-term patterns of movement between local populations, their many complications make these analyses a difficult business. Foremost among these issues is untangling the effects of local selection from those of movement, both of which contribute to patterns of genetic variation within and among populations. We suggest Dobson et al. (1999) and Steinberg and Jordan (1998) and the references on interpretation of genetic data they contain as good entries into this literature.

HOW TO QUANTIFY MOVEMENT: PATTERNS IN SITE OCCUPANCY Finally, simply knowing which patches of suitable habitat in the landscape are currently occupied or how occupancy has changed through time can (with many additional assumptions) yield information about both colonization rates and population extinction probabilities. The basic idea is to use information on site occupancy to infer how colonization probabilities and extinction rates vary as a function of the distances between habitat patches, their sizes, and other factors, (such as components of habitat quality) (Sjögren-Gulve and Hanski 2000). Estimating these functions is part and parcel of the patch-based multi-site PVA approach, keeping track of occupancy of sites and the "birth" of whole populations, rather than individual movements. To use this approach, you need data from one or preferably many years about the occupancy of a large number of suitable habitat patches, as well as the information to classify the distances between all pairs of patches, their areas, and their habitat qualities. Key problems in using these methods are making sure what unoccupied habitat truly *is* habitat, being careful to include *all* patches of suitable habitat in the field surveys, and the confounding of dispersal and environmental correlation, as we discussed earlier in the chapter. Even more problematic for any really rare species is having a large enough sample of sites to powerfully estimate both extinction and colonization rates and how they vary with distance, patch size, and surrogates of patch quality. We will discuss how to implement these methods in Chapter 11.

A SUMMARY OF MOVEMENT RATE ESTIMATION Estimating movement rates for rare species virtually always requires us to make the best of a bad situation. Each of the methods outlined above has drawbacks. Capture-recapture estimates, although they are based on observed dispersal between sites, are highly specific to the peculiarities of the landscape in which movement was measured, and may not generalize to other landscapes or to the same landscape after it has been modified by human or other impacts. Similarly, genetic and patch-based approaches, which determine average movement rates over long timescales, give no information to extrapolate how movement may change with alteration of intervening habitats or other landscape changes. On the other hand, movement estimates based on individual behavior can quantify how individual movement rules change in degraded habitat or at habitat boundaries but may suffer from unreliable or unrealistic extrapolation from short-term, small-scale data to longer-term, landscape-level outcomes. Finally, all the methods are subject to a common problem: the longest distance movements are very rare and highly sporadic but are also likely to be the most important for connecting populations at the scales we care about.

Of course, it may be that no data at all are available to estimate movement rates. In this case, data on movements of related, often more common, species may help to estimate a range of plausible movement rates. And, even with no direct data, you can often make a defensible, if qualitative, distinction between three situations: movement is high enough that essentially all sites support portions of a single population; movement is essentially or completely nonexistent; or, movement rates are intermediate and require careful evaluation. Below, we use this categorization of different movement patterns to help guide approaches to the analysis of multi-site viability.

A Schematic Breakdown of Multi-Site Situations

We have just outlined the basic types of data needed to fully analyze the viability of a multi-site population, but it is very rare indeed that we will have all these categories of information. Normally, we have poor or no estimates of at least some of this information (e.g., estimates of population growth at only one site out of five, some estimate of movement rates, and no information on variation or correlation in the site-specific growth rates). Therefore, we should carefully consider when we really need to know, or even to guess, all these rates in order to understand multi-site viability in different situations. This requires a consideration of the relative viability of populations at different sites, and also a consideration of how the joint forces of movement and correlation in dynamics influence viability. Our goal is to create a breakdown of quantitatively different types of multi-site population structures. In making this categorization, we'll use some of the multi-site scenarios proposed by Harrison (1991), but also elaborate upon these categories.

The first question to ask is whether one population is of overwhelming importance for overall viability. In other words, is one population essentially invulnerable to extinction over the time scale on which we are making conservation decisions, while all others are at substantial risk? If so, then the populations represent an extreme *mainland-island* or *source-sink* situation (Pulliam 1988, Harrison 1989), in which the viability of the single best population (the *mainland* or *source*) is of paramount importance for overall viability, while the others (the *islands* or *sinks*) are of minor importance.[7] Although it may be useful to analyze the extinction risk of these risk-prone populations to understand how to maintain or improve their status (especially if they harbor genetically unique individuals), their fate over the timescale of a typical viability assessment will probably have little effect on the fate of the total population. This is likely the case for the population of mountain lions in the Santa Ana Mountains of California. Results of PVA models constructed by Beier (1993) predicted that extinction of this small population would be strongly influenced by extinction of the much larger population in the Palomar Mountains, but any reciprocal effect is highly unlikely. Similarly, Harrison (1989) found that even though many small sites supported ephemeral populations of the Bay checkerspot butterfly, a single much larger population appeared to be the source of almost all immigrants to the small sites, and was likely to be relatively impervious to extinction. In these cases, movement behaviors and correlations in growth rates can be important in understanding the viability of the extinction-prone sites, but these complexities are probably not important in assessing overall viability. In sum, if one (or more) of the populations would have a very low extinction risk in isolation, then it is probably safe to forgo a multi-site analysis and conclude that the risk of total extinction is also very low.

The more usual situation when conducting a PVA is that all populations in the constellation of sites being considered have a significant risk of quasi-extinction. If so, then we can proceed with a breakdown of different types of population dynamics according to the magnitude of correlations in population growth or vital rates and the probabilities of movement among sites (Table 10.3). This breakdown is useful in emphasizing that many multi-site situations are really quite simple demographically and thus can be adequately treated without the full range of complex information needed for a complete spatial analysis.

[7]Strictly speaking, *mainland-island* and *source-sink* apply to situations in which at least some movement among populations occurs. But even if there is no movement, one population may be far more viable than the others.

TABLE 10.3 A classification of multi-site scenarios

Correlations in population-specific vital rates	*Movement rates*		
	Essentially none	*Low to medium*	*High*
Significantly negative	1A. Separate populations, contrasting environmental drivers; multiple populations strongly beneficial	2A. Highly effective metapopulation	3A. Multiple sites, very different habitat effects, one population
Not different from zero	1B. Separate populations, uncorrelated fates; multiple populations beneficial	2B. Somewhat effective metapopulation	3B. Multiple sites, somewhat different habitat effects, one population
Significantly positive	1C. Separate popuations, shared fates; multiple populations not very effective	2C. Ineffective metapopulation	3C. One population

No Movement between Populations (Table 10.3: Cases 3A, 3B, 3C)

If all populations have some risk of extinction, then we should ask if any significant amount of dispersal is likely to occur between them. When the chances of successful dispersal are so remote that they cannot be relied upon to occur over a reasonable time horizon for management, we need only consider the possible impact of correlations among sites. If there is no correlation in the fluctuations experienced by separate populations, then the probability of multi-site extinction is simply the product of the extinction risks in each population (as discussed in the simple clapper rail analysis that we performed earlier). However, common events will often impact different populations of the same species at the same time, leading to negative (Case 1A) or, more commonly, positive (Case 1C) correlations in population growth or vital rates among sites. In these cases, we can build models that are essentially a set of single-site PVAs with correlated dynamics, while still avoiding most of the complexity of real spatial dynamics.

High Movement Rates between Populations (Table 10.3: Cases 3A, 3B, 3C)

At the other extreme, physically isolated habitat areas may be intimately connected to each other by frequent movement of individual animals or plants. In these cases, regardless of the degree of correlation between demography in different sites, it is best to think of the set of sites as supporting a single population. If sites differ substantially in the means and variances of the vital rates, or if the rates are negatively correlated with those in other sites (Case 3A), the population is essentially using more than one type of habitat, but high movement will quickly mix individuals among habitats. In these cases, single site PVA methods will usually be adequate to estimate viability, although individuals may need to be categorized by their temporary location (say, breeding habitat) as well as by stage, size, or age. The question is, how high must movement rates be to make a single population analysis defendable? There is no right answer to this question, but a reasonable approach is to ask if in each bout of reproduction the progeny of individuals living in different sites are thoroughly mixed. For animals, we can also ask if individuals frequently move back and forth between sites, thereby experiencing conditions in multiple areas over their own lifetimes. If either of these tests for mixing is met, then the individuals in multiple sites are so intimately linked that there is probably little reason, for the purposes of PVA, to try to distinguish separate populations in different locations. Rather, there is one population utilizing multiple locations.

Low to Medium Movement Rates between Populations (Table 10.3: Cases 2A, 2B, 2C)

The most complicated cases are those in which movement rates are high enough to be important for recolonization and for "rescue" of faltering populations, but are still low enough that the majority of individuals live out their lives at a single site and the majority of their progeny are quite likely to stay put as well. In this case, both movement rates and the correlation in demography between populations are important to consider in a multi-site PVA. Barring situations in which some local populations are impervious to extinction, Cases 2A through 2C can fairly be said to represent true metapopulations. However, these three types of metapopulations are not all equal in the effectiveness with which they ensure multi-site viability. If correlation in demography is low, and especially if it is negative (Case 2A), then there is great potential for separate populations to buttress each other's viability. We call this a highly effective metapopulation because there are good chances of rescue and recolonization of some populations by immigrants from others, a process that can sustain the entire assemblage in the face of substantial extinction risk to any one population. Whether correlations are strongly negative or near zero, these situations are the most complicated to analyze, because viability will be strongly influenced by both movement and correlation.

If there are quite similar, correlated responses of demography at different sites (Case 2C), then a collection of populations will still act as a metapopulation, but as a much less effective one: when one local population has been depressed by a series of environmentally unfavorable years, others in its vicinity are also likely to be depressed and so will not be pumping out enough dispersers to provide rescue. Similarly, with highly correlated local dynamics, populations are likely to become extinct simultaneously, so that recolonization is unlikely to happen. Basically, high correlations can spoil the substantial insurance against multi-site extinction that classic metapopulation dynamics are often envisioned as providing. A particular concern raised by a consideration of positive correlation is how the spatial arrangement of sites will influence overall viability. Virtually all metapopulation models assume that there are no correlations in the dynamics of neighboring populations, and thus they suggest that proximity of populations will increase metapopulation viability via increased movements (Hanski et al. 1996, Moilanen 1999, 2000, Sjögren-Gulve and Hanski 2000). However, if spatial proximity increases both movement and correlation in growth rates, and if positive correlation enhances the probability of joint extinction, this issue is much more complicated, and some mixture of sites that includes distant as well as closely adjacent populations is likely to provide greater overall safety.

Summary: Using Occam's Razor in Multi-Site PVAs

Two main points come out of this classification of spatial dynamics. The first is that in many cases, we may not need to grapple directly with "space" or movement in a multi-site situation. If you have reason to think that movement is either extremely low or extremely high, you can safely use PVA methods that ignore any explicit treatment of space and movement. This is all for the good, as getting accurate estimates of movement rates is quite difficult; we are aware of only a handful of PVAs for which sufficient data were available to confidently estimate movement parameters. A second point is that the strong emphasis on metapopulation dynamics in PVA (indeed in conservation biology in general) is, in our opinion, somewhat misplaced. Although multi-site situations with metapopulation dynamics can and do occur, the cases in which they can powerfully stave off extinction represent only a small subset of all the combinations of movement and correlation patterns likely to exist. Inferring that safety will come through metapopulation effects, or even that a metapopulation analysis is needed or would be useful, should only come after a careful consideration of the inherent extinction risks separate populations face, the movement rates between populations, and especially (because it tends to be ignored) correlation in the fates of different populations.

In the following chapter, we will present the details of applying demographic, count-based, and patch-based approaches to the different multi-site situations in Table 10.3.

11

Methods of Viability Analysis for Spatially Structured Populations

In Chapter 10 we discussed the basic ways in which temporal correlation and movement can influence viability assessments, as well as how they can be measured. We then used the strength of correlations and the frequency of movement between local populations to create a classification that can help to simplify viability assessments for many multi-site situations. We now turn to actual methods that can be used to perform multi-site PVAs, dependent both on the classification given in Table 10.3 and on the type of population-level data available (count-based, demographic, or patch-based). Recognizing the usual data limitations found in multi-site situations, we will try to highlight ways to make multi-site PVA models useful even when only poor or incomplete information is available. In multi-site situations, it is important to focus not just on the likelihood of overall extinction but also on the relative value of different populations within the larger archipelago of sites. In many cases, conclusions about the overall viability of a suite of sites are likely to have low accuracy, but inferences about the relative importance of different populations may be much more robust (and useful). This emphasis parallels the ideas in previous sections of the book about the value of PVA in ranking the viability of different populations (using count-based methods), in determining the contributions of different life history stages and vital rates to population growth, and in evaluating the efficacy of alternative management interventions (using demographic methods).

Patch-Based Approaches

We begin with patch-based models because they use the very simplest type of multi-site data. Recall that patch-based measurements track only the presence or absence of populations in local sites of suitable habitat. At this level of resolution, the two sets of rates that govern multi-site viability are the extinction probabilities and colonization rates for each local site. In discussing the use of patch-based models we will consider only situations in which there is thought to be modest movement between local populations (i.e., Cases 2A, 2B, and 2C in Table 10.3). Although it is theoretically possible to use patch-based models for viability assessment in the absence of movement, actually estimating the extinction rates for such a situation is difficult to imagine in most real circumstances. If movement is not occurring, then any local extinction will be final, and cannot be "undone" by recolonization. However, to estimate local extinction probabilities for a patch-based model, we must see a reasonable number of extinctions. Together, these mean that the number of local populations must be declining fairly rapidly. In this situation, it is hard to see why we would need a quantitative model to tell us that the total population is in serious danger of extinction.

A patch-based approach is also unlikely to be appropriate when movement rates are high, since in this case true population dynamics (changes in numbers of individuals) will be much more important in governing overall viability and are likely to be poorly approximated by local extinction and colonization probabilities. Thus, patch-based models are most often used for situations with moderate to low movement, corresponding to more or less "classic" metapopulations. We will begin by explaining how to estimate extinction and colonization rates using patch-based data and then discuss how to simulate the resulting models to make inferences about multi-site viability.

Structure and Fitting of Patch-Based Models

The metapopulation models most often used by theoreticians assume that all local populations have identical extinction rates, and that they are all equidistant from one another (e.g., Harrison and Quinn 1989, Frank and Wissel 1998). These assumptions make for elegantly simple models, but they are too far removed from the realities of specific multi-site situations to be of practical use. Although not quite as restrictive in their assumptions, many of the more recent elaborations of metapopulation models still make it difficult to include site-by-site differences in colonization and extinction, again limiting their utility for our purposes.

The most useful metapopulation frameworks for viability analysis are incidence function models, first applied to metapopulations by Hanski (1991, 1994), and the logically similar logistic regression models for colonization and extinction rates (Sjögren-Gulve and Ray 1996, Moilanen 1999, 2000, Kindvall 2000). However, even these methods may be of quite limited value in most conservation situations, given that the number of sites and number of years of data are typically quite small. But we will get to these problems in describing how to fit the models. First, we describe their structure and assumptions in more detail.

THE INCIDENCE FUNCTION MODEL The basic idea of both incidence function and logistic regression models is to use patterns of patch occupancy over time and space (i.e., "incidence") to estimate E_i, the probability of extinction for habitat patch i when it is occupied, and C_i, the probability that patch i becomes colonized when it is unoccupied, taking into account the patch's habitat area, habitat quality, distance to other populations, and other potentially important characteristics. Importantly, the models allow these causal factors to include both metapopulation features, such as the number and proximity to other populations, which alter colonization and extinction rates via dispersal probabilities, and local features such as habitat area and quality, which can influence the ease of successful colonization and the probability of extinction through effects on local population dynamics.

To apply an incidence function model to data, we must assume specific functional forms for the effects of the causal factors (e.g., patch area and quality, and proximity to other populations) on E_i and C_i. Typical forms are (Hanski et al. 1996, Sjögren-Gulve and Hanski 2000):

$$E_i = \frac{e}{A_i^x} \qquad C_i = \frac{M_i^2}{M_i^2 + y^2} \tag{11.1}$$

For the extinction function, e is the maximum extinction probability and is decreased by increasing habitat area (A_i), with the strength of the area effect governed by the parameter x. Colonization is governed by a site-specific number of arriving migrants, M_i, and a parameter y that determines the scaling of realized colonization as a function of M_i. If the number of arriving migrants is zero, the colonization probability is also zero, but as the number of migrants increases, the colonization probability first increases and then asymptotes at 1. The number of migrants, M_i, depends on the probabilities of other patches having extant populations, their population sizes if they are occupied (which is assumed to be proportional to the area of each patch), the rate at which dispersers leave patches, and the distances from each other patch to focal patch i:

$$M_i = \sum_{j \neq i}^{N} \beta A_j^b \exp(-\alpha D_{ij}) p_j \tag{11.2}$$

In words, the number of migrants arriving at site i depends on the sum over all $N - 1$ other sites of a per-unit-area migrant production rate, β, multiplied by each site's area (scaled by the parameter b to allow for nonlinearity) and discounted by its distance, D_{ij} (scaled by α), all conditional on each site actually harboring a population (since $p_j = 0$ if site j is empty and 1 if it is occupied).[1]

[1]Note that although we have given the usual explanations of what M_i and β represent, neither term has such a clear biological meaning. In particular, estimated values of M_i should not be thought of as literally the number of migrants arriving per year at a site, but rather as a relative measure of the chance that migrants may successfully reach a site.

Part of Hanski's original idea of using the incidence function for metapopulations involved the ability to combine E_i and C_i into a single expression that predicts the static pattern of presence–absence data across many sites, thus allowing estimates of the parameters in E_i and C_i to be easily found for a particular metapopulation. For example, the probability that site i is occupied in any one year is predicted to be

$$J_i = \frac{C_i}{C_i + E_i(1 - C_i)} \tag{11.3}$$

in cases where rescue effects are thought to occur[2]. Using this expression for J_i, a snap-shot of the pattern of patch occupancy (with no direct data on colonization or extinction events) can be used to estimate the parameters governing C_i and E_i. However, doing so generally means assuming that the occupancy patterns seen in the field are very near their equilibrium values: the number of populations in the metapopulation is neither declining nor increasing on average, and neither is it fluctuating very much around some dynamic equilibrium. These assumptions may not be grievously wrong for very large metapopulations with low rates of colonization and extinction in stable sets of habitat patches, but for most conservation situations (in which we often expect metapopulations to be small and in decline) they are very unlikely to be good approximations to reality. Worse yet, estimating parameters under an assumption of stability practically guarantees that you will not detect evidence that a metapopulation is declining. Because of these problems, most recent uses of the incidence function model use both static occupancy data and data on actual extinction and colonization (turnover) events, or use turnover data alone.

Although the standard forms for the extinction and colonization rate functions (Equations 11.1 and 11.2) take into account effects arising from the distances between sites and their areas, they assume very particular patterns in these effects and they also do not account for a myriad of other factors likely to be of importance for many species. In particular, different aspects of habitat quality, including the presence or absence of food species, competitors, predators, and parasites, do not appear in the usual formulation. While additional factors can be added (e.g., Moilanen and Hanski 1998), the very specific structure of the incidence function equations can make the formulation and testing of more detailed models cumbersome, especially as specially written computer routines are required to find the best parameter values. For these reasons, many recent authors have turned to logistic regression methods to understand metapopulation dynamics.

[2]Specifically, the term $(1 - C_i)$ in the denominator causes the baseline extinction probability (E_i) to be reduced by migrants from other populations, with realized extinction rate equal to $E_i(1 - C_i)$.

LOGISTIC REGRESSION As with incidence functions, logistic regression models estimate the site-specific probabilities of extinction and colonization (Sjögren-Gulve and Ray 1996). However, these models are based on generalized logit functions, which can be used to predict the value of any probability (bounded by zero and one) as a function of various independent variables. For example, the probability of extinction could be modeled as:

$$E_i = \frac{\exp(a + bA_i + cI_i)}{1 + \exp(a + bA_i + cI_i)} \tag{11.4}$$

where A_i is area, as for the incidence function model, and I_i is an isolation measure, such as the harmonic mean[3] distance to other extant populations; a, b, and c are fitted coefficients governing the magnitude and shape of the effects of A_i and I_i on extinction risk. More generally, logit functions for extinction and colonization can be written as $E_i = \exp(u_e)/[(1 + \exp(u_e)]$, and $C_i = \exp(u_c)/[(1 + \exp(u_c)]$, where u_e and u_c are linear functions of any number of continuous or categorical explanatory variables.

In addition to patch area and isolation, many other factors can be put into u_e and u_c, including measures of habitat quality, year of observation as a categorical variable, latitude and longitude, and interactions between any of these factors. Effects of year and geographic position are especially important, because they allow us to estimate temporal variability and spatial gradients in extinction and colonization rates, respectively. In thinking about the formulation of logistic regression models, you should note that because u_e and u_c are always exponents, the linear form of their terms is in fact multiplicative, meaning that the effects of the independent variables on extinction or colonization rates can take on a wide variety of forms, depending on the fitted parameter values. One can also use polynomial functions (e.g., by including the square and cube of patch area as independent variables) to allow further flexibility in the fitted relationships. This flexibility, with no need to fix *a priori* the basic functional forms of the dependence of E_i and C_i on any of the independent variables, is a key advantage of this approach to metapopulation modeling relative to the incidence function approach.

---

[3]The harmonic mean is the inverse of the average reciprocal of a set of numbers. For example, the harmonic mean of the two numbers x_1 and x_2 is $1/\{[(1/x_1) + (1/x_2)]/2\} = 2x_1x_2/(x_1 + x_2)$. The harmonic mean weighs small numbers even more strongly than does the geometric mean, making it an appealing way to combine distances relevant for dispersal, since the nearest sites are likely to have by far the greatest role in providing dispersers. Thus, regardless of the number of distant populations, a few short distances to neighboring populations will result in a small harmonic mean distance, and thus a high colonization probability or a low extinction risk (thanks to rescue effects).

Unlike incidence function models, logistic regression models have generally been used to make parameter estimates based separately on colonization and extinction rates. This is largely because doing so permits use of standard statistical packages to make maximum likelihood estimates of the fitted coefficients in a logistic regression. However, it is also possible to fit these coefficients in the extinction and colonization probabilities to occupancy data, as has frequently been done with incidence function models. We now turn to the questions of why and how to fit either model to different types of data.

FITTING PATCH-BASED MODELS Finding the coefficient values that best predict a given data set is done with maximum likelihood methods and Information Criterion analysis for both logistic regression and incidence function models. Because of its advantages, we will only describe these steps for the logistic regression model, although for most of what we now present, the procedure is identical for the incidence function model. In Chapter 4 we presented the basic steps of maximum likelihood fitting, and we proceed to use it here assuming that you have gleaned the basic idea of the approach from our earlier description.

As we described in Chapter 10, to obtain the necessary data to fit a patch-based model, we must survey a set of patches of suitable habitat at least once, and preferably two or more times, and record which patches are occupied at each survey. We must also quantify characteristics of each patch that we suspect might influence its probability of colonization and extinction. At a minimum, we need to record the location and size of each patch, but other features (e.g., indices of habitat quality) may also be useful to include in the analysis. The survey should include *all* patches of suitable habitat in a region, because estimating the extinction and colonization probabilities of each patch requires us to account for the potential influence of *all* other occupied patches in the metapopulation. Moreover, the surveys must include a reasonable number of occupied and unoccupied patches,[4] with higher numbers needed if extinction or colonization is rare, since we must observe multiple extinction and colonization events in order to estimate model parameters. In short, the data requirements for this PVA method are not trivial.

We begin with the usual case of separately fitting the parameters of the E function to data on extinction events and those of the C function to data on colonization events. To begin, let's assume that only two surveys have been conducted, and let O represent the number of patches that were occupied at

[4]The same increases in the variability of binomial outcomes that create demographic stochasticity in survival rates at small population sizes (Chapter 2) will make estimates of extinction and colonization rates unreliable with small numbers of sites. What a "reasonable" number is has not been carefully examined, but at least ten initially occupied and ten unoccupied sites probably are an absolute minimum, even if turnover is rapid and several years of survey data are collected.

the initial survey (so that $N - O$ patches were initially unoccupied, where N is the total number of patches surveyed). To estimate extinction parameters, we only use information from the O initially occupied patches. The log-likelihood for a particular E function (with specific coefficient values in u_e) is the sum of a set of log-transformed probabilities :

$$\log L_E = \sum_{i=1}^{O} \left[f_i \log(E_i) + (1 - f_i) \log(1 - E_i) \right] \tag{11.5}$$

where f_i is the binary fate of the population in patch i (1 if it has gone extinct by the second survey and zero if not)[5] and E_i is the model's predicted extinction probability for patch i. To estimate parameters in the colonization functions, we use only the $N - O$ patches that were unoccupied at the initial survey. The log-likelihood formula for the function C, $\log L_C$, is nearly identical to Equation 11.5, except that we replace E_i with C_i (and its component parameters and independent variables), we replace O with $N - O$, and f_i now equals 1 if patch i has become colonized by the second survey and zero if not. In this explanation we have assumed data from a single time period, but using data from multiple intervals is perfectly acceptable. In this case the fate of each site in each time period is treated as a separate term in the summation in Equation 11.5 (which is then a sum of logged probabilities over both sites and years). If you are using data on turnovers from two or more intervals, you should include a categorical variable for time period in the u_e and u_c functions.

As with any likelihood fitting, the goal is to find the values for the coefficients that provide the best fit to the observed data. This is equivalent to maximizing the log-likelihood of the E and C functions, or, as is more usually done, minimizing the negative log-likelihoods. Most statistical programs now perform maximum likelihood fitting for logistic regression models, alleviating the need for special programs to do parameter estimation. In addition, using the negative log-likelihoods for each logistic regression model, we can compute AIC_c values for models that do and do not include specific effects (say, an effect of isolation on the extinction probability) to determine whether there is support for including these complexities.

Harrison et al.'s (2000) study of five serpentine seep plants provides an example of such an analysis. In particular, Harrison et al. applied a logistic regression to extinction data to show that increasing isolation (measured as the harmonic mean of the distances to the three nearest sites) strongly increased extinction probability, while increasing distance from human-

[5]Since with the logistic regression model we can directly include rescue effects in E_i (e.g., by including a distance term among the independent variables), there is no need to include the C_i function when estimating extinction rates. See Moilanen 1999 (Equation 7) for the corresponding, more complicated likelihood formula for an incidence function model with rescue effects.

caused disturbances resulted in a weak decrease in extinction risk. As we have emphasized in Chapter 10, these results can either be interpreted as the result of rescue effects or as a reflection of spatial correlation in fates due to correlated habitat qualities or temporal variations, without any direct interaction between the populations. This is not an easy problem to solve, but it is a key one to keep in mind when interpreting the results of logistic regression and incidence function models. Ecologists have usually thought of these models as testing for movement effects, but they are actually forms of spatial auto-regression, and as such don't give direct information on the underlying *cause* of associations between colonization or extinction and the occupancy of nearby sites (Augustin et al. 1996, 1998). To search for spatial correlation patterns more thoroughly, we also suggest conducting a more explicit analysis of spatial correlations via join-count methods, as we described in Chapter 10 for this same example (Figure 10.5, Box 10.1). That analysis found evidence for movement and/or spatial environmental correlations at multiple spatial scales in the data of Harrison and colleagues. The advantage of join-count statistics as an exploratory tool is their ability to look for relationships at multiple spatial scales without any presumption of a smooth functional form for these correlations (e.g., a decline with increasing distance), in contrast to either logistic regression or incidence function models.

Fitting separate logistic regression models for E_i and C_i to extinction and colonization data has the distinct advantage of allowing us to use readily available software packages. However, there are some circumstances in which it may be advantageous to perform more complex model fitting. First, using only simple linear combinations of variables in u_e and u_c prevents the inclusion of functions such as M_i (Equation 11.2) that allow a more detailed estimation of how the size and distance of neighboring patches will influence extinction or colonization rates. It is simple to include more complex functions like this within u_e and u_c, but doing so then requires a fitting routine that can handle any arbitrary model structure.

A second, more difficult issue is whether or not to use information on the initial pattern of patch occupancies in addition to data on subsequent colonization and extinction events when fitting E and C. Although it is clear that fitting from static occupancy data alone can be quite misleading (Moilanen 1999, 2000), to ignore this information entirely is to throw away a great deal of *potentially* useful data. This is particularly true if few turnover events were observed, in which case nearly all the useable information in the data comes from the static occupancy pattern. However, whether occupancy information actually *is* useful depends on the metapopulation's history and what we know or are willing to assume about it. When one uses patch occupancy data to fit the E and C functions, the usual assumption made is that the occupancy pattern reflects a dynamic equilibrium between colonization and extinction, and thus that the current number and arrangement of occupied sites can tell us a great deal about the mean rates of extinctions and colonizations that govern the metapopulation. Even though this is not a reasonable assumption in many conservation settings, it may well be in others, as when we analyze a set of populations on undisturbed sites in anticipation of future impacts. In this case, using occupancy data can be extremely valuable in honing estimates of the parameters in E and C (Moilanen 1999), forcing the model to predict not only estimates of turnover that match the data, but also the outcome of these events for the metapopulation as a whole. However, don't forget that using occupancy information in this way will also make it much less likely that we will arrive at parameter estimates predicting either metapopulation declines or increases, even if such a trend is actually occurring. Thus, using occupancy data and therefore assuming stability should never be done as a matter of course.

In sum, despite the need for caution, it can be desirable to use occupancy data along with extinction and colonization information when fitting metapopulation models for some species and situations. Moilanen (1999, 2000) has recently critiqued past methods of fitting models to occupancy data and suggested a better approach for using occupancy data and extinction–colonization data simultaneously, based on the assumption of a dynamically stable metapopulation (with all its attendant problems). Moilanen concentrates on incidence function models when delving into specifics, but the approach works for any extinction and colonization function, and we will present his methods in the context of logistic regression models.

We can summarize the occupancy information about a set of sites in year t with a vector $\mathbf{O}(t)$, where the elements of the vector are ones for patches with extant populations and zeros for patches without populations. With two years of data, we want to fit extinction and colonization functions that together maximize both $P[\mathbf{O}(1)]$, the probability of seeing the observed pattern of occupancy in the first year, and $P[\mathbf{O}(2) \mid \mathbf{O}(1)]$, the probability of seeing the observed occupancy pattern in the second year given the pattern of occupancy in the first. In general, if we have multiple years of occupancy data, the overall probability of seeing the series of occupancies, given functions E and C, is:

$$P(data \mid E,C) = P[\mathbf{O}(1)]\, P[\mathbf{O}(2) \mid \mathbf{O}(1)]\, P[\mathbf{O}(3) \mid \mathbf{O}(2)] \cdots \qquad (11.6)$$

But note that $P[\mathbf{O}(2) \mid \mathbf{O}(1)]$ must logically be determined by the separate probabilities of the observed pattern of extinctions from years 1 to 2 and the observed pattern of colonizations over the same interval. Therefore, all of the terms on the right-hand side of Equation 11.6 except for the probability of seeing the initial occupancy pattern are, in fact, composed of the separate extinction and recolonization probabilities we've already considered, giving:

$$P(data \mid E,C) = P[\mathbf{O}(1)]\, P[\mathbf{C}(1)]\, P[\mathbf{E}(1)]\, P[\mathbf{C}(2)]\, P[\mathbf{E}(2)] \cdots \qquad (11.7)$$

where $\mathbf{C}(t)$ and $\mathbf{E}(t)$ are vectors of zeros and ones that describe the colonization fates of initially empty sites and the extinction fates of initially occupied sites from t to $t + 1$, respectively. $P[\mathbf{C}(1)]$ is the probability of seeing the observed set of colonization events, $\mathbf{C}(1)$, from years 1 to 2, and $P[\mathbf{E}(1)]$ is the same for the set of observed extinction events, $\mathbf{E}(1)$. To estimate the log-likelihood of seeing the entire data set, we take the logarithm of both sides of Equation 11.7 and then substitute in the log-likelihoods for each year of colonization and extinction data:

$$\log L(data \mid E,C) = \log L\left[\mathbf{O}(1) \mid E,C\right] + \sum_{t=2}^{T} \log L_E(t) + \log L_C(t) \qquad (11.8)$$

We know how to estimate $\log L_E(t)$ and $\log L_C(t)$, the log-likelihoods for the extinction and colonization functions (i.e., we apply Equation 11.5), so we only have to estimate the likelihood of seeing the initial occupancy pattern, given particular parameters for E and C. Direct calculation of this probability is feasible in theory, but in practice it is impossible for even a moderate number of sites (Moilanen 1999). The most obvious way to estimate the probability of seeing a pattern of occupancy is to simulate a metapopulation governed by a particular set of parameters in the E and C functions and use the frequency of years with exactly the correct occupancy pattern as the probability of that model. However, there are so many possible combinations of occupied and unoccupied patches that again this approach is rarely feasible. Moilanen's solution to this problem is to simulate a metapopulation for a long time, using the observed initial occupancy pattern as the starting condition, and then to calculate the probability of getting from each year's simulated occupancy pattern to the observed occupancy pattern in one time-step, which simply relies on each year's set of E_i and C_i probabilities (presumably discarding an initial period of the simulation to reduce effects of the starting conditions). Taking these many probabilities of getting to the observed occupancy pattern as an indication of the likelihood of seeing that pattern, we can then calculate the overall likelihood of seeing the initial occupancy pattern given the governing E and C functions:

$$L[\mathbf{O}(1) \mid E,C] = \frac{1}{Z}\sum_{t=1}^{Z} P[\mathbf{X}(t+1) = \mathbf{O}(1) \mid \mathbf{X}(t)] \qquad (11.9)$$

where $\mathbf{X}(t)$ is the simulated occupancy pattern in year t and Z years of simulated occupancy data are used. We substitute the log of this likelihood into Equation 11.8 to arrive at an overall log-likelihood for observed occupancies, extinctions, and colonizations for a given set of parameter values. Although we don't pursue it here, Moilanen also suggested using this same method to make predictions of extinction and colonization (and hence occupancy) patterns when there are years with no data in the midst of a series of observations of occupancies.

Two notes of caution must be sounded about using Equation 11.9. First, Moilanen points out that using it can give predicted extinction and colonization rates that are considerably higher than the correct ones. Presumably this is because, without a *very* large sample of $\mathbf{X}(t)$ occupancy distributions, most or all of the distributions are quite far from the single initial observed set of occupancies. If this is the case, the model may be fitting less for colonization and extinction rates that result in $\mathbf{X}(t)$ being very close to $\mathbf{O}(1)$ and more for rates that are high enough to increase the probability of "leaping" from poor predicted $\mathbf{X}(t)$ distributions to the observed one. Moilanen suggests limiting the range of possible colonization and extinction parameters so that the total population turnover (transitions from extant to extinct or visa versa) is not much above that actually observed. We could do so by instructing the search routine used to fit the model to assign a very low likelihood to parameter combinations that yield a total number of turnovers higher than observed. While this "fix" allows use of the method with low numbers of simulated years, we suggest the less ad hoc approach of using more and more simulated years until the parameter estimates converge to steady values. The second caution is that the use of a single long time series to generate the many replicate $\mathbf{X}(t)$ distributions needed for this method means that the method will favor parameter values that predict extremely high viability, with no chance of random extinction of the metapopulation over even thousands of years. Since this is an unreasonable assumption for many situations, it is generally better to run many shorter simulations, each with $\mathbf{O}(1)$ as the starting condition, using only one or a few $\mathbf{X}(t)$ distributions from the end of each of these replicates.

We have described the basic idea of Moilanen's method, but it is worth noting that it could also be extended to assumptions about patch occupancy patterns other than that of a fairly stable dynamic equilibrium. In particular, if we know that all patches were occupied at some point in the past, we can estimate $\log L[\mathbf{O}(1) \mid E,C]$ from simulations that embody the process of getting from complete occupancy to the initial pattern observed in the occupancy data. For example, we might know that 20 years ago a currently fragmented habitat was continuous, and then in a short amount of time, most of the habitat was lost. With this knowledge, we could assume that all current habitat patches were occupied 20 years ago, and we can run many replicate 19-year simulations with these initial conditions to estimate $\log L[\mathbf{O}(1) \mid E,C]$ from a modification of Equation 11.9, using only the last year of each replicate to calculate the probability of reaching the current occupancy pattern. This approach allows much more flexibility in the use of occupancy data to estimate the parameters of E and C, but it also requires considerable prior knowledge of how occupancy patterns probably changed before your own sampling began.

With Moilanen's method in hand, we have a way to judge the fit of a model to initial occupancy, colonization, and extinction data. However, this set of estimation methods must be used in conjunction with some minimization algorithm that will search for the best values for the coefficients in u_e and u_c, given the likelihood criterion being used (see the Appendix in Chapter 4). Minimization routines differ drastically in their efficiency and in how reliably they identify the best global solution, especially when there is a stochastic element in the calculated log-likelihoods, as there is in Moilanen's method. If you plan to do a large number of parameter searches, it will be worth your while to investigate several options for good search algorithms and read about how each works (Press et al. 1995).[6]

In Boxes 11.1 and 11.2 we give two linked programs that find the best fit parameters for Harrison et al.'s study of *Delphinum uliginosum*, using the `fminsearch` minimization routine that is a standard function in MATLAB.[7] While we do not include code to find the confidence limits around each parameter estimate, this is a wise extra step in model-fitting (Hilborn and Mangel 1997). For this example, we fit logistic regression models for extinction and colonization, with the following forms for u_e and u_c:

$$u_e = a_e + \beta_e \sum_{j \neq i}^{N} \exp(-\alpha D_{ij}) p_j, \qquad u_c = a_c + \beta_c \sum_{j \neq i}^{N} \exp(-\alpha D_{ij}) p_j \quad (11.10)$$

where a_e and a_c are coefficients that set the baseline values of the extinction and colonization probabilities (compare to a in Equation 11.4), and the other parameters have the same meanings as in Equation 11.2. We don't include any effect of patch size on extinction or colonization rates, based upon Harrison et al.'s results that show no such effects.[8] Furthermore, we fit the same α parameter, which controls the rate of decline in influence of neighboring patches with distance, for both colonization and extinction. Both this shared parameter in u_e and u_c and the nonlinear effect of distance mean that we cannot use standard logistic regression routines to fit this model. The programs in Boxes 11.1 and 11.2 are written to enable the use of both occupancy patterns and turnover data in estimating model fit (Equation 11.8). However, the assumption of stability is not appropriate for *Delphinum uliginosum* since min-

[6]Members of one of our labs (Doak's) became so frustrated with the free minimization (solver) programs that work in conjunction with MATLAB that we finally paid good money for what has turned out to be an excellent set of routines called TOMLAB, which we can strongly recommend.

[7]Our goal in Boxes 11.1 and 11.2 is to illustrate the general approach of estimating colonization and extinction parameters by minimizing the negative log likelihood. In reality, the `fminsearch` routine does not do a good job of converging on the maximum likelihood parameter estimates for this data set, even when we do not include the stochastic estimation of $\log L[\mathbf{O}(1) \mid E,C]$ in the calculation. Minimization routines from TOMLAB do rapidly converge on a good solution when $\log L[\mathbf{O}(1) \mid E,C]$ is not included in the likelihood function. Moilanen recommends using simulated annealing methods to fit his model (Moilanen 1999, Press et al. 1995).

[8]But note that to make the code more generally useful, we have included area effects in the structure of the programs in Boxes 11.1 and 11.2.

BOX 11.1 ***A MATLAB program to find the maximum likelihood parameter values for a logistic regression model of metapopulation dynamics. This program calls on the function in Box 11.2, which does most of the actual calculations.***

```
% Program logregB.m This program uses the function logregA.m
% and the MATLAB function fminsearch to find maximum likelihood
% estimates for a logistic regression model of metapopulation
% dynamics.
% As written below, this function is tailored to the particular
% example used in Morris and Doak Chapter 11 (Delphinium
% uliginosum data from the study of Harrison et al. 2000), but
% it contains the code to include effects, such as site sizes,
% that are not used in this example, and also can be modified
% to use more years of data or more explanatory variables.
clear all;
global FF1 FF2 AA sitenum Dists AAmx NNL randtime ...
 mintime bestparams bestNLL; % define global variables

%****************User-defined Parameters*******************
% The program is set up to use two years of data.
% Load the stored information on each site.
% Each row of data should be one site.
% The columns of variables in the data file should be:
%   X-coordinate of site,
%   Y-coordinate,
%   occupancy state in first year(0=unoccupied, 1=occupied)
%   occupancy state in second year
%   independent variable columns (e.g., size of site)
load delugA.txt;
siteinfo = delugA; % assign data to the siteinfo variable
randtime = 1; % the number of replicate random simulations
% used to calculate the likelihood of initial occupancy pattern
mintime = 20; % number of initial years to discard in
% simulations to get rid of transitory dynamics

% There are two arrays of initial parameter values:
% 1) eeparams stores the values of the extinction parameters
%    ae, betae, se, alpha, and b, used in the expression
%    Ei=ae + se*Ai + betae*sum{(exp(-alpha*Dij)*pj*Aj^b}
%    where Ai is the area of patch i, Dij is the distance from
%    patch i to patch j, pj indexes occupancy of patch j
%    (1=occupied, 2=unoccupied), and the sum is taken over
```

BOX 11.1 (continued)

```
%    all patches other than patch i.
% 2) ccparams stores the values of the colonization parameters
%    ac, betac, and sc in the expression:
%     Ci=ac + sc*Ai + betac*sum{(exp(-alpha*Dij)*pj*Aj^b}
% Note that both expressions use the same values for the
% parameters alpha and b. Also, se, sc, and b (governing site
% size effects) are not used for Delphinium uliginosum.
eeparams = [0.233   0.000000  -2.862   5.85   0.000000];
ccparams = [-1.108     0.000000   0.653];
% IMPORTANT NOTE: starting values can have drastic effects on
%  the final parameter values to which a search routine
% converges. It is very important to try different starting
% values in eeparams and ccparams to verify your results.
%*************************************************************

rand('state',sum(100*clock)); randn('state',sum(100*clock));
% Making variables to use in calculations:
XX = siteinfo(:,1);        % latitude in km
YY = siteinfo(:,2);        % longitude in km
FF1 = siteinfo(:,3);       % occupancy state in year 1 (0,1)
FF2 = siteinfo(:,4);       % occupancy state in year 2 (0,1)
AA = siteinfo(:,5);        % size of site
sitenum = length(XX);      % number of sites
x1 = XX(:,[ones(1,sitenum)]);      % making matrices
x2 = x1';                          % of x
y1 = YY(:,[ones(1,sitenum)]);      % and y values
y2 = y1';
Dists = sqrt((x1-x2).^2 + (y1-y2).^2); % dist's between sites
AAmx = AA(:,[ones(1,sitenum)])';       % matrix of sizes
% combine parameters into one array to pass to fminsearch:
params = [eeparams,ccparams];
bestparams=params; bestNLL=10000; % these help to track results
% Now, call the MATLAB minimization routine fminsearch to find
% the best values for logregA, a function that calculates the
% negative log likelihood of the data given the parameters.
fminsearch('logregA',params);  % call fminsearch
disp('Locally optimum parameter values:');
disp('extinction parameters:');
disp([params(1:5)']);
disp('Locally optimum parameter values:');
disp([params(6:8)']);
```

BOX 11.2 ***A MATLAB function that provides an estimate of the log-likelihood of a set of occupancy, extinction, and colonization data, given a set of parameters provided by the program in Box 11.1***

```
function NLL=logregA(params)
% This function gives the likelihood of a logistic regression
% model for a metapopulation, given a particular set of
% parameter values.
% This function is called by a program (logregB) that fits the
% maximum likelihood values for the parameters.
% This function is tailored to the particular example
% used in Morris and Doak Chapter 11 (Delphinium
% uliginosum data from the study of Harrison et al. 2000)

%****************User-defined Parameters*******************
global FF1 FF2 AA sitenum Dists AAmx FF1mx NNL randtime ...
 mintime bestparams bestNLL;
ee = params(1:5)';                  % separate the extinction
cc = [params(6:8)',params(4:5)']; % and colonization parameters
totProbff1 = 0;
simcount = 0;
%*********************************************************
% setting these parameters equal to 0 eliminates size effects:
ee(2)=0; cc(2)=0; ee(5)=0; cc(5)=0; % remove to include size

for reps = 1:randtime  % multiple simulations of the metapop'n
% On the first time through, it predicts the probability of
% seeing the second year of data - other simulations are for
% predictions of the first year occurrence pattern.
 FFnow = FF1;       % set initial occupancies
 for tt=1:(mintime +1) % time loop

%----------Start of basic calculations--------------------
 % matrix of occupancies for all neighbors:
 FFnowmx = FFnow(:,[ones(1,sitenum)])';
 % Calculation of neighbor effects for extinction probability:
 bit1=-ee(4)*Dists; % first part of extinction equation
 bit1(find(bit1>100))=100; % check for values too big or small
 bit1(find(bit1<-100))=-100;
 connmxe = exp(bit1).*(AAmx.^(ee(5))).*FFnowmx;
 connmxe = connmxe-diag(diag(connmxe)); % remove site's self
 connmxe(find(connmxe>20))=20; % check for too big or small
```

BOX 11.2 ***(continued)***

```
connmxe(find(connmxe<(-20)))=-20;
% summed, dist.-discounted areas of neighbors:
conne = sum(connmxe')';

% Calculate neighbor effects for colonization probability:
bit1=-cc(4)*Dists;
bit1(find(bit1>100))=100;
bit1(find(bit1<-100))=-100;
connmxc = exp(bit1).*(AAmx.^(cc(5))).*FFnowmx;
connmxc = connmxc-diag(diag(connmxc));
connmxc(find(connmxc>20))=20;
connmxc(find(connmxc<(-20)))=-20;
% summed, dist.-discounted areas of neighbors:
connc = sum(connmxc')';

% make the linear fns ue and uc to go into logit fns:
uue = ee(1) + ee(2)*AA + ee(3)*conne;
uuc = cc(1) + cc(2)*AA + cc(3)*connc;
% initialize logistic equations:
loge=zeros(sitenum,1); logc=zeros(sitenum,1);

% find the predicted probabilities of ext and col:
uue(find(uue >10))=10; uue(find(uue<-10))=-10;
loge = exp(uue)./(exp(uue)+1);
uuc(find(uuc >10))=10; uuc(find(uuc<-10))=-10;
logc = exp(uuc)./(exp(uuc)+1);

% calculate Negative Log-Likelihood for the observed
% extinctions and colonizations:
if ((tt == 1) & (reps ==1))
   LLec=FFnow.*( (1-FF2).*log(loge) +FF2.*log(1-loge) ) ...
   + (1-FFnow).*( (FF2).*log(logc) +(1-FF2).*log(1-logc) );
   NLLec = -sum(LLec); % neg LL of extinctions and col.'s
end; % if

% randomly make next year's simulated occupancy:
transprobs=rand(sitenum,1);
FFnew = FFnow.*(ceil(transprobs-loge)) + ...
             (1-FFnow).*(ceil(logc-transprobs));

%----------End of basic calculations----------------------
```

BOX 11.2 *(continued)*

```
 % calculate the probability of going from the simulated
 if tt == mintime +1 % occupancy to the initial real one:
    logProbff1= sum(FFnew.*( (1-FF1).*log(loge) + ...
      FF1.*log(1-loge) ) + (1-FFnew).*( (FF1).*log(logc) ...
      +(1-FF1).*log(1-logc)));
    totProbff1 = totProbff1 +exp(logProbff1);
    simcount = simcount+1;
 end; % tt == mintime + 1
 FFnow = FFnew; % advance the simulated occupancies
 end; % tt loop over the multi-year simulation
end; % reps loop

% negative log-likelihood of initial occupancy data
NLLint=-log(totProbff1/simcount);
% to use both initial occupancy and ext/col data for fitting:
% NLL= NLLec+NLLint; % final, total negative log-likelihood
% to use only ext/col data for fitting:
NLL= NLLec; % final negative log-likelihood
disp(params') % displays each set of parameter values used
disp([NLLec,NLLint, NLL]); % displays the 3 NLL values
if NLL <bestNLL % keeps a running tab of best estimates found
    bestNLL = NLL; % remove ';' on this line to show estimates
    bestparams=params';
end;
```

ing activities destroyed some seeps between the two census periods and may have had other less direct impacts as well (Harrison et al. 2000). Therefore, we fit u_e and u_c based upon only extinction and colonization events, using Equation 11.8 but setting $\log L[\mathbf{O}(1) \mid E,C] = 0$. One last point to make is that we have fit the probabilities of extinction and colonization as decadal rather than annual rates. Even though it would be possible to fit the best annual rates given only beginning and ending censuses (Moilanen 1999), it is easier and more direct to fit them for the actual time interval over which the data were collected.

The maximum likelihood parameter estimates for *D. uliginosum* show the expected effects of distance on both extinction and colonization, with extinction probability being negatively influenced by proximity of neighboring populations ($\beta_e = -2.86$) and colonization being positively affected ($\beta_c = 0.653$).

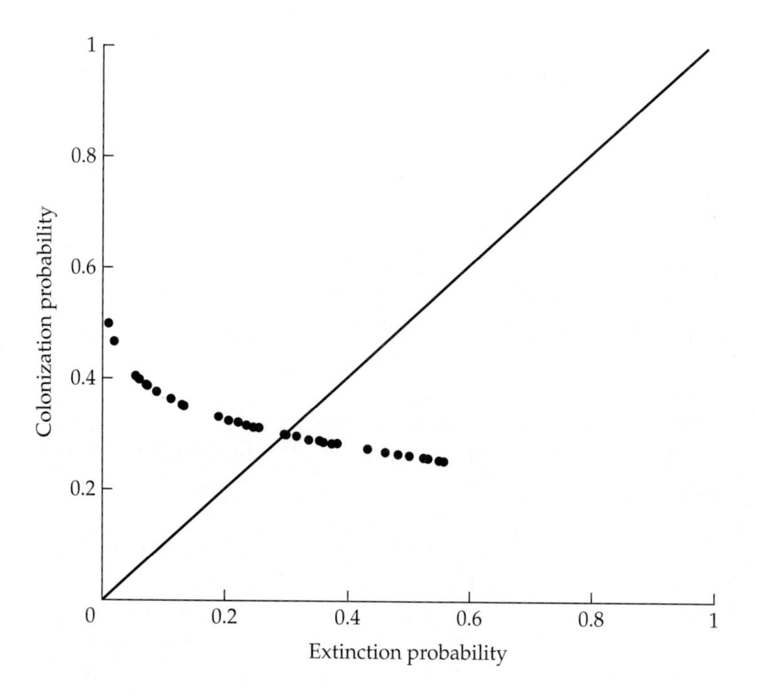

Figure 11.1 Predicted extinction and colonization probabilities for 56 seep sites that provide habitat for *Delphinium uliginosum*. Each seep is shown as a single point. Probabilities are calculated based upon fitted logistic regression parameters, the distances between sites, and the observed pattern of occupancy in the 1980s. The 45° line indicates equal probabilities of colonization and extinction

The distance exponent is large ($\alpha = 5.85$), indicating that the effect of neighboring populations on the extinction and colonization probabilities falls off rapidly with distance. Together with the two constant terms ($a_e = 0.234$ and $a_c = -1.11$) we can now use these parameter estimates, the distances between seeps, and the initial pattern of occupancy to estimate the probabilities of extinction or colonization for each site following the 1980s survey, giving a rough feeling for the viability and importance of each site (Figure 11.1); the large number of sites with higher predicted colonization than extinction probabilities suggests a highly viable metapopulation. Furthermore, we can use the distance of a site above the line of equal extinction and colonization probabilities as a rough gauge of its likely importance for metapopulation viability. However, because each site's rates will fluctuate with its neighbors' occupancy and will influence its neighbors in turn, this graphic is at best a very rough way to determine metapopulation viability or population importance (see Crone et al. 2001). A better way is to simulate the metapopulation over time.

Simulation of Patch-Based Models

With parameter estimates in hand it is a simple matter to simulate the expected dynamics of patch occupancy over time (Box 11.3). The basic idea is simply to use expressions such as Equations 11.4 and 11.10 to estimate each site's chances of extinction (if there is a population there) or coloniza-

BOX 11.3 ***A program to simulate a logistic regression model for patch-based metapopulation dynamics.***

```
% Program logregextsim.m. This does stochastic simulations to
% estimate the extinction CDF for a patch-based metapopulation
% model. All the data file requirements, parameter definitions,
% and simulation methods are the same as for logregA and
% logregB in Boxes 11.2 and 11.1, so comments are kept to a
% minimum here.
clear all;
%****************User-defined Parameters*******************
load delugA.txt; % get basic information file for the
siteinfo=delugsmall; % metapopulation to simulate
simreps = 500; % how many simulations to do
maxtime = 200; % number of years to replicate
% Values for parameters (same order and meaning as in logregA)
eeparams = [0.233908  0.000000  -2.862026 5.849953  0.000000];
ccparams = [-1.107957 0.000000   0.652810];
%*******************************************************
rand('state',sum(100*clock));
XX = siteinfo(:,1);        % latitude in km
YY = siteinfo(:,2);        % longitude in km
FF1 = siteinfo(:,3);       % occupancy state in year 1 (0,1)
FF2 = siteinfo(:,4);       % occupancy state in year 2 (0,1)
AA = siteinfo(:,5);        % size of site
sitenum = length(XX);      % number of sites
x1 = XX(:,[ones(1,sitenum)]);  x2 = x1';
y1 = YY(:,[ones(1,sitenum)]);  y2 = y1';
Dists = sqrt((x1-x2).^2 + (y1-y2).^2);
AAmx = AA(:,[ones(1,sitenum)])'; % size matrix
ee = eeparams(1:5)';        cc = [ccparams,eeparams(4:5)];
simcount = 0;               PrExt = zeros(1,maxtime);

for reps = 1:simreps  % loop of simulations
 FFnow = FF2; % set initial occupancy as it was at last census
 disp(reps);
 for tt=1:maxtime % time loop

 % Next, to run, insert the lines in the program in Box 11.2
 % that are between the same two comment lines as follow here:
 %----------Start of basic calculations--------------------
 %----------End of basic calculations----------------------
```

BOX 11.3 *(continued)*

```
 % check for extinction
 if isempty(find(FFnew>0))
    PrExt(tt) = PrExt(tt) +1;   end;
 extantsites(tt,reps) = sum(FFnew); % stores number occupied
 FFnow = FFnew;

 end; % tt loop
end; % reps loop

CDFExt = cumsum(PrExt/simreps);
disp('Below is mean and standard deviation')
Disp('of number of pop"s each year');
disp([mean(extantsites')', std(extantsites')']);
disp('And now, the extinction time CDF');
figure;
plot(CDFExt);
```

tion (if there isn't) at each time step, based upon its own characteristics (e.g., size) and the characteristics (e.g., size and distance) of its neighboring sites (Sjögren-Gulve and Ray 1996). Then, to determine a site's actual fate, a uniform random number between zero and one is drawn and compared with the probability of each fate. Although this simulation method is straightforward, it has two underlying assumptions. First, if done as just described, this method simulates the equivalent of demographic stochasticity, with whole populations playing the role that individuals play in the usual definition of demographic stochasticity (see Chapters 2 and 4), but it does not include temporal variability in extinction and colonization probabilities due to environmental changes. If multiple years of occupancy data are available, environmental stochasticity can be estimated and then included in the simulation, either by random draws of entire sets of parameter values based on a set of discrete year effects or as continuous variation (see Moilanen 1999). Second, the simulations are based on the assumption that the estimated effects of neighboring sites on occupancy patterns are due to direct interactions and not to any spatial or spatio-temporal correlations driven by shared environmental conditions. Although we can't usually test this assumption, it is good to keep it in mind when thinking about the output of metapopulation simulations.

To simulate the *D. uliginosum* metapopulation, we started with the occupancy pattern in the 1990s and used the parameter estimates derived above.

For the current collection of seep sites, the probability of metapopulation extinction over 200 years is estimated as zero, and the number of occupied sites in any year is very stable at 28.0 ± 3.6 (mean ±1 S.D.). Next, we tested the influence on these results of the 13 sites with the highest ratio of current colonization to extinction probabilities, all with colonization probabilities greater than twice their extinction probabilities.[9] To see whether these sites were critical in maintaining metapopulation viability, we removed them from the metapopulation simulation and reran the analysis. Surprisingly, the metapopulation is still stable even without these 13 sites, with no extinctions over 200 years and the number of extant populations stable at 15.5 ± 3.1. Recall that we did not use the initial occupancy pattern when estimating the colonization and extinction parameters, so the conclusion of high metapopulation viability is not the result of an *a priori* assumption that the metapopulation is in a dynamic equilibrium, a problem we discussed earlier. These results parallel those of Harrison and Ray (2002) who also found that this population would remain viable even with large reductions in the number of habitat sites. This example illustrates that at times metapopulations may be quite robust even to major perturbations, and that patch-based models can be a way to test for these responses. Given these results, it is not surprising that even though it specializes on an extremely rare habitat, *D. uliginosum* is relatively common on serpentine seeps in this study area (Harrison et al. 2000).

Count-Based Approaches

In general, count-based approaches to multi-site PVAare much more useful than patch-based methods, in part due to their greater flexibility and in part because they can be applied to more widely available data. However, to use these models, we need estimates of the mean, variance, and covariance in population growth rates for local populations, as well as the probabilities of movement (if any) between local populations. Ideally, some estimate of density dependence—at the minimum, a local carrying capacity—should also be available. With these estimates in hand, we can perform viability analysis for situations with or without movement and with or without correlation between populations, as we now describe.

The heart of a count-based model is a transition matrix, but one that governs population counts at different sites, rather than numbers of individuals in different classes in a single population. For the clapper rail example introduced in the last chapter (Harding et al. 2001b), a simple density-independent transition matrix for the three populations can be written as:

---

[9]This illustrates one way that patch-based models can be used in a management context, enabling us to ask, for example, whether allowing suburban development to destroy these seeps would endanger the metapopulation's chance of persistence.

$$\begin{bmatrix} n_M(t+1) \\ n_F(t+1) \\ n_L(t+1) \end{bmatrix} = \begin{bmatrix} \lambda_{M,t} & 0 & 0 \\ 0 & \lambda_{F,t} & 0 \\ 0 & 0 & \lambda_{L,t} \end{bmatrix} \begin{bmatrix} n_M(t) \\ n_F(t) \\ n_L(t) \end{bmatrix} \tag{11.11}$$

where $n_M(t)$is the number of rails in the Mowry population at time t, and $\lambda_{M,t}$ is the annual population growth rate for Mowry from t to $t + 1$ (and similarly for the Faber and Laumeister populations). Only the diagonal elements of the matrix are nonzero, meaning that the model assumes no birds move from one population to another. This corresponds to the best guess about the biology of the species in its now-isolated patches of marsh habitat. Thus the total population size across the three marshes is the sum of three completely independent populations. In this situation, the results of a formal matrix analysis, such as those presented in Chapters 7, 8, and 9 for demographic matrix models, are difficult to interpret or even misleading. Using eigenvalues and eigenvectors to estimate a meaningful overall growth rate for a matrix or to perform sensitivity analyses of this growth rate rely on the assumption that there is some probability of moving between every pair of classes represented in the model (Caswell 2001), which is not the case here. Therefore, we must resort to simulations to predict how total numbers and overall extinction risk are likely to change over time.

To simulate the model, we need to account for the mean, variance, and correlations of growth rates in each population. To do so, we can either use the machinery explained in Chapter 8 for generating projection matrices with correlated vital rates, or randomly draw sets of estimated population growth rates for all populations from different years, just as we drew entire projection matrices at random in Chapter 7. Box 11.4 presents a program to do the former, estimating extinction risks and the means and variances of population growth rates over time. Note that in simulating growth rates, we use the correlation structure and μ and σ^2 for each marsh (based on the values in Tables 10.1 and 10.2) to first generate $\log\lambda_t$ values for each site and then convert these to λ_t values for use in Equation 11.11.

Setting a quasi-extinction threshold for a multi-site model can be done in several ways. For the clapper rails, we set a lower limit of 20 birds in a single marsh as the quasi-extinction level for that marsh, with quasi-extinction occurring for the total population once all three populations have hit this threshold. With no movement between marshes, this is a more reasonable threshold than would be the total number of birds across all marshes, an approach that would make sense if there were fairly frequent movement between local populations. To build a more realistic model, we can also add negative density dependence. Although there is no direct evidence of density dependence in the rail data (Harding et al. 2001b), rails clearly cannot reach an infinite population density. Therefore, we set a cap at the highest density ever seen in any marsh (1.6 birds/ha). Using this density and the

BOX 11.4 ***A stochastic simulation for a count-based multi-site PVA. This program is a modification of single-population matrix models developed in Chapter 8.***

```
% MultiSiteCount. A program to do a multi-site stochastic
% PVA simulation using count data.
% NOTES:
% 1. This model is made to generate random simulations of a
% stochastic  model, generating estimates of the mean and SD of
% realized growth rates and of extinction probabilities over a
% specified time horizon.
% 2. The model is specified as the means, variances, and
% correlations of population growth rates.  The model assumes
% that log growth rates are entered
% so they can be simulated as normals, then transformed.
clear all;
%****************Simulation Parameters****************
% The parameters for population growth are mu
% and sig^2 values for three populations, then the
% prob. of leaving, and the prob. of reaching another site.
% Data for the California clapper rail; Harding et al. 2001b
vrmeans= [0.043 -0.002 0 0.2 0.5] ;     % means
vrvars=[0.051 0.041 0.051 0 0 ];        % variances
% correlations of growth rates:
corrmx1 = ...
 [    1.000        0.995   0.896   0    0;
      0.995        1.000   0.938   0    0;
      0.896        0.938   1.000   0    0;
      0            0       0       1    0
      0            0       0       0    1];
% you must also have a separate m-file that takes a vector of
% vital rate, vrs, and creates the matrix elements with them:
makemx='railmxdef';
n0= [70 26 33]';      % initial population sizes of each pop'n
Nmax = [286 60 58]; % maximum numbers in each population
Ne = [20 20 20];      % quasi-extinction thresholds for each pop'n
                      % (applied separately to each population)
tmax = 100;           % number of years to simulate
np = 5;               % number of vital rates
popnum  = 3;          % number of local populations
runs = 1000;          % how many trajectories to do
%*******************************************************
```

BOX 11.4 *(continued)*

```
rand('state',sum(150*clock)); % M12 is used for correlations
[W,D] = eig(corrmx1);  M12 = W*(sqrt(abs(D)))*W';
vrs = vrmeans;              eval(makemx);
[uu,lam1] = eig(mx);        lam1s = max(lam1);
[lam0,iilam] = max(lam1s);  % dominant eigenvalue
uvec = uu(:,iilam)/sum(uu(:,iilam));   % dominant right vector

results = [];
PrExt = zeros(tmax,1);     % extinction time tracker
logLam = zeros(runs,1);    % tracker of stoch-log-lambda values
stochLam = zeros(runs,1);  % tracker of stochastic lambda values
for xx = 1:runs;
 if round(xx/5) == xx/5 disp(xx); end;
 nt = n0; % start at initial population size vector
 extinct = zeros(1,popnum); % vector of extinction recorders
 for tt = 1:tmax
      m = randn(1,np);
      rawelems = (M12*(m'))';    % correlated std. normals
      vrs = vrmeans + sqrt(vrvars).*rawelems;
      eval(makemx); % make a matrix with the new vrs values
      nt = mx*nt;   % multiply by the population vector
      nt = min([nt';Nmax])'; % applying population cap
      if sum(extinct) < popnum  % these two ifs check for
       for nn = 1:popnum        % extinction and count it
              if nt(nn) <= Ne(nn) extinct(nn) = 1;  end;
       end;
       if sum(extinct) == popnum PrExt(tt)=PrExt(tt) +1; end;
      end; % if
 end % tt
 logLam(xx) = (1/tmax)*log(sum(nt)/sum(n0));
 stochLam(xx) = (sum(nt)/sum(n0))^(1/tmax);
end; % xx
CDFExt = cumsum(PrExt./runs);
disp('This is the deterministic lambda value');    disp(lam0);
disp('And this is the mean stochastic lambda');
disp(mean(stochLam));
disp('Below is mean and standard deviation of log lambda');
disp(mean(logLam));        disp(std(logLam));
disp('Next is plotted a histogram of logLams'); Hist(logLam);
disp('And now,the extintion time CDF'); figure; plot(CDFExt);
```

area of each marsh, we set maximum population sizes of 286, 60, and 58 for the Mowry, Faber, and Laumeister populations, respectively.

The results of the rail model confirm what we might have guessed from Table 10.1: Extinction risk for the three populations is relatively low in the near-term but increases to 5% after only 20 years and over 10% after 60 years (Figure 11.2A). To check how much this result depends on the estimated correlations between growth rates at the three sites, we reran the simulations with all correlations between growth rates set to zero. Removing these positive correlations approximately halved the multi-site extinction probability over the first 40 years, with converging CDF functions at longer timespans (Figure 11.2A). Thus correlations among the population growth rates can have a strong effect on the short-term extinction risk. Knowing that the Mowry marsh has the highest mean growth rate (Table 10.1) it is also important to ask how much this one population determines the overall level of safety of the entire set of populations. We therefore reran the simulations without Mowry, and obtained the not-surprising result that multi-site viability is quite low for the Faber and Laumeister populations by themselves (Figure 11.2B).

We can easily add more complexity to this simplest of multi-site models. For the rails, there is, in fact, some small possibility of movement between marshes and management could certainly be undertaken to enhance movement rates. The simplest way to add movement is with a constant probability d of an individual dispersing from a marsh, along with a constant probability a of arriving at each other marsh if a bird does initiate dispersal:

$$\begin{bmatrix} n_M(t+1) \\ n_F(t+1) \\ n_L(t+1) \end{bmatrix} = \begin{bmatrix} (1-d)\lambda_{M,t} & da & da \\ da & (1-d)\lambda_{F,t} & da \\ da & da & (1-d)\lambda_{L,t} \end{bmatrix} \begin{bmatrix} n_M(t) \\ n_F(t) \\ n_L(t) \end{bmatrix} \quad (11.12)$$

The way we've entered dispersal into the model assumes that movement happens after the census period, but before much mortality or reproduction occurs. (For example, the population growth rate for the Mowry population, $\lambda_{M,t}$, is multiplied by the $(1-d)n_M(t)$ birds that did not disperse away from that site.) We also assume that the measured population growth rates (Table 10.1) do not include loss or gain to a population from dispersal and that all marshes are effectively equidistant from one another. (An alternative would be to use separate a parameters that are proportional to the distances between pairs of marshes.)

To run this new model, we need some estimates for the dispersal parameters. Since we have no data at all with which to estimate a or d, we ran simulations for several different values of each and compared the resulting extinction probabilities. Increasing values of a always improve viability, but dispersal initiation was never favored, even for models with complete sur-

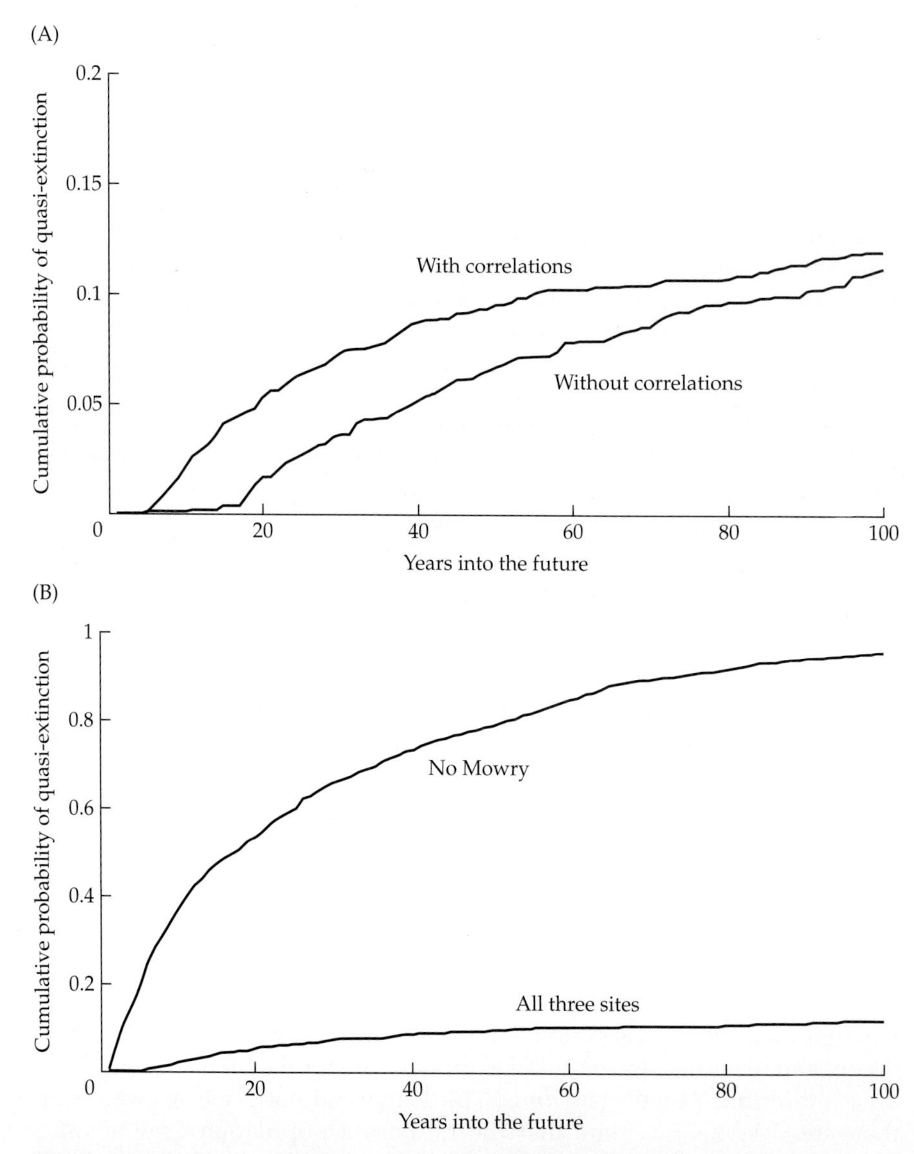

Figure 11.2 Extinction time cumulative distribution functions for count-based simulations of the California clapper rail. (A) Extinction probabilities with and without the estimated correlation structure in population-specific growth rates (Tables 10.1 and 10.2). (B) Extinction results based on all three populations and on simulations that eliminated the Mowry Marsh population from consideration. Note the different scales of the y-axes.

vival of all dispersing birds (Figure 11.3). The result reflects the much better demographic performance of birds at the Mowry population, such that dispersal away from this site is never beneficial to overall viability. Thus

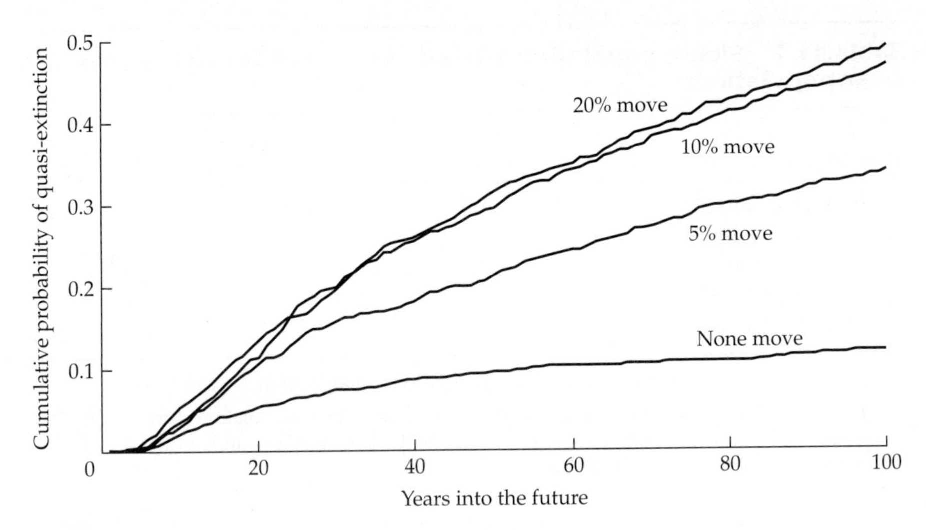

Figure 11.3 Extinction time cumulative distribution functions for count-based simulations of the California clapper rail, with different hypothetical levels of dispersal, *d*. Results are shown for models that assume no mortality during dispersal ($a = 0.5$).

enhancing dispersal among sites is not an automatically advantageous management strategy, especially when some populations are demographic sinks. In part this result is also due to the high correlations in growth rates of the three sites, making any potential benefits of dispersal too weak to offset the loss of birds from Mowry.

As we discussed in Chapter 10, count-based approaches can also be adapted to multi-site situations with even less information than we have for the clapper rail. A common situation is one in which we have census data from only one or two sites, but estimates of the current population sizes in several other locations. If we suspect that all of the sites are sufficiently isolated from one another that there is neither movement among them nor environmentally driven correlations in their population growth rates, this situation corresponds to Case 1B in Table 10.3. Even with so little information, we can gain some understanding of the value of preserving multiple populations if we are willing to assume that the population parameters measured in the censused sites apply to all the populations. As we discussed in Chapter 10, this assumption must be applied with great caution; if most populations are actually doing better or worse than the censused population(s), our analysis can be highly inaccurate.

As an example of such an analysis, we return to the grizzly bear. The Yellowstone population was the focus of our attention in Chapter 3, but there are actually five known grizzly bear populations in the Lower 48 United States and adjacent parts of Canada. In 1993, the United States Fish and Wildlife Service estimated population sizes ranging from 306 in the Northern Continental Divide Ecosystem to five in the North Cascades (Table 11.1). Since each of

Table 11.1 Single population viability estimates for five grizzly bear populations[a]

Population	*Population size*	*Probability of quasi-extinction (500 years)*[b]
Northern Divide	306	7.1×10^{-8}
Greater Yellowstone Ecosystem	236	1.66×10^{-7}
Selkirk Mountains	25	0.000258
Cabinet/Yaak Mountains	15	0.001374
North Cascade Mountains	5	0.04995

[a]Population size estimates are from the United States Fish and Wildlife Service (1993).

[b]The estimated quasi-extinction risk shown for the Yellowstone population differs from that estimated in Chapter 3 because here we use total population numbers and, more importantly, we use a quasi-extinction threshold of 2 rather than 20.

these areas is distinct and fairly isolated from the others, we will proceed with an analysis that assumes both that there is no movement between the populations (which is almost certainly true) and that there is no correlation between the growth rates of the populations. This second assumption is probably not correct; there is good evidence of widespread environmental factors, including weather and masting of white-bark pines (Mattson et al. 1992) that will create at least some correlations in the fates of bear populations in the inter-mountain West. However, since we can't clearly evaluate the strength of such correlations, we will proceed here with the simplifying assumption of totally independent fates, knowing that such correlations (if they exist) will render our analysis optimistic. To estimate the viability of each population, we used the Dennis et al. method described in Chapter 3. Since none of the other four grizzly populations have enough data to estimate μ and σ^2, we instead use estimates from the extensive Yellowstone data set (see Table 3.1): $\hat{\mu} = 0.02134$ and $\hat{\sigma}^2 = 0.01305$. With these values, we used Equation 3.5 to calculate the probability of extinction at 500 years for each population and assumed an extinction threshold of 2 bears[10] (Table 11.1). This extinction threshold is unrealistically low, but we use it here so that the North Cascade population is not considered to be extinct from the start.

Given our estimates of the extinction risk of each population, we can ask about the advantages for overall viability of protecting sequentially smaller populations. Recall that if populations are truly independent, then the probability of all of them becoming extinct is simply the product of the probabilities that each of them do. For the 500-year time-span, protecting the

[10]Note that we have used the total number of bears here, rather than the number of adult females as in Chapter 3.

Yellowstone population in addition to that of the Northern Divide decreases the probability of extinction from 7.1×10^{-8} (for the Northern Divide alone) to the product of the two populations' extinction probabilities: $(7.1 \times 10^{-8}) \times (1.7 \times 10^{-7}) = 1.2 \times 10^{-14}$, a very small number. Although this dramatic gain in safety may not seem surprising, there is also a substantial lessening of extinction risk from adding the much smaller Selkirk population to the first two: $(7.1 \times 10^{-8}) \times (1.7 \times 10^{-7}) \times (2.6 \times 10^{-4}) = 3.1 \times 10^{-18}$. Figure 11.4 illustrates that adding the remaining two populations to the portfolio of sites yields diminishing benefits. (Note the logarithmic scale of the y-axis.)

We might also ask whether the three smallest populations together can guarantee as little risk of extinction as either of the two largest populations. The answer is no: The three small populations together have a combined extinction risk more than double that of either of the two larger populations alone: $(0.04995) \times (0.001374) \times (2.6 \times 10^{-4}) = 3.6 \times 10^{-7}$. However, in terms of the total number of individuals involved, this is a surprisingly small extinction risk, since these three populations together harbor only 45 individuals, less than one-fifth the number in either of the two large populations. This example clearly illustrates the potential advantages of preserving multiple populations if their fates are independent.

Most uses of count-based methods for multi-site PVAs will probably not be much more complicated than the examples we have used here. These approaches have the virtue of simplicity, but we must turn to demographic methods in order to incorporate more detailed biological information into a multi-site PVA.

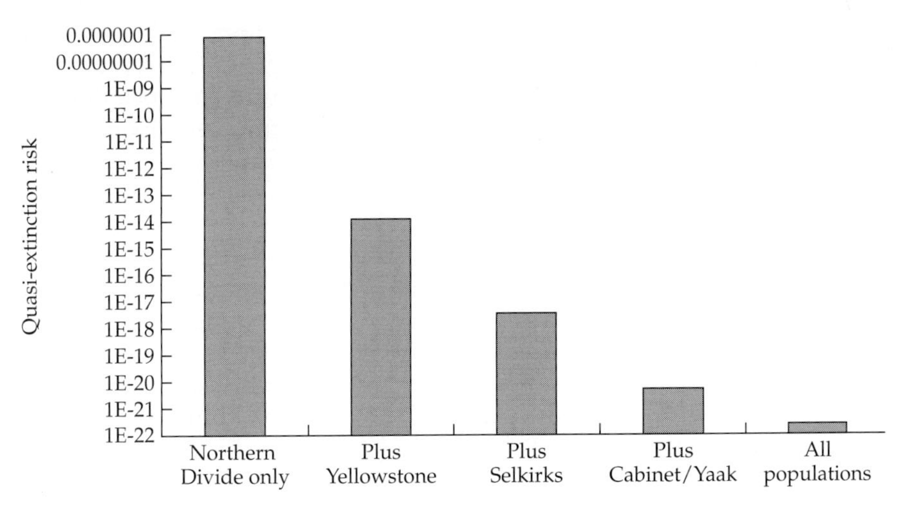

Figure 11.4 Extinction risk for sets of grizzly bear populations over 500 years. From left to right, each bar plots the overall extinction probability obtained by adding sequentially smaller populations to a collection of protected areas.

Demographic Approaches

To use demographic approaches for multi-site viability analysis, you need estimates of demographic rates for each local population as well as class-specific estimates of movement rates between each pair of populations. In addition, correlations among vital rates within and across populations can be critical. With these estimates, there are two major demographic approaches that can be used to address viability questions. First are extensions of projection matrix methods (Chapters 6–9), which are closely parallel to the count-based multi-site models we have just presented. Basically, these are models that only differ from single population matrix models in that they classify individuals by location as well as by age, size, or stage. The second class of demographic models consists of more elaborate, individually based simulations. When abundant data exist, this second way of exploring viability questions may be favored because of its ability to better incorporate complicated movement processes and the effects of complex habitat patterns. In this section, we first discuss in some detail how to construct matrix-based models and then give a brief overview of individually based simulation approaches.

Multi-Site Matrix Models

Probably the majority of published multi-site PVAs fall into this class of models (e.g., Wootton and Bell 1992, McCarthy 1996, Lahaye et al. 1994, Lindenmayer and Possingham 1996, Root 1998). Even though it is not always clear from reading descriptions of their construction or results, these models are all enlarged versions of a basic projection matrix model that includes both class- and site-specific demographic rates and also movement probabilities, which again are usually class- and often site-specific. As for count-based multi-site models, analysis of these models is usually by simulation, since density dependence and other violations of simple matrix structure will often preclude the more elegant analysis methods we use for many single-site matrix models (Chapters 7–9).

To illustrate the basic structure of this class of models, we will construct a model for *Coryphantha robbinsorum*, the endangered cactus introduced in Chapter 10. Schmalzel et al. (1995) collected data to estimate a four-class demographic model for *C. robbinsorum* in three neighboring sites, and collected data for a total of four time intervals. The four classes are seeds (class 1), small and large juveniles (classes 2 and 3), and adults (class 4), resulting in the following projection matrix **A** for a single site, for a census conducted directly after seed production[11]:

[11]In Schmalzel et al. 1995, the matrix was incorrectly constructed without the effect of adult survival on the reproduction element of the matrix, but this makes very little difference in the results.

$$\mathbf{A} = \begin{pmatrix} 0 & 0 & 0 & f_4 s_4 \\ s_1 & s_2(1-g_2) & 0 & 0 \\ 0 & s_2 g_2 & s_3(1-g_3) & 0 \\ 0 & 0 & s_3 g_3 & s_4 \end{pmatrix} \tag{11.13}$$

where s_i, g_i, and f_i are the survival, growth, and fecundity estimates for class i, respectively.

To make a multi-site model for all three sites that does not include movement, we just make a grand matrix **G** that contains three single-site matrices **A, B**, and **C**, one for each of the three sites and zeros for all transitions involving movement between sites:

$$\mathbf{G} = \begin{pmatrix} 0 & 0 & 0 & a_{1,4} & 0 & 0 & 0 & 0 & 0 & 0 & 0 & 0 \\ a_{2,1} & a_{2,2} & 0 & 0 & 0 & 0 & 0 & 0 & 0 & 0 & 0 & 0 \\ 0 & a_{3,2} & a_{3,3} & 0 & 0 & 0 & 0 & 0 & 0 & 0 & 0 & 0 \\ 0 & 0 & a_{4,3} & a_{4,4} & 0 & 0 & 0 & 0 & 0 & 0 & 0 & 0 \\ 0 & 0 & 0 & 0 & 0 & 0 & 0 & b_{1,4} & 0 & 0 & 0 & 0 \\ 0 & 0 & 0 & 0 & b_{2,1} & b_{2,2} & 0 & 0 & 0 & 0 & 0 & 0 \\ 0 & 0 & 0 & 0 & 0 & b_{3,2} & b_{3,3} & 0 & 0 & 0 & 0 & 0 \\ 0 & 0 & 0 & 0 & 0 & 0 & b_{4,3} & b_{4,4} & 0 & 0 & 0 & 0 \\ 0 & 0 & 0 & 0 & 0 & 0 & 0 & 0 & 0 & 0 & 0 & c_{1,4} \\ 0 & 0 & 0 & 0 & 0 & 0 & 0 & 0 & c_{2,1} & c_{2,2} & 0 & 0 \\ 0 & 0 & 0 & 0 & 0 & 0 & 0 & 0 & 0 & c_{3,2} & c_{3,3} & 0 \\ 0 & 0 & 0 & 0 & 0 & 0 & 0 & 0 & 0 & 0 & c_{4,3} & c_{4,4} \end{pmatrix} \tag{11.14}$$

where a, b, and c terms denote the non-zero elements of **A**, **B**, and **C**, respectively (e.g., $a_{4,4} = s_4$ for population a). To simulate the set of three populations, we use means and variances for each rate (Table 11.2), plus the full correlation matrix (which we don't show), and then perform simulations like those explained in Chapter 8 (Box 11.5). As with the rail example, we have set site-specific quasi-extinction thresholds: 20 juvenile and adult plants (i.e., we have not counted the seed class). We also set a cap of 100 on the sum of juveniles and adults in each population. We set the initial numbers of plants in each class and site at the numbers in the final census conducted in 1993 (Schmalzel et al. 1995). Although there is a fair amount of variability in most of the demographic rates, there is a low probability of extinction over a 100-year time period (Figure 11.5). Furthermore, the total number of plants is predicted to show very little variability from year to year, suggesting very stable population dynamics in the future (and assuming removal of plants by human collec-

Table 11.2 Estimated vital rates for *Coryphantha robbinsorum* for each of three sites[a,b]

Site	s_1	s_2	s_3	s_4	g_2	g_3	f_4
a	0.0173 (0.0175)	0.8545 (0.0524)	0.9875 (0.025)	0.9692 (0.0386)	0.2145 (0.1304)	0.1411 (0.0528)	33 (0)
b	0.0028 (0.0017)	0.6197 (0.1924)	0.9645 (0.0416)	0.9852 (0.0295)	0.1834 (0.2135)	0.2466 (0.1711)	33 (0)
c	0.0073 (0.0057)	0.767 (0.0905)	0.9135 (0.0512)	0.9563 (0.0361)	0.4376 (0.3198)	0.3348 (0.071)	33 (0)

Vital rate (column group spanning s_1 through f_4)

[a]Data analyzed from Schmalzel et al. 1995.

[b]Values are the means and standard deviations over four annual time periods.

tors—the main threat to this species—does not change; Schmalzel et al. 1995). This safety appears to result mostly from the very high growth rates of population *a*, with an estimated deterministic growth rate of $\lambda = 1.09$. Still, removing this population and simulating just the other two does not appreciably increase the risk of extinction, suggesting that populations *b* and *c*, though less viable, actually are in aggregate quite safe from extinction (Figure 11.5).

We do not present sensitivity analysis results here for separate demographic rates, but the methods are the same as those used for stochastic single site matrix models (Chapter 9). You simply alter each mean, variance, or correlation parameter in turn by a small amount, simulate each new set of

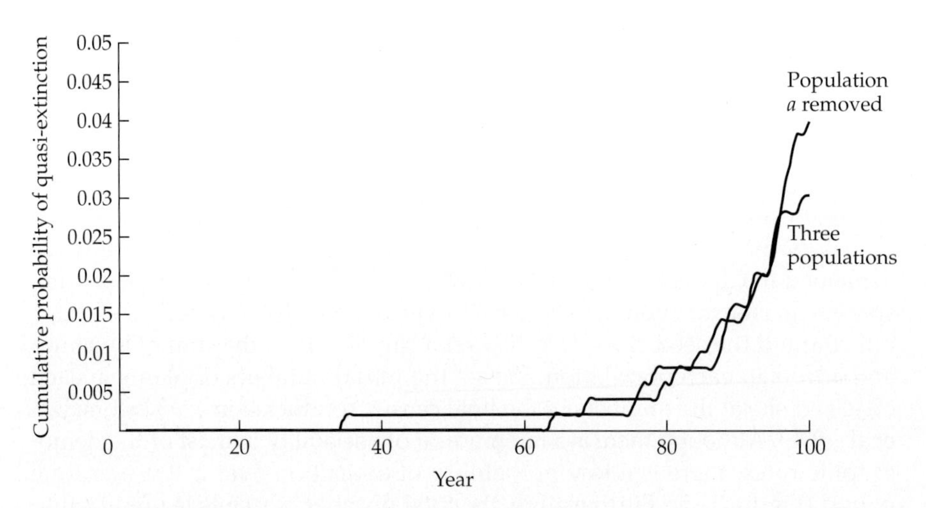

Figure 11.5 Extinction time cumulative distribution functions for two simulations of *Coryphantha robbinsorum* populations, one with all three populations and one with only the two populations with lower growth rates.

BOX 11.5. ***A MATLAB program to perform demographic, multi-site simulations. Much of the machinery needed for this program is identical to that in the vital rate-based simulation models presented previously, and in this code we refer to pieces of VitalSim.m in Box 8.12 that need to be inserted to make a fully functional program.***

```
% DemoMetaSim. A program to do a multi-site stochastic
% demography simulation with correlations and a choice of
% different distributions for the vital rates.
% The code is now tailored to data for Coryphantha robbinsorum
% see Schmalzel et al. 1995
clear all;
%****************Simulation Parameters**********************
% the order of the variables is survival and growth rates from
% youngest to oldest, for each of the three sites (a,b,c),
% and then the fecundity used in all the matrices:
% s1a s2a   s3a   s4a   g2a   g3a   s1b   s2b   s3b   s4b
% g2b g3b   s1c   s2c   s3c   s4c   g2c   g3c   ff
load coryrates.txt; % contains the values for each rate
                    % (columns) for each year of data (rows)
vrates = coryrates; % assign data to the generic rate variable
vrtypes= [ones(1,18),3]; % see VitalSim.m for explanation
vrmins=zeros(1,19); vrmaxs=zeros(1,19);
% Initial distribution across stages and sites:
n0 = [720, 26, 13, 24, 920, 6, 10, 23, 980, 12, 20, 31]';
makemx='makecorymx'; % assign matrix definition file: this
%  makes a separate matrix for each population (mx1 mx2 mx3)
%  and a matrix for all pop'ns, mx, same as G in Equ. 11.14
Ncap = [100, 100, 100]; % cap on juveniles and adults in pop'ns
Ne = [20 20 20]; % quasi-extinction threshold for each pop'n
tmax = 100;         % number of years to simulate;
np = 19;            % number of vital rates
dims = 12;          % dimensions of total multi-site matrix;
runs = 500;         % how many trajectories
popnum = 3;         % number of separate populations
dimpop = 4;         % number of classes per population
%*********************************************************
vrmeans = mean(vrates);   % means of vital rates
vrvars=diag(cov(vrates)); % variances of vital rates
corrmatrix = corrcoef(vrates);   % correl. matrix for rates
corrmatrix(:,19) = 0; % specific to this matrix: the last
```

BOX 11.5 *(continued)*

```
corrmatrix(19,:) = 0; % parameter has no variance
corrmatrix(19,19) = 1;
for pp = 1:popnum % total each pop'n size w/o seeds
   popNstart(pp) = sum(n0((2+dimpop*(pp-1)):dimpop*pp)); end;
randn('state',sum(100*clock)); rand('state',sum(150*clock));
[W,D] = eig(corrmatrix);  M12 = W*(sqrt(abs(D)))*W';

% INSERT HERE: the section in VitalSim.m to make or retrieve
% sets of beta and stretched beta values to use in simulations:
% replace np+np2 in this section with np.

% find the eigenvalues for each of the three populations:
vrs=vrmeans;                  eval(makemx);
[uu,lam1] = eig(mx1);         lam1a = max(max(lam1));
[uu,lam1] = eig(mx2);         lam1b = max(max(lam1));
[uu,lam1] = eig(mx3);         lam1c = max(max(lam1));
% initialize result trackers:
PrExt = zeros(tmax,1);              logLam = zeros(runs,popnum);
stochLam = zeros(runs,popnum);

for xx = 1:runs;
 disp(xx);
 nt = n0;    % initialize population sizes
 extinct = zeros(1,popnum); % vector of extinction recorders
 for tt = 1:tmax
      m = randn(1,np);
      yrxy = (M12*(m'))';
% INSERT HERE: the lines contained in the loop in VitalSim.m:
% [for yy = 1:(np+np2)], assigning values for each rate,
% each year. In this loop, replace np+np2 with np
       eval(makemx); % make a matrix with new vrs values.
       nt = mx*nt;   % multiply by population vector
       for pp = 1:popnum % enforce population cap
        popNs(pp) = sum(nt((2+dimpop*(pp-1)):dimpop*pp));
        if popNs(pp) > Ncap(pp)
             nt((2+dimpop*(pp-1)):dimpop*pp) = ...
                   nt((2+dimpop*(pp-1)):dimpop*pp)/Ncap(pp);
        end;
       end; % for pp
       if sum(extinct) < popnum % check for extinction
        for nn = 1:popnum
```

BOX 11.5 *(continued)*

```
            if popNs(nn) <= Ne(nn) extinct(nn) = 1;  end;
         end;
         if sum(extinct) == popnum PrExt(tt) = PrExt(tt) +1;
         end;
       end;
 end % tt
logLam(xx,:) = (1/tmax)*log(popNs./popNstart);
stochLam(xx,:) = (popNs./popNstart).^(1/tmax);
end; % xx
CDFExt = cumsum(PrExt./runs); % make the extinction CDF
disp('deterministic lambdas for the three populations:');
display([lam1a, lam1b,lam1c]);
% INSERT HERE: the final 6 lines in VitalSim.m to show results
```

vital rates, and examine the effects on the population growth rate or extinction risk. Because the methodology is identical to what we have already described, we refer you to Chapter 9 for the details of these analyses. We also note that a complete analysis should include estimation of confidence intervals for the extinction risk, for example using bootstrap methods (Ellner and Fieberg 2002).

Adding movement between populations follows the same general approach we used for the count-based rail model. However, in a demographic model, there are likely to be quite different movement rates for different life stages. For *C. robbinsorum*, only seeds can move; unfortunately, there are no data available on seed dispersal distances. However, to illustrate the inclusion of movement, let's assume that we can estimate specific probabilities of seed dispersal, m_{ij}, from each site j to site i. Including this movement results in a new multi-site matrix:

$$\mathbf{G} = \begin{pmatrix} 0 & 0 & 0 & a_{1,4}(1-m_{ba}-m_{ca}) & 0 & 0 & 0 & b_{1,4}m_{ab} & 0 & 0 & 0 & c_{1,4}m_{ac} \\ a_{2,1} & a_{2,2} & 0 & 0 & 0 & 0 & 0 & 0 & 0 & 0 & 0 & 0 \\ 0 & a_{3,2} & a_{3,3} & 0 & 0 & 0 & 0 & 0 & 0 & 0 & 0 & 0 \\ 0 & 0 & a_{4,3} & a_{4,4} & 0 & 0 & 0 & 0 & 0 & 0 & 0 & 0 \\ 0 & 0 & 0 & a_{1,4}m_{ba} & 0 & 0 & 0 & b_{1,4}(1-m_{ab}-m_{cb}) & 0 & 0 & 0 & c_{1,4}m_{bc} \\ 0 & 0 & 0 & 0 & b_{2,1} & b_{2,2} & 0 & 0 & 0 & 0 & 0 & 0 \\ 0 & 0 & 0 & 0 & 0 & b_{3,2} & b_{3,3} & 0 & 0 & 0 & 0 & 0 \\ 0 & 0 & 0 & 0 & 0 & 0 & b_{4,3} & b_{4,4} & 0 & 0 & 0 & 0 \\ 0 & 0 & 0 & a_{1,4}m_{ca} & 0 & 0 & 0 & b_{1,4}m_{cb} & 0 & 0 & 0 & c_{1,4}(1-m_{ac}-m_{bc}) \\ 0 & 0 & 0 & 0 & 0 & 0 & 0 & 0 & c_{2,1} & c_{2,2} & 0 & 0 \\ 0 & 0 & 0 & 0 & 0 & 0 & 0 & 0 & 0 & c_{3,2} & c_{3,3} & 0 \\ 0 & 0 & 0 & 0 & 0 & 0 & 0 & 0 & 0 & 0 & c_{4,3} & c_{4,4} \end{pmatrix} \tag{11.15}$$

To analyze a matrix with the possibility of movement between all sites, we can use simulations as above, but we can also use the analytical tools of formal matrix analysis (Chapter 7, 8, and 9) so long as we ignore other complications, such as density dependence in any rates. Wootton and Bell (1992) provide a clear example of the use and analysis of just such a simple multi-site matrix model for two populations of peregrine falcons linked by dispersal.

Although we've discussed only quite simple models in this section, more complicated multi-site demographic models have been developed for many species. The most common additions to these models are some form of density dependence, including both Allee effects, negative density dependence in vital rates, and density-dependent movement behaviors (e.g., Lindenmayer and Possingham 1996). Furthermore, many analyses take into account both males and females, with different vital rates and movement behaviors for each (e.g., Beier 1993, McCarthy 1996). Even though these additions may at first seem difficult, it is very simple to add any of these complexities to the basic structure of the stochastic matrix simulations that are the foundation of this entire family of models. Indeed, the only real challenge to including them in a model is obtaining the necessary parameter estimates.

Individually Based Simulations

Although the matrix-based approach to multi-site PVAs can include a great deal of complexity, it is an awkward tool with which to analyze the effects of dispersal, landscape structure, and habitat pattern on viability. Matrix models take the net result of what may be highly complex movement behavior to arrive at simple probabilities of going from any one site to another. By demarcating a small number of habitat areas or types and tracking population numbers on these discrete areas, we simplify the real complexity of interdigitated and intergrading habitat types and the complexity of real movement paths and barriers or modifiers of movement.

The solution to these limitations is to make models that track individuals through space, allowing them to experience location-specific fates and to move according to rules that involve their current state (e.g., hunger level or need to find mates) as well as their reactions to complex spatial stimuli. These spatially explicit, individually based simulation models are often tied to habitat maps generated by Geographic Information Systems (GIS) in an effort to reflect the real arrangements of particular landscapes (e.g., Akçakaya and Atwood 1997, Walters et al. 2002).

The idiosyncratic, highly tailored nature of individually based models makes it difficult and unproductive for us to give a general overview here. Instead, we describe the main features of one such model for the red-cockaded woodpecker, an endangered species of the Southeastern United States (Letcher et al. 1998, Walters et al. 2002). Letcher et al. constructed a spatially explicit, individually based model in order to incorporate the complex social interactions and the effects of habitat fragmentation that are both important for this bird. In the model, territories are randomly arrayed on the landscape according to rules governing the average territory density, the degree of spatial clumping of territories, and the spatial extent of the population. Red-cockaded woodpeckers live in social groups, and each individual in the model has a sex and social status that, along with the numbers and status of the individuals with which it interacts, determines its behavior and fate. For example, the oldest nonbreeding "helper" bird in a territory inherits its breeding status if the same-sex breeder dies, and helper birds can also compete for breeding vacancies in any territory within 3 km of their own. Dispersal uses fairly simple rules; a disperser moves in a randomly chosen direction until it dies, finds a breeding position, or leaves the edge of the modeled landscape. In addition to accounting for movement and social interactions (which create density dependence), the model also simulates demographic and environmental stochasticity. A key thing to note is that the structure of this model allows it to incorporate the multiple effects that habitat quantity and distribution have on individual performance, in part mediated by social interactions. The ability to simulate these effects is important because assessment of habitat preservation and restoration strategies is the key issue for this species.

This example conveys both the advantages and the potential disadvantages of using individually based simulations to perform multi-site PVA. Clearly, there is a potential to include more biological realism, including nuances of animal behavior and of habitat pattern and changes to habitats that management might affect. Indeed, this desire to make models as realistic as possible is the main force driving the construction of individually based simulation models. However, countering this benefit are serious data limitations. A great deal of information exists on the red-cockaded woodpecker, but this is not the case for most PVA targets. We've discussed this tradeoff before; it is especially critical to individually based simulations, since they are constructed to allow much greater complexity than are any other type of PVA. It is therefore particularly important when building them to remember that a single, simple parameter for, say, movement will often yield a much more accurate understanding of viability than will modeling a suite of complicated movement rules, but doing so poorly (Ruckelshaus et al. 1997, 1999, Mooij and DeAngelis 1999).

Even though these ideas do not lead to any clear rules about when to use and when not to use spatially explicit, individually based PVA models, we hope that they give you pause when deciding what type of multi-site PVA to construct. A desire to include important biological processes in your model must be weighed against the data you have to make the PVA accurate and well-grounded. If you do decide to use a spatially explicit simulation, your next decision is whether to build your own model from scratch or to use one of the many canned programs of this type. We do not present code to do these complex simulations. Many models of this type are so tied to particular systems that they are not easily modified to work for others. Of the programs that are available to perform spatially explicit, individually based PVAs, Schumaker's PATCH model is probably the best.[12] PATCH is designed explicitly for species that are territorial as adults. One reason we recommend it is the effort that Schumaker has taken to make the movement rules used in the model clear and simple, allowing a user to test the sensitivity of model results to these often poorly estimated parameters. For clear instructions on how to construct your own individually based simulations from scratch using the C programming language, see Wilson 2000.

Using Multi-Site PVAs with Care

The amount and kind of data available will determine the multi-site model you are able to construct; however, for any of these approaches it is important to assess how sensitive PVA results are to the specific parameter values used. For many multi-site situations, the crucial question is the importance of protecting some populations versus others. In other cases, management of intervening habitat for increased dispersal may be the key issue. The answers to these questions depend on all the parameters governing the dynamics of each population and the movements and correlations between them. As the number of populations becomes large, and as the complexity of the models increases, carefully assessing the sensitivity of results to the handful, dozens, or hundreds of parameters moves from being irritating to difficult to impossible. How should the practitioner deal with this difficult issue in order to generate believable results?

One answer is to always build simple models—or at least the simplest that capture the basic, important dynamics of the populations being considered. Although we argued for the "simple is better" philosophy in Chapter 1, a multi-site situation can still result in strikingly complex models. Even the simple multi-site demographic model for *Coryphantha* presented in this chapter has 22 mean parameter estimates, along with 18 variance and 153 correlation estimates. To be careful, but practical, in checking the reliability of results, we suggest prioritizing a sensitivity analysis in two ways. First, alter in concert large sets of parameters that you suspect will have little effect on the qualitative results to confirm (hopefully) that their particular values are not key to the results you see. For example, running the *C. robbinsorum* model with all correlations set to zero and then again with them all set to 1 would quickly test whether these admittedly poor estimates matter very much for the questions being asked. Second, priori-

[12]PATCH can be downloaded from: http://www.epa.gov/wed/pages/models/patch/patchmain.htm.

tize explorations of particular rates based upon your knowledge of the general life history of the species involved and general sensitivity patterns in the single population case. For long-lived species, population growth rates are often highly sensitive to annual adult survival, and this is likely to be true for multi-site models as well. Similarly, movement rates to and from the most viable population in a suite of interconnected sites are probably more critical than movement between less robust populations. Even though these rules are hardly infallible, they can often help to make analyses of multi-site PVA more fruitful, and hence make the final results more convincing.

12

Criticisms and Caveats: When to Perform (and When Not to Perform) a Population Viability Analysis

In the preceding chapters, we have reviewed in detail how to construct population viability analyses using three different types of data: census counts of the number of individuals in a population or population subset; demographic data on the vital rates of individuals that differ in size, age, or stage; and data pertinent to multiple populations, including information about presence/absence of populations at suitable sites, differences in population growth rates or vital rates among sites, the amount of movement, and the degree of correlation in vital rates among populations. Throughout, we have tried to lay out the advantages of taking a quantitative approach to the conservation of threatened and endangered populations, emphasizing how PVA models help us not only to assess the risk of extinction but also to identify the most promising management strategies. However, along the way we have also tried to point out the many factors that make it difficult to come up with precise statements about population viability, most notably the sparse amount and noisy nature of the data available for most threatened populations. Indeed, a number of authors have raised important criticisms and caveats about applying quantitative population models to rare species (see the following section). Given these limitations, and given that population viability analysis is only one way to arrive at conservation decisions (others include identifying and preserving critical habitat for species of interest, "rules of thumb" for the number and size of populations to conserve, expert opinions regarding population health, and extrapolation based on historical extinctions of similar species.), it becomes important to raise the question: When will performing a population viability analysis be justified, and when will it not?

In this final chapter, we look at the question of when to perform—and when not to perform—a population viability analysis. We begin by reviewing the most important criticisms that have been raised about the quantitative methods described in this book. In our view, these critiques should always be kept sharply in focus whenever a PVA is constructed, so that we are less likely to be misled by the biases and limitations of our analyses. Indeed, as we've discussed in conjunction with each analysis method we've presented, incorporating and exploring the uncertainties about parameter values, model structure, and predictions are key steps in performing a PVA.

However, in more than a few cases, biases and limitations due to poor information will be so severe as to render performing a PVA unproductive at best and detrimental at worst. After all, when data limitations are extreme, using a quantitative analysis invites us to view as mathematically rigorous a viability assessment that is really not much better than a guess. However, even though for some populations and questions we may not be able to take a quantitative approach, we do *not* believe that quantitative analyses are in general invalidated by these problems. It is important to remember that the alternatives to some type of quantitative analysis are unsubstantiated "expert" opinions, sweeping generalities, and politically motivated priority-setting. Thus, we remain sanguine about PVA's current and future use as a conservation tool, while we are also genuinely concerned that users of PVA not be blind to the important limitations raised by its critics. To drive this concern home, we end the chapter and the book by making some general recommendations and cautions that should apply to all population viability assessments.

Criticisms and Critiques of PVA

Although many concerns about population viability analysis have been voiced in the conservation biology literature, here we boil them down into four general and important criticisms. In the following sections, we both describe these four criticisms and present a defense of the usefulness of PVA in the light of each.

Criticism 1: Too Few Data Are Available upon which to Base Quantitative Risk Assessments for Rare Species

Rare species are difficult to study, but as conservation biologists we are interested in them precisely because they are rare. Thus PVA practitioners will constantly face the "Catch-22" that the species and populations most urgently in need of viability assessments will typically be those with the fewest available data. Sparse data can impart two weaknesses, bias and imprecision, to estimates of population parameters and to the extinction risk metrics based upon them. Probably the most important bias is in estimates of the temporal variability of population growth rates and vital rates; because extremes in environmental conditions are unlikely to be seen in only a few years, estimates of variability based on limited data are likely to be downwardly biased (Pimm and Redfearn 1988, Arino and Pimm 1995). Sparse data also result in imprecision in population parameters and thus in viability measures such as the long-term population growth rate or the probability of quasi-extinction. Unless population growth rates or vital rates are nearly constant over time, short data sets will produce wide confidence intervals on extinction probabilities or population growth rates due to sampling variation (see Chapters 3 and 5), even if population size or vital rate estimates are accurate (Dennis et al. 1991, Taylor 1995, Ludwig 1999). Fieberg and Ellner (2000) looked at this problem in another way. Using the expression for the cumulative probability of extinction for a density-independent population based upon a diffusion approximation (i.e., Equation 3.5), they calculated the number of years of count data that would be needed to estimate the probability of quasi-extinction by year T to within a margin of error of ± 0.2. (That is, if the true probability is 0.3, the estimate should lie between 0.1 and 0.5.) Even when the year-to-year variability in the population growth rate is relatively small, achieving even this modest level of precision requires $5T$ to $10T$ census counts. In other words, we must have 5 to 10 times as many years of census data as the number of years into the future we wish to predict. Even for the relatively long 39-year census of the Yellowstone grizzly bear (see Table 3.1), this level of precision is compatible with predicting extinction risk at most only about 8 years into the future.

A sensible way to proceed given the inherent imprecision of viability estimates is to realize that we are usually less interested in calculating *absolute* measures of viability than we are in assessing *relative* risks (Ralls and Taylor 1997, Beissinger and Westphal 1998). Often, knowing an exact, quantitative value of a population's long-term growth rate is of lesser importance than obtaining a simpler, more qualitative assessment of whether the population will tend to grow or decline (i.e., whether μ or $\log \lambda_s$ is positive or negative). Moreover, we usually care less about knowing the exact extinction probability of a single population than we do about knowing which of two or more populations is more likely to go extinct by a given future time, so that we can identify which population is in more urgent need of management. Similarly, a precise estimate of the rate of population decline is of less use to us than a reliable assessment of the relative efficacy of two or more management strategies for slowing or reversing that decline. Elderd et al. (2002) found that, provided the true value of μ is not fairly close to zero, ten years of census data are sufficient to yield a correct estimate of the sign of μ in a high proportion of simulated data sets generated from an underlying size-structured population model. Moreover, they found that a similar amount of data would allow us to do a fairly reliable job of ranking the growth rates of multiple populations in which the true values of μ varied from slightly positive to substantially negative. Of course, if the true values of μ are very close in all the populations, it will be difficult to accurately rank them using so few data points. The bottom line is that the imprecision that characterizes extinction risk metrics does not necessarily compromise viability assessments that are more qualitative but that nevertheless can provide extremely useful conservation guidance. Nonetheless, even relative measures of viability should be scrutinized for the effects of uncertainty (Ellner and Fieberg 2002).

Criticism 2: Observation Errors Will Compromise the Usefulness of Even Moderately Long Data Sets

As we saw in Chapters 5 and 8, observation errors can inflate estimates of the variability in population growth rates and vital rates. Even though methods such as the ones we reviewed in Chapters 5 and 8 have been developed to reduce observation error biases, this reduction often comes at a cost of reduced precision in parameter estimates (Ludwig 1999; also see the increasing range of corrected estimates with increasing sampling variation in Figure 5.2C). As a result, Ludwig (1999) has argued that measures of extinction risk based on those parameter estimates will frequently have such broad confidence limits as to be of very little value in assessing true population vulnerability.

Of course, whether viability assessments will be rendered hopelessly inaccurate by the bias and imprecision introduced by observation errors will depend on the magnitudes of those errors. To explore this question, Meir and Fagan (2000) intentionally added either biases (i.e., systematic under- or overcounts) or random errors of different magnitudes to simulated sets of population counts of moderate length (15 or 30 years) produced by "true" models with no, weak, or strong density dependence. They then asked how large the biases or random errors could be before the true probability of quasi-extinction could no longer be predicted to a given degree of accuracy[1] by a viability analysis that assumed there were no observation errors (such as the density-independent analysis described in Chapter 3). Over all models, degrees of environmental stochasticity, and lengths of data sets, they found that quite large systematic biases (e.g., consistent 20% undercounts or 400% overcounts) would not seriously compromise the accuracy of the estimated probability of quasi-extinction. More modest random errors (e.g., a distribution in which two-thirds of the observed counts fall within 15% above or below the true counts) could substantially degrade the estimated

[1]Specifically, Meir and Fagan considered an accurate assessment to be achieved if the absolute difference between the estimated and the true probability of quasi-extinction was no more than 10% higher for a data set contaminated with observation errors than for the corresponding uncontaminated data set.

extinction risks when density dependence was strong, particularly when the population's intrinsic growth rate was low and environment stochasticity was at an intermediate level,[2] but the effect of such errors was far less when density dependence was weak or absent. Importantly, Meir and Fagan combined their simulation studies with estimates of the magnitude of observation error from the small number of field studies in which both exhaustive population counts and estimates of population size (e.g., from capture-recapture studies) were obtained. They concluded that for most of the field studies, the level of observation error would not cause serious difficulty in estimating extinction risk, except possibly when combined with strong density dependence.

As was the case for criticisms based on data limitations described above, the take home message of this section is that observation errors do not *automatically* doom viability estimates to be hopelessly inaccurate; rather, we should be aware of the situations in which observation errors are likely to be especially problematic (i.e., observation errors of large magnitude combined with intermediate levels of environmental variation, near-zero log population growth rates, and strong density dependence), as well as the situations in which such errors will be comparatively unimportant. Furthermore, it is well to keep in mind that the whole debate over the amount and quality of data needed for a PVA has centered on the accuracy of the estimated extinction probabilities, which are probably the most inaccurate output of a PVA model. In contrast, many PVAs focus more on questions of population growth, sensitivity analysis for management, or the relative viability of multiple populations, all of which are likely to have more robust answers with less information.

Criticism 3: PVA Models Omit Too Many Factors That Are Likely to Affect Population Viability

A commonly voiced criticism of PVA (and indeed of the use of models in ecology in general) is that the models that form the basis for most analyses, including those described in this book, are highly simplified representations of true biological populations, and they consequently omit many ecological, economic, and political factors that can affect population viability. Aspects of this criticism are discussed in several recent reviews of PVA (Boyce 1992, Groom and Pascual 1998, Beissinger and Westphal 1998, Reed et al. 1998,

---

[2]In Meir and Fagan's simulations, a highly variable environment resulted in most population trajectories going extinct; added observation errors did not prevent a standard viability analysis (i.e., one that ignores observation error) from predicting so. In a nearly constant environment, few of their simulated trajectories went extinct (i.e., their modeled populations tended to grow), and again moderate observation errors did not cause a standard analysis to yield the wrong prediction (in this case, extinction).

Coulson et al. 2001). The features that are most frequently omitted from PVA models (and therefore the absence of which is most frequently lamented) are genetic factors, density dependence, infrequent catastrophes (and bonanzas), and temporal trends in environmental conditions due to invasive species, declining availability of critical habitat, and ongoing changes in abiotic factors (e.g., altered weather patterns caused by climate change).

The essential truth of the claim that omitting such factors may compromise the veracity of risk assessments is unquestionable. Nevertheless, we can envision no more sensible approach to predicting extinction risks than to adhere to the modeling philosophy we articulated in the first chapter of this book, namely to keep PVA models simple. When data are limited, it is well known that, as model complexity increases, the increased difficulty of obtaining precise parameter estimates will quickly come to outweigh any perceived advantages of enhanced biological realism (Ludwig and Walters 1985, Burnham and Anderson 1998). This basic fact must be remembered whenever we are tempted to expand a simple PVA by including factors such as genetics, complex density dependence, catastrophes, or environmental change. For example, while the temptation to include genetic processes such as inbreeding may be great, we will usually lack the data to link the amount of inbreeding to its effects on vital rates (Lande 2002, Allendorf and Ryman 2002); indeed, genetic data of any kind are lacking for the majority of endangered species (Schemske et al. 1994, Tear et al. 1995). The same caution applies to estimating the frequency and severity of catastrophes. Similarly, attempting to estimate the carrying capacity or the Allee threshold by fitting a complex density-dependent model to a small number of data points is usually counterproductive. Finally, attempting to include changing conditions in a viability assessment may serve only to compound the uncertainty inherent in projecting future population size under the status quo with the uncertainties involved in projecting future states of the environment. With the perpetual data shortage that conservation biology faces, perhaps the best we can do is to follow the advice of the mathematician and philosopher Alfred North Whitehead, who advised us to "seek simplicity, but distrust it." For PVA, this means making simple models, but also understanding and acknowledging the potential ways that this simplicity may influence our results (see Recommendation 5 below).

Criticism 4: PVA Models Are Rarely Validated

Several authors have pointed out that population viability analyses usually fail to test the adequacy of the model before using it to make predictions about future population status (Reed et al. 1998, Beissinger and Westphal 1998). A more rigorous modeling protocol would require that, before it is used in a predictive fashion, the model be "validated" by testing its ability to predict data other than those used to estimate the model parameters (McCarthy et al. 2001). In many cases, such validation will be impossible, but sometimes it may be feasible. One way to validate the model is to use one subset of the data (e.g., the first half of a series of census counts) to parameterize the model and another subset (e.g., the second half of the counts) to validate its predictive accuracy (Brook et al. 2000a; also see Gerber et al. 1999). Although the probability of quasi-extinction by a given future time is one of the most desirable predictions of a PVA model, it is difficult to test a model's accuracy at predicting extinction risk, for the following reason. We can think of the predicted probability of quasi-extinction as the proportion of all possible population trajectories that should fall below the quasi-extinction threshold at or before a future time horizon. Yet the data set presents only a single population trajectory, which either will or will not have hit the threshold by the end of the census period. An alternative is to test the model's ability to predict the degree of population decline or the population growth rate over the second half of the data set. Taking this approach, Brook et al. (2000a) concluded that PVA models using five widely available software packages did a good job of predicting the aggregate population dynamics seen in 21 high-quality data sets from vertebrate species (mostly birds and mammals). However, Ellner et al. (2002) point out that even when models predict the correct proportion of a suite of species declining by a certain amount, the predictions for any one species, which is the usual focus of a PVA, may still be highly imprecise. Other approaches that have been taken to validate PVA models are to use data from one population to fit the model and then test its ability to predict data from other populations (McCarthy and Broome 2000) and to use information on patterns of occupancy across a set of habitat patches to test multi-site PVA models parameterized with data collected elsewhere (Lindenmayer et al. 1999, 2000, McCarthy et al. 2000).

The criticism that PVA models are rarely validated is often phrased in another way. Most viability analyses adopt only a single model, and assume that it is the best description of the biological population. Yet we rarely know *a priori* which is the best model. As we have touched on in several places, methods to test alternative models (e.g., ones with or without density dependence) should be used whenever possible. Unfortunately, different models fit to the same data do not always reach the same conclusions regarding population viability. For example, Pascual et al. (1997) found that different models described data from the Serengeti wildebeest population equally well, yet made quite divergent predictions about the probability of future quasi-extinction and about how extinction risk would be affected by harvesting. A number of studies have found that different PVA software packages, which make different assumptions about the biology of the focal species and include different potential threats, yield different predictions about population viability when fit to essentially the same set of data (Mills et al. 1996, Brook et al. 1997, Brook et al. 2000b).

Unfortunately, thoroughly validating a PVA model before using it to predict population viability will always be difficult to achieve, not the least because dividing an already limited data set into "parameterization" and "validation" subsets will only exacerbate uncertainty in parameter estimates. When enough data are available to do so, an attempt to validate the PVA model should certainly be undertaken. In addition, careful attention should be paid to the possibility that different models that all adequately fit the available data can make different predictions about population viability and the consequences of competing management strategies (see Recommendation 6 below).

General Recommendations and Cautions for Conducting a Population Viability Analysis

In our opinion, the criticisms we discussed in the preceding section do not deliver a fatal blow to the entire endeavor of population viability analysis. The conditions under which PVAs should or should not be constructed is an active area of research, as work proceeds on how best to correct or minimize bias and uncertainty. In part this situation reflects the youth and activity of the field of PVA, with changing methods and frequent reassessment of how best to proceed. Nonetheless, the concerns raised by skeptics pose important cautions that must be heeded whenever a PVA is performed. We close the book with eight general recommendations to encourage viability assessments to proceed usefully but cautiously in light of these criticisms.

Recommendation 1: Avoid Conducting a Formal PVA if the Amount of Available Data is Truly Sparse

The very act of constructing a PVA model can have important benefits, even when too few data exist to fully parameterize it (e.g., it can help to organize our thoughts about the dynamics of the focal population and can indicate what future data collection efforts should be given priority). Still, we also need to avoid the "fallacy of illusory precision" (Reed et al. 1998) that an unsubstantiated model might create. That being said, it is difficult to define precisely a threshold amount of data at which it becomes worthwhile to conduct a formal PVA. As a tentative guideline, we recommend that ten years of census data be viewed as a minimum amount for conducting a count-based PVA. The results of Elderd et al. (2002) and Meir and Fagan (2000) discussed above suggest that this amount of data stands a reasonable chance of providing at least a qualitative sense of population viability (i.e., whether a population is growing or declining, and whether one population is more viable than another), even in the face of moderate observation error. Although a similar amount of data (or even more) would be desirable for a demographic PVA that is used to predict extinction risk, a threshold of ten years of data would preclude nearly every demographic PVA that has ever been performed. Thus bowing to the reality that demographic data sets will usually be shorter than count-based data sets, we suggest that four years of data, yielding three annual transitions on which to base the means, variances, and covariances of the vital rates, be viewed as a minimum threshold for performing a stochastic demographic PVA. However, even less data can be used to perform sensitivity analyses on deterministic models in order to prioritize management actions and understand threats. For a species with fairly stable vital rates, data on just one transition can be used to build a useful deterministic model for this purpose. For multi-site PVAs, the minimum amount of data depends on the type of model being used. Logistic regression models typically require data from at least two presence/absence surveys, and incidence function models should probably conform to the same rule. The number of sites needed to perform either analysis should exceed 20 at a bare minimum (see Chapter 11). Stochastic matrix models that include migration among subpopulations should probably adhere to the four-year rule. For spatially explicit individual-based models, which include a plethora of parameters that must be estimated, the bar must be set substantially higher than for any other kind of PVA model in terms of both the types and amounts of data needed.

Recommendation 2: Do Not Present Estimates of Population Viability That Are Unaccompanied by Confidence Intervals

Given the uncertainty inherent in estimating model parameters with few data and the omnipresence of observation error, it is critical that measures of population viability be delivered with an estimate of their associated uncertainty. Confidence intervals can be obtained by sampling parameter values from their analytic probability distributions (Dennis et al. 1991; see Chapter 3), by using nonparametric bootstrapping techniques (Ellner and Fieberg 2002), or with the aid of Bayesian approaches (Ludwig 1996b, Wade 2002). Extremely broad confidence intervals provide evidence that sufficient data do not yet exist to justify a formal viability analysis (see Recommendation 1). However, even these can be quite useful in a management context. A PVA that forcefully brings home the message that we simply don't have enough information to predict future risk accurately can serve a very useful role in averting risky decisions that increase threats to a population.

Recommendation 3: View Viability Metrics as Relative or Qualitative, Not Absolute, Gauges of Population Status

Given the wide confidence intervals that will typically accompany population viability metrics (see Recommendation 2), point estimates become relatively meaningless. Thus rather than making the goal of a PVA the estimation of the exact long-term growth rate or the exact probability of future extinction, we should seek more modest objectives, such as the direction of population growth (increase or decrease) or the ranking of extinction risks for a set of two or more populations (Beissinger and Westphal 1998). We agree with Taylor (1995) that in most cases, PVA is probably too blunt an instrument to accurately classify species according to precise criteria of extinction risk, such as those proposed by the World Conservation Union (Mace and Lande 1991, World Conservation Union 1994).

Recommendation 4: Do Not Try to Project Population Viability Far into the Future

In light of the results of Fieberg and Ellner (2000), it is foolhardy to try to predict the probability that a population will avoid extinction for many centuries into the future, as many PVAs attempt to do. The confidence intervals around the probability of extinction quickly come to encompass the entire range from zero to one as the time horizon moves farther into the future. Furthermore, complications such as negative density dependence that we often can't include in a PVA take on more and more importance over longer time horizons. Thus, simpler, better-justified models are best used to make relatively short-term predictions. Although as conservation biologists we would all like to see that the species we are charged with protecting will still be in existence a long time into the future, seeking to achieve long-term persistence by predicting and managing for persistence over a series of shorter time intervals is more in keeping with both the demands of statistical rigor and the realities of political and economic constraints.

Recommendation 5: Always Consider How Potential Determinants of Population Viability That Have Been Omitted from the Model Might Cause Risk Assessments to Be Optimistic or Pessimistic

The "keep it simple" modeling philosophy argues that we not try to include in our models factors about which we have few data or none. Nonetheless, we can still keep in mind how the factors omitted, if they are truly important for the population under study, would increase or decrease the risk of extinction relative to the calculated viability (Chapter 3). It is usually clear whether omitting a certain factor will cause the analysis to be optimistic (i.e., to underestimate the extinct risk) or pessimistic (i.e., to overestimate the risk). For example, omitting rare catastrophes or the increased expression of inbreeding depression as population size becomes small will render the estimated risk optimistic. On the other hand, failure to correct the upward bias in the estimated variance of the population growth rate introduced by observation error, or neglecting migration among populations, would cause the extinction risk estimate to be overly pessimistic. Although simply knowing that a viability estimate is likely to be optimistic or pessimistic is certainly less satisfying that being able to include the missing factor and compute a revised estimate of viability, the fact that we can anticipate the *direction* in which our estimate would be in error allows us to use qualitative information (such as the discovery that there is *some* migration among populations, even if we lack quantitative estimates of its magnitude) to better assess viability.

Recommendation 6: Never Base Management on Estimates of the Probability of Absolute Extinction

As we stated in Chapter 2, a PVA should never attempt to estimate the chance that a population will go completely extinct (i.e., hit a population size of zero for hermaphroditic species or one for species with separate sexes). Apart from the practical fact that when such an outcome occurs, it is too late to save the population, setting a higher "quasi-extinction" threshold allows us to partially account for several factors, such as inbreeding depression at low population size, demographic stochasticity, and Allee effects, about which we will usually lack sufficient information to include in the model explicitly. Thus a quasi-extinction threshold is another way to enhance biological realism without resorting to an overly complex model (see Recommendation 5).

Recommendation 7: Whenever Possible, Consider Multiple Models to Address "Model Uncertainty"

When we don't have enough data to validate the PVA model, we can at least try to fit multiple models to the available data. In this way, we acknowledge our uncertainty as to which model best describes the real population. Although we recognize that by advocating the use of Information Criteria (see Chapters 4 and 11) we are suggesting that PVA modelers attempt to identify the single best model, we also recognize that in some cases more than one model will produce as good or nearly as good a fit to the available data (Pascual et al. 1997). In such cases, we agree with Pascual et al. (1997) and Groom and Pascual (1998) that the best approach is to use all well-fitting models to assess viability and management options. For example, a precautionary approach would argue for using a combined estimate of extinction risk across several models, weighted by the support each model has from the data (Burnham and Anderson 1998), or even for using only the most pessimistic extinction risk estimate across the suite of models considered. On the other hand, if a single management strategy is consistently predicted to provide the greatest improvement in viability for all models considered, we would have greater confidence that our management recommendations are relatively insensitive to model uncertainty.

Recommendation 8: Consider a Population Viability Analysis to Be a Work in Progress, Not the Final Word

The last point we wish to make is that constructing a PVA is like building a house—the work is never really completed. Additional data are constantly being accumulated, in part thanks to basic monitoring for rare and endangered species. As a result, the quality of previously performed viability assessments is likely to improve as additional data are collected and incorporated into the model. More data allow us to improve the accuracy of the estimates for parameters currently in the model, to reconsider basic assumptions we made in constructing the PVA, and to incorporate more factors into the model.

As more data accumulate, not only will previous PVAs be improved, but the number of species and populations that can be subjected to quantitative analysis is likely to increase. The hope is that PVAs will have "strength in numbers." That is, by building up a library of viability analyses, we may eventually be able to place rough bounds on population parameters (e.g., μ and σ^2, see Chapter 3) for groups of species that share similar life histories or inhabit similar environments. We could then ask: "If we know only the current size of a population of interest, but we assume its parameters to be similar to those of ecologically similar species, what is its relative risk of extinction?" With such comparative approaches, viability assessments for relatively well-studied populations may aid us in making decisions about other, less-studied populations. For initial attempts at developing such generalizations, see Fagan et al. (2001) and Heppell (1998).

Closing Remarks

To a large extent, the above critiques of PVAs and our recommendations are just statements of common sense. We wish to use PVAs in the first place because we have a limited ability to understand population dynamics and to predict the future. Of necessity, this means that we are building imperfect representations to make uncertain predictions, and to most powerfully understand viability problems we must acknowledge these limitations. However, while it is crucial to recognize the short-comings of PVA, we also believe that the standard against which the contribution of PVA to conservation biology should be measured is conservation planning that uses no formal analysis whatsoever. PVA methods provide ways to use the information we do have to the maximum extent possible to understand and manage for viability. Although expert opinions and other sources of information can help to guide management, PVA plays a unique role in sharpening our understanding and in combining diverse, quantitative information into unified predictions that no other tool can provide.

Appendix: Mathematical Symbols Used in This Book

Symbol[a]	**Definition**
log	Natural logarithm
exp	Exponential; $\exp(x) = e^x$, where $e \approx 2.71828$ is the base of natural logarithms
π	Ratio of the circumference of a circle to its diameter ($\pi \approx 3.14159$)
$\frac{\partial f}{\partial x}$	Partial derivative of f (a function of several variables, including x) with respect to x; the derivative of f with respect to x, treating all other variables in f as constants.
N_t	Population size in year t
λ	Population growth rate
λ_t	Population growth rate in year t
λ_A	Arithmetic mean of the population growth rate
λ_G	Geometric mean of the population growth rate
μ	Arithmetic mean of the log population growth rate
$\hat{u}$	Estimated value of μ
σ^2	Environmentally driven variance of the log population growth rate
$\hat{\sigma}^2$	Estimated value of σ^2
PDF	Probability density function
CDF	Cumulative distribution function
N_c	Current population size
N_x	Population size at the quasi-extinction threshold
d	The difference between log N_c and log N_x
T	Time horizon for calculating extinction probability
$g(t \mid \mu,\sigma^2,d)$	The extinction time PDF (probability density for quasi-extinction occurring at time t given the values of μ, σ^2, and d)
$G(T \mid \mu,\sigma^2,d)$	The extinction time CDF (the probability that quasi-extinction occurs at or before time T, given the values of μ, σ^2, and d)

[a]Symbols are listed more or less in the order of their first use in the book. Thus they proceed from symbols used primarily in count-based models, to those used in demographic PVAs, to those used in multi-site simulations. Many symbols appear frequently throughout the entire book, while others arise and are used in only a single chapter.

Symbol	Definition
$\hat{G}$	Estimated value of G
$\Phi(z)$	Standard normal CDF (the probability that $x \leq z$, where x is drawn from a normal distribution with a mean of zero and a variance of 1)
q	Number of estimates of population growth rate (number of censuses minus 1)
t_q	Duration of census (years)
$t_{\alpha,q-1}$	Critical value of the two-tailed Student's t distribution with a significance level of α and $q - 1$ degrees of freedom
$SE(\hat{\mu})$	Standard error of the estimate of μ
$\chi^2_{\alpha,q-1}$	100αth percentile of the chi-square distribution with $q - 1$ degrees of freedom
K	Carrying capacity, or population ceiling
r	Average log population growth rate when N_t is very small relative to K
θ	Parameter controlling the shape of negative density dependence in the theta logistic equation
ε_t	Environmentally caused deviation between observed and predicted log population growth rate in year t
$\overline{T}$	Mean time to reach the quasi-extinction threshold
V_r	Residual variance (mean squared deviation between observed log population growth rates and those predicted by a density-dependent model; also equals the maximum likelihood estimate of the environmentally-driven variance in the log population growth rate)
L	Likelihood (probability of the data given a model)
$\log L_{\max}$	Maximum log likelihood
AIC_c	Corrected Akaike Information Criterion
p	Number of parameters in a model (as used in Chapter 4)
A	Population density at which potential per-capita reproduction is one-half its maximum value in an Allee effect model (as used in Chapter 4)
β	Strength of negative density dependence in offspring survival in Allee effect model (as used in Chapter 4)
C_{it}	Contribution of individual i to population growth in year t
$\overline{C}_t$	Mean contribution across individuals in year t
m_t	Number of individuals whose contributions to population growth are known in year t
$V_d(t)$	Variance in individual contributions to population growth in year t
ρ	Correlation coefficient between population growth rates in adjacent years
z_t	Random number drawn in year t from a standard normal distribution (i.e., with a mean of zero and a variance of one)
σ^2_{corr}	Estimate of the environmental variance in (i.e., σ^2) corrected for sampling variation
R_t	Running sum of census counts beginning in year t

Symbol	Definition
f_j	Fertility rate of individuals in class j in a structured population (a vital rate)
s_j	Survival rate of individuals in class j in a structured population (a vital rate)
$g_{i,j}$	Rate of transition of individuals from class j to class i in a structured population (a vital rate)
s	Number of classes in a projection matrix
$\mathbf{A}$	Population projection matrix (constant)
a_{ij}	The element in row i and column j of the matrix $\mathbf{A}$
$\mathbf{A}(t)$	Population projection matrix in year t
$\mathbf{n}(t)$	Population vector in year t
$N(t)$	Total size of a structured population in year t (sum of the elements of $\mathbf{n}(t)$)
λ_1	Dominant eigenvalue of a constant projection matrix (also, the ultimate growth rate of a structured population in a constant environment)
$\mathbf{w}$	Stable distribution vector (dominant right eigenvector of a constant projection matrix)
$\mathbf{v}$	Vector of reproductive values (dominant left eigenvector of a constant projection matrix)
S_{ij}	Sensitivity of λ_1 to changes in matrix element a_{ij}
$\mathbf{v}' * \mathbf{w}$	Scalar product of $\mathbf{v}$ and $\mathbf{w}$ (sum of the pairwise products of the elements of $\mathbf{v}$ and $\mathbf{w}$)
$\log \lambda_s$	Stochastic log growth rate for a structured population
$\overline{\mathbf{A}}$	The arithmetic mean of a set of temporally varying projection matrices
$\bar{a}_{ij}$	Element in row i, column j of $\overline{\mathbf{A}}$
$\bar{\lambda}_1$	The dominant eigenvalue of $\overline{\mathbf{A}}$
$\bar{S}_{ij}$	Sensitivity of $\bar{\lambda}_1$ to changes in $\bar{a}_{ij}$
$\mathrm{Cov}(a_{ij}, a_{kl})$	Covariance of matrix elements a_{ij} and a_{kl}
$\mathrm{Var}(a_{ij})$	Variance of matrix element a_{ij}
τ^2	Sum of all pairwise products of covariances and sensitivities of a_{ij} and a_{kl}
$\tau^2 / \bar{\lambda}_1^2$	Approximate variance of $\log \lambda_s$
$r(x,y)$	Correlation coefficient between variables x and y
$V_c(v_i)$	Variance estimate for rate v_i, corrected for sampling error
$P(k \mid x,y)$	Probability of event k, given (conditional on) the values of events x and y
$Beta(a,b)$	The value of a beta function with parameters a and b
$\Gamma(a)$	The value of a gamma function with parameter a
$g_{\geq i,j}$	Rate of transition of individuals from class j to class i *or any larger class* in a structured population (a vital rate)
$F(x \mid a,b)$	Value of a cumulative distribution function for the value x, given distribution parameters a and b (the form of the distribution must also be specified)

Symbol	Definition
$\mathbf{C}$	Matrix of correlation coefficients between all pairs of vital rates underlying the elements of a population matrix
$\mathbf{C}^{1/2}$	Square-root of **C**, used to generate correlated vital rates
$\mathbf{M}$	Matrix of within- and between-year correlations between vital rates comprising the elements of a population matrix
S_{r_i}	Sensitivity of λ_1 to changes in a vital rate r_i
E_{r_i}	Elasticity of λ_1 to changes in a vital rate r_i
S_x^s	Stochastic sensitivity of λ_s to the mean matrix element, mean vital rate, or covariance term denoted by x
E_x^s	Stochastic elasticity of λ_s to the mean matrix element, mean vital rate, or covariance term denoted by x
A_i	Area of habitat patch i, used in metapopulation models
M_i	Number of migrants arriving to habitat patch i
E_i	Extinction probability for habitat patch i
C_i	Colonization probability for habitat patch i
J_i	Probability that habitat patch i is occupied

Literature Cited

Numbers in brackets at the end of each reference indicate the chapter or chapters in which the work is cited.

Åberg, P. 1992a. A demographic study of two populations of the seaweed *Ascophyllum nodosum*. Ecology 73: 1473–1487. [7]

Åberg, P. 1992b. Size-based demography of the seaweed *Ascophyllum nodosum* in a stochastic environment. Ecology 73: 1488–1501. [7]

Abramowitz, M. and I. Stegun. 1964. *Handbook of Mathematical Functions, with Formulas, Graphs and Mathematical Tables.* National Bureau of Standards, Washington, DC. (reprinted by Dover Publications, New York) [3, 8]

Akçakaya, H. R. 1997. RAMAS/metapop: Viability analysis for stage-structured metapopulations. Version 2. Applied Biomathematics, Setauket, NY. (also see http: //www.ramas.com/) [1]

Akçakaya, H. R. and J. L. Atwood. 1997. A habitat-based metapopulation model of the California gnatcatcher. Conservation Biology 11: 422–434. [10, 11]

Akçakaya, H. R. and S. Ferson. 1992. RAMAS/space: Spatially structured population models for conservation biology. Applied Biomathematics, Setauket, NY. (also see http: //www.ramas.com/). [1]

Allendorf, F. W. and N. Ryman. 2002. The role of genetics in population viability analysis. In S. Beissinger and D. McCullough (eds.). *Population Viability Analysis* (in press). University of Chicago Press, Chicago. [2, 12]

Allendorf. F. W., D. Bayles, D. L. Bottom, K. P. Currens, C. A. Frissell, D. Hankin, J. A. Lichatowich, W. Nehlsen, P. C. Trotter and T. H. Williams. 1997. Prioritizing Pacific salmon stocks for conservation. Conservation Biology 11: 140–152. [1]

Amarasekare, P. 1998a. Interactions between local dynamics and dispersal: insights from single species models. Theoretical Population Biology 53: 44–59. [4]

Amarasekare, P. 1998b. Allee effects in metapopulation dynamics. American Naturalist 152: 298–302. [4]

Arino, A. and S. L. Pimm. 1995. On the nature of population extremes. Evolutionary Ecology 9: 429–443. [3, 12]

Armbruster, P. and R. Lande. 1993. A population viability analysis for African elephant (*Loxodonta africana*): How big should reserves be? Conservation Biology 7: 602–610. [1, 2]

Augustin, N. H., M. A. Mugglestone and S. T. Buckland. 1996. An autologistic model for the spatial distribution of wildlife. Journal of Animal Ecology 33: 339–347. [11]

Augustin, N. H., M. A. Mugglestone and S. T. Buckland. 1998. The role of simulation in modelling spatially correlated data. Environmetrics 9: 175–196. [11]

Bailey, E. A. and P. J. Mock. 1998. Dispersal capability of the California gnatcatcher: A landscape analysis of distribution data. Western Birds 29: 351–360. [10]

Beier, P. 1993. Determining minimum habitat areas and habitat corridors for cougars. Conservation Biology 7: 94–108. [2, 10, 11]

Beier, P. 1996. Metapopulation models, tenacious tracking, and cougar conservation. Pages 293–324 in D. R. McCullough (ed.), *Metapopulations and Wildlife Conservation*. Island Press, Washington, DC. [10]

Beissinger, S. 1995. Modeling extinction in a periodic environment: Everglades water levels and snail kite population viability. Ecological Applications 5: 618–631. [7]

Beissinger, S. and M. I. Westphal. 1998. On the use of demographic models of population viability in endangered species management. Journal of Wildlife Management 62: 821–841. [12]

Benton, T. G., C. T. Lapsley and A. P. Beckerman. 2001. Population synchrony and environmental variation: An experimental demonstration. Ecology Letters 4: 236–243. [10]

Beverton, R. J. H. and S. J. Holt. 1957. On the Dynamics of Exploited Fish Populations. United Kingdom Ministry of Agriculture, Fisheries, and Food Fishery Investment Series 2, no. 19. [2]

Birkhead, T. R. 1977. The effect of habitat and density on breeding success in the common Guillemot (*Uria aalge*). Journal of Animal Ecology 46: 751–764. [4]

Bishop, Y. M. M., S. E. Fienberg and P. W. Holland. 1975. *Discrete Multivariate Analysis: Theory and Practice*. M. I. T. Press, Cambridge, MA. [6]

Bjørnstad, O. N., J.-M. Fromentin, N. C. Stenseth and J. Gjøsæter. 1999. Cycles and trends in cod populations. Proceedings of the National Academy of Sciences USA 96: 5066–5071. [5]

Bouzat, J. L., H. A. Lewin and K. N. Paige. 1998. The ghost of genetic diversity past: Historical DNA analysis of the greater prairie chicken. American Naturalist 152: 1–6. [2]

Boyce, M. S. 1992. Population viability analysis. Annual Review of Ecology and Systematics 23: 481–506. [12]

Brook, B. W., L. Lim, R. Harden and R. Frankham. 1997. Does population viability analysis software predict the behaviour of real populations? A retrospective study on the Lord Howe Island Woodhen *Ticholimnas sylvestris* (Sclater). Biological Conservation 82: 119–128. [12]

Brook, B. W., J. J. O'Grady, A. P. Chapman, M. A. Burgman, H. R. Akçakaya and R. Frankham. 2000a. Predictive accuracy of population viability analysis in conservation biology. Nature 404: 385–387. [12]

Brook, B. W., M. A. Burgman and R. Frankham. 2000b. Differences and congruencies between PVA packages: The importance of sex ratio for predictions of extinction risk. Conservation Ecology (online at http://www.consecol.org/vol4/iss1/art6) [12]

Brown, J. H. and A. Kodric-Brown. 1977. Turnover rates in insular biogeography: Effect of immigration on extinction. Ecology 58: 445–449. [10]

Bullock, J. M. and R. T. Clarke. 2000. Long-distance seed dispersal by wind: Measuring and modelling the tail of the curve. Oecologia 124: 506–521. [10]

Burgman, M. A. and B. B. Lamont. 1992. A stochastic model of the viability of *Banksia cuneata* populations: Environmental, demographic, and genetic effects. Journal of Applied Ecology 29: 719–727. [2]

Burgman, M. A., S. Ferson and H. R. Akçakaya. 1993. *Risk Assessment in Conservation Biology*. Chapman and Hall, London. [4, 8]

Burnham, K. P. and D. R. Anderson. 1998. *Model Selection and Inference: A Practical Information-Theoretic Approach*. Springer, New York. [4, 12]

Bustamante, J. 1996. Population viability analysis of captive and released bearded vulture populations. Conservation Biology 10: 822–831. [1]

Carlin, B. P., N. G. Polson and D. S. Stoffer. 1992. A Monte Carlo approach to nonnormal and nonlinear state-space modeling. Journal of the American Statistical Association 87: 493–500. [5]

Carpenter, S. R., K. L. Cottingham and C. A. Stow. 1994. Fitting predator-prey models to time series with observation errors. Ecology 75: 1254–1264. [5]

Carroll, R., C. Augspurger, A. Dobson, J. Franklin, G. Orians, W. Reid, R. Tracy, D. Wilcove and J. Wilson. 1996. Strengthening the use of science in achieving the goals of the Endangered Species Act: An assessment by the Ecological Society of America. Ecological Applications 6: 1–11. [12]

Caswell, H. 1978. A general formula for the sensitivity of population growth rate to changes in life history parameters. Theoretical Population Biology 14: 215–230. [7, 9]

Caswell, H. 1996. Second derivatives of population growth rate: Calculation and applications. Ecology 77: 870–879. [9]

Caswell, H. 2001. *Matrix Population Models: Construction, Analysis and Interpretation*. 2nd Ed. Sinauer Associates, Sunderland, MA. [1, 6, 7, 8, 9, 11]

Caswell, H., R. Naiman and R. Morin. 1984. Evaluating the consequences of reproduction in complex salmonid life cycles. Aquaculture 43: 123–143. [9]

Caswell, H., S. Brault, A. Read and T. Smith. 1998. Harbor porpoise and fisheries: An uncertainty analysis of incidental mortality. Ecological Applications 8: 1226–1238. [1]

Caswell, H., M. Fujiwara and S. Brault. 1999. Declining survival probability threatens the North Atlantic right whale. Proceedings of the National Academy of Sciences, USA 96: 3308–3313. [2]

Caughley, G. 1994. Directions in conservation biology. Journal of Animal Ecology. 63: 215–244. [2]

Coulson, T., G. M. Mace, E. Hudson and H. Possingham. 2001. The use and abuse of population viability analysis. Trends in Ecology and Evolution 16: 219–221. [12]

Courchamp, F. T. Clutton-Brock and B. Grenfell. 1999. Inverse density dependence and the Allee effect. Trends in Ecology and Evolution 14: 405–410. [2, 4]

Cox, D. R. and H. D. Miller. 1965. *The Theory of Stochastic Processes.* Chapman and Hall, London. [3]

Crone, E. E., D. F. Doak and J. Pokki. 2001. Ecological influences on the dynamics of a field vole metapopulation. Ecology 82: 831–843. [10, 11]

Crooks, K. R., M. A. Sanjayan and D. F. Doak. 1998. New insights on cheetah conservation through demographic modeling. Conservation Biology 12: 889–895. [9]

Crouse, D., L. Crowder and H. Caswell. 1987. A stage-based population model for loggerhead sea turtles and implications for conservation. Ecology 68: 1412–1423. [1, 6, 7]

Crowder, L., D. Crouse, S. Heppell and T. Martin. 1994. Predicting the impact of turtle excluder devices on loggerhead sea turtle populations. Ecological Applications 4: 437–445. [1]

de Kroon, H., A. Plaisier, J. van Groenendael and H. Caswell. 1986. Elasticity: The relative contribution of demographic parameters to population growth rate. Ecology 67: 1427–1431. [9]

de Valpine, P. and A. Hastings. 2002. Fitting population models with process noise and observation error. Ecological Monographs 72: 57–76. [5]

Dennis, B. 1989. Allee effects: Population growth, critical density, and the chance of extinction. Natural Resource Modeling 3: 481–538. [4]

Dennis, B. and M. L. Taper. 1994. Density dependence in time series observations of natural populations: Estimation and testing. Ecological Monographs 64: 205–224. [3, 4]

Dennis, B., P. L. Munholland and J. Michael Scott. 1991. Estimation of growth and extinction parameters for endangered species. Ecological Monographs 61: 115–143. [3, 4, 5, 12]

Dixon, P. M. 2001. The bootstrap and the jackknife: describing the precision of ecological studies. Pages 267–288 in S. Scheiner and J. Gurevitch (eds.). *Design and Analysis of Ecological Experiments*, 2nd ed. Oxford University Press, Oxford. [4]

Doak, D. F. 1989. Spotted owls and old growth logging in the Pacific Northwest USA. Conservation Biology 3: 389–396. [1]

Doak, D. F. 1995. Source-sink models and the problem of habitat degradation: general models and applications to the Yellowstone grizzly bear. Conservation Biology 9: 1370–1379. [1]

Doak, D. F., P. Kareiva and B. Klepetka. 1994. Modeling population viability for the desert tortoise in the Western Mojave Desert. Ecological Applications 4: 446–460. [1, 2, 8]

Dobson, A. P., J. P. Rodriguez, W. M. Roberts and D. S. Wilcove. 1997. Geographic distribution of endangered species in the United States. Science 275: 550–553. [10]

Dobson, A., K. Ralls, M. Foster, M. Soule, D. Simberloff, D. F. Doak, J. Estes, L. S. Mills, D. Mattson, R. Dirzo, H. Arita, Ryan, E. Norse, R. Noss and D. Johns. 1999. Landscape viability equals landscape connectivity: Reconnecting fragmented landscapes. In M. E. Soulé and J. Terborgh (eds.). *Continental Conservation: Scientific Foundations of Regional Reserve Networks.* Island Press. [10]

Donovan. T. M., F. R. Thompson III., J. Faaborg and J. A. Probst. 1995. Reproductive success of migratory birds in habitat sources and sinks. Conservation Biology 9: 1380–1395. [10]

Downer, R. 1993. *GAPPS II User Manual. Version 1.3.* Applied Biomathematics, Setauket, NY. [1]

Draper, N. and H. Smith. 1981. *Applied Regression Analysis.* 2nd Ed. John Wiley and Sons, New York. [3]

Easterling, M. R., S. P. Ellner and P. M. Dixon. 2000. Size-specific sensitivity: Applying a new structured population model. Ecology 81: 694–708. [6]

Eberhardt, L. L. and D. B. Siniff. 1988. Population model for Alaska Peninsula sea otters. U.S. Department of Interior, Minerals Management Service, OCS Study, MMS 88–0091. 94 pp. [6]

Eberhardt, L. L., R. R. Knight and B. Blanchard. 1986. Monitoring grizzly bear population trends. Journal of Wildlife Management 50: 613–618. [3, 6]

Eberhardt, L. L., B. M. Blanchard and R. R. Knight. 1994. Population trend of the Yellowstone grizzly bear as estimated from reproduction and survival rates. Canadian Journal of Zoology 72: 360–363. [1]

Edwards, A. W. F. 1972. Likelihood: An Account of the Statistical Concept of Likelihood and its Application to Scientific Work. Cambridge University Press, Cambridge. [4]

Ehrlich, P. R., D. S. Dobkin and D. Wheye. 1988. *The Birder's Handbook: A Field Guide to the Natural History of North American Birds*. Simon and Schuster, New York. [2]

Eldered, B., P. Shahani and D. F. Doak. 2002. The problems and potential of count-based PVA. In C. Brigham and M. Schwartz (eds.). *Population Viability Analysis for Plants* (in press). Spring-Verlag, Berlin. [1, 3, 12]

Ellner, S. P. and J. Fieberg. 2002. Using PVA for management in light of uncertainty: Effects of habitat, hatcheries, and harvest on salmon viability. Ecology, in press. [3, 4, 11, 12]

Ellner, S. P., J. Fieberg, D. Ludwig and C. Wilcox. 2002. Precision of population viability analysis. Conservation Biology, in press. [12]

Elzinga, C. L., D. W. Salzer and J. W. Willoughby. 1998. Measuring & Monitoring Plant Populations. U.S. Dept. of the Interior, Bureau of Land Management, National Applied Resource Sciences Center. Washington DC. [5]

Engen, S., O. Bakke and A. Islam. 1998. Demographic and environmental stochasticity: Concepts and definitions. Biometrics 54: 840–846. [4, 8]

Fagan, W. F., E. Meir, J. Prendergast, A. Folarin and P. Kareiva. 2001. Characterizing population vulnerability for 758 species. Ecology Letters 4: 132–138. [12]

Falconer, D. S. 1981. *Introduction to Quantitative Genetics*. 2nd Ed. Longman, New York. [2]

Ferrière, R., F. Sarrazin, S. Legendre and J. P. Baron. 1996. Matrix population models applied to viability analysis and conservation: Theory and practice using the ULM software. Acta Oecologica 17: 629–656. [1]

Ferson, S. 1994. RAMAS/stage: Generalized stage-based modeling for population dynamics. Applied Biomathematics, Setauket, NY (online: http: //www.ramas.com/). [2]

Ferson, S. and M. A. Burgman. 1995. Correlations, dependency bounds and extinction risks. Biological Conservation 73: 101–105. [2]

Fieberg, J. and S. P. Ellner. 2000. When is it meaningful to estimate an extinction probability? Ecology 81: 2040–2047. [2, 3, 12]

Fieberg, J. and S. P. Ellner. 2001. Stochastic matrix models for conservation and management: A comparative review of methods. Ecology Letters 4: 244–266. [8]

Fishman, G. S. 1973. *Concepts and Methods in Discrete Event Digital Simulation*. Wiley and Sons, New York. [5, 8]

Foley, P. 1994. Predicting extinction times from environmental stochasticity and carrying capacity. Conservation Biology 1: 124–137. [4]

Forsman, E. D., S. DeStefano, M. G. Raphael and R. J. Gutiérrez. 1996. *Demography of the Northern Spotted Owl*. Studies in Avian Biology 17, Cooper Ornithological Society, Camarillo, CA. [1]

Fowler, C. W. and J. D. Baker. 1991. A review of animal population dynamics at extremely reduced population levels. Reports of the International Whaling Commission 41: 545–554. [2]

Fox, G. A. and B. E. Kendall. 2002. Demographic stochasticity and the variance reduction effect. Ecology, *in press*. [2]

Frank, K. and C. Wissel. 1998. Spatial aspects of metapopulation survival: From model results to rules of thumb for landscape management. Landscape Ecology 13: 363–379. [11]

Frost, C. 1990. *Hudsonia montana* final report: Effects of fire, trampling, and interspecies competition, 1985–1989. Report submitted to the North Carolina Department of Agriculture Plant Conservation Program. [6]

Gabriel, W. and R. Bürger. 1992. Survival of small populations under demographic stochasticity. Theoretical Population Biology 41: 44–71. [4]

Gaillard, J., M. Festa-Bianchet and N. G. Yoccoz. 1998. Population dynamics of large herbivores: variable recruitment with constant adult survival. Trends in Ecology and Evolution 13: 58–63. [2]

Gaona, P., F. Ferreras and M. Delibes. 1998. Dynamics and viability of a metapopulation of the endangered Iberian lynx (*Lynx pardinus*). Ecological Monographs 68: 349–370. [8]

Gerber, L. R., D. P. DeMaster and P. M. Kareiva. 1999. Gray whales and the value of monitoring data in implementing the U.S. Endangered Species Act. Conservation Biology 13: 1215–1219. [1, 3, 12]

Gibbs, J. P. 2000. Monitoring populations. Pages 213–252 in L. Boitani and T. K. Fuller (eds.). *Research*

Techniques in Animal Ecology: Controversies and Consequences. Columbia University Press, New York. [3]

Gibbs, J. P., H. L. Snell and C. E. Causton. 1999. Effective monitoring for adaptive wildlife management: Lessons from the Galapágos Islands. Journal of Wildlife Management 63: 1055–1065. [5]

Gilpin, M. E. and F. J. Ayala. 1973. Global models of growth and competition. Proceedings of the National Academy of Sciences (USA) 70: 3590–3593. [4]

Gilpin, M. E. and I. Hanski. 1991. *Metapopulation Dynamics: Empirical and Theoretical Investigations*. Academic Press, New York. [10]

Gilpin, M. E. and M. E. Soulé. 1986. Minimum viable populations: Processes of species extinction. Pages 19–34 in M. E. Soulé (ed.). *Conservation Biology: The Science of Scarcity and Diversity*. Sinauer Associates, Sunderland, MA. [2]

Ginzberg, L. R., L. B. Slobodkin, K. Johnson and A. G. Bindman. 1982. Quasi-extinction probabilities as a measure of impact on population growth. Risk Analysis 2: 171–181. [2]

Goodman, D. 1987. The demography of chance extinction. Pages 11–34 in M. E. Soulé (ed.). *Viable Populations for Conservation*. Cambridge University Press, Cambridge. [2]

Goodman, D. 2002. Predictive Bayesian PVA: A logic for listing criteria, delisting criteria, and recovery plans. In S. Beissinger and D. McCullough (eds.). *Population Viability Analysis* (in press). University of Chicago Press, Chicago.

Gould, W. R. and J. D. Nichols. 1998. Estimation of temporal variability of survival in animal populations. Ecology 9: 2531–2538. [8]

Grant, A. and T. G. Benton. 2000. Elasticity analysis for density-dependent populations in stochastic environments. Ecology 81: 680–693. [9]

Greenwood, J. J. D. 1996. Basic techniques. Pages 11–110 in W. J. Sutherland (ed.). *Ecological Census Techniques: A Handbook*. Cambridge University Press, Cambridge. [5]

Groom, M. 1998. Allee effects limit population viability of an annual plant. American Naturalist. 151: 487–496. [2]

Groom, M. and M. Pascual. 1998. The analysis of population persistence: An outlook on the practice of population viability analysis. Pages 4–27 in P. L. Fiedler and P. M. Karieva (eds.) *Conservation Biology for the Coming Decade*. Chapman and Hall, New York. [12]

Gross, K. G., J. R. Lockwood, C. Frost and W. F. Morris. 1998. Modeling controlled burning and trampling reduction for conservation of *Hudsonia montana*. Conservation Biology 12: 1291–1301. [2, 6, 7, 8]

Haddad, N. M. 1999. Corridor and distance effects on interpatch movements: A landscape experiment with butterflies. Ecological Applications 9: 612–622. [10]

Haddad, N. M. and K. A. Baum. 1999. An experimental test of corridor effects on butterfly densities. Ecological Applications 9: 623–633. [10]

Halley, J. M. 1996. Ecology, evolution, and 1/f noise. Trends in Ecology and Evolution 11: 33–37. [2, 4]

Halley, J. M. and W. E. Kunin. 1999. Extinction risk and the 1/f family of noise models. Theoretical Population Biology 56: 215–230. [4]

Halliday, T. 1978. *Vanishing Birds: Their Natural History and Conservation*. Holt, Rinehart & Winston, New York. [2]

Hanski, I. 1991. Single-species metapopulation dynamics: Concepts, models, and observations. Pages 17–38 in M. E. Gilpin and I. Hanski (eds.). *Metapopulation Dynamics: Empirical and Theoretical Investigations*. Academic Press, New York. [11]

Hanski, I. 1994. A practical model of metapopulation dynamics. Journal of Animal Ecology 63: 151–162. [11]

Hanski, I. 1999. *Metapopulation Ecology*. Oxford University Press, Oxford. [10]

Hanski, I., J. Alho and A. Moilanen. 2000. Estimating the parameters of survival and migration of individuals in metapopulations. Ecology 81: 239–251. [10]

Hanski, I., A. Moilanen, T. Pakkala and M. Kuussaari. 1996. The quantitative incidence function model and persistence of an endangered butterfly metapopulation. Conservation Biology 10: 578–590. [10, 11]

Harding, E. K., E. E. Crone, B. D. Elderd, J. M. Hoekstra, A. J. McKerrow, J. D. Perrine, J. Regetz, L. J. Rissler, A. G. Stanley, E. L. Walters, and the NCEAS Habitat Conservation Plan Working Group. 2001a. The scientific foundations of habitat conservation plans: A quantitative assessment. Conservation Biology 15: 488–500. [1, 2, 10, 11]

Harding, E. K., D. F. Doak and J. D. Albertson. 2001b. Evaluating the effectiveness of predator control for the non-native red fox. Conservation Biology 15: 1114–1122. [1, 2, 10, 11]

Hargrove, W. W., R. H. Gardner, M. G. Turner, W. H. Romme and D. G. Despain. 2000. Simulating fire patterns in heterogeneous landscapes. Ecological Modelling 135: 243–263. [1-]

Haroldson, M. 1999. Bear monitoring and population trend: unduplicated females. Pages 3–17 in C. C. Schwartz and M. A. Haroldson (eds.). Yellowstone grizzly bear investigations: Annual report of the Interagency Grizzly Bear Study Team, 1999. U.S. Geological Survey,

Bozeman, MT. 128 pp. (www.nrmsc.usgs.gov/research/igbstpub. htm). [3]

Harris, R. B., L. H. Metzgar and C. D. Bevins. 1986. *GAPPS: Generalized Animal Population Projection System, Version 3. 0.* Montana Cooperative Wildlife Research Unit, University of Montana, Missoula, MT. [1]

Harrison, S. 1989. Long-distance dispersal and colonization in the bay checkerspot butterfly, *Euphydryas editha bayensis*. Ecology 70: 1236–1243. [10]

Harrison, S. 1991. Local extinction in a metapopulation context: An empirical evaluation. Biological Journal of the Linnaean Society 42: 73–88. [10]

Harrison, S. and J. F. Quinn. 1989. Correlated environments and the persistence of metapopulations. Oikos 56: 293–298. [11]

Harrison, S. and C. Ray. 2002. Plant population viability and metapopulation-level processes. In S. Beissinger and D. McCullough (eds.). *Population Viability Analysis* (in press). University of Chicago Press, Chicago. [11]

Harrison, S., J. F. Quinn, J. F. Baughman, D. D. Murphy and P. R. Ehrlich. 1991. Estimating the effects of scientific study on two butterfly populations. American Naturalist 137: 227–243. [4]

Harrison, S., J. Maron and G. Huxel. 2000. Regional turnover and fluctuation in populations of five plants confined to serpentine seeps. Conservation Biology 14: 769–79. [10, 11]

Hartl, D. L. and A. G. Clark. 1997. *Principles of Population Genetics.* 3rd Ed. Sinauer Associates, Sunderland, MA. [2]

Hastings, N. and J. Peacock. 1974. *Statistical Distributions.* Butterworths, London. [8]

Hedrick. P. W. and P. S. Miller. 1992. Conservation genetics: Techniques and fundamentals. Ecological Applications 2: 30–46. [2]

Hedrick, P. W., R. C. Lacy, F. W. Allendorf. and M. E Soulé. 1996. Directions in conservation biology: Comments on Caughley. Conservation Biology 10: 1312–1320. [2]

Heino, M. 1998. Noise, colour, synchrony and extinctions in spatially structured populations. Oikos 83: 368–375. [4]

Heppell, S. S. 1998. Application of life-history theory and population model analysis to turtle conservation. Copeia 1998: 367–375. [12]

Heppell, S. S. and L. B. Crowder. 1996. Analysis of a fisheries model for harvest of hawksbill sea turtles (*Eretmochelys imbricata*). Conservation Biology 10: 874–880. [1]

Heppell, S. S., J. R. Walters and L. B. Crowder. 1994. Evaluating management alternatives for red-cockaded woodpeckers: A modeling approach. Journal of Wildlife Management 58: 479–487. [6, 7]

Heppell, S. S., L. B. Crowder and D. T. Crouse. 1996. Models to evaluate headstarting as a management tool for long-lived turtles. Ecological Applications 6: 556–565. [7, 9]

Heppell, S., H. Caswell and L. B. Crowder. 2000. Life histories and elasticity patterns: Perturbation analysis for species with minimal demographic data. Ecology 81: 654–665. [7]

Heyer, W. C. (ed.). 1994. *Measuring and Monitoring Biological Diversity: Standard Methods for Amphibians.* Smithsonian Institution Press, Washington, DC. [5]

Hilborn, R. and M. Mangel. 1997. *The Ecological Detective: Confronting Models with Data.* Princeton University Press, Princeton, NJ. [4, 8, 11]

Hitchcock, C. L. and C. Gatto-Trevor. 1997. Diagnosing a shorebird local population decline with a stage-structured population model. Ecology 78: 522–534. [6, 7]

Holmes, E. E. 2001. Estimating risks in declining populations with poor data. Proceedings of the National Academy of Science USA 98: 5072–5077. [2, 3, 5]

Holmes, E. E. and W. F. Fagan. In press. Validating population viability analysis for corrupted data sets. Ecology. [5]

Howells, O. and G. Edwards Jones. 1997. A feasibility study of reintroducing wild boar *Sus scrofa* to Scotland: Are existing woodlands large enough to support minimum viable populations? Biological Conservation 81: 77–89. [1]

Hurvich, C. M. and C-L. Tsai. 1989. Bias of the corrected AIC criterion for underfitted regression and time series models. Biometrika 78: 499–509. [4]

Johst, K. and C. Wissel 1997. Extinction risk in a temporally correlated fluctuating environment. Theoretical Population Biology 52: 91–100. [4]

Kalinowski, S. T., P. W. Hedrick and P. S. Miller. 2000. Inbreeding depression in the Speke's gazelle captive breeding program. Conservation Biology 14: 1375–1384. [2]

Karevia, P. 1990. Population dynamics in spatially complex environments: Theory and data. Philosophical Trans. Royal Society of London B 330: 175–190. [7, 9]

Kareiva, P., M. Marvier and M. McClure. 2000. Population persistence time: Estimates, models and mechanisms. Ecological Applications 7: 107–117. [7, 9]

Kendall, B. E. 1998. Estimating the magnitude of environmental stochasticity in survivorship data. Ecological Applications 8: 184–193. [2, 4, 8]

Kendall, B. E. and G. A. Fox. 2002. Variation among individuals reduces demographic stochasticity. Conservation Biology, in press. [2]

Kendall, B. E., O. N. Bjornstad, J. Bascompte, T. H. Keitt and W. F. Fagan. 2000. Dispersal, environmental correlation, and spatial synchrony in population dynamics. American Naturalist 155: 628–636. [10]

Kindvall, O. 2000. Comparative precision of three spatial realistic simulation models of metapopulation dynamics. Ecological Bulletin 48: 101–110. [11]

Knight, R. R. and L. L. Eberhardt. 1985. Population dynamics of Yellowstone grizzly bears. Ecology 66: 323–334. [1]

Koenig, W. D. 1998. Spatial autocorrelation in California land birds. Conservation Biology 12: 612–620. [10]

Krebs, C. J. 1989. *Ecological Methodology*. Harper & Row, New York. [5]

Kuusaari, M., I. Saccheri, M. Camara and I. Hanski. 1998. Allee effect and population dynamics in the Glanville fritillary butterfly. Oikos 82: 384–392. [2]

Lacy, R. C., K. A. Hughes and P. S. Miller. 1995. *VORTEX Users Manual, Version 7*. IUCN/SSC Conservation Breeding Specialist Group, Apple Valley Michigan, USA (also see http: //pw1. netcom. com/~rlacy/vortex. html). [1]

Lahaye, W. S., R. J. Gutierrez and H. R. Akçakaya. 1994. Spotted owl metapopulation dynamics in Southern California. Journal of Animal Ecology 63: 775–785. [10, 11]

Lamberson, R. H., K. McKelvey, B. R. Noon and C. Voss. 1992. A dynamic analysis of northern spotted owl viability in a fragmented forest landscape. Conservation Biology 6: 505–512. [1]

Lamberson, R. H., B. R. Noon and K. S. McKelvey. 1994. Reserve design for territorial species: The effects of patch size and spacing on the viability of the Northern spotted owl. Conservation Biology 8: 185–195. [1]

Lamont, B. R., P. G. L. Klinkhamer, E. T. F. Witkowski. 1993. Population fragmentation may reduce fertility to zero in *Banksia goodii*: A demonstration of the Allee effect. Oecologia 94: 446–450. [2]

Lande, R. 1988a. Demographic models of the Northern Spotted Owl (*Strix occidentalis caurina*). Oecologia 75: 601–607. [1]

Lande, R. 1988b. Genetics and demography in biological conservation. Science 241: 1455–1460. [1]

Lande, R. 1993. Risks of population extinction from demographic and environmental stochasticity and random catastrophes. American Naturalist 142: 911–927. [4]

Lande, R. 1998. Demographic stochasticity and Allee effect on a scale with isotropic noise. Oikos 83: 353–358. [4]

Lande, R. 2002. Incorporating stochasticity in Population Viability Analysis. In S. Beissinger and D. McCullough (eds.). *Population Viability Analysis* (in press). University of Chicago Press, Chicago. [2, 4, 12]

Lande, R. and S. H. Orzack. 1988. Extinction dynamics of age-structured populations in a fluctuating environment. Proceedings of the National Academy of Sciences USA 85: 7418–7421. [3, 7]

Lande, R., S. Engen and B.-E. Saether. 1994. Optimal harvesting, economic discounting and extinction risk in fluctuating populations. Nature 372: 88–90. [4]

Lande, R., S. Engen and B.-E. Saether. 1999. Spatial scale of population synchrony: Environmental correlation versus dispersal and density regulation. American Naturalist 154: 271–281. [10]

Law, R. 1983. A model for the dynamics of a plant population containing individuals classified by age and size. Ecology 64: 224–230. [6]

Lebreton, J. D., K. P. Burnham, J. Clobert and D. R. Anderson. 1992. Modeling survival and testing biological hypotheses using marked animals: A unified approach with case studies. Ecological Monographs 62: 67–118. [6]

Lefkovitch, L. P. 1965. The study of population growth in organisms grouped by stages. Biometrics 21: 1–18. [6]

Legendre, S., J. Clobert, A. P. Møller and G. Sorci. 1999. Demographic stochasticity and social mating systems in the process of extinction of small populations: The case of passerines introduced to New Zealand. American Naturalist 153: 449–463. [4]

Lele, S., M. L. Taper and S. Gage. 1998. Statistical analysis of population dynamics in space and time using estimating functions. Ecology 79: 1489–1502. [10]

Leslie, P. H. 1945. On the use of matrices in certain population mathematics. Biometrika 33: 183–212. [6]

Letcher, B. H., J. A. Priddy, J. R. Walters and L. B. Crowder. 1998. An individual-based, spatially-explicit simulation model of the populations dynamics of the endangered red-cockaded woodpecker, *Picoides borealis*. Biological Conservation 86: 1–14. [11]

Levin, S. A. and C. P. Goodyear. 1980. Analysis of an age-structured fishery model. Journal of Mathematical Biology 9: 245–274. [8]

Levins, R. 1969. Some demographic and genetic consequences of environmental heterogeneity for biological control. Bulletin of the Entomological Society of America 15: 237–240. [10]

Lewis, M. A. and P. Kareiva. 1993. Allee dynamics and the spread of invading organisms. Theoretical Population Biology 43: 141–158. [4]

Lewontin, R. C. and D. Cohen. 1969. On population growth in a randomly varying environment. Proceedings of the National Academy of Sciences USA 62: 1056–1060. [2]

Lindenmayer, D. B. and H. P. Possingham. 1996. Ranking conservation and timber management options for Leadbeater's possum in Southeastern Australia using population viability analysis. Conservation Biology 10: 235–251. [1, 10, 11]

Lindenmayer, D. B., M. A. Burgman, H. R. Akçakaya, R. C. Lacy and H. Possingham. 1995. A review of the generic computer programs ALEX, RAMAS/space and VORTEX for modeling the viability of wildlife metapopulations. Ecological Modeling 82: 161–174. [1]

Lindenmayer, D. B., I. Ball, H. Possingham, M. A. McCarthy and M. L. Pope. 1999. A landscape-scale test of the predictive ability of a spatially explicit model for Population Viability Analysis. Journal of Applied Ecology 38: 36–48. [12]

Lindenmayer, D. B., R. C. Lacy and M. L. Pope. 2000. Testing a simulation model for population viability analysis. Ecological Applications 10: 580–597. [12]

Ludwig, D. 1996a. The distribution of population survival times. American Naturalist 147: 506–526. [12]

Ludwig, D. 1996b. Uncertainty and the assessment of extinction probabilities. Ecological Applications 6: 1067–1076. [12]

Ludwig, D. 1999. Is it meaningful to estimate a probability of extinction? Ecology 80: 298–310. [2, 3, 5, 12]

Ludwig, D. and C. J. Walters. 1981. Measurement errors and uncertainty in parameter estimates for stock and recruitment. Canadian Journal of Fisheries and Aquatic Sciences 38: 711–720. [2, 5]

Ludwig, D. and C. J. Walters. 1985. Are age-structured models appropriate for catch-effort data? Canadian Journal of Fisheries and Aquatic Sciences 42: 1066–1072. [12]

Ludwig, D., C. J. Walters and J. Cooke. 1988. Comparison of two models and two estimation methods for catch and effort data. Natural Resource Modeling 2: 457–498. [5]

Mace, G. M. and R. Lande. 1991. Assessing extinction threats: Towards a reevaluation of IUCN threatened species categories. Conservation Biology 5: 148–157. [2, 12]

Mangel, M. and C. Tier. 1993. A simple, direct method for finding persistence times of population and applications to conservation problems. Proceedings of the National Academy of Sciences USA 90: 1083–1086. [2, 3, 4]

Mangel, M. and C. Tier. 1994. Four facts every conservation biologist should know about persistence. Ecology 75: 607–614. [2]

Marshall, K. and G. Edwards Jones. 1998. Reintroducing capercaillie (*Tetrao urogallus*) into southern Scotland: identification of minimum viable populations at potential release sites. Biodiversity and Conservation 7: 275–296. [1]

Maschinski, J., R. Frye and S. Rutman. 1997. Demography and population viability of an endangered plant species before and after protection from trampling. Conservation Biology 11: 990–999. [2]

Mattson, D. J., B. M. Blanchard and R. R. Knight. 1992. Yellowstone grizzly bear mortality, human habituation, and whitebark pine seed crops. Journal of Wildlife Management 56: 432–442. [11]

McCarthy, M. A. 1996. Extinction dynamics of the helmeted honeyeater: Effects of demography, stochasticity, inbreeding, and spatial structure. Ecological Modelling 85: 151–163. [11]

McCarthy, M. A. 1997. The Allee effect, finding mates, and theoretical models. Ecological Modelling 103: 99–102. [4]

McCarthy, M. A. and L. S. Broome. 2000. A method for validating stochastic models of population viability: A case study of the mountain pygmy-possum (*Burramys parvus*). Journal of Animal Ecology 69: 599–607. [12]

McCarthy, M. A., D. B. Lindenmayer and H. P. Possingham. 2000. Testing spatial PVA models of Australian treecreepers (Aves: Climacteridae) in fragmented forest. Ecological Applications 10: 1722–1731. [12]

McCarthy, M. A., H. P. Possingham, J. R. Day and A. J. Tyre. 2001. Testing the accuracy of population viability analysis. Conservation Biology 15: 1030–1038. [12]

McGarrahan, E. 1997. Much-studied butterfly winks out on Stanford preserve. Science 275: 479–480. [4]

Meir, E. and W. F. Fagan. 2000. Will observation error and biases ruin the use of simple extinction models? Conservation Biology 14: 148–154. [12]

Menges, E. 1990. Population viability analysis for an endangered plant. Conservation Biology 4: 52–62. [1, 10]

Menges, E. S. and R. W. Dolan. 1998. Demographic variability of populations of *Silene regia* in midwestern prairies: Relationships with fire management, genetic variation, geographic location, population size and isolation. Journal of Ecology 86: 63–78. [10]

Menges, E. S. and D. R. Gordon. 1996. Three levels of monitoring intensity for rare plant species. Natural Areas Journal 16: 227–237. [1]

Meretsky, V. J., N. F. R. Snyder, S. R. Beissinger, D. A. Clendenen and J. W. Wiley. 2000. Demography of the California condor: implications for reestablishment. Conservation Biology 14: 957–967. [2]

Middleton, D. A. J. and R. M. Nisbet. 1997. Population persistence time: Estimates, models and mechanisms. Ecological Applications 7: 107–117. [10]

Middleton, D. A. J., A. R. Veitch and R. M. Nisbet. 1995. The effect of an upper limit to population size on persistence time. Theoretical Population Biology 48: 277–305. [4]

Mills, L. S. and P. E. Smouse. 1994. Demographic consequences of inbreeding in remnant populations. American Naturalist 144: 412–431. [1, 2]

Mills, L. S. and F. W. Allendorf. 1996. The one-migrant-per-generation rule in conservation and management. Conservation Biology 10: 1509–1518. [10]

Mills, L. S., S. G. Hayes, C. Baldwin, M. J. Wisdom, J. Citta, D. J. Mattson and K. Murphy. 1996. Factors leading to different viability predictions for a grizzly bear data set. Conservation Biology 10: 863–873. [1, 12]

Mills, L. S., D. F. Doak and M. Wisdom. 1999. The reliability of conservation actions based upon elasticities of matrix models. Conservation Biology 13: 815–829. [9]

Moilanen, A. 1999. Patch occupancy model of metapopulation dynamics: Efficient parameter estimation using implicit statistical inference. Ecology 80: 1031–1043. [10, 11]

Moilanen, A. 2000. The equilibrium assumption in estimating the parameters of metapopulation models. Journal of Animal Ecology 69: 143–153. [10, 11]

Moilanen, A. and I. Hanski. 1998. Metapopulation dynamics: Effects of habitat quality and landscape structure. Ecology 79: 2503–2515. [10, 11]

Moloney, K. A. 1986. A generalized algorithm for determining category size. Oecologia 69: 176–180. [6]

Mooij, W. M. and D. L. DeAngelis. 1999. Error propagation in spatially explicit models: A reassessment. Conservation Biology 13: 930–933. [11]

Morales, J. M. 1999. Viability in a pink environment: why "white noise" models can be dangerous. Ecology Letters 2: 228–232. [4]

Moritz, M. A. 1997. Analyzing extreme disturbance events: Fires in Los Padres National Forest. Ecological Applications 7: 1252–1262. [10]

Morris, W. F., D. F. Doak, M. Groom, P. Kareiva, J. Fieberg, L. Gerber, P. Murphy and D. Thomson. 1999. *A Practical Handbook for Population Viability Analysis*. The Nature Conservancy, Washington, DC. [3, 10]

Morris, W. F., P. L. Bloch, B. R. Hudgens, L. C. Moyle and J. R. Stinchcombe. 2002. Population viability analysis in endangered species recovery planning: Past use and recommendations for future improvement. Ecological Applications, in press. [1]

Myers, R. A., N. J. Barrowman, J. A. Hutchings and A. A. Rosenberg. 1995. Population dynamics of exploited fish stocks at low population levels. Science 269: 1106–1108. [4]

Nantel, P., D. Gagnon and A. Nault. 1996. Population viability analysis of American Ginseng and Wild Leek harvested in stochastic environments. Conservation Biology 10: 608–621. [1, 6]

National Research Council. 1995. *Science and the Endangered Species Act*. National Academy Press, Washington, DC.

Nature Conservancy, The. 1997. *Conservation by Design: A Framework for Mission Success*. Arlington, VA. [1]

Nault, A. and D. Gagnon. 1993. Ramet demography of *Allium tricoccum*, a spring ephemeral, perennial forest herb. Journal of Ecology 81: 101–119. [6]

Nichols, J. D., J. R. Sauer, K. H. Pollock and J. B. Hestbeck. 1996. Estimating transition probabilities for stage-based population projection matrices using capture-recapture data. Ecology 73: 306–312. [6]

Novaro, A. J., K. H. Redford and R. E. Bodmer. 2000. Effect of hunting in source-sink systems in the Neotropics. Conservation Biology 14: 713–721. [10]

Nowell, K. and P. Jackson. 1996. *Wild Cats: Status Survey and Conservation Action Plan*. International Union for Conservation of Nature and Natural Resources (IUCN), Gland, Switzerland. [8]

Nunney, L. 1993. The influence of mating system and overlapping generations on effective population size. Evolution 47: 1329–1341. [2]

Nunney, L. and D. R. Elam. 1994. Estimating the effective population size of conserved populations. Conservation Biology 8: 175–184. [2]

Okubo, A. and S. A. Levin. 1989. A theoretical framework for data analysis of wind dispersal of seeds and pollen. Ecology 70: 329–338. [10]

Olea, P. P. 2001. Postfledging dispersal in the endangered Lesser Kestrel, *Falco naumanni*. Bird Study 48: 110–115. [10]

Pascual, M. A., P. Kareiva and R. Hilborn. 1997. The influence of model structure on conclusions about the viability and harvesting of Serengeti wildebeest. Conservation Biology 11: 966–976. [12]

Pease, C. M. and D. J. Mattson 1999. Demography of the Yellowstone grizzly bears. Ecology 80: 957–975. [1]

Perry, G. L. W., A. D. Sparrow and I. F. Owens. 1999. A GIS-supported model for the simulation of the spatial structure of wildland fire, Cass Basin, New Zealand. Journal of Applied Ecology 36: 502–518. [10]

Petit, S., A. Moilanen, I. Hanski and M. Baguette. 2001. Metapopulation dynamics of the bog fritillary butterfly: movements between habitat patches. Oikos 92: 491–500. [10]

Pfister, C. A. 1998. Patterns of variance in stage-structured populations: Evolutionary predictions and ecological implications. Proceedings of the National Academy of Sciences USA 95: 213–218. [2]

Pimm, S. L. and A. Redfearn. 1988. The variability of animal populations. Nature 334: 613–614. [3, 12]

Pokki, J. 1981. Distribution, demography and dispersal of the field vole, *Microtus agrestis* (L.) in the Tvarminne archepeligo, Finland. Acta Zoologica Fennica 164: 1–48. [10]

Pollard, E., K. H. Lakhani and P. Rothery. 1987. The detection of density-dependence from a series of annual censuses. Ecology 68: 2146–2155. [3]

Porneluzi, P. A. and J. A. Faaborg. 1999. Season-long fecundity, survival, and viability of ovenbirds in fragmented and unfragmented landscapes. Conservation Biology 13: 1151–1161. [10]

Possingham, H. P. and I. Davies. 1995. ALEX: A model for the viability analysis of spatially structured populations. Biological Conservation 73: 143–150. [1]

Powell, R. A., J. W. Zimmerman, D. E. Seaman and J. F. Gilliam. 1996. Demographic analyses of a hunted black bear population with access to a refuge. Conservation Biology 10: 224–234. [1]

Press, W. H., B. P. Flannery, S. A. Teukolsky and W. T. Vetterling. 1995. *Numerical Recipes In C: The Art Of Scientific Computing*. Cambridge University Press, Cambridge. [4, 11]

Press, D., D. F. Doak and P. Steinberg. 1996. The role of local governments in rare species protection. Conservation Biology 10: 1538–1548. [8, 10]

Pulliam, H. R. 1988. Sources, sinks, and population regulation. American Naturalist 132: 652–661. [10]

Ralls, K. J. and B. L. Taylor. 1997. How viable is population viability analysis? Pages 228–235 in S. T. A. Pickett, R. S. Ostfeld, M. Shachak and G. E. Likens (eds.). *The Ecological Basis of Conservation*. Chapman and Hall, New York. [12]

Ralls, K. J., J. D. Ballou and A. R. Templeton. 1988. Estimates of lethal equivalents and the cost of inbreeding in mammals. Conservation Biology 2: 185–193. [2]

Ralls, K., D. P. DeMaster and J. A. Estes. 1996. Developing a criterion for delisting the Southern sea otter under the U.S. Endangered Species Act. Conservation Biology 10: 1528–1537. [2]

Ranta, E., V. Kaitala, J. Lindström and H. Linden. 1995. Synchrony in population dynamics. Proceedings of the Royal Society of London B 262: 113–118. [10]

Ranta, E., V. Kaitala and J. Lindström. 1998. Spatial dynamics of populations. In J. Bascompte and R. V. Soleì (eds.), *Modeling Spatiotemporal Dynamics in Ecology*. Springer-Verlag, Berlin. [10]

Ratner, S., R. Lande and B. R. Roper. 1997. Population viability analysis of Spring chinook salmon in the South Umpqua River, Oregon. Conservation Biology 11: 879–889. [7, 8, 9]

Ratsirarson J, J. A. Silander and A. F. Richard. 1996. Conservation and management of a threatened Madagascar palm species, *Neodysis decaryi* Jumelle. Conservation Biology 10: 40–52. [1]

Reed, J. M., D. D. Murphy and P. F. Brussard. 1998. Efficacy of population viability analysis. Wildlife Society Bulletin 26: 244–251. [12]

Richards, S. A., H. P. Possingham and J. Tizard. 1999. Optimal fire management for maintaining community diversity. Ecological Applications 9: 880–892. [10]

Ricker, W. E. 1954. Stock and recruitment. Journal of the Fisheries Research Board of Canada 11: 559–623. [4]

Ricketts, T. H. 2001. The matrix matters: Effective isolation in fragmented landscapes. American Naturalist 158: 87–99. [10]

Ripa, J. and M. Heino. 1999. Linear analysis solves two puzzles in population dynamics: The route to extinction and extinction in coloured environments. Ecology Letters 2: 219–222. [4, 8]

Ripa, J. and P. Lundberg. 1996. Noise colour and the risk of population extinctions. Proceedings of the Royal Society of London B 263: 1751–1753. [4, 8]

Ripa, J. and P. Lundberg. 2000. The route to extinction in variable environments. Oikos 90: 89–96. [8]

Robinson, S. K., F. R. Thompson and T. M. Donovan. 1995. Regional forest fragmentation and the nesting success of migratory birds. Science 267: 1987–1990. [10]

Root, K. V. 1998. Evaluating the effects of habitat quality, connectivity, and catastrophes on a threatened species. Ecological Applications 8: 845–865. [11]

Roughgarden, J. 1975. A simple model for population dynamics in stochastic environments. American Naturalist 109: 713–736. [4]

Roughgarden, J. 1997. *Primer of Ecological Theory*. Prentice-Hall, Englewood Cliffs, NJ. [1]

Royama, T. 1992. *Analytical Population Dynamics*. Chapman and Hall, London. [4]

Ruckelshaus, M., C. Hartway and P. Kareiva. 1997. Assessing the data requirements of spatially explicit dispersal models. Conservation Biology 11: 1298–1306. [11]

Ruckelshaus, M., C. Hartway and P. Kareiva. 1999. Dispersal and landscape errors in spatially explicit population models: A reply. Conservation Biology 131: 1223–1224. [11]

Saether, B.-E. and S. Engen. 2002. Including uncertainties in Population Viability Analysis using population prediction intervals. In S. Beissinger and D. McCullough (eds.). *Population Viability Analysis* (in press). University of Chicago Press, Chicago. [4]

Saether, B.-E., R. H. Ringsby and E. Roskaft. 1996. Life history variation, population processes and priorities in species conservation: Towards a reunion of research paradigms. Oikos 77: 217–226. [4]

Saether, B.-E., S. Engen, A. Islam, R. McCleery and C. Perrins. 1998a. Environmental stochasticity and extinction risk in a population of a small songbird, the great tit. American Naturalist 151: 441–450. [2, 4]

Saether, B.-E., S. Engen, J. Swenson, O. Bakke and F. Sandegren. 1998b. Assessing the viability of Scandinavian brown bear, *Ursus arctos*, populations: The effects of uncertain parameter estimates. Oikos 83: 403–416. [2, 4]

Saether, B.-E., J. Tufto, S. Engen, K. Jerstad, O. W. Rostad and J. E. Skatan. 2000a. Population dynamical consequences of climate change for a small temperate songbird. Science 287: 854–856. [4]

Saether, B.-E., S. Engen, R. Lande, P. Arcese and J. N. M. Smith. 2000b. Estimating the time to extinction in an island population of song sparrows. Proceedings of the Royal Society of London B. 267: 621–626. [4]

SAS. 1990. *SAS/STAT User's Manual. Version 6*. 4th Ed. SAS Institute, Cary, NC.

Scheiner, S. M. and J. Gurevitch (eds.). 1993. *Design and Analysis of Ecological Experiments*. Chapman and Hall, New York. [10]

Schemske, D. W. and R. Lande. 1985. The evolution of self-fertilization and inbreeding depression in plants: Two empirical observations. Evolution 39: 41–52. [2]

Schemske, D. W., B. C. Husband, M. H. Ruckelhaus, C. Goodwillie, I. M. Parker and J. G. Bishop. 1994. Evaluating approaches to the conservation of rare and endangered plants. Ecology 75: 584–606. [2, 12]

Schmalzel, R. J., F. W. Reichenbacher and S. Rutman. 1995. Demographic study of the rare Coryphantha robbinsorum (Cactaceae) in Southeastern Arizona. Madroño 42: 332–348. [10, 11]

Schmutz, J. A., R. F. Rockwell and M. R. Petersen. 1997. Relative effects of survival and reproduction on the population dynamics of emperor geese. Journal of Wildlife Management 61: 191–201. [9]

Schultz, C. B. and E. E. Crone. 2001. Edge-mediated dispersal behavior in a prairie butterfly. Ecology 82: 1879–1892. [10]

Shaffer, M. L. 1978. Determining minimum viable population sizes: A case study of the grizzly bear (*Ursus arctos* L.). Ph. D. Dissertation, Duke University, Durham, NC. [1]

Shaffer, M. L. 1981. Minimum population sizes for species conservation. Bioscience 31: 131–134. [1]

Shaffer, M. L. and F. B. Sampson. 1985. Population size and extinction: A note on determining critical population size. American Naturalist 125: 144–152. [1]

Sjögren-Gulve, P. and C. Ray. 1996. Large scale forestry extirpates the pool frog: using logistic regression to model metapopulation dynamics. In D. R. McCullough (ed.). *Metapopulations and Wildlife Conservation*. Island Press, Washington, DC. [11]

Sjögren-Gulve, P. and I. Hanski. 2000. Metapopulation viability analysis using occupancy models. Ecological Bulletins 48: 53–71. [10, 11]

Snedecor, G. W. and W. G. Cochran. 1980. *Statistical Methods*. 7th Ed. Iowa State University Press, Ames, IA. [3]

Sokal, R. R. and F. J. Rohlf. 1995. *Biometry*. 3rd Ed. W. H. Freeman, New York. [6]

Sokal, R. R. and N. L. Oden. 1978a. Spatial autocorrelation in biology. 1. Methodology. Biological Journal of the Linnaean Society 10: 199–228. [10]

Sokal, R. R. and N. L. Oden. 1978b. Spatial autocorrelation in biology. 2. Some biological implications and four applications of evolutionary and ecological interest. Biological Journal of the Linnaean Society 10: 229–249. [10]

South, A. S. Rushton and D. Macdonald. 2000. Simulating the proposed reintroduction of the European beaver (*Castor fiber*) to Scotland. Biological Conservation 93: 103–116. [1]

Southwood, T. R. E. 1978. *Ecological Methods.* 2nd Ed. Chapman and Hall, New York. [5]

Stamps, J. A. 1988. Conspecific attraction and aggregation in territorial species. American Naturalist 131: 329–347. [10]

Stamps, J. A. and V. V. Krishnan. 1990. The effect of settlement tactics on territory sizes. American Naturalist 135: 527–546. [10]

Steele, J. H. 1985. A comparison of terrestrial and marine ecological systems. Nature 313: 355–358. [2, 4]

Steenbergh, W. F. and C. H. Lowe. 1983. Ecology of the Saguaro. III. Growth and Demography. National Park Service Scientific Monograph Number 17. U.S. Dept. of Interior, National Park Service, Washington, DC. [2]

Steinberg, E. K. and C. E. Jordan. 1998. Using molecular genetics to learn about the ecology of threatened species: The allure and illusion of measuring genetic structure in natural populations. Pages 440–460 in P. L. Fiedler and P. M. Kareiva (eds.), *Conservation Biology for the Coming Decade*, 2nd Ed. Chapman and Hall, New York. [10]

Stephan, T. and C. Wissel. 1994. Stochastic extinction models discrete in time. Ecological Modelling 75–76: 183–192. [4]

Stephens, P. A. and W. J. Sutherland. 1999. Consequences of the Allee effect for behaviour, ecology, and conservation. Trends in Ecology and Evolution 14: 401–405. [2]

Sutcliffe, O. L., C. D. Thomas and D. Moss. 1996. Spatial synchrony and asynchrony in butterfly population dynamics. Journal of Animal Ecology 65: 85–95. [10]

Sutherland, W. J. (ed). 1996. *Ecological Census Techniques: A Handbook.* Cambridge University Press, Cambridge. [5]

Swanson. B. J. 1998. Autocorrelated rates of change in animal populations and their relationship to precipitation. Conservation Biology 12: 801–808. [10]

Taylor, B. 1995. The reliability of using population viability analysis for risk classification of species. Conservation Biology 9: 551–558. [12]

Tear, R. H., J. M. Scott, P. H. Hayward and B. Griffith. 1993. Status and prospects for success of the Endangered Species Act. : A look at recovery plans. Science 262: 976–977. [2]

Tear, R. H., J. M. Scott, P. H. Hayward and B. Griffith. 1995. Recovery plans and the Endangered Species Act: Are criticisms supported by data? Conservation Biology 9182–195. [12]

Thomas, J., E. D. Forsman, J. B. Lint, E. C. Meslow, B. R. Noon and J. Verner. 1990. A conservation strategy for the northern spotted owl. Report of the Interagency Scientific Committee to address the conservation of the northern spotted owl. Portland: USDA Forest Service; USDI Bureau of Land Management/Fish and Wildlife Service/National Park Service. [1]

Thompson, W. L., G. C. White and C. Gowan. (eds.). 1998. *Monitoring Vertebrate Populations.* Academic Press, New York. [5]

Tufto, J., B.-E. Saether, S. Engen, J. E. Swenson and F. Sandegren. 1999. Harvesting strategies for conserving minimum viable populations based on World Conservation Union criteria: Brown bears in Norway. Proceedings of the Royal Society of London B 266: 961–967. [1]

Tufto, J., B.-E. Saether, S. Engen, P. Arcese, K. Jerstad, O. W. Rostad and J. N. M. Smith. 2000. Bayesian meta-analysis of demographic parameters in three small, temperate passerines. Oikos 88: 273–281. [4]

Tuljapurkar, S. 1982. Population dynamics in variable environments. III. Evolutionary dynamics of *r*-selection. Theoretical Population Biology 21: 141–165. [7]

Tuljapurkar, S. and S. H. Orzack. 1980. Population dynamics in variable environments. I. Long-run growth rates and extinction. Theoretical Population Biology 18: 314–342. [7]

Turchin, P. 1998. *Quantitative Analysis of Movement: Measuring and Modeling Population Redistribution in Animals and Plants.* Sinauer Associates, Sunderland, MA. [10]

United States Fish and Wildlife Service. 1993. *Grizzly Bear Recovery Plan (and Summary).* U.S. Department of Agriculture, Bethesda, MD. [3]

van Tienderen, P. H. 1995. Life cycle trade-offs in matrix population models. Ecology 76: 2482–2489. [8]

Vandermeer, J. 1978. Choosing category size in a stage projection matrix. Oecologia 32: 199–225. [6]

Veit, R. R. and M. A. Lewis. 1996. Dispersal, population growth, and the Allee effect: Dynamics of the house finch invasion of eastern North America. American Naturalist 148: 255–274. [4]

Vos, P., E. Meelis and W. J. ter Keurs. 2000. A framework for the design of ecological monitoring programs as a tool for environmental and nature management. Environmental Monitoring and Assessment 61: 317–344. [5]

Wade, P. R. 2000. Bayesian methods in conservation biology. Conservation Biology 14: 1308–1316. [12]

Wade, P. R. 2002. Bayesian population viability analysis. In S. Beissinger and D. McCullough (eds.). *Population Viability Analysis* (in press). University of Chicago Press, Chicago. [12]

Walters, J. R. 1990. Red-cockaded woodpeckers: A 'primitive' cooperative breeder. Pages 67–101 in P. B. Stacey and W. D. Koenig (eds.), *Cooperative Breeding in Birds: Long-Term Studies of Ecology and Behavior.* Cambridge University Press, Cambridge. [6]

Walters, J. R., P. D. Doerr and J. H. Carter, III. 1988. The cooperative breeding system of the red-cockaded woodpecker. Ethology 78: 275–305. [6]

Walters, J. R., L. B. Crowder and J. A. Priddy. 2002. Population viability analysis for red-cockaded woodpeckers using an individual-based model. Ecological Applications, in press. [11]

Weiss, S. B., D. D. Murphy and R. R. White. 1988. Sun. slope, and butterflies: Topographic determinants of habitat quality for *Euphydryas editha*. Ecology 69: 1486–1496. [10]

Weiss, S. B., D. D. Murphy, P. R. Ehrlich and C. F. Metzler. 1993. Adult emergence phenology in checkerspot butterflies: The effects of macroclimate, topoclimate, and population history. Oecologia 96: 261–270. [10]

Westemeier, R. L., J. D. Brawn, S. A. Simpson, T. L. Esker, R. W. Jansen, J. W. Walk, E. L. Kershner, J. L. Bouzat and K. N. Paige. 1998. Tracking the long-term decline and recovery of an isolated population. Science 282: 1659–1698. [2]

White, G. C. 2000. Population viability analysis: data requirements and essential analyses. Pages 288–331 in L. Boitani and T. K. Fuller (eds.), *Research Techniques in Animal Ecology: Controversies and Consequences.* Columbia University Press, New York. [4, 8]

White, G. C., A. B. Franklin and T. M. Shenk. 2002. Estimating parameters of PVA models from data on marked animals. In S. Beissinger and D. McCullough (eds.). *Population Viability Analysis* (in press). University of Chicago Press, Chicago. [5, 6]

Whitmore, G. A. and V. Seshadri. 1987. A heuristic derivation of the inverse Gaussian distribution. American Statistician 41: 280–281. [3]

Widén, B. 1993. Demographic and genetic effects on reproduction as related to population size in a rare, perennial herb, *Senecio integrifolius* (Asteraceae). Biological Journal of the Linnaean Society 50: 179–195. [2, 4]

Williams, L. R., A. A. Echelle, C. S. Toepfer, M. G. Williams and W. L. Fisher. 1999. Simulation modeling of population viability for the leopard darter (Percidae: *Percina patherina*). The Southwestern Naturalist 44: 470–477. [2]

Wilson, D. E. (ed.). 1996. *Measuring and Monitoring Biological Diversity: Standard Methods for Mammals.* Smithsonian Institution Press, Washington, DC. [5]

Wilson, W. G. 2000. *Simulating Ecological and Evolutionary Systems in C.* Cambridge University Press, Cambridge. [11]

Wisdom, M. J. and L. S. Mills. 1997. Sensitivity analysis to guide population recovery: Prairie-chickens as an example. Journal of Wildlife Management 61: 302–312. [9]

Wisdom, M. J., L. S. Mills and D. F. Doak. 2000. Life stage simulation analysis: Estimating vital rate effects on population growth for species conservation. Ecology 81: 628–641. [9]

Wootton, J. T. and D. A. Bell. 1992. A metapopulation model for the Peregrine Falcon in California: viability and management strategies. Ecological Applications 2: 307–321. [11]

World Conservation Union. 1994. IUCN Redlist Categories. 40th Meeting of the IUCN Council, Gland, Switzerland. (online at http: //www.iucn.org/themes/ssc/redlists/ssc-rl-c.htm) [12]

Young, T. P. 1994. Natural die-offs of large mammals: implications for conservation. Conservation Biology 8: 410–418. [2]

Index